The Clostridia:

Molecular Biology and Pathogenesis

The Clostridia:
Molecular Biology and Pathogenesis

edited by

Julian I. Rood
Department of Microbiology,
Monash University, Melbourne, Australia

Bruce A. McClane
Department of Molecular Genetics and Biochemistry,
University of Pittsburgh, Pittsburgh, Pennsylvania, USA

J. Glenn Songer
Department of Veterinary Science,
University of Arizona, Tucson, Arizona, USA

Richard W. Titball
Defence Evaluation and Research Agency,
Porton Down, CBD Salisbury, Wiltshire, UK

ACADEMIC PRESS
San Diego London Boston
New York Sydney Tokyo Toronto

This book is printed on acid-free paper.

Academic Press, Inc.
525 B Street, Suite 1900, San Diego, California 92101–4495, USA
http://www.apnet.com

Academic Press Limited
24–28 Oval Road, London NW1 7DX, UK
http://www.hbuk.co.uk/ap/

ISBN 0–12–595020–9

Library of Congress Cataloging-in-Publication Data

The Clostridia: Molecular biology and pathogenesis / edited by
Julian Rood . . . [et al.].
p. cm.
Includes index.
ISBN 0–12–595020–9 (alk. paper)
1. Clostridium. 2. Clostridium diseases—Molecular aspects.
I. Rood, Julian.
QR201.C54C56 1997
616′.01464—DC21

96–37807
CIP

A catalogue record for this book is available from the British Library

Typeset by J&L Composition Ltd, Filey, North Yorkshire
Printed and bound in Great Britain by The University Press, Cambridge

97 98 99 00 01 02 EB 9 8 7 6 5 4 3 2 1

Cover illustration
Sporulating culture of *C. perfringens* stained using the method of Schaeffer and Fulton. Courtesy of Richard Titball.

About the Editors

JULIAN I. ROOD is a Reader in Microbiology at Monash University in Melbourne, Australia. He is a Fellow of both the Australian Society for Microbiology and the American Academy of Microbiology. His research career has focused on the development of the genetics of *Clostridium perfringens*, the use of molecular genetics to study the regulation of toxin production and the pathogenesis of clostridial myonecroses, and the genetic analysis of clostridial antibiotic resistance determinants.

BRUCE A. McCLANE is currently an Associate Professor in the Department of Molecular Genetics and Biochemistry at the University of Pittsburgh School of Medicine, (Pittsburgh, Pennsylvania). His primary research interest concerns the role of toxins in bacterial pathogenesis, with a particular emphasis on understanding the action and genetics of *Clostridium perfringens* enterotoxin.

J. GLENN SONGER is Professor of Veterinary Science and Microbiology at the University of Arizona (Tucson, Arizona). His research is directed to understanding the pathogenesis of infections by Gram-positive bacteria, including *Clostridium perfringens*, *Corynebacterium pseudotuberculosis* and *Actinomyces pyogenes*. In addition, he has interests in the application of molecular and immunologic methods in diagnostic veterinary microbiology.

RICHARD W. TITBALL is the Head of Microbiology at the Defence Evaluation and Research Agency, Porton Down. He has worked on the pathogenic clostridia for the past eight years. His other research interests include the development of vaccines against other bacterial pathogens, and he holds a number of patents in the area.

Contents

D PATHOGENESIS OF INFECTIONS

E MEMBRANE ACTIVE TOXINS AND ENZYMES

F NEUROTOXINS

G ENTEROTOXINS

H OTHER TOXINS AND EXTRACELLULAR ENZYMES

A colour plate appears between pages 110 and 111

Contributors

Stephen S. Arnon
California Department of Health Services, 2151 Berkeley Way, Room 716, Berkeley, CA 94704–1011, USA

Klaus Aktories
Institut für Pharmakologie und Toxicologie, Albert-Ludwigs-Universität Freiburg, Herman-Herderstrasse 5, D–79104 Freiburg, Germany

Jimmy D. Ballard
Microbiology and Molecular Genetics, 200 Longwood Ave, D1–502, Harvard Medical School, Boston, MA 02115, USA

J. Edward Brown
US Army Medical Research Institute of Infectious Diseases, Fort Detrick, MD 21702, USA

Amy E. Bryant
Veterans Affairs Medical Center, Infectious Diseases Section, 500 West Fort Street, Boise, ID 83702, USA and University of Washington School of Medicine, Seattle, WA, USA

Bruno Canard
Unité Génétique Moléculaire Bactérienne, Institut Pasteur, 28 Rue du Docteur Roux, 75724 Paris Cedex 15, France

Robert J. Carman
TechLab Inc., 1861 Pratt Drive, Blacksburg, VA 24060–6364, USA

Stewart T. Cole
Unité de Génétique Moléculaire Bactérienne, Institut Pasteur, 28 Rue du Docteur Roux, 75724 Paris Cedex 15, France

Renée E. Collie
Department of Molecular Genetics and Biochemistry, University of Pittsburgh School of Medicine, E1240 Biomedical Science Tower, Pittsburgh, PA 15261, USA

Tom Davis
Department of Molecular Microbiology, Centre for Applied Microbiology and Research, Porton Down, Salisbury, Wiltshire SP4 0JG, UK

Michael Elmore
Department of Molecular Microbiology, Centre for Applied Microbiology and Research, Porton Down, Salisbury, Wiltshire, SP4 0JG, UK

Dale N. Gerding
Department of Medicine, Northwestern University Medical School, Chicago, IL, USA

Ian Henderson
Department of Molecular Microbiology, Centre for Applied Microbiology and Research, Porton Down, Salisbury, Wiltshire SP4 0JG, UK

Fred Hofmann
Institut für Pharmakologie und Toxicologie, Albert-Ludwigs-Universität Freiburg, Herman-Herderstrasse 5, D–79104 Freiburg, Germany

Eric A. Johnson
Department of Food Microbiology, Toxicology and Bacteriology, University of Wisconsin, 1925 Willow Drive, Madison, WI 53706, USA

Stuart Johnson
Medicine Service, VA Lakeside Medical Center, 333 East Huron, Chicago, IL 60611, USA

Ingo Just
Institut für Pharmakologie und Toxicologie, Albert-Ludwigs-Universität Freiburg, Herman-Herderstrasse 5, D–79104 Freiburg, Germany

John F. Kokai-Kun
E1240 Biomedical Science Tower, University of Pittsburgh School of Medicine, Pittsburgh, PA 15261, USA

Ronald G. Labbé
Food Microbiology Laboratory, Department of Food Science, University of Massachusetts Amherst, Amherst, MA 01003, USA

Gregor W. Lawrence
Queensland Institute of Medical Research, Post Office Royal Brisbane Hospital, Queensland 4029, Australia

Sophie E.C. Leary
Defence Evaluation and Research Agency, CBD Porton Down, Salisbury, Wiltshire SP4 0JG, UK

David M. Lyerly
Department of Biochemistry and Anaerobic Microbiology, Virginia Polytechnic Institute and State University, Blacksburg, VA, USA

Dena Lyras
Department of Microbiology, Monash University, Clayton, Victoria 3168, Australia

Bruce A. McClane
Department of Molecular Genetics and Biochemistry, University of Pittsburgh School of Medicine, Pittsburgh, PA 15261, USA

Stephen B. Melville
Department of Microbiology and Immunology, University of Tennessee School of Medicine, 858 Madison Avenue, Memphis, TN 38163, USA

Nigel P. Minton
Department of Molecular Microbiology, Centre for Applied Microbiology and Research, Porton Down, Salisbury, Wiltshire SP4 0JG, UK

J. Scott Moncrief
Department of Biochemistry and Anaerobic Microbiology, Virginia Polytechnic Institute and State University, Blacksburg, VA, USA

Cesare Montecucco
Centro CNR Biomembrane and Dipartimento di Scienze Biomediche, Università di Padova Via Trieste 75, 35121 Padova, Italy

Akinobu Okabe
Department of Microbiology, Faculty of Medicine, Kagawa Medical University, 1750–1, Ikenobe, Miki-cho, Kita-gun, Kagawa 761–07, Japan

Petra Oyston
Chemical and Biological Defence Establishment, Porton Down, Salisbury, Wiltshire SP4 0JG, UK

Dean Payne
Chemical and Biological Defence Establishment, Porton Down, Salisbury, Wiltshire SP4 0JG, UK

Sylvie Perelle
Laboratoire des Toxines, Institut Pasteur, 28 Rue du Docteur Roux, 75724 Paris Cedex 15, France

Michel R. Popoff
Laboratoire des Toxines, Institut Pasteur, 28 Rue du Docteur Roux, 75724 Paris Cedex 15, France

Fred A. Rainey
DSMZ-German Collection of Microorganisms and Cell Cultures GmbH, Mascheroder Weg 1b, D–38124 Braunschweig, Germany

N.-J. Remi Shih
Department of Chemical Engineering and Material Science, University of California, Davis, CA 95616–5294, USA

Peter Roggentin
Biochemisches Institut, Christian-Albrechts-Universität, Olshausenstrasse 40, D–24098 Kiel, Germany

Julian I. Rood
Department of Microbiology, Monash University, Clayton 3168, Australia

Roland Schauer
Biochemisches Institut, Christian-Albrechts-Universität, Olshausenstrasse 40, D–24098 Kiel, Germany

Giampietro Schiavo
Imperial Cancer Research Fund, PO Box 123, 44 Lincoln's Inn Fields, London WC2A 3PX, UK

Jörg Selzer
Institut für Pharmakologie und Toxicologie, Albert-Ludwigs-Universität Freiburg, Herman-Herderstrasse 5, D–79104 Freiburg, Germany

Tohru Shimizu
Department of Microbiology, Institute of Basic Medical Sciences, University of Tsukuba, Tsukuba Ibaraki 305, Japan

J. Glenn Songer
Department of Veterinary Science, The University of Arizona, Tucson, AZ 85721, USA

Erko Stackebrandt
DSMZ-German Collection of Microorganisms and Cell Cultures GmbH, Mascheroder Weg 1b, D–38124 Braunschweig, Germany

Dennis L. Stevens
Veterans Affairs Medical Center, Infectious Diseases Section, 500 West Fort Street, Boise, ID 83702, USA and University of Washington School of Medicine, Seattle, WA, USA

Richard W. Titball
Defence Evaluation and Research Agency, CBD Porton Down, Salisbury, Wiltshire SP4 0JG, UK

Rodney K. Tweten
Department of Microbiology and Immunology, University of Oklahoma Health Sciences Center, Oklahoma City, OK 73190, USA

Tracy D. Wilkins
Department of Biochemistry and Anaerobic Microbiology, Virginia Polytechnic Institute and State University, Blacksburg, VA, USA.

E. Diane Williamson
Chemical and Biological Defence Establishment, Porton Down, Salisbury, Wiltshire SP4 0JG, UK

Preface

The objective of this book is to provide a state of the art reference work which describes the latest knowledge in an exciting and quickly developing area of research. The stimulus arose from the First International Conference on the Molecular Genetics and Pathogenesis of the Clostridia, which was held in Rio Rico, Arizona, USA, in January 1995. That meeting brought together over 120 microbiologists, geneticists, biochemists, cell biologists, clinicians and veterinarians from 11 different countries to discuss the latest developments in this rapidly moving field. Exciting research progress was described in areas as diverse as the structure, function and regulation of extracellular toxins, the pathogenesis of clostridial infections, and the mapping and genetic analysis of clostridial genomes. The underlying theme was that major advances in our understanding of these fundamental processes were being made by applying the techniques of modern molecular biology. The structure of the meeting was such that the fundamental sessions on clinical aspects of clostridial diseases and on the molecular genetics of the clostridia were followed by sessions which used a system-based, rather than an organism-based, format to discuss the latest developments in the pathogenesis of clostridial infections and toxaemias. This book follows the same general format. However, we would like to emphasize that it does not represent the proceedings of the Rio Rico meeting. Many of the authors who have written chapters in this book were not at Rio Rico and several of the chapters involve topics that were not discussed at that meeting. The authors have written comprehensive reviews of their respective topics rather than presenting their own findings. It is our hope that the book will reach a wider audience than those with very specific research interests in the clostridia.

We believe that the text will be invaluable for researchers interested in the molecular genetics and pathogenesis of the clostridia and that it will stimulate many younger scientists to devote their considerable talents to the study of these fascinating and complex pathogens. We also hope that this book will be of value as a reference tool for undergraduate and postgraduate science, medical and veterinary students and for academics involved in teaching molecular genetics and pathogenesis to these students. If the book has achieved these objectives, credit must be given to the enthusiasm and energy that was devoted to this project by the many leading clostridial researchers who willingly contributed their time and effort in writing the various chapters. We are indebted not only to the authors but to our many colleagues who have acted as reviewers of the individual chapters. Finally, we thank Dr Tessa Picknett of Academic Press who has guided the editors throughout the project and been a major source of advice, assistance, and enthusiasm.

Julian I. Rood
Bruce A. McClane
J. Glenn Songer
Richard W. Titball

A: General Biology

Chapter 1

Phylogenetic Relationships

ERKO STACKEBRANDT AND FRED A. RAINEY

DSMZ-German Collection of Microorganisms and Cell Cultures GmbH, Mascheroder Weg 1b, D-38124 Braunschweig, Germany

I. INTRODUCTION

The genus *Clostridium*, traditionally defined as containing Gram-positive, anaerobic rod-shaped and endospore-forming bacteria, constitutes a phylogenetically incoherent genus. However, in contrast to the traditionally defined genus *Pseudomonas*, where members are of polyphyletic origin, found in different subphyla of the phylum (class) Proteobacteria (most of them being already assigned to new genera), members of *Clostridium* can still be considered descendants of a common ancestor that emerged early in the evolution of Gram-positive bacteria. The formation of endospores appears to be a monophyletic trait that is expressed in a few genera of the *Clostridium/Bacillus* subphylum of Gram-positive bacteria, i.e., *Clostridium*, *Bacillus*, *Desulfotomaculum* and *Heliobacterium*,[1] *Sarcina*,[2] *Sporomusa*,[3] and a few new genera that were recently described for certain members of the genera *Clostridium* and *Bacillus* (see Section IV). Loss of spore formation and changes in morphology led to the evolution of phenotypes that were clearly recognizable as being different from spore-forming species. Consequently, they were described as novel taxa, such as the anaerobic members of the genera *Peptococcus*, *Peptostreptococcus* and

The Clostridia: Molecular Biology and Pathogenesis
ISBN 0-12-595020-9

Ruminococcus which are related to species of *Clostridium*, or the aerobic members of the genera *Planococcus*, *Caryophanon*, *Kurthia*, *Filibacter* and *Staphylococcus*, which are related to members of *Bacillus*.

The emphasis on a few morphological, physiological and ultrastructural traits as a basis for classification was accepted while phylogenetic data on the intra- and inter-generic relationships were missing. However, by applying the polyphasic approach to classification, the phylogenetic position of an isolate provides the starting point for phenotypic and genomic characterization to delineate the taxon from neighboring taxa. However, the elucidation of genomic relationships and the determination of the phylogenetic position of a strain is often more rapid than the determination of phenotypic properties. This may lead to the situation where the phylogenetic affiliation of a species to a different genus is not immediately supported by the presence of phenotypic properties characteristic of that genus. Similarly, the description of a new genus for a misclassified species, justified by its phylogenetic position, must await investigation of the phenotypic properties necessary to support the uniqueness of this taxon in a polyphasic approach to classification. The reason for the lack of these properties is obvious. The phenotypic properties of an isolate are usually determined only in comparison with the characteristics of other members of the genus to which the isolate is thought to belong. After more than 20 years of phylogenetic analysis, taxonomists are still confronted with the finding that phenotypically defined genera represent collections of remotely related species which are either members of other genera or which form the nuclei of novel genera.

II. PHYLOGENETIC ANALYSES OF EVOLUTIONARILY CONSERVED MACROMOLECULES

Of the many macromolecules contained in a bacterial cell, only a few have been identified as suitable phylogenetic markers. These molecules are the products of historical events during the course of evolution, and their structures at all levels of information reflect their evolutionary history. For example, the number of sequence differences among proteins, and genes coding for rRNA and proteins, and also the specific properties of their secondary structures, are reflections of their phylogenies. The advantage of working with such genes is their intrinsically chimeric nature; they are composites of regions of varying degrees of sequence conservation. Within one molecule, positions or regions that have different levels of conservation are informative for the analysis of different phylogenetic levels. While highly variable positions are most useful in the determination of close relationships, the more conserved regions report on earlier events in evolution, hence on remote and ancient relationships.

Consequently, it is possible to recognize pairs or groups of genes/proteins that have been derived from a common ancestor.

Results of phylogenetic studies using ribosomal RNA genes, and those coding for the β-subunit of the ATPase and the protein chain elongation factor EF-Tu,[4] have demonstrated a high degree of correlation in the branching patterns obtained. This finding indicates not only the genetic stability of these genes but also the ability of these markers to reflect the phylogeny of the majority of housekeeping genes – hence the phylogeny of the organisms themselves. As a consequence, recently-evolved organisms can be ranked according to their evolutionary history (preferably by 16S rDNA sequence analysis)[5] and physiological, biochemical and morphological characters can be superimposed onto the phylogenetic tree. It can then be decided whether a given trait evolved only once in evolution (monophyletic) or whether it evolved independently in several different lineages (polyphyletic). For example, multilayered peptidoglycan has evolved at least two or three times (deinococci and Gram-positive bacteria of the *Clostridium* and actinomycete subphyla; see below), while endospore formation appears to be a monophyletic property (albeit lost in several lineages during evolution).

III. RECOGNITION OF THE PHYLOGENETIC HETEROGENEITY OF THE GENUS *CLOSTRIDIUM*

The history of the taxonomic dissection of the genus *Clostridium* parallels the history of the use of ribosomal RNA/DNA in phylogenetic studies. It covers the period from 1975 to the present and will not be completed until (i) all validly described species have been subjected to 16S rDNA analysis, (ii) the taxonomic affiliation of invalidly described *Clostridium* species has been determined, and (iii) taxonomic conclusions have been drawn and used to redescribe those clusters that fall outside the confines of the redefined genus *Clostridium* as novel taxa.

The pioneering work of Johnson and Francis[6] who investigated 56 *Clostridium* species by determination of ribosomal ribonucleic acid homologies led to the recognition of several groups. Group I contained 30 species (including the type species, *Clostridium butyricum*) of low DNA G+C content (21–28%) which fell into 10 subgroups Less closely related to members of Group I were 11 species of Group II (*C. lituseburense* and relatives) which possessed a similarly low DNA G+C content. The six species of Group III organisms (*C. sticklandii* and relatives) showed low rRNA similarities to members of Groups I and II. All species with a low DNA G+C content were only distantly related to five *Clostridium* species with a DNA G+C content above 40 mol%, e.g., *C. innocuum*, *C.*

sphenoides, *C. indolis*, *C. barkeri* and *C. oroticum* (the interrelationship of which was not determined for the last four species).

This study clearly demonstrated that the genus *Clostridium* was composed of a wide range of phylogenetically remotely related species. However, the intermixing of *Clostridium* phenotypes with non-*Clostridium* phenotypes could not be recognized because only members of the genus *Clostridium* were included in the reassociation experiments. The extent to which this occurred was first recognized by analysis of the cumulative database of partial 16S rRNA sequences.[7–11] To the surprise of taxonomists, many species of *Clostridium* were found to be more closely related to species of different genera which contained a mixture of clostridial and non-clostridial phenotypes, e.g., spherical form and non-spore-forming (e.g. *Ruminococcus* and *Peptococcus*), or rod-shaped form and non-spore-forming (e.g., *Eubacterium*).

IV. A BRIEF OVERVIEW OF THE PHYLOGENETIC STRUCTURE OF THE GENUS *CLOSTRIDIUM*

Almost complete 16S rRNA/DNA sequences have been determined for about 100 of the 120 validly described species of the genus *Clostridium*. This enormous progress was only made possible by the introduction of polymerase chain reaction (PCR)-mediated amplification of the gene coding for the 16S rRNA.[12–18] The compilation of data by Collins *et al.*[12] and recent updates in the ARB ('A software environment for sequence data') program (retrievable from arb@mikro.biologie.tu-muenchen.de) demonstrate the enormous heterogeneity of the genus. Rather than repeating here the published detailed phylogenetic trees and the extensive discussion of the relationships between the *Clostridium* species, we have summarized the data by displaying the groups defined by Collins *et al.*[12] (Figure 1.1).

Comparison of the data presented by Johnson and Francis[6] and Collins *et al.*[12] shows that the core of the genus, defined as Group I in both studies, correlates well in species composition, but it appears that the rRNA reassociation approach failed to unambiguously determine certain remote interspecific relationships. While the high degree of relatedness between *C. botulinum* (types A, B and F proteolytic), *C. putrificum* and *C. sporogenes* (subgroup I-F), and also between *C. botulinum* types C and D and *C. novyi* is observed in both studies, the four species of rRNA Group I-K are found either in rDNA similarity cluster II (*C. histolyticum* and *C. limosum*) or are located at different positions within cluster I (*C. malenominatum* and *C. subterminale*). The allocation of species to rRNA groups is indicated in Tables 1.1 and 1.2. It is obvious that rRNA Group II is especially heterogeneous, containing mainly members of Cluster XI, but also

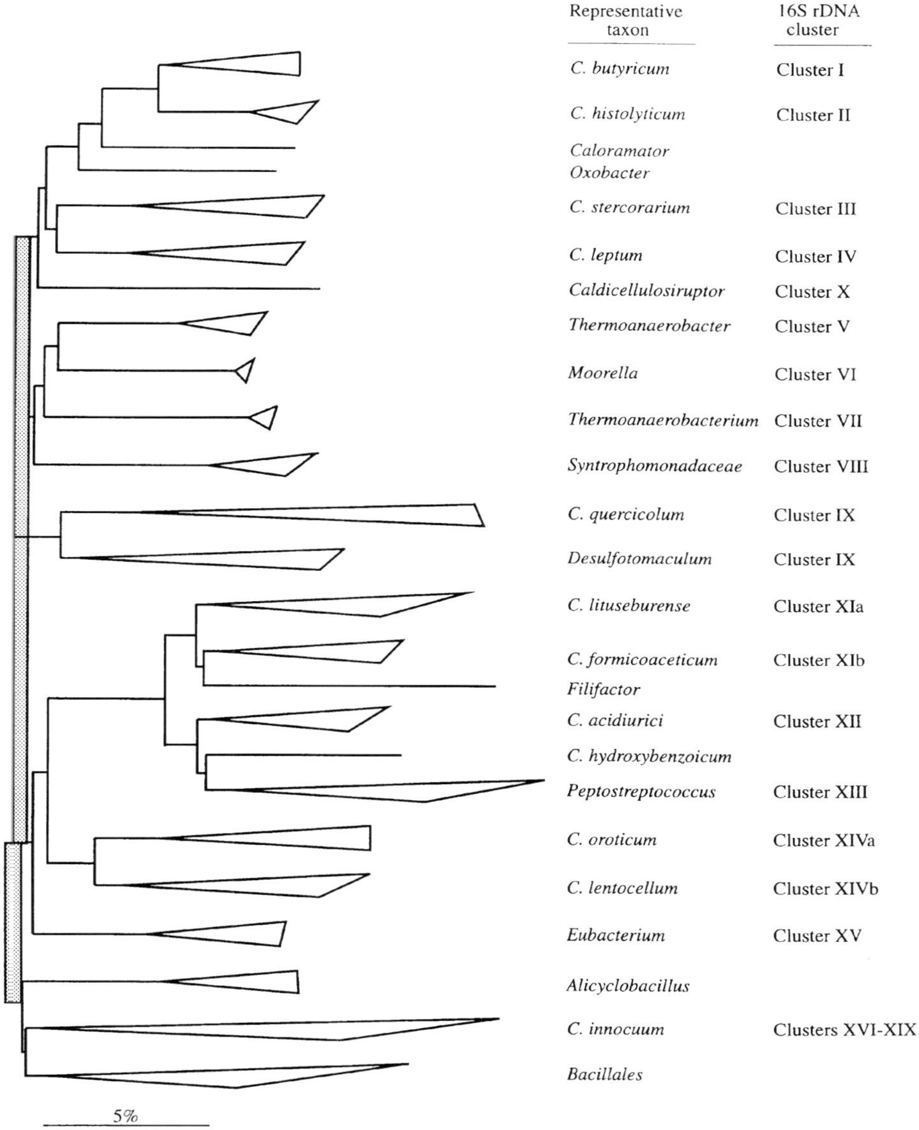

Figure 1.1 Phylogenetic dendrogram summarizing the compilation of the currently available 16S rRNA/DNA sequence data for the members of the genus *Clostridium* and related taxa. Clusters as defined by Collins *et al.*[12] The scale bar represents 5% sequence divergence.

species of Clusters I (*C. tetani*), III (*C. cellobioparum*), and XIX (*C. rectum*). Many species described since 1975 fall mostly into novel clusters, e.g., clusters II, IV, and XIVb. Some of these species, which are well characterized with respect to diagnostic phenotypic characters, have later been assigned to novel

Table 1.1. Distribution of *Clostridium* species groups in the 16S rDNA Clusters I and II[a]

Cluster Ia	Cluster Id	Cluster Ih
C. absonum P[b]	*C. coliagenovorans*	*C. cochlearium*
C. acetobutylicum (I)[c]	*C. sardiniensis*	*C. malenominatum* (I) P
C. aurantibutyricum (I)		*C. tetani* (II) MP
C. barati (I) P	Cluster Ie	*C. tetanomorphum*
C. beijerinckii (I)	*C. botulinum* types C, D (I) MP	
C. botulinum (types B, E, F	*C. grantii*	Cluster Ii
(non-proteolytic) MP	*C. homopropionicum*	*C. kluyveri*
C. butyricum (I)	*C. novyi* (I) MP	*C. ljungdahlii*
C. carnis (I) P	*C. thermopalmarium*	*C. magnum*
C. celatum	*C. thermobutyricum*	*C. pasteurianum* (I)
C. paraputrificum (I) P	'Eubacterium delafieldii'	*C. scatologenes* (I)
C. perfringens MP		*C. tyrobutyricum* (I)
C. puniceum	Cluster If	
C. quinii	*C. argentinense* (I) P	Cluster II
'C. kainantoi'	*C. estertheticum*	*C. histolyticum* (I) MP
Sarcina ventriculi	*C. subterminale* (I) P	*C. limosum* (I) P
Sarcina maxima		*C. proteolyticum*
Eubacterium budayi	Cluster Ig	
Eubacterium nitritogenes	*C. botulinum* types A, F	Species assigned to groups I
E. moniliforme	(proteolytic) (I)	and II by Johnson and
	C. oceanicum (I)	Francis[13]
Cluster Ib	*C. putrificum* (I) P	*C. chauvoei* (I) MP
C. cellulovorans	*C. sporogenes* (I) P	*C. septicum* (I) MP
		C. haemolyticum (I) P
Cluster Ic		*C. tertium* (II) P
C. algidicarnis		
C. cadaveris (I) P		
C. fallax (I) P		
C. intestinalis		

[a] Within each cluster the *Clostridium* species are listed alphabetically. For a detailed intracluster analysis see Collins *et al.*[12] and ARB.
[b] P, pathogenic; MP, major pathogenic.
[c] Numbers in brackets refer to rRNA groups as defined by Johnson and Francis.[6]

genera such as *C. villosum* to *Filifactor*, *C. fervidus* to *Caloramator*, *C. pfennigii* to *Oxobacter* and *C. oxalicum* to *Oxalophagus*.[12] Other *Clostridium* species have been transferred either to new genera, such as *C. thermoaceticum* and *C. thermoautotrophicum* to *Moorella*, or to established genera as new combinations (comb. nov.), such as *C. barkeri* to *Eubacterium*, *C. thermocopriae* to *Thermoanaerobacter*, *C. thermosaccharolyticum* to *Thermoanaerobacterium* or *C. durum* to *Paenibacillus*.[12]

As pointed out above, the apparent homogeneity of the genus *Clostridium*, as defined by rRNA operon cistron similarities, was confused when species from other genera were included in the phylogenetic analysis. Some of the *Clostridium* groups now contain more non-*Clostridium*-type organisms than *Clostridium* species (Table 1.2). The phylogenetic structure of one of these

Table 1.2. Distribution of *Clostridium* species in the 16S rDNA Clusters III–XVIII[a] and indication of relationship of non-*Clostridium* species

Cluster III
C. aldrichii
C. cellobioparum (II)[c]
C. cellulolyticum
C. papyrosolvens
C. stercorarium
C. termitidis
C. thermocellum
C. thermolacticum
Acetivibrio cellulolyticus
Bacteroides cellulosolvens

Cluster IV
C. cellulosi
C. leptum
C. sporosphaeroides
Ruminococcus bromii

Cluster IX
C. quercicolum
Acetonema longum
Sporomusa

Cluster XIa
C. bifermentans (II) P[d]
C. difficile (II) MP
C. ghoni (II) P – *Eubacterium tenue*
C. glycolicum (II) P
C. irregularis
C. litorale
C. lituseburense (II)
C. mangenotii (II)
C. mayombei
C. paradoxum
C. sordellii (II) MP
C. sticklandii
C. thermoalcaliphilum
Eubacterium acidaminophilum
Peptostreptococcus anaerobius

Cluster XIb
'C. aminobutyricum'
C. felsineum
C. formicoaceticum (II)
C. halophilum
Filifactor villosus

Cluster XII
C. acidiurici (II)
'C. filamentosum'
C. hastiforme P
C. purinolyticum (II)
Eubacterium angustum
'Eubacterium thermomarinus'
Tissierella species

Cluster XIII b[b]
C. hydroxybenzoicum
Helcococcus kunzii
Peptostreptococcus (several species)

Cluster XIVa
(for non *Clostridium* species see Figure 1.2)
C. aerotolerans
C. aminophilum
C. aminovalericum (III)
C. celerecrescens
C. clostridiiforme P
C. coccoides
C. herbivorans
C. nexile
C. oroticum (III) P
C. polysaccharolyticum
C. populeti
C. sphenoides (III) P
C. symbiosum P
C. xylanolyticum

Cluster XIVb
C. colinum P
C. lentocellum
C. neopropionicum
C. piliforme
C. propionicum
Eubacterium yurii
'Epulopiscium'

Cluster XVI
C. innocuum P
Eubacterium biforme
Eubacterium cylindroides
Eubacterium dolichum
Eubacterium tortuosum
Streptococcus pleomorphus

Cluster XVIII
C. ramosum P
C. spiroforme
Mycoplasmas

Cluster XIX
C. rectum (II)
Fusobacterium nucleatum
Leptotrichia buccalis
Propionigenium modestum

Continued

Table 1.2. Continued

Additional group I (associated with group IV)	
Eubacterium desmolans	
Eubacterium plautii	Additional group II
Eubacterium siraeum	*Eubacterium brachy*
Ruminococcus albus	*Eubacterium nodatum*
Ruminococcus callidus	*Eubacterium timidum*
Ruminococcus flavefaciens	

[a] Within each cluster the species are listed alphabetically. For a detailed intracluster analysis see Collins *et al.*[12] and ARB.
[b] Not defined by Collins *et al.*,[12] see Figure 1.1.
[c] Numbers in brackets refer to rRNA groups as defined by Johnson and Francis.[6]
[d] P, pathogenic; MP, major pathogenic.

highly intermixed groups, group XIV as defined by Collins *et al.*,[12] is shown in Figure 1.2. Only one of the *Clostridium* species, *C. sphenoides*, which was found to be unrelated to other species of this genus, had been included in the rRNA similarity study of Johnson and Francis.[6] At present, this group contains the type strains of 11 *Clostridium* species and members of the genera *Eubacterium*, *Ruminococcus*, *Acetitomaculum*, *Roseburia*, *Butyrivibrio* and *Lachnospira*. Species of the last three genera have been reported to stain Gram-negative.[19] The only subgroup that contains mainly *Clostridium* species centers around *C. sphenoides* and the species of this cluster are united in having G+C contents in the range 38–42%. In addition to the above-mentioned extreme heterogeneity of the genus *Clostridium*, all of the intermixed genera which contain more than a single species are also heterogeneous. One of the two species of *Butyrivibrio*, *B. fibriosolvens*, is remotely related to *C. aminophilum*, while the other, *B. crossotus* is distantly related to *Eubacterium saburreum*. The phylogenetic heterogeneity of the genus *Ruminococcus* has recently been demonstrated independently by Rainey and Janssen[20] and Willems and Collins.[21] While four species, including the type species *R. flavefaciens*, cluster in group IV, the others are found in group XIVa. Within the latter group, only two of the *Ruminococcus* species, *R. hansenii* and *R. obeum*, cluster together, while the other three species constitute distinct lines of descent. The phylogenetic heterogeneity of the genus *Eubacterium* is not surprising as this genus has been defined as a dumping ground for anaerobic, asporogenous, rod-shaped Gram-positive bacteria lacking additional diagnostic features. While some of the *Eubacterium* species, such as *E. combesii*, *E. fosser*, and *E. lentum*, have been found to group with the actinomycetes and their relatives, other species are members of *Clostridium* groups I, IV, IX, XI, XII and XIV. Of the 13 *Eubacterium* species falling within

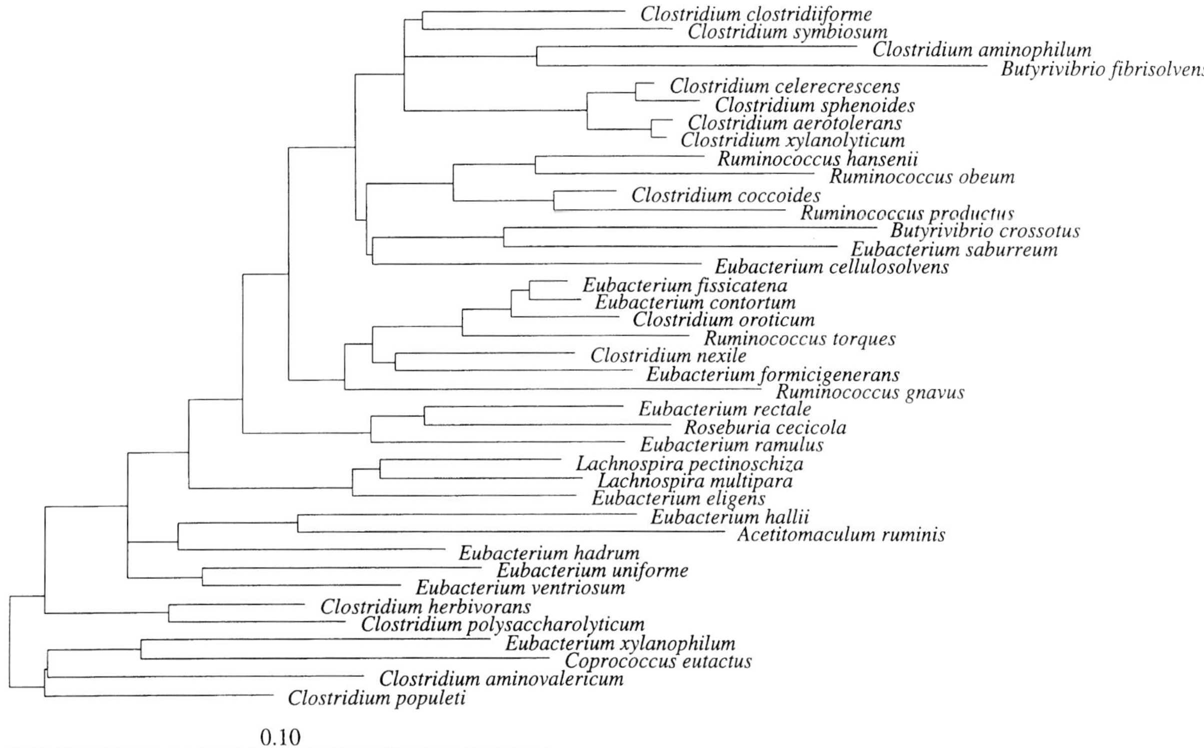

Figure 1.2 Dendogram indicating the phylogenetic heterogeneity within the single phylogenetic cluster XIV as defined by Collins *et al.*[12] The phylogenetic dendrogram was constructed using the neighbor joining method. The scale bar represents 10 nucleotide substitutions per 100 nucleotides.

group XIV, two species are closely related to *C. oroticum*, one species is moderately related to *C. nexile*, while the other species are related to varying degrees either among themselves or to non-*Clostridium* species. In no other group of *Clostridium* species is the lack of significance of endospore formation in classification more impressively demonstrated than in group XIV. Loss of spore formation, reduction of peptidoglycan layers and changes in morphology have apparently occurred during the evolution of *Clostridium* species immediately following the first occurrence of the clostridial phenotype.

V. THE ORIGIN OF THE ANAEROBIC ROD-SHAPED SPORE-FORMING GRAM-POSITIVE BACTERIA

The 16S rRNA/DNA sequence analysis has revealed that members of the genus *Clostridium* branch deep within the subphylum of Gram-positive bacteria, the members of which are characterized by a DNA G+C content of less than 55 mol%. Their appearance in the phylogenetic tree pre-dates the evolution of the facultative and aerobic descendants of the clostridial genotype and phenotype such as the bacilli, lactobacilli and staphylococci. This order of evolution of the respiration chain parallels the situation in certain other major sublines of descent, in which the anaerobic phenotypes appear to be more ancient than the aerobic metabolism. Examples are found in the actinomycete subphylum (e.g., *Bifidobacterium* versus mycelium-forming actinomycetes), the *Bacteroides* (anaerobic) and *Cytophaga* (aerobic) phylum, or within the subclasses of *Proteobacteria* where the anaerobic purple sulfur- and non sulfur-photosynthetic organisms represent deep-rooting lineages.

The presence of Gram-positive peptidoglycan structure in members of the phylogenetically neighboring *Clostridium/Bacillus* lineage and actinomycetes lineage has been interpreted as the divergence of a Gram-positive ancestral organism into two phylogenetic lineages.[9] Support for this theory came from the phylogenetic placement of Gram-positive staining cyanobacteria close to the root of the Gram-positive organisms, in contrast to the deinococci which are related to the genus *Thermus*.[1] This hypothesis of common origin has been challenged by two findings. First, the common ancestry of the *Clostridium/Bacillus* lineage and the actinomycete lineages cannot be recovered from the analysis of all phylogenetic markers, and the unity of these two subphyla is not significant. For example, analyses of 16S rDNA has revealed that the order of divergence of the main lines of descent is not constant but changes as a function of the particular set of sequences included in the analyses.[22] Similarly, analyses of 23S rDNA and the elongation factor Tu (EF-Tu) show the two subphyla of Gram-positive bacteria to be separated.[4] Second, the most deeply branching species in the *Clostridium/Bacillus* lineage are not Gram-positive but possess a Gram-negative cell wall. This finding was originally observed in members of *Sporomusa*, *Selenomonas* and *Megasphaera*[3] and in *Heliobacterium*[1]; this group now contains a significantly larger number of genera, including *Zymophilus*, *Pectinatus*, *Dialister*, and *Johnsonella*. Of special interest in this context is the report of an outer membrane in members of the endospore-forming genus *Sporomusa*. Thus, the topology of the *Sporomusa* cell wall does not appear to be a reduction of the Gram-positive wall but rather an ancestral feature of a Gram-negative (pre-proteobacterial) ancestor.

However, even the elucidation of phenotypic properties of the hypothetical ancestor is complicated by the fact that the branching order of deep-rooting lineages is not significant and may change with the inclusion of more sequences. Furthermore, the deeply branching Gram-negative staining species *Acidaminococcus* and *Phascolarctobacterium* are intermixed with Gram-positive staining species of the genera *Desulfotomaculum*, *Desulfitobacterium* and *Peptococcus niger*. The close relatedness between the Gram-negative staining *Tissierella* and *Clostridium hastiforme*,[23] and between *Roseburia*, *Catonella* and *Johnsonella*[24] species and members of Cluster XIVa indicates that a significant reduction of the Gram-positive cell wall may occur at any stage of the evolution of the peptidoglycan. No Gram-negative descendant of a putative Gram-negative ancestor has as yet been detected in the actinomycete subphylum.

Of the four major groups of bacteria that possess a Gram-positive cell wall, that containing *Deinococcus* and its relatives evolved early and cannot be considered the ancestral type of the cyanobacteria and the clostridia/actinomycetes. The phylogenetic position of the cyanobacteria based on 16S rDNA oligonucleotide catalogues needed to be revised as analysis of nearly complete sequences did not confirm this finding (sequence analysis program ARB, see above).[22,25] It thus appears that the wall structure of the oxygenic phototrophic bacteria also evolved independently from that of clostridia and actinomycetes. The origin of the multilayered peptidoglycan of the clostridia and the actinomycetes can only be discussed meaningfully when the relatedness of these two groups has been elucidated. While no Gram-negative ancestor of the actinomycetes line of descent that forms a common root with the clostridial lineage is known, one can assume an independent origin of the actinomycete wall.

VI. PROBLEMS ASSOCIATED WITH RECLASSIFICATION

The need to use information derived from the primary and higher-order structure of evolutionary conserved macromolecules in the determination of the genealogy of living matter has been recognized since the pioneering work of Zuckerkandl and Pauling.[26] A phylogenetically based classification for prokaryotes began with the recognition that more than any of the previously analysed macromolecules, the structure of the ribosomal RNA species offers a most convincing insight into their genealogy. In addition to revealing evolutionary aspects of biochemistry, ulstrastructure and ecology, the backbone of the phylogenetic tree of life has allowed microbial taxonomists to fulfil the expectations of a polyphasic classification system. The strategy of the modern classification system includes, first, the determination of the phylogenetic position of an organism and, second, the use of discriminating genomic, chemotaxonomic, physiological and structural features for the delineation of taxa. As a result of this approach, the last 20 years have seen a dramatic rearrangement of taxa, mainly above the species level. The taxonomic changes have perturbed those scientists in the fields of plant microbiology and veterinary and human medical microbiology, who have been used to a traditional taxonomic name for decades. For severe human pathogenic organisms, the traditional name has often been conserved over scientifically more sound taxonomic affiliations and names, as in the case of the retention of the genomically indistinguishable taxa *Escherichia* and *Shigella*, *Mycobacterium bovis* and *M. tuberculosis* and *Yersinia pestis* and *Y. pseudotuberculosis*. A similar situation is seen for *C. botulinum*, known to constitute a phylogenetically very heterogeneous species. If this species is restricted to the type strain ATCC 25763 (toxin types A, B, and F, proteolytic),[27] the phylogenetically unrelated strains of this species need to be reclassified. As these include strains responsible for botulism in humans (type E, proteolytic) and animals (types C and D) this scientifically logical step would certainly cause significant problems in communication and diagnosis in the field of medical microbiology, and should not be done without prior consultation with medical microbiologists.

As outlined above, a similar situation is now seen within the genus *Clostridium*. The most important aspect is the high phylogenetic relatedness between the type species of *Sarcina*, *S. ventriculi*, the type species of *Clostridium*, *C. butyricum* and other members of Cluster I. As pointed out by Willems and Collins,[2] subsequent application of the now widely accepted use of 16S rDNA similarities for merging and splitting higher taxa would lead to the reclassification of *Clostridium* species as *Sarcina* species, because the latter genus[28] has priority over the genus *Clostridium*.[29] While such a move would be in agreement with the Bacteriological Code[30] it would probably not be accepted by most microbiologists. For such far-reaching and dramatic changes, with their striking impact on nomenclature, the Code provides the possibility of conserving the name *Clostridium* and the type species *C. butyricum* (*nomen periculosum*; Rule 56a).

The genus *Clostridium* contains 35 species which are considered pathogenic. Most of these species have been investigated by 16S rDNA sequence analysis

(Figure 1.3). About half of these pathogenic species are members of rDNA clusters I and II, including those considered the major pathogenic clostridia,[31] i.e., *C. barati*, *C. botulinum*, *C. novyi*, *C. perfringens*, and *C. tetani* (cluster I), as well as *C. septicum* and *C. chauvoei* (phylogeny determined by Johnson and Francis[6]), and *C. histolyticum* (cluster II). If the genus *Clostridium* was restricted to clusters I and II, the generic assignment for these pathogens would not be affected. Unless considered *nomen periculosa*, the names of other major and minor pathogens should be changed when the clusters to which they belong phylogenetically are reclassified. These include *C. difficile* and *C. sordellii* and some minor pathogens in cluster XI, and several pathogens of clusters XII, XIV, and XVIII. The distribution of pathogens in the rDNA clusters is indicated in Tables 1.1 and 1.2.

For the purpose of diagnosis and identification of bacterial species, including members of *Clostridium*, ribosomal RNA is of proven value. As part of the ribosome, rRNA occurs as naturally amplified molecules, and hence provides an ideal target for labeled oligonucleotide probes. Analysis of variable stretches of the 23S rDNA of *C. botulinum* strains that produce different types of neurotoxins has confirmed the genomic heterogeneity of this species, evident in the significant differences in metabolic properties,[32] rRNA cistron similarity studies[6] (see above), DNA reassociation experiments[33] and 16S rDNA sequence analysis.[13,14] Probes developed against the variable 23S rDNA regions of *C. botulinum* were found to reliably identify (i) proteolytic strains with toxins of types A, B, and F as well as related species, (ii) non-proteolytic strains with toxins B and E, (iii) strains with toxin D, and *C. argentinense* (the former *C. botulinum*) with toxin G.[27] Using a similar approach, the phylogenetic coherence of the species *C. perfringens* has been confirmed.[34] For identification of pathogenic strains, 16Sr DNA based determination of phylogenetic identity is unable to identify the toxin type or type group, as the genealogies of ribosomal RNA genes and toxin genes are probably not coupled. Certain toxin genes (C and D) of *C. botulinum*, are contained in the genome of temperate bacteriophages[35,36] and certain non-pathogenic *Clostridium* strains can be transformed to express a toxin gene.[36,37] A non-virulent strain that has lost its toxin gene(s) may still be identified as a nonpathogenic strain of *C. botulinum* by analysis of 16S rDNA, while the sole application of electroimmunodiffusion and enzyme-linked immunosorbent assays, able to identify the toxin type, will most likely misdetermine the phylogenetic identity of the organism. In the design of probes based on 16S rDNA sequence data for identification of *Clostridium* species or strains, two points must be considered. First, the quality of the sequence data currently available in the public databases needs careful assessment; in many cases the 100–150 nucleotides of the 5′ end is missing or of low quality, yet this region contains variable regions useful as probe target sites. The second point relates to the

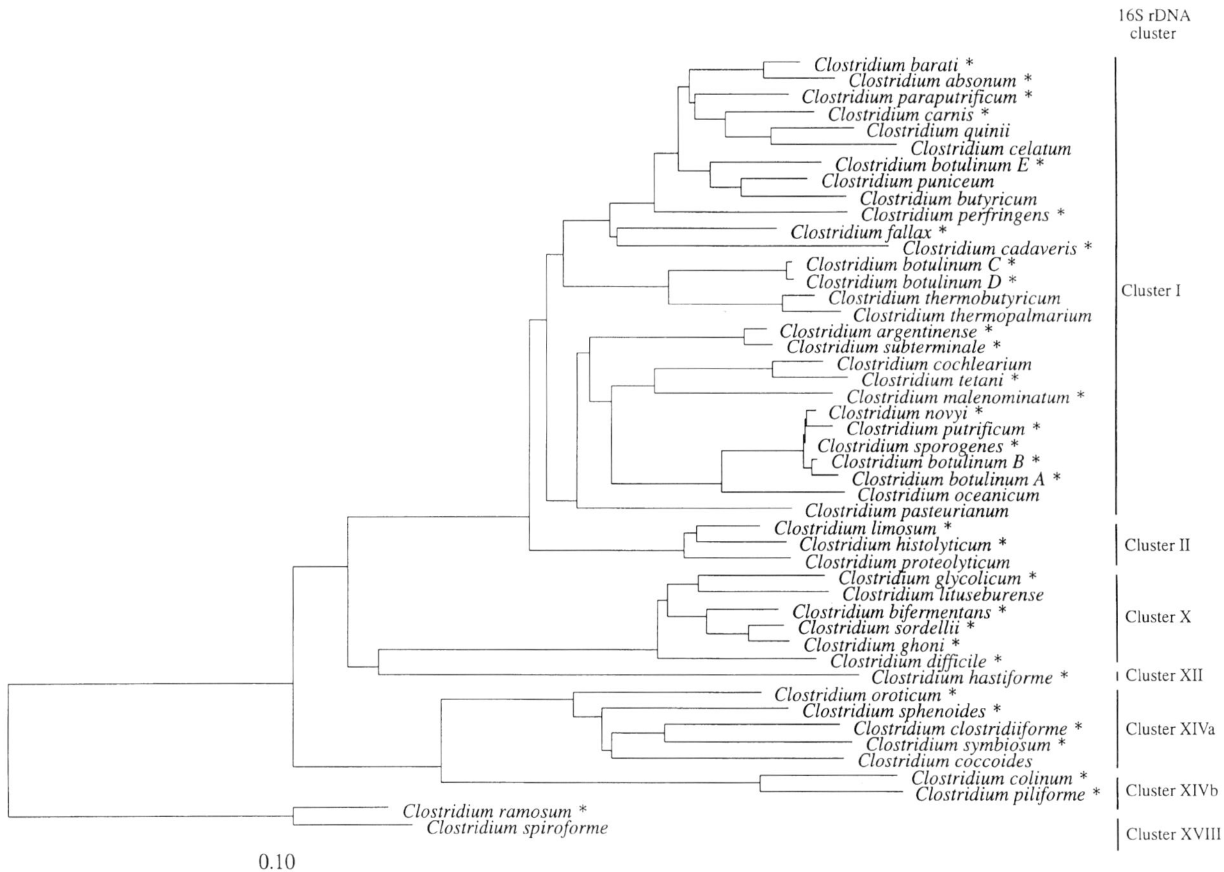

Figure 1.3 Phylogenetic dendrogram indicating the relationships of the pathogenic *Clostridium* species*. Clusters as defined by Collins *et al.*[12] The scale bar represents 10 nucleotide substitutions per 100 nucleotides.

presence of multiple rRNA operons, which can differ in sequence composition within one strain. This has been demonstrated in *Clostridium paradoxum*[38] and in members of the genus *Desulfotomaculum* (C. Spröer, N. Ward-Rainey, F. Rainey, H. Hippe, and E. Stackebrandt, unpublished data). The variability of operon sequence composition between strains of the same species has yet to be investigated.

The future taxonomy of the genus *Clostridium* and related taxa now awaits the completion of the 16S rDNA sequence database for at least the type strains of all valid species, a comprehensive phylogenetic analysis of all available sequence data, and the initiation of a polyphasic approach in the hope of finding additional taxonomic markers to complement the phylogenetically based classification.

VII. REFERENCES

1. Woese C.R., Debrunner-Vossbrinck B.A., Oyaizu H., Stackebrandt E. and Ludwig W. (1985) Gram-positive bacteria: possible photosynthetic ancestry. *Science* **229**: 762–765.
2. Willems A. and Collins M.D. (1994) Phylogenetic placement of *Sarcina ventriculi* and *Sarcina maxima* within group I *Clostridium*, a possible problem for future revision of the genus *Clostridium*. *Int J Syst Bacteriol* **44**: 591–593.
3. Stackebrandt E., Pohla H., Kroppenstedt R.M., Hippe H. and Woese C.R. (1985) 16S rRNA analysis of *Sporomusa*, *Selenomonas*, and *Megasphaera*: on the phylogenetic origin of Gram-positive Eubacteria. *Arch Microbiol* **143**: 270–276.
4. Schleifer K.H. and Ludwig W. (1994) Molecular taxonomy: classification and Identification. In Priest F.G. and Goodfellow M. (eds), *Bacterial Diversity and Systematics*, pp. 1–15. Plenum Press, New York.
5. Woese C.R. (1987) Bacterial evolution. *Microbiol Rev* **51**: 221–271.
6. Johnson J.L. and Francis B.S. (1975) Taxonomy of the clostridia: ribosomal ribonucleic acid homologies among the species. *J Gen Microbiol* **88**: 229–244.
7. Cato E.P. and Stackebrandt E. (1989) Taxonomy and phylogeny. In Minton N.P. and Clarke D.J. (eds), *Clostridia*, pp. 1–26. Plenum Press, New York.
8. Fox G.E. and Stackebrandt E. (1987) The application of 16S rRNA cataloguing and 5S rRNA sequencing in microbial systematics. In Colwell R.R. and Grigorova R. (eds), *Methods in Microbiology* Vol. 19, pp. 405–458. Academic Press, Orlando.
9. Fox G.E., Stackebrandt E., Hespell R.B., Gibson J., Maniloff J., Dyer T., Wolfe R.S., Balch W., Tanner R., Magrum L., Zablen L.B., Blakemore R., Gupta R., Luehrsen K.R., Bonen L., Lewis B.J., Chen K.N. and Woese C.R. (1980) The phylogeny of prokaryotes. *Science* **209**: 457–463.
10. Tanner R.S., Stackebrandt E., Fox G.E. and Woese C.R. (1981) A phylogenetic analysis of *Acetobacterium woodii, Clostridium barkeri, Clostridium butyricum, Clostridium lituseburense, Eubacterium limosum* and *Eubacterium tenue*. *Curr Microbiol* **5**: 35–38.
11. Tanner R.S., Stackebrandt E., Fox G.E. and Woese C.R. (1982) A phylogenetic analysis of anaerobic eubacteria capable of synthesizing acetate from CO_2. *Curr Microbiol* **7**: 127–132.
12. Collins M.D., Lawson P.A., Willems A., Cordoba J.J., Fernandez-Garayzabal J., Garcia P., Cai J., Hippe H. and Farrow J.A.E. (1994) The phylogeny of the genus *Clostridium*: proposal of five new genera and eleven new species combinations. *Int J Syst Bacteriol* **44**: 812–826.
13. Hutson R.A., Thompson D.E. and Collins M.D. (1993) Genetic interrelationships of saccharolytic

Clostridium botulinum types B, E and F and related clostridia as revealed by small subunit rRNA gene sequences. *FEMS Microbiol Lett* **108**: 103–110.

14. Hutson R.A., Thompson D.E., Lawson P.A., Schocken-Itturino R.P., Böttger E.C. and Collins M.D. (1993) Genetic interrelationships of proteolytic *Clostridium botulinum* types A, B, and F and other members of the *Clostridium botulinum* complex as revealed by small-subunit rRNA gene sequences. *Ant Leeuwenhoek J Microbiol* **64**: 273–283.
15. Rainey F.A. and Stackebrandt E. (1993) 16S rDNA analysis reveals phylogenetic diversity among the polysaccharolytic clostridia. *FEMS Microbiol Lett* **113**: 125–128.
16. Rainey F.A., Ward N.L., Morgan H.W., Toalster R. and Stackebrandt E. (1993) Phylogenetic analysis of anaerobic, thermophilic bacteria: aid for their reclassification. *J Bacteriol* **175**: 4772–4779.
17. Rainey F.A., Donnison A.M., Janssen P.H., Saul D., Rodrigo A,. Bergquist P.L., Daniel R.M., Stackebrandt E. and Morgan H.W. (1994) Description of *Caldicellulosiruptor saccharolyticus* gen. nov., spec. nov: an obligately anaerobic, extremely thermophilic, cellulolytic bacterium. *FEMS Microbiol Lett* **120**: 263–266.
18. Rainey F.A., Janssen P.H., Morgan H.W. and Stackebrandt E. (1994) A biphasic approach to the determination of the phenotypic and genotypic diversity of some anaerobic, cellulolytic, thermophilic, rod-shaped bacteria. *Ant Leeuwenhoek J Microbiol* **64**: 341–355.
19. Hespell R.B. (1992) The genera *Butyrivibrio*, *Lachnospira*, and *Roseburia*. In Balows A., Trüper H.G., Dworkin M., Harder W. and Schleifer K-H. (eds), *The Prokaryotes*, pp. 2022–2033. Springer-Verlag, New York.
20. Rainey F.A. and Janssen P.H. (1985) Phylogenetic analysis by 16S ribosomal DNA sequence comparison reveals two unrelated groups of species within the genus *Ruminococcus*. *FEMS Microbiol Lett* **129**: 69–74.
21. Willems A. and Collins M.D. (1995) Phylogenetic analysis of *Ruminococcus flavefaciens*, the type species of the genus *Ruminococcus*, does not support the reclassification of *Streptococcus hansenii* and *Peptostreptococcus productus* as ruminococci. *Int J Syst Bacteriol* **45**: 572–575.
22. Van de Peer Y., Neefs J.-M., De Rijk P., De Vos P. and De Wachter R. (1994) About the order of divergence of the major bacterial taxa during evolution. *Syst Appl Microbiol* **17**: 32–38.
23. Farrow J.A.E., Lawson P.A., Hippe H., Gauglitz U. and Collins M.D. (1995) Phylogenetic evidence that the Gram-negative nonsporulating bacterium *Tisseriella* (*Bacteroides*) *praecuta* is a member of the *Clostridium* subphylum of the Gram-positive bacteria and description of *Tisseriella creatinini*. sp. nov. *Int J Syst Bacteriol* **45**: 436–440.
24. Willems A. and Collins M.D. (1995) Evidence for the placement of the Gram-negative *Catonella morbi* (Moore and Moore) and *Johnsonella ignava* (Moore and Moore) within the *Clostridium* subphylum of the Gram-positive bacteria on the basis of 16S rRNA sequences. *Int J Syst Bacteriol* **45**: 855–857.
25. Maidak B.L., Larsen N., McCaughey J., Overbeck R., Olsen G.J., Fogel K., Blandy J. and Woese C.R. (1994) The ribosomal database project. *Nucl Acids Res* **22**: 3483–3487.
26. Zuckerkandl E. and Pauling L. (1965) Molecules as documents of evolutionary history. *J Theor Biol* **8**: 357–366.
27. Rönner S.G.E. and Stackebrandt E. (1994) Further evidence for the genetic heterogeneity of *Clostridium botulinum* as determined by 23S rDNA oligonucleotide probing. *Syst Appl Microbiol* **17**: 180–188.
28. Goodsir J. (1842) History of a case in which a fluid periodically ejected from the stomach contained vegetable organisms of an undescribed form. With a chemical analysis of the fluid, by George Wilson. *Edinb Med Surg J* **57**: 430–443.
29. Prazmowski A. (1880) Untersuchung über die Entwicklungsgeschichte und Ferment-wirkung einiger Bakterien-Arten. Inaugural Dissertation Hugo Voigt, Leipzig, Germany.
30. LaPage S.P., Sneath P.H.A., Lessel E.F., Skerman V.B.D., Seeliger H.P.R. and Clark W.A. (1992)

International Code of Nomenclature of Bacteria. American Society for Microbiology, Washington, D.C.

31. McClane B.A. (1995) ASM first international conference on the molecular genetics and pathogenesis of the clostridia. *ASM News* **61**: 465–468.
32. Holdeman L.V. and Brooks J.B. (1970) Variation among strains of *Clostridium botulinum* and related clostridia. In Herzberg M. (ed.), *Proceedings of the Botulism Conference*, pp. 278–286, UJNR Conference on Toxic Microorganisms, U.S. Govt. Printing Office, Washington, D.C.
33. Suen J.C., Hatheway C.L., Steigerwalt A.G., Brenner D.J. (1988) *Clostridium argentinense* sp. nov.: a genetically homogeneous group composed of all strains of *Clostridium botulinum* toxin type G and some nontoxic strains previously identified as *Clostridium subterminale* and *Clostridium hastiforme. Int J Syst Bacteriol* **38**: 375–381.
34. Rönner S.G.E. and Stackebrandt E. (1994) Identification of *Clostridium perfringens* by 16S and 23S rRNA oligonucleotide probes. *Syst Appl Microbiol* **17**: 425–432.
35. Ecklund M.W., and Poysky F.T. (1974) Interconversion of type C and D strains of *Clostridium botulinum* by specific bacteriophages. *Appl Microbiol* **27**: 251–258.
36. Ecklund M.W., Poysky F.T., Meyers J.A., and Pelroy G.A. (1974) Interspecies conversion of *Clostridium botulinum* type C to *Clostridium novyi* type A by bacteriophage. *Science* **186**: 456–458.
37. Oguma K., Iida H., Shiozaki M. and Inoue K. (1976) Antigenicity of converting phages obtained from *Clostridium botulinum* types C and D strains. *Infect Immun* **13**: 855–860.
38. Rainey F.A., Ward-Rainey N., Janssen P.H., Hippe H. and Stackebrandt E. (1996) *Clostridium paradoxum* DSM 7303 contains multiple 16S rRNA genes with heterogeneous intervening sequences. *Microbiol* **142**: 2087–2095.

Chapter 2

Physiology of Sporulation of Clostridia

RONALD G. LABBÉ AND N-J. REMI SHIH*

Food Microbiology Laboratory, Department of Food Science, University of Massachusetts, Amherst, MA 01003, USA and,
**Department of Chemical Engineering and Material Science, University of California Davis, Davis, CA 95616–5294, USA*

I. INTRODUCTION

Exposure of clostridia to conditions of nutrition depletion activates alternate metabolic pathways. Sporulation is the ultimate example of this adaptation response, which begins at the onset of stationary phase. Unlike most adaptive responses, bacterial sporulation involves several metabolic pathways associated with morphological, physiological and biochemical changes and can affect end products associated with energy metabolism. Several changes involve metabolites associated with stationary phase and sporulation. However, not all are sporulation-related or specific. Studies of the sporulation process at the molecular level have just begun in the genus *Clostridium* but are extensive in the case of aerobic spore-formers, *Bacillus subtilis* in particular.[1,2] This chapter presents some aspects of the current knowledge about how sporulation affects the metabolism of the clostridia and focuses on the methods that promote *in vitro* sporulation and the regulation of sporulation-associated products. Not surprisingly, most studies of sporulation of clostridia have focused on species of medical or economic importance.

The Clostridia: Molecular Biology and Pathogenesis
ISBN 0–12–595020–9

II. MORPHOLOGICAL CHANGES DURING SPORULATION

The morphological events during sporulation of *Clostridium* species are similar to those occurring during sporulation of *Bacillus* species. That is, they are divided into seven stages (0–VII) based on electron microscopic studies and associated physical and biochemical properties.[3–7] Stage VII is associated with the autolysis of the sporangium and release of the mature spore. The sporulation process is thought to be induced by nutrient deprivation which causes a drop in the GTP pool.[8]

Vegetative and sporulating cells of the same *Clostridium* species often exhibit distinct shapes. For example, in a glucose-rich medium, even in stationary phase, *Clostridium perfringens* cells are relatively short, whereas in a starch-based sporulation medium, they are longer. Although this is not the case in all clostridia, the change of morphology provides a useful tool in laboratory observations. Sporulation is also fundamental in the classification of the clostridia. In addition, in the case of one species, *C. perfringens*, an enterotoxin is produced during the sporulation process. From a practical viewpoint the production of bacterial spores also minimizes storage and handling requirements.

III. *IN VITRO* SPORULATION

Although sporulation generally is initiated by a lack of nutrients, clostridia need an energy source for macromolecular synthesis during sporulation and the cells may die if one or more nutrients are completely absent. However, not all *Clostridium* species can sporulate well in laboratory media because conditions for their sporulation can be highly specific. For example, *Clostridium cellulolyticum* sporulates following cellobiose starvation. A level of 30% sporulation occurs if cells are starved during the exponential phase of growth, otherwise no spores or very few spores are observed.[9]

High sporulation levels of certain *Clostridium* species such as *C. perfringens*, are notoriously difficult to achieve, which has resulted in numerous modifications of defined and complex media. Several reagents such as caffeine, theophylline, isobutylmethylxanthine, and papaverine in the methylxanthine family have been reported to enhance sporulation in the clostridia although their effects are somewhat strain-specific and concentration-dependent.[10–13] The effect of sporulation enhancement by these compounds is characterized by reduced growth due to their ability to inhibit macromolecular synthesis.[14] Because the compounds in the methylxanthine family are purine analogs, Setlow and Sacks[15] concluded that their mechanisms of action are similar to that of purine deprivation in *B. subtilis*, namely, a drop in the GTP pool. Therefore, it is not surprising that guanosine (1 mM) can suppress sporulation and increase vegetative growth of *C. perfringens*.[16] Human bile salts also enhance the sporulation of some *Clostridium* species, although an excess amount of human bile salts may result in poor growth and result in little sporulation.[17]

Rapidly metabolizable carbohydrates such as glucose are generally avoided in sporulation media because they are vigorously fermented by certain saccharolytic species of *Clostridium*, resulting in considerable acid production.[18,19] In fact, improved sporulation of *C. perfringens* has been reported by adjusting the initial pH value of the sporulation medium to 7.8 with sodium carbonate,[20] which contributes to the buffering capacity of the medium.[21]

The optimal temperature for sporulation of *C. perfringens* is between 35 and 40°C. Although a few strains of this species can sporulate at high temperatures (>40°C),[22] they require the exogenous addition of α-amylase to sporulate in starch-based media[23] since amylase production by these strains is minimal at higher temperatures.[24,25] Addition of exogenous *C. perfringens* enterotoxin has also been reported to hasten the sporulation process of this organism[26] although the mechanism is unknown.

The specific carbohydrate composition of the sporulation media also affects the ability to sporulate. For example, sporulation of *C. perfringens* is inhibited by high concentrations (>15 mM) of glucose, maltose, mannose, lactose and sucrose but is unaffected by the presence of high amounts (>30 mM) of ribose, galactose and fructose.[27] In the case of *Clostridium botulinum* a level of 270 mM glucose decreases sporulation from 80% to 30%.[28] However, some strains of *C. botulinum* require glucose for growth.[29] A high concentration of glucose also inhibits sporulation of *Clostridium thermosaccharolyticum*[30] but stimulates sporulation in *Clostridium acetobutylicum*.[31] Ababouch and Busta[32] have reported that an increase in glucose concentration (from 0.2% to 1%) delays the onset of sporulation in *Clostridium sporogenes* and results in a tenfold increase in the number of spores (per millilitre).

Starch and dextrin are used as carbohydrate sources in sporulation media of *C. perfringens*.[33,34] This organism can sporulate in the presence of 8 mM glucose when soluble starch and dextrin are included in a defined medium. The type of starch is also of importance, with soluble starch resulting in higher spore levels than potato, corn or arrowroot starch.[35] Substitution of starch with raffinose can also enhance sporulation in some strains of *C. perfringens*.[36]

Supplementation of complex sporulation medium with 1 mM Mn^{2+} and Zn^{2+} enhances sporulation of *C. botulinum* and results in a one-log increase in spores. However, Fe^{2+} is less effective at the same concentration[37] and less sporulation of *C. botulinum* is observed if Cu^{2+} (0.5–1.0 mM) is present in the medium together with Zn^{2+}.[38]

IV. MOLECULAR REGULATION OF SPORULATION

It has long been suggested that the sporulation processes in *Bacillus* spp. and *Clostridium* spp. share similar molecular mechanisms. Nevertheless, little attention has been paid to the genetic analysis of the latter, despite its considerable biotechnical potential.[39,40]

In *Bacillus* spp. phosphorylation of the regulatory protein Spo0A controls the initiation of sporulation and also changes in gene expression that occur during transition into stationary phase.[41,42] Homologues of the *B. subtilus spo0A* and *spoIVB* genes have been identified in six *Clostridium* species with three highly conserved regions within the effector domain.[43] Such homologs were absent in several nonsporulating bacteria closely related to *B. subtilus*. Disruption of the *spo0A* homolog in *C. acetobutylicum* by integrative recombination produces a Spo^- phenotype and abolishes the solvent biosynthesis properties of the parent strain.

Small cytoplasmic RNA (scRNA) is a stable and abundant RNA of approximately 270 nucleotides from *B. subtilus*.[44] It is a member of the signal-recognition particle (SRP)RNA family and consists of four domains based on secondary structure with domains I and II necessary for formation of heat-resistant spores.[45] Recent evidence indicates that scRNA of *C. perfringens* is of a size similar to that of *B. subtilus* scRNA with about 70% primary sequence homology.[46] Functional analysis showed that *C. perfringens* scRNA allows the formation of heat-resistant spores in a scRNA-depleted *B. subtilus* strain. This suggests that conservation of domains I and II, needed for efficient sporulation, has conferred a selective advantage throughout eubacterial evolution.

Finally, there is evidence that some of the more complex regulatory features of sporulation in *B. subtilus* are conserved in *C. acetobutylicum*. Transposon insertions can occur at 12 or more chromosomal sites which normally express the developmental signals which induce solvent production and sporulation.[47] The variety of solvent classes observed in such mutants suggest several levels of control of solventogenesis in *C. acetobutylicum*. In *B. subtilus* four sporulation-specific sigma factors, σ^E, σ^F, σ^G, and σ^K, are involved in the sporulation process. Homologous genes for these sigma factors are present in *C. acetobutylicum*, and have been cloned and sequenced.[48,49] Hybridization analysis indicates the presence of σ^H, σ^D and *spoIID* homologs, and again indicates the evolutionary conservation of molecular mechanisms of sporulation in members of the extremely diverse *Bacillus* and *Clostridium* genera.[48,49]

V. METABOLITES ASSOCIATED WITH THE SPORULATION PROCESS

A Regulation of enzymes during sporulation

Clostridia synthesize certain compounds and enzymes when they sporulate, but not all are sporulation-specific or related to sporulation. Some sporulation media contain complex carbohydrates such as starch or dextrin. Amylolytic action during sporulation in such media contributes to cell growth and sporulation by providing metabolizable, short-chain carbon sources. α-Amylase synthesis may be regulated by sporulation.[27] Cells of *C. perfringens* at stages II–III of sporulation are committed to complete sporulation and fail to synthesize α-amylase when transferred to a medium optimized for α-amylase synthesis. Conversely, α-amylase synthesis decreases when vegetative cells are induced to sporulate.[50] In *C. perfringens*, eight amylolytic enzymes, which can be divided into two distinct groups according to the immunological and biochemical properties of the α-amylase they produce, are excreted during vegetative cell growth. Sporulation affects the amount of these eight amylolytic enzymes in the culture supernatant fluid (Figure 2.1). The intensification of amylase bands 1, 7, and 8, and the disappearance of bands 5 and 6 is observed when *C. perfringens* sporulates.[27] Synthesis of the highest levels of a α-amylase require the presence of a small amount (6–10 mM) of a simple sugar while higher levels inhibit sporulation.[50]

It has been proposed that in Gram-positive bacteria the regulation of amylolytic enzymes involves a global regulatory mechanism. A *trans*-acting factor, CcpA, encoded by the *ccpA* gene, and closely related to the Lac and Gal repressors, can bind to a responsive element in many genes encoding proteins that are regulated by mechanisms similar to that of α-amylase.[51] Similar to *ccpA*, a regulatory gene (*regA*) has been identified in *C. acetobutylicum*. This gene encodes a RegA protein that contains a DNA-binding domain homologous to CcpA in *B. subtilis*.[52] A mutation in the *ccpA* gene allows sporulation and amylase production of *B. subtilis* in the presence of excess glucose.[53,54]

A chymotrypsin-like intracellular protease has been isolated from sporulating *C. perfringens* cells and may function in the processing of enterotoxin but is less important in spore coat protein synthesis.[55] The relationship between protease levels and sporulation has also been studied using Spo^- mutants of *C. perfringens*. Protease activity of Spo^- mutants is reduced by 87–88% compared with wild type. The wild type also has drastically reduced protease levels when grown under conditions that repress sporulation. A serine-type protease rather than metallo-type is associated with sporulation.[56] Like *C. perfringens*, *C. cellulolyticum* strains lacking proteolytic activity also fail to induce sporulation.[9]

The nitrogenase activity of *C. butyricum* is reportedly low in vegetative cells, increases at Stage III of sporulation (forespore formation) and then decreases

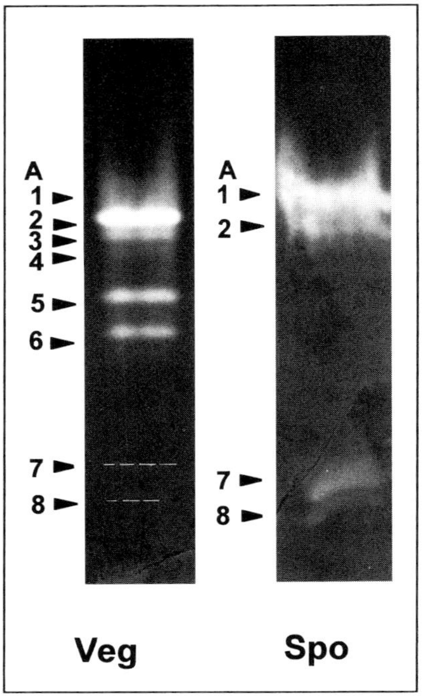

Figure 2.1 Distribution of extracellular amylolytic activities in the *Clostridium perfringens* NCTC 8679 culture fluid during vegetative growth (Veg) (mid-log phase) and sporulation (Spo). Amylases were precipitated by ammonium sulfate, dialysed, concentrated by ultrafiltration and separated by polyacrylamide gel electrophoresis under native conditions. Zymograms of amylase activity were prepared by immersing gels in soluble starch followed by a brief exposure to iodine. Minor amylase activity was enhanced by white dotted lines: reproduced from Shih.[25]

when sporulation continues.[57] Similar observations have been reported in *Clostridium pasteurianum*.[58]

B. Toxins and proteins

Some *C. perfringens* type A strains produce an enterotoxin (CPE) which causes food poisoning[59] and which is discussed at length in Chapter 19. The amount of enterotoxin increases dramatically during sporulation.[18,60–62] Although the enterotoxin gene, *cpe*, has been cloned in *Escherichia coli*, the amount of enterotoxin produced by *E. coli* is far less than that produced by *C. perfringens* during sporulation.[63] The enterotoxin was proposed to be a component of the spore coat.[64] However, recent studies indicate that spore coat proteins are not immunologically identical to CPE.[65] The presence of trace amounts of enterotoxin-like protein in vegetative cells of *C. perfringens* has been reported.[62,63]

Thus, CPE is unlikely to be a component of *C. perfringens* spores. By ligating the *cpe* promoter to the *E. coli gusA* gene, which encodes β-glucuronidase, it has been shown that the induction of *cpe* gene expression is an early stationary phase event.[66] The enterotoxin is induced between Stage 0 and Stage III of sporulation; mutants blocked at later stages still produce the toxin.[67] Because virtually all enterotoxin-negative strains also sporulate it appears more likely that the *cpe* gene plays no role in sporulation but is activated by transcriptional factors which also control sporulation genes. Results of on-going studies of *cpe* regulation at the transcriptional level may provide explanations for the high levels of enterotoxin produced during sporulation of this organism (see Chapter 27).

An inclusion body associated with the sporulation of enterotoxin-positive strains of *C. perfringens* has been reported (Figure 2.2a). It possesses biological activity and is composed of *C. perfringens* enterotoxin.[68,69] Another type of inclusion body is present in vegetative cells of entertoxin-positive and enterotoxin-negative strains of *C. perfringens*, especially when they are incubated above 40°C (Figure 2.2b), and, as viewed by phase-contrast microscopy, is remarkably similar in appearance to a small spore.[70] It differs from the inclusion

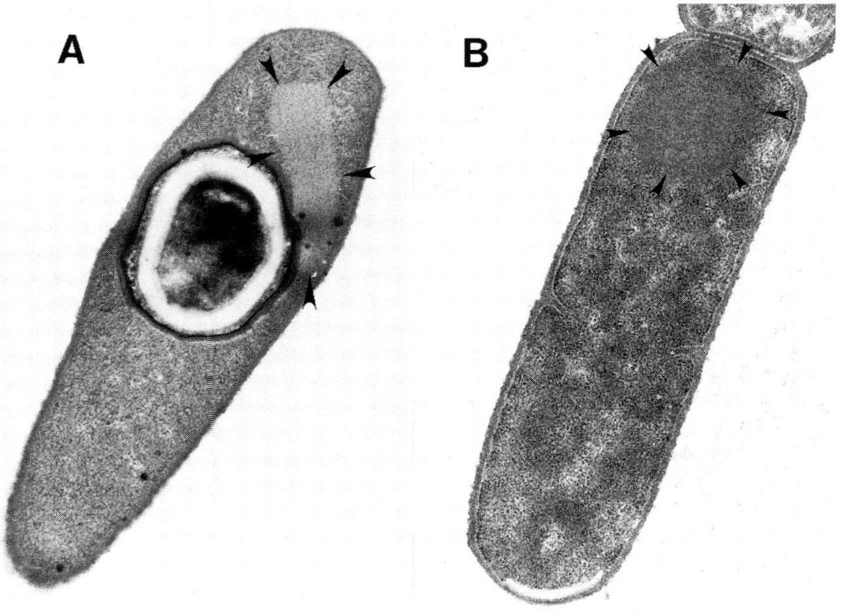

Figure 2.2 Electron micrographs of *Clostridium perfringens* inclusion bodies. A, Sporulation-associated inclusion body (NCTC 8679): reproduced from Löffler and Labbé,[68] with permission. B, Inclusion body from vegetative cells (NCTC 8238): reproduced from Garcia-Alvarado *et al.*,[70] with permission.

body associated with sporulation in that it is refractile when observed by phase-contrast microscopy, is not surrounded by vacuoles, and is composed of nonenterotoxin-related protein.

Unlike *C. perfringens* enterotoxin, the production of *Clostridium difficile* toxin A (also an enterotoxin) is not sporulation-specific.[71] However, the protein toxins of *Clostridium bifermentans*, a novel class of insecticidal toxins, are synthesized during sporulation.[72] *C. botulinum* type C_2 neurotoxin is reportedly associated with sporulation, whereas type A and E toxins are produced by vegetative cultures.[8]

Like *Bacillus* species, clostridia synthesize several small, acid-soluble spore proteins (SASPs) which account for about 20% of the spore protein during sporulation and are degraded during germination.[73,74] The amino acid sequences of two major SASPs (SASP-α and SASP-β) of *C. bifermentans* are almost identical and are very similar to those of *C. perfringens*.[74]

C. Acids, solvents and other byproducts

Many clostridia produce solvents such as ethanol, acetone or butanol by converting the organic acids produced from carbohydrate sources in the medium. *Clostridium acetobutylicum* excretes acetic and butyric acids which cause the pH of the medium to fall below 5.0 at the onset of the stationary phase of growth, although the intracellular pH of *C. acetobutylicum* is unaffected.[75] The acids are converted to acetone, butanol and ethanol after prolonged incubation as a detoxification process in response to the accumulation of the acid end-products.[76] Although sporulation also is observed at the time of organic acid accumulation, it is not a prerequisite for solvent production.[77,78] The overall yield of solvents depends on the total amount of acidic products converted from the carbon source in the growth medium rather than the degree of sporulation of *C. acetobutylicum*.[79] At a low initial glucose concentration, *C. acetobutylicum* is unable to produce acetone and butanol because of the low levels of acetic and butyric acids excreted.[80] In contrast, high-sporulating strains of *Clostridium felsineum* have been reported which produce less butyric and propionic acid but twice as much butanol than low-sporulating strains.[81]

It has been recently noted that sterile, concentrated supernatants of *C. perfringens* can promote sporulation of certain homologous and heterologous strains of this organism.[82] The effect of this sporulation factor is concentration-dependent. The substance responsible for this reaction is still unidentified but it is heat- and acid-stable, and possesses a molecular weight of less than 5000. Although its physiological role is unclear it is produced by both vegetative and sporulating cultures of *C. perfringens* and also by enterotoxin-positive and enterotoxin-negative strains.

In summary, the conditions of nutrition deprivation that induce sporulation also induce other metabolic responses. It is not surprising that regulatory factors expressed during stationary phase can affect metabolic pathways. However, details of the mechanisms of sporulation in the genus *Clostridium* and their sporulation-associated metabolites have yet to be clarified. The multiple regulatory events controlling vegetative cell growth, sporulation and enzyme and solvent production and excretion in *Clostridium* species indicate the diversity of this genus.

VI. REFERENCES

1. Errington J. (1993) *Bacillus subtilis* sporulation: Regulation of gene expression and control of morphogenesis. *Microbiol Rev* **57**: 1–33.
2. Hoch J.A. (1993) *spo0* genes, the phosphorelay, and initiation of sporulation. In Sonenshein A.L. and Hoch J.A. (eds), Bacillus subtilis *and other Gram-positive bacteria*, pp. 747–755. American Society for Microbiology, Washington, D.C.
3. Shakhbanov A.A. (1980) Ultrastructure of *Clostridium septicum. J Mikrobiol Epidemiol Immunobiol* **7**: 43–47.
4. Johnstone K. and Holland K.T. (1977) Ultrastructural changes during sporulation of *Clostridium bifermentans. J Gen Microbiol* **100**: 217–220.
5. Labbé R.G. and Rufner R. (1980) Ultrastructure of sporulating cells of *Clostridium perfringens* type A grown in the presence of raffinose. *Can J Microbiol* **26**: 1153–1157.
6. Long S., Jones D.T. and Woods D.R. (1983) Sporulation of *Clostridium acetobutylicum* P262 in a defined medium. *Appl Environ Microbiol* **45**: 1389–1393.
7. Ryter A. (1965) Etude morphologique de la sporulation de *Bacillus subtilis. Ann Inst Pasteur* **108**: 40–60.
8. Marks C.L. and Freese E. (1987) Aspects of carbohydrate metabolism related to sporulation and germination. In Reizer J. and Peterkofsky A. (eds), *Sugar Transport and Metabolism in Gram-positive Bacteria*, pp. 295–332. Ellis-Horwood, Chichester.
9. Gehin A., Gelhaye E., Raval G. and Petidemange H. (1995) *Clostridium cellulolyticum* viability and sporulation under cellobiose starvation conditions. *Appl Environ Microbiol* **61**: 868–871.
10. Sacks L.E. and Thompson P.A. (1977) Increased spore yields of *Clostridium perfringens* in the presence of methylxanthines. *Appl Environ Microbiol* **34**: 189–193.
11. Labbé R.G. and Nolan L.L. (1981) Stimulation of *C. perfringens* enterotoxin formation by caffeine and theobromine. *Infect Immun* **34**: 50–54.
12. Sacks L.E. (1982) Stimulation of sporulation of *Clostridium perfringens* by papaverine. *Experientia* **38**: 246–247.
13. Craven S.E. and Blankenship L.C. (1982) Effect of purine derivatives, papaverine hydrochloride, and imidazole on enterotoxin formation by *Clostridium perfringens* type A. *Can J Microbiol* **28**: 851–859.
14. Labbé R.G. and Nolan L.L. (1987) Inhibition of macromolecular synthesis by caffeine in *Clostridium perfringens. Can J Microbiol* **33**: 589–592.
15. Setlow P. and Sacks L.E. (1983) Cyclic AMP is not detectable in *Clostridium perfringens. Can J Microbiol* **29**: 1228–1230.
16. Sacks L.E. (1983) Influence of carbohydrates on growth and sporulation of *Clostridium perfringens* in a defined medium with or without guanosine. *Appl Environ Microbiol* **46**: 1169–1175.

17. Heredia N.L., Labbé R.G., Rodriquez M.A. and Garcia-Alvarado J.S. (1991) Growth, sporulation and enterotoxin production by *C. perfringens* type A in the presence of human bile salts. *FEMS Microbiol Lett* **84**: 15–22.
18. Labbé R.G. (1989) *Clostridium perfringens*. In Doyle M. (ed.) *Foodborne Bacterial Pathogens*, pp. 191–234. Marcel Dekker, New York.
19. Labbé R.G. and Rey D.K. (1979) Raffinose increases sporulation and enterotoxin production by *Clostridium perfringens* type A. *Appl Environ Microbiol* **37**: 1196–1200.
20. Harmon S. and Kautter S. (1986) Improved media for sporulation and enterotoxin production by *C. perfringens*. *J Food Prot* **49**: 706–711.
21. Craven S.E. (1988) Increased sporulation of *Clostridium perfringens* in a medium prepared with the prereduced anaerobically sterilized technique or with carbon dioxide or carbonate. *J Food Prot* **51**: 700–706.
22. Garcia-Alvarado J.S., Rodriguez M.A. and Labbé R.G. (1992) Sporulation and enterotoxin production by *Clostridium perfringens* type A at 37 and 43°C. *Appl Environ Microbiol* **58**: 1411–1414.
23. Garcia-Alvarado J.S., Rodriguez M.A. and Labbé R.G. (1992) Influence of elevated temperature on starch hydrolysis by enterotoxin-positive and enterotoxin-negative strains of *Clostridium perfringens*. type A. *Appl Environ Microbiol* **58**: 326–330.
24. Shih N-J. and Labbé R.G. (1995) Purification and characterization of an extracellular α-amylase from *Clostridium perfringens* Type A. *Appl Environ Microbiol* **61**: 1776–1779.
25. Shih N.-J. (1995) Isolation, characterization, and regulation of amylases from *Clostridium perfringens*. PhD Dissertation, University of Massachusetts, Amherst.
26. Dillon M.E. and Labbé R.G. (1989) Stimulation of growth and sporulation of *C. perfringens* by its homologous enterotoxin. *FEMS Microbiol Lett* **59**: 221–224.
27. Shih N.-J. and Labbé R. (1996) Characterization and distribution of amylases during vegetative cell growth and sporulation of *Clostridium perfringens*. *Can J Microbiol* **42**: 628–633.
28. Emeruwa A.C. and Hawirko R.Z. (1975) Effect of cyclic AMP on catabolite repressed bacterial sporogenesis of an anaerobe. *Arch Microbiol* **105**: 67–71.
29. Witmer M.E. and Johnson E.A. (1988) Development of improved defined media for *Clostridium botulinum* serotypes A, B, and E. *Appl Environ Microbiol* **54**: 753–759.
30. Woods D.R. and Jones D.T. (1986) Physiological responses of *Bacteroides* and *Clostridium* strains to environmental stress factors. *Adv Microbiol Physiol* **28**: 1–64.
31. Long S., Jones D.T. and Woods D.R. (1984) Initiation of solvent production, clostridial stage formation and endospore formation in *Clostridium acetobutylicum* P262. *Appl Microbiol Biotech* **20**: 256–261.
32. Ababouch L. and Busta F.F. (1985) Adaptation of biphasic culture technique to the sporulation of *Clostridium sporogenes* PA 3679: Effect of glucose concentration on spore yield and heat resistance. *Food Microbiol* **2**: 107–114.
33. Sacks L.E. and Thompson P.A. (1978) Clear defined medium for the sporulation of *Clostridium perfringens*. *Appl Environ Microbiol* **35**: 405–410.
34. Duncan C.L. and Strong D.H. (1968) Improved medium for sporulation of *Clostridium perfringens*. *Appl Microbiol* **47**: 1172–1174.
35. Labbé R.G., Somers E. and Duncan C.L. (1976) Influence of starch source on sporulation and enterotoxin production by *Clostridium perfringens* type A. *Appl Environ Microbiol* **31**: 455–457.
36. Labbé R.G. and Rey D.K. (1979) Raffinose increases sporulation and enterotoxin production by *Clostridium perfringens* type A. *Appl Environ Microbiol* **37**: 1196–1200.
37. Kihm D.J., Hutton M.T., Hanlin J.H. and Johnson E.A. (1990) Influence of transition metals added during sporulation on heat resistance of *Clostridium botulinum* 113 B spores. *Appl Environ Microbiol* **56**: 681–685.
38. Kihm D.J., Hutton M.T., Hanlin J.H. and Johnson E.A. (1988) Zinc stimulates sporulation in *Clostridium botulinum* 113B. *Curr Microbiol* **17**: 193–198.

39. Woods D.R. (ed.) (1993) *The Clostridia and Biotechnology*. Butterworth-Heinemann, Stoneham, MA.
40. Woods D. (1993) The genetic engineering of microbial solvent production. *Trends Biotechnol* **13**: 259–264.
41. Olmeda G., Ninfa E., Stock J. and Youngman P. (1990) Novel mutations that alter the regulation of sporulation in *Bacillus subtilus*: evidence that phosphorylation of regulatory protein Spo0A controls the initiation of sporulation. *J Mol Biol* **215**: 359–372.
42. Perego M., Cole S., Burbulys D., Trach K. and Hoch J. (1989) Characterization of the gene for a protein kinase which phosphorylates the sporulation-regulatory proteins Spo0A and Spo0F. *J Bacteriol* **171**: 6187–6196.
43. Brown D., Ganova-Raeva L., Green B., Wilkinson S., Young M. and Youngman P. (1994) Characterization of *spo0A* homologues in diverse *Bacillus* and *Clostridium* species identifies a probable DNA-binding domain. *Mol Microbiol* **14**: 411–426.
44. Struck J., Vogel D., Ulbrich N. and Erdmann V. (1988) The *Bacillus subtilus* scRNA is related to 4.5S RNA from *Escherichia coli*. *Nucl Acids Res* **16**: 2719.
45. Nishiguchi M., Honda K., Amikura R., Nakamura K. and Yamane K. (1994) Structural requirements of *Bacillus subtilus* small cytoplasmic RNA for cell growth, sporulation, and extracellular enzyme production. *J Bacteriol* **176**: 157–165.
46. Nakamura K., Hashizume E., Shibata T., Nakamura Y., Mala S. and Yamane K. (1995) Small cytoplasmic RNA (scRNA) gene from *Clostridium perfringens* can replace the gene for the *Bacillus subtilus* scRNA in both growth and sporulation. *Microbiol* **141**: 2965–2975.
47. Mattsson D.M. and Rogers P (1994) Analysis of Tn916-induced mutants of *Clostridium acetobutylicum* altered in solventogenensis and sporulation. *J Ind Microbiol* **13**: 258–268.
48. Sauer U., Treuner A., Buchholz M., Santangelo J. and Durre P. (1994) Sporulation and primary sigma factor homologous genes in *Clostridium acetobutylicum*. *J Bacteriol* **176**: 6572–6582.
49. Wong J., Sass C. and Bennett G. (1995) Sequence and arrangement of genes encoding sigma factors in *Clostridium acetobutylicum*. *J Bacteriol* **176**: 6572–6582.
50. Shih N-J. and Labbé R.G. (1994) Effect of glucose on sporulation and extracellular amylase production by *Clostridium perfringens* type A in a defined medium. *Current Microbiol* **29**: 163–169.
51. Hueck C.J. and Hillen W. (1995) Catabolite repression in *Bacillus subtilus*: A global regulatory mechanism for the Gram-positive bacteria? *Mol Microbiol* **15**: 395–401.
52. Davison S.P., Santangelo J.D., Reid S.J. and Woods D.R. (1995) A *Clostridium acetobutylicum* regulator gene (*regA*) affecting amylase production in *Bacillus subtilis*. *Microbiol* **141**: 989–996.
53. Henkin T.M., Grundy F.J., Nicholson W.L. and Chambliss G.H. (1991) Catabolite repression of α-amylase gene expression in *Bacillus subtilis* involves a *trans*-acting gene product homologous to the *Escherichia coli lacI* and *galR* repressors. *Mol Microbiol* **5**: 575–584.
54. Chambliss G.H. (1993) Carbon source-mediated catabolite repression. In Sonnenshein A.L., Hoch J.A. and Losick R. (eds), Bacillus subtilis *and other Gram Positive Bacteria: Biochemistry, Physiology and Molecular Genetics*, pp. 213–219, American Society for Microbiology, Washington, D.C.
55. Park K.B. and Labbé R.G. (1990) Purification and characterization of intracellular protease of *Clostridium perfringens* type A. *Can J Microbiol* **37**: 19–27.
56. Löffler A. and Labbé R.G. (1983) Intracellular protease during sporulation and enterotoxin formation by *Clostridium perfringens* type A. *Curr Microbiol* **8**: 187–190.
57. Savalov A.I. (1980) Regulation of nitrogenase activity and biosynthesis in *Clostridium butyricum*. *Mikrobiologiya* **49**: 427–432.
58. Vallespinos F. and Kleiner D. (1980) Selective inactivation of nitrogenase and gluctamate synthase during sporulation of (nitrogen fixing) *Clostridium pasteurianum*. *J Gen Microbiol* **117**: 543–545.
59. McDonel L.J. (1979) *Clostridium perfringens* toxin (type A, B, C, D, E). *Pharmacol Ther* **10**: 617–635.
60. Labbé R.G. (1980) Relationship between sporulation and enterotoxin production in *Clostridium perfringens* type A. *Food Technol* **34**: 88–90.

61. Ryu S. and Labbé R.G. (1992) Kinetics of spore coat protein synthesis by *Clostridium perfringens* type A. *Curr Microbiol* **25**: 35–40.
62. Granum P.E., Telle W., Olsvik O. and Stavn A. (1984) Enterotoxin formation by *Clostridium perfringens* during sporulation and vegetative growth. *Int J Food Microbiol* **1**: 43–49.
63. Czeczulin J.R., Hanna P.C. and McClane B.A. (1993) Cloning, nucleotide sequencing, and expression of the *Clostridium perfringens* enterotoxin gene in *Escherichia coli. Infect Immun* **61**: 3429–3439.
64. Frieben W.R. and Duncan C.L. (1973) Homology between enterotoxin protein and spore structural protein in *Clostridium perfringens* type A. *Eur J Biochem* **39**: 393–401.
65. Ryu S. and Labbé R.G. (1993) A 48 kilodalton enterotoxin-related protein from *Clostridium perfringens* vegetative and sporulating cells. *Int J Food Microbiol* **17**: 343–348.
66. Melville S.B., Labbé R. and Sonenshein A.L. (1994) Expression from the *Clostridium perfringens cpe* promoter in *C. perfringens* and *Bacillus subtilis. Infect Immun* **62**: 5550–5558.
67. Duncan C.L., Strong D.H. and Sebald M. (1972) Sporulation and enterotoxin production by mutants of *Clostridium perfringens. J Bacteriol* **110**: 378–391.
68. Löffler A. and Labbé R.G. (1985) Isolation of an inclusion body from sporulating, enterotoxin-positive *Clostridium perfringens. FEMS Microbiol* **27**: 143–147.
69. Löffler A. and Labbé R.G. (1986) Characterization of a parasporal inclusion body from sporulating, enterotoxin-positive *Clostridium perfringens* type A. *J Bacteriol* **165**: 542–548.
70. Garcia-Alvarado J., Labbe R. and Rodriguez M. (1992) Isolation of inclusion bodies from vegetative *Clostridium perfringens*: partial purification of a 47 kDa inclusion protein. *J Appl Bacteriol* **73**: 157–162.
71. Ketley J.M., Mitchell T.J., Haslam S.C., Stephen J., Candy D.C. and Burdon D.W. (1986) Sporogenesis and toxin A production by *Clostridium difficile. J Med Microbiol* **22**: 33–38.
72. Nicholas L., Charles J.L., de Barjac H. (1993) *Clostridium bifermentans* serovar malaysia: Characterization of putative mosquito larvicidal proteins. *FEMS Microbiol Lett* **113**: 23–28.
73. Granum D.E., Richardson M. and Blom H. (1987) Isolation and amino acid sequence of an acid-soluble protein from *Clostridium perfringens* spores. *FEMS Microbiol Lett* **42**: 225–230.
74. Cabrera-Martinez R.M., Mason J.M., Setlow B., Waites W.M. and Setlow P. (1989) Purification and amino acid sequence of two small acid-soluble proteins from *Clostridium bifermentans* spores. *FEMS Microbiol Lett* **61**: 139–143.
75. Huang L,. Forsberg C.W. and Gibbins L.N. (1986) Influence of external pH and fermentation products on *Clostridium acetobutylicum* intracellular pH and cellular distribution of fermentation products. *Appl Environ Microbiol* **51**: 1230–1234.
76. Jones D. and Woods D (1986) Acetone–butanol fermentation revisited. *Microbiol Rev* **50**: 484–524.
77. Jones D.T., van der Westhuzen A., Long S., Allcock E.R., Reid S.J. and Woods D.J. (1982) Solvent production and morphological changes in *Clostridium acetobutylicum. Appl Environ Microbiol* **43**: 1434–1439.
78. Babb B.L., Collett H.J., Reid H.J. and Woods D.R. (1993) Transposon mutagenesis of *Clostridium acetobutylicum* P262. Isolation and characterization of solvent and metronidazole resistant mutants. *FEMS Microbiol Lett* **114**: 343–348.
79. Awang G.M., Ingledew W.M. and Jones D.T. (1992) The effect of fermentable carbohydrate on sporulation and butanol production by *Clostridium acetobutylicum* P262. *Appl Microbiol Biotech* **38**: 12–16.
80. Monot F, Martin J.M., Petitdemange H. and Gary R. (1982) Acetone and butanol production by *Clostridium acetobutylicum* in a synthetic medium. *Appl Environ Microbiol* **44**: 1318–1324.
81. Avrova N.P., Zubko I.K. and Alekseeva E.G. (1981) Fermentation products and the pectolytic enzymatic activity in *Clostridium felsineum* strains differing in intensity of sporulation. *Mikrobiogiya* **50**: 458–460.
82. Shih N-J. and Labbé R.G. (1996) Sporulation promoting ability in *Clostridium perfringens* culture fluids. *Appl Environ Microbiol* **62**: 1441–1443.

B: Genome Organization and Molecular Genetics

Chapter 3

Extrachromosomal Virulence Determinants in the Clostridia

ERIC A. JOHNSON
Department of Food Microbiology, Toxicology and Bacteriology, University of Wisconsin, 1925 Willow Drive, Madison, WI 53706, USA

I. INTRODUCTION

Virulence factors are those components of an organism whose loss (usually by mutation or shedding of an unstable genetic element) specifically impairs pathogenicity but not viability.[1] The two major classes of virulence factors are toxins and surface molecules. Virulence factors are generally highly polygenic traits that are often affected by aspects of cellular physiology such as nutritional requirements, growth rate, temperature, and response to host factors. With certain exceptions, rarely does the virulence of an organism depend on a single gene product but the capacity to cause disease usually involves a balanced genome whereby many traits contribute to the virulence.

The clostridia produce more toxins than any other bacterial genus,[2,3] and pathogenic clostridia are usually identified on the basis of the characteristic toxins they produce. More than 20 toxins and other extracellular proteins contributing to virulence such as spreading factors and proteolytic enzymes have been identified in *Clostridium* spp.[4] Two of the toxins, botulinum and tetanus toxins, are the most powerful poisons known – lethal doses of botulinum and

The Clostridia: Molecular Biology and Pathogenesis
ISBN 0-12-595020-9

Table 3.1. Virulence factors associated with extrachromosomal genetic factors or mobile genetic elements in clostridia

Clostridium species	Virulence factor	Extrachromosomal genetic element	References
C botulinum types C and D	C_3 toxin	Nonintegrative lysogenic phage	25
C. perfringens	CPE	Large plasmid (animal isolates)	66
C. botulinum types C and D	BoNT/C and BoNT/D	Nonintegrative lysogenic phage	18,19,22,23
C. butyricum	BoNT/E	Lysogenic phage and/or plasmid	47,48
C. botulinum type G (*C. argentinense*)	BoNT/G	Plasmid	40,41
C. botulinum types C and D	HA/C,D	Nonintegrative lysogenic phage	24
C. botulinum type G (*C. argentinense*)	HA, NTNH	Plasmid	41
C. tetani	TeNT	Large plasmid	56,57
C. novyi	α-Toxin	Lysogenic phage	17,26
C. perfringens	ε-Toxin	Plasmid	64,70,71
C. perfringens	λ-Toxin	Plasmid	66

Abbreviations: BoNT, botulinum neurotoxin; CPE, *C. perfringens* enterotoxin; HA, hemagglutinin; NTNH, nontoxic nonhemagglutinin; TeNT, tetanus neurotoxin; see text for details.

tetanus toxins for various animals are ≤1 ng per kg.[5] These toxins act by selectively blocking neurotransmission of presynaptic nerve terminals in the peripheral and central nervous systems. Other toxins of clostridia have other modes of action such as degradation of tissue, hemolysis, causing diarrhea, or inducing a hyperactive immune response in the host, as described in other chapters of this book.

The genes coding for several clostridial toxins are located on nonintegrative lysogenic bacteriophages or on plasmids (Table 3.1). Others are believed to be located on the chromosome and have not been shown to be associated with lysogenic bacteriophages, or are located on lysogenic bacteriophages that reside extrachromosomally in the cytoplasm. The presence of genes for virulence factors on extrachromosomal elements often results in phenotypic properties such as genetic instability and the capacity for toxinogenicity to be dispersed by horizontal gene transfer to nontoxigenic organisms. These genotypic and phenotypic properties are consistent with contributing to the adaptation of a bacterium to a specific disease-causing niche without being essential for viability.

Genes encoding clostridial toxins that exist at least temporally on extrachromosomal virulence plasmids are those of *C. botulinum* type G, enterotoxigenic

C. perfringens, and *C. tetani*. The genes for *C. botulinum* types C and D botulinum toxins are carried on pseudolysogenic bacteriophages, in which the relationship between the bacteriophage and chromosomal DNA is unstable; these phages reside mainly extrachromosomally in the cytoplasm. The genetic location and function of extrachromosomal virulence genes in clostridial species are described below.

II. *CLOSTRIDIUM BOTULINUM*

Clostridium botulinum produces seven serologically distinct neurotoxins (BoNT) (A–G) as well as a binary cytotoxin (C_2) and an ADP-ribosylating toxin (C_3). The genetic basis and mode of action of *C. botulinum* neurotoxins are described in Chapters 17 and 18. In this chapter, the association of *C. botulinum* virulence genes with extrachromosomal genetic elements is described. Neurotoxins of *C. botulinum* are produced as protein complexes consisting of BoNT, hemagglutinin (HA), and nontoxic nonhemagglutinin (NTNH).[6–8] With the exception of types C and D, strains of *C. botulinum* generally produce only one toxin type. A few strains have been demonstrated to produce two toxin types, of which one is produced at a much higher titer than the other.[4] Recently, a large number of type A strains were reported to possess the gene for type B botulinum toxin, although type B toxin was not expressed in all but one of the 43 strains examined.[9] The surprisingly high frequency of the silent BoNT/B gene in *C. botulinum* strains producing only type A BoNT raised the important question of why the gene is not expressed and also supported the hypothesis that the BoNT/B gene may be associated with a transferable genetic element.[10]

The genes encoding the botulinum toxin complexes occur as clusters[8,10] in *C. botulinum*. The genes for NTNH and BoNT are probably transcribed as a single polycistronic message, with transcription initiating in the NTNH gene and continuing through the BoNT gene. The HA gene encoding hemagglutinin is transcribed in the opposite direction to the NTNH gene. A gene encoding a lysin is located downstream from the BoNT gene in *C. botulinum* type A.[11,12]

Almost all strains of *C. botulinum* and related nontoxigenic organisms examined have been shown to harbor bacteriophages inducible by mitomycin C, ultraviolet light, or other treatments.[13–17] Inoue and Iida[18] demonstrated a relationship between toxigenicity and lysogenicity in *C. botulinum* type C Stockholm. Nontoxigenic mutants were readily isolated after treatment of vegetative cells with acridine orange, and were shown to be nonlysogenic.[18] Phage infection in broth and up to 96% conversion to toxigenicity was achieved by infection of a nonlysogenic strain by a lysate prepared from a toxigenic strain. Toxin conversion by the filtrate was not destroyed by trypsin or DNase, and converted strains stably gained the ability to produce toxin, supporting a stable phage infection. In 1971, Eklund *et al.*[19] independently showed that bacteriophages converted nontoxigenic cultures of *C. botulinum* type C to a toxigenic state; they were also able to obtain turbid plaques on solid media by the agar layer method.[20]

Shortly after these experiments were reported, C_2 toxin was demonstrated in culture filtrates of *C. botulinum* type C,[17,21] but the gene encoding C_2 toxin was not associated with specific bacteriophages.[17] Similar procedures demonstrated that the genes for BoNT/D,[22,23] HA,[24] and C_3 exoenzyme[25] in *C. botulinum* types C and D are located on lysogenic phages. Hemagglutinin production can be transmitted separately or concomitantly with toxin production.[24] The *tox*$^+$ phages from *C. botulinum* types C and D have a broad host range and can interconvert type C to type D or vice versa. Infection of *C. botulinum* types C and D with phages induced from *Clostridium novyi* enable the recipient to produce the lethal alpha toxin of *C. novyi*.[17,26]

The life cycle of *tox*$^+$ phages involves a pseudolysogenic relationship in which the prophage state is readily disturbed and the phage genome is believed to reside extrachromosomally in the cytoplasm. The life cycle of *tox*$^+$ phages of *C. botulinum* types C and D has not been elucidated at the molecular level, and the extrachromosomal location is based on indirect evidence. Oguma *et al.*[24,27] distinguished two classes of *tox*$^+$ phages according to host range and antigenicity. The *tox*$^+$ phages resemble the T-even phages in morphology except that the *tox*$^+$ phages have a much longer tail. The DNA content of the *tox*$^+$ phages is 70–100 kb smaller than most T-even phages. The genomic nucleic acid of the *tox*$^+$ phages was digested by DNase but not RNase. Hybridization studies indicated that homologous sequences exist among the converting *tox*$^+$ phages. However, hybridization experiments were not performed with chromosomal DNA throughout the logarithmic growth and stationary phases of the host bacterium to determine if the phage integrates into the chromosomal DNA at any period during the growth cycle. Hence, the conclusion that the phages permanently reside extrachromosomally is based solely on their instability in the host and not from molecular evidence. The mechanism by which the *tox*$^+$ phages transduce different complements of genes is not known, but recombination with the chromosome could explain this ability. Several phages and plasmids transiently integrate into the chromosome of bacteria[28] by site-specific recombination which is catalysed by recombinases of the integrase family. It would be of value to determine if the *tox*$^+$ phages transiently integrate in the chromosome and to identify the genes involved in their life cycle to elucidate the host–parasite relationship. Other phages possessing toxin genes such as those found in *Corynebacterium diphtheriae* exist in a variety of states, including a stable lysogenic state and in a pseudolysogenic 'carrier' state.[29,30]

Following the demonstration of phages harboring genes for BoNT/C and BoNT/D production, efforts were made to demonstrate a relationship between phage and toxin production in other serotypes of *C. botulinum*.[17] Although phages are abundantly present in virulent clostridia, their involvement in toxin production has not been demonstrated in *C. botulinum* types A, B, E, F, and G.[17] Experimental difficulties have been encountered including curing strains of phages and especially in developing indicator strains and infectious lysates that can form plaques on agar plates. The evidence that bacteriophages or plasmids are not involved in toxinogenicity has been indirectly based on continued toxin production in strains cured of phages and plasmids after treatment with such agents as mitomycin C and acridine orange. The clostridia often carry six or more different phages and several plasmids and it is difficult to ascertain whether all phages and plasmids have been cured or if stable integration of plasmids or phage genomes into the chromosome has occurred. Hence, it cannot be stated unequivocally at this time that the genes for botulinum toxins are not associated with phages in serotypes A, B, E, and F.

Several plasmids have been demonstrated in *C. botulinum* but they have not been demonstrated to affect virulence or to code for known gene products.[31–33] Strom *et al.*[32] screened *C. botulinum*, *C. sporogenes*, *C. subterminale*, and *C. botulinum*-like organisms for plasmids. Fifty-six per cent of 68 toxigenic *C. botulinum* strains contained one or more plasmids, ranging in mass from 31 to 121 kb. The largest plasmids of 65 and 121 kb were found in a type Bf and type G strain, respectively. Difficulties were encountered in identifying plasmids in several strains owing to DNase activity and poor lysis of the cells. Of 15 nontoxigenic *C. botulinum*-like strains examined, 40% also possessed one or more plasmids ranging from 2.5 to 37 kb. No phenotypic functions including toxigenicity could be attributed to the plasmids. Ferreira *et al.*[33] identified one or more plasmids in 54% of the *C. botulinum* type A, B, E, and F strains examined. The plasmids ranged in mass from 5.1 to around 103 kb, but it was difficult to cure the plasmids from two type A strains using acridine orange and elevated growth temperatures. Weickert *et al.*[34] demonstrated the presence of plasmids in 12 strains of *C. botulinum* type A and seven strains of *C. sporogenes*. Nine strains had one or more plasmids, and several strains showed a large plasmid of 91 kb, but it was not consistently recovered. Toxigenicity of *C. botulinum* type A was not correlated with the presence of a plasmid. The gene for BoNT/A has been reported to be present on the chromosome from cloning and sequencing experiments.[35,36] However, these experiments used total genomic DNA which could contain contaminating DNA from plasmids and therefore DNA of lysogenic phages. Using a transposon Tn*916* mutagenesis system[37] and genome analysis by pulsed-field gel electrophoresis (PFGE),[38] the genes encoding BoNT/A, HA, and NTNH were mapped to relatively large chromosomal

DNA fragments in *C. botulinum* type A[38] and in a strain possessing a silent B gene.[10] These fragments mapped to fragments ⩾72 kb, larger than any plasmids in the strains. In nontoxigenic mutants isolated by Tn*916* mutagenesis, a large deletion was detected that resulted in concomitant loss of the entire genes for BoNT/A, NTNH, and HA.[38,39] The chromosomal macrorestriction fragments originally harboring the BoNT, NTNH, and HA genes were also decreased in size by around 70 kb. These results suggest that the toxin complex genes in *C. botulinum* type A are chromosomally located but unstable and may therefore be associated with mobile genetic elements.

Of the seven serotypes of neurotoxigenic strains, only the gene encoding type G has been clearly demonstrated to reside on a plasmid. Evidence for plasmid-mediated toxicity was suggested when a culture of *C. botulinum* type G ceased production of BoNT/G and a bacteriocin when it was cultured at 44°C.[40] The plasmid location of the genes for BoNT/G, NTNH, and HA was directly demonstrated when a 114 kb plasmid purified from *C. botulinum* type G was shown to hybridize to a BoNT/G gene probe.[42] The plasmid, but not chromosomal DNA, from the toxigenic strain provided a template for PCR to amplify an oligonucleotide of the same size (*c* 1.3 kb) as the L chain of the BoNT/G gene. Zhou *et al.*[41] also demonstrated that the genes for the BoNT/G, NTNH, and HA were linked in a cluster and had the orientation previously found in type C.[8,12,42–44]

An interesting aspect of the genes encoding BoNT and other virulence genes is their association with unstable and potentially mobile genetic elements. Investigations suggest that the gene encoding C_3 botulinal toxin is associated with a 21.5 kb DNA fragment resembling a mobile DNA element[43,44] of the Tn*554* family of transposons.[45] The isolation of neurotoxigenic *C. butyricum* and *C. baratii* strains[46] indicates that the toxin genes have been transferred to these normally nontoxigenic clostridia. Hauser *et al.*[47] reported that a large extrachromosomal plasmid was present in a toxigenic *C. butyricum* strain. When plasmid or chromosomal DNA was used as the template for polymerase chain reaction (PCR) using primers corresponding to DNA sequences at the 5′ and 3′ termini of the BoNT/E light chain, amplification products were obtained from plasmid (strong signal) and chromosomal DNA (weaker signal). The chromosomal amplification product was attributed to contamination with plasmid DNA. The isolated plasmid was not reported to react directly with a toxin gene probe. These results are suggestive of, but not conclusive for, plasmid localization of the type E BoNT gene since PCR amplification was required to detect the presence of the toxin gene.

Zhou *et al.*[48] reported that a toxin gene probe hybridized directly to chromosomal DNA of toxigenic strains of *C. butyricum* and to *C. botulinum* type E but not to plasmid DNA. All toxigenic and most nontoxigenic strains were lysogenized by a prophage on the chromosome. Prophages of toxigenic strains, irrespective of species, had related or identical DNA sequences which differed from the genomes of prophages found in nontoxigenic strains. The prophages were located close to the toxin gene on the chromosome. DNA from the phages induced by mitomycin C and purified by cesium chloride density centrifugation did not hybridize with the toxin gene probe. These results suggested that a mixed population of phage was induced, and that the phage carrying the toxin gene was present at low levels and was undetectable by direct hybridization. Phage DNA but not chromosomal DNA purified from toxigenic strains served as a PCR template to amplify the toxin gene. This amplification was not negated by first incubating the phage particles with DNase and RNase, indicating that the toxin gene was sheltered in the phage particle.

These findings supported natural interspecies transfer of the BoNT/E gene from *C. botulinum* to *C. butyricum*. Interspecies transfer of the toxin gene could not be obtained in the laboratory when *C. botulinum* type E or phage lysates from type E were used as the gene donor. Interspecies transfer of the toxin gene did occur when a nontoxigenic *C. botulinum* type E-like strain was used as a recipient and phage lysates induced from toxigenic *C. butyricum* together with phage lysates from nontoxigenic *C. butyricum* were used as gene donors. The toxin gene was detected in about 1 in 10^4 colonies obtained by plating the recipient strain. Hence, the gene for BoNT/E was associated with a prophage and transfer to a nontoxigenic strain appeared to result from transduction by a defective phage which required a helper phage for infectivity.

BoNT gene transfer is also supported by the isolation of *C. baratii* strains that produce type F botulinal toxin. BoNT gene transfer to *C. baratii* appears to have occurred several times, because toxigenic *C. baratii* strains have been isolated from humans with intestinal botulism in distinct geographic locations within the USA.[49,50] Natural interspecies transfer of toxin genes could be significant for the food industry and medical microbiology. However, understanding the mechanisms of toxin gene transfer is hindered by the rudimentary knowledge of molecular genetics and gene transfer in *C. botulinum* and by the lack of a positive genetic selection for the BoNT phenotype. Genetic transfer of genes encoding neurotoxins is also supported by the isolation of clostridial strains that produce more than the BoNT serotype or that harbor the genes for more than one serotype, yet express only one serotype.[10,47]

III. *CLOSTRIDIUM TETANI*

Clostridium tetani produces two toxin proteins: a hemolysin, referred to as tetanolysin, and its potent neurotoxin (TeNT). The production of the hemolysin is not associated with toxin production because strains that have lost the ability to produce tetanus toxin still synthesize tetanolysin. Early investigations showed that synthesis of TeNT was an unstable character in *C. tetani*[51] and that strains varied considerably in production. Mueller and colleagues found that TeNT production was highly dependent on the strain and its maintenance.[52] Instability of TeNT production and frequent occurrence of nontoxic derivatives suggested that the gene encoding TeNT was associated with an unstable genetic element. Nontoxigenic derivatives of *C. tetani* were isolated by treatment with acridine orange, *N*-methyl-*N*′-nitrosoguanidine, rifampicin or ultraviolet light.[53] The frequency of nontoxigenic colonies was 0.8–3.2%, which is much higher than expected for mutations in bacterial chromosomes. Hara *et al.*[53] predicted that genetic factors controlling tetanus toxin production were associated with extra-chromosomal elements – specifically a plasmid, or possibly with a nonintegrative lysogenic bacteriophage. Attempts to demonstrate a relationship between toxigenicity and bacteriophages were unsuccessful.[53–55]

Laird *et al.*[56] found that 52 toxigenic strains possessed a single large plasmid while all nontoxigenic derivatives did not carry the plasmid. Direct evidence of the presence of the tetanus toxin structural gene on a large plasmid was presented by Finn *et al.*,[57] who isolated a plasmid and showed that a pool of oligonucleotides prepared on the basis of a six amino acid stretch near the amino terminus of fragment C hybridized to plasmid DNA. The oligonucleotides did not hybridize to plasmids from nontoxigenic strains. Eisel *et al.*[58] definitively showed that the entire structural gene for TeNT was located on a plasmid. They determined the nucleotide sequence of the TeNT structural gene that had been subcloned and expressed in *Escherichia coli*. Overlapping DNA fragments encompassing the entire structural gene for TeNT hybridized with plasmid DNA isolated from the Massachusetts strain of *C. tetani*.

It is interesting that although the structural gene for TeNT is on a plasmid, and TeNT is entirely responsible for virulence, since humans or animals immunized against toxoided TeNT are protected against tetanus, the expression of the toxin is regulated in *C. tetani* by nutritional and environmental factors. TeNT is regulated by the concentration of iron in the medium,[52] by the presence of a peptide containing histidine,[59] and by nitrogen (glutamate) repression of toxin synthesis.[60] These findings suggest that when the *tox*$^{+}$ plasmid is present in *C. tetani*, the expression of the gene for tetanus toxin is regulated by chromosomal genes. Mueller and Miller[52] isolated a strain in which tetanus toxin production was not repressed by iron and that produced 1–2 million mouse LD_{50} per ml, or close to 5% of total cellular protein as TeNT. This level of protein is comparable to that obtained in high-producing strains of *C. botulinum* type A which can produce 1–4 million LD_{50} per ml.[61]

IV. *CLOSTRIDIUM PERFRINGENS*

Clostridium perfringens produces a variety of toxins that are lethal to animals and also an enterotoxin that causes diarrhea in humans and necrotic enteritis in domestic animals.[62] Plasmids were suggested to be linked to enterotoxin and beta toxin production in strains of *C. perfringens*.[63] Canard *et al.*[64] used PFGE and gene mapping to investigate genomic diversity and virulence in 10 strains of *C. perfringens* (see Chapter 4). They identified three polymorphic regions of chromosomal DNA, one of which was associated with the enterotoxin gene, *cpe*. Another variable region spanned the virulence gene, *plc*, which encodes the cytolytic α-toxin, a phospholipase C. They also demonstrated that the gene for epsilon toxin, a protease-activated lethal toxin, resides on large plasmids that differ in size in different strains. The comparative genome mapping showed that the chromosome had a similar organization among *C. perfringens* strains isolated from different geographical locations and over a time period exceeding half a century. However, three polymorphic regions were detected, which were linked to virulence genes. Variable regions were also linked to rRNA operons, suggesting that one mechanism of genomic diversification could involve homologous recombination between *rrn* genes, as reported in *E. coli* and *Salmonella enterica* ser. typhimurium.[65]

In a subsequent study[66] PFGE and gene probes were used to investigate the *cpe* location in a larger collection of *C. perfringens* strains. The *cpe* gene encoding the enterotoxin was demonstrated to be associated with the chromosome in human isolates, but with a large plasmid in isolates from livestock. The chromosomal *cpe* gene was linked to a repetitive DNA sequence and an open reading frame that may be part of an insertion sequence related to IS*200* which was initially identified in *S.* typhimurium.[67] The *cpe* gene on the plasmid was linked to a copy of the transferable element IS*1151* (see Chapter 6). In both locations, *cpe* may be associated with mobile genetic elements that could lead to instability or horizontal transfer to nontoxigenic clostridia. Young and Cole[68] provided evidence that *C. perfringens* has engaged in gene transfer with *Enterococcus faecalis* and *Clostridium difficile*. Cornillot *et al.*[66] also proposed that infection of *cpe*-negative *C. perfringens* initially occurred by a temperate phage that could also exist episomally, or alternatively by introduction of a transposon similar to Tn*554* from *Staphylococcus aureus*.[45]

The *etx* gene encoding the major virulence factor, the epsilon toxin, also resided on the same plasmid as the *cpe* gene. The *etx* gene is linked to IS*1151*.[66,69] These results provide an explanation of the relatively rare phenotype of enterotoxin production in randomly sampled *C. perfringens* strains. Recently, lambda toxin from *C. perfringens* was purified and shown to be a thermolysin-like metalloprotease.[70] The gene encoding the lambda toxin or caseinase, was originally reported to reside on a 2.1 kb plasmid in strain ATCC 3626B.[71] Using DNA hybridization with probes prepared from the cloned *lam* gene, the gene was localized to a 70 kb plasmid. Further investigation should elucidate whether *lam* resides on different plasmids depending on the strain.

V. OTHER CLOSTRIDIA

Other toxigenic clostridia include the species *Clostridium difficile*, *Clostridium sordellii*, *Clostridium novyi*, *Clostridium haemolyticum*, *Clostridium chauvoei*, *Clostridium septicum*, *Clostridium histolyticum*, *Clostridium spiroforme*, and *Clostridium colinum*.[4,62] Except for *C. difficile*, where toxin genes are chromosomally located, and *C. novyi* which is closely related to *C. botulinum* types C and D, detailed investigations of the genetic locations of virulence factors have not been carried out. It can be expected that future investigations will reveal an association of many virulence factors in these species with episomal or mobile genetic elements.

VI. CONCLUDING REMARKS

The determination of the genetic location of virulence genes of clostridia has been greatly facilitated by cloning of the structural genes, sequence analysis, and the use of DNA probes to investigate the location and mobility of the genes. The use of PFGE, gene probes and other molecular approaches has demonstrated that virulence genes in clostridia are frequently associated with unstable genetic elements such as plasmids, bacteriophages, transposons, and possibly other classes of mobile genetic elements. The study of the location, genetic transfer, and expression of virulence genes is in its infancy and has been investigated in only a few species. Many questions remain to be answered including the mechanisms governing instability of toxin production, the movement of virulence genes between extrachromosomal and chromosomal locations, horizontal transfer to other microbial species, and ultimately the function of the virulence genes in the host bacterium. It is expected that molecular approaches to the analysis of virulence genes will reveal further examples of association with extrachromosomal genetic elements and shed light on the evolution and biology of toxin production in the clostridia.

VII. REFERENCES

1. Wood W.B. Jr. and Davis B.R. (1980) Host–parasite relations in bacterial infections. In Davis B.R., Dulbecco R., Eisen H.N. and Ginsberg H.S. (eds), *Microbiology*, 3rd edn, pp. 552–571. Harper and Row, Philadelphia, PA.
2. van Heyningen W.E. (1950) *Bacterial Toxins.* Blackwell Scientific Publications, Oxford.
3. Smith L.D.S. and Williams B.L. (1984) *The Pathogenic Anaerobic Bacteria.* C. Thomas, Springfield, IL.
4. Hatheway C.L. (1990) Toxigenic clostridia. *Clin Microbiol Rev* **3**: 66–98.
5. Gill D.M. (1982) Bacterial toxins: a table of lethal amounts. *Bacteriol Rev* **46**: 86–94.
6. Sugiyama H. (1980) *Clostridium botulinum* neurotoxin. *Microbiol Rev* **44**: 419–448.
7. Sakaguchi G. (1983) *Clostridium botulinum* toxins. *Pharmacol Ther* **19**: 165–194.
8. Oguma K., Fujinaga Y. and Inoue E. (1995) Structure and function of *Clostridium botulinum* toxins. *Microbiol Immunol* **39**: 161–168.
9. Franciosa G., Feirreira J.L. and Hatheway C.L. (1994) Detection of type A, B, and E botulism neurotoxin genes in *Clostridium botulinum* and other *Clostridium* species by PCR: evidence of unexpressed type B toxin genes in type A toxigenic organisms. *J Clin Microbiol* **32**: 1911–1917.
10. Hutson R.A., Zhou Y., Collins M.D., Johnson E.A., Hatheway C.L. and Sugiyama H. (1996) Genetic characterization of *Clostridium botulinum* type A containing silent type B neurotoxin gene sequences. *J Biol Chem* **271**: 10786–10792.
11. Minton N.P. (1995) Molecular genetics of clostridial neurotoxins. In Montecucco C. (ed.), *Clostridial Neurotoxins*, pp. 161–194. Springer, Berlin.
12. Fujinaga Y., Inoue K., Shimazaki S., Tomochika K., Tsuzuki K., Fujii N., Watanabe T., Ohyama T., Takeshi K., Inoue K. and Oguma K. (1994) Molecular construction of *Clostridium botulinum* type C progenitor toxin and its gene organization. *Biochem Biophys Res Commun* **205**: 1291–1298.
13. Inoue K. and Iida H. (1968) Bacteriophages of *Clostridium botulinum. J Virol* **2**: 537–540.
14. Eklund M.W., Poysky F.T and Boatman E.S. (1969) Bacteriophages of *Clostridium botulinum* types A, B, E, and F and nontoxigenic strains resembling type E. *J Virol* **3**: 270–274.
15. Dolman C.E. and Chang E. (1972) Bacteriophages of *Clostridium botulinum. Can J Microbiol* **18**: 67–76.
16. Sugiyama H. and King G.J. (1972) Isolation and taxonomic significance of bacteriophages for non-proteolytic *Clostridium botulinum. J Gen Microbiol* **70**: 517–525.
17. Eklund M.W., Poysky F.T. and Habig W.H. (1989) Bacteriophages and plasmids in *Clostridium botulinum* and *Clostridium tetani* and their relationship to production of toxins. In Simpson L.L. (ed.) *Botulinum Neurotoxin and Tetanus Toxin*, pp. 25–51. Academic Press, San Diego, CA.
18. Inoue K. and Iida H. (1970) Conversion of toxigenicity in *Clostridium botulinum* type C. *Jap J Microbiol* **14**: 87–89.
19. Eklund M.W., Poysky F.T., Reed S.M. and Smith C.A. (1971) Bacteriophage and toxigenicity of *Clostridium botulinum* type C. *Science* **172**: 480–482.
20. Adams M.H. (1959) *Bacteriophages.* Wiley Interscience, New York.
21. Jansen B.C. (1971) The toxic antigenic factors produced by *Clostridium botulinum* types C and D. *Onderstepoort J Vet Res* **38**: 93–98.
22. Inoue K. and Iida H. (1971) Phage conversion of toxigenicity in *Clostridium botulinum* types C and D. *Jpn J Med Sci* **24**: 53–56.
23. Eklund M.W., Poysky F.T. and Reed S.M. (1972) Bacteriophages and the toxigenicity of *Clostridium botulinum* type D. *Nature New Biol* **235**: 16–18.
24. Oguma K. Iida H. and Shiozake M. (1976) Phage conversion to hemagglutinin production in *Clostridium botulinum* types C and D. *Infect Immun* **14**: 597–602.
25. Popoff M.R., Hauser D., Boquet P., Eklund M.W. and Gill D.M. (1991) Characterization of the C_3

gene of *Clostridium botulinum* types C and D and its expression in *Escherichia coli*. *Infect Immun* **59**: 3673–3679.

26. Eklund M.W., Poysky F.T., Meyers J.A. and Pelroy G.A. (1974) Interspecies conversion of *Clostridium botulinum* type C to *Clostridium novyi* by bacteriophage. *Science* **186**: 456–458.
27. Oguma K., Iida H., Shiozaki M. and Inoue I. (1976) Antigenicity of converting phages obtained from *Clostridium botulinum* types C and D. *Infect Immun* **13**: 855–860.
28. Campbell A.M. (1992) Chromosomal insertion sites for phages and plasmids. *J Bacteriol* **174**: 7496–7499.
29. Barksdale L. (1971) The gene, *tox*$^+$, of *Corynebacterium diphtheriae*. In Monod J. and Borek E. (eds), *Of Microbes and Life*, pp. 215–232. Columbia University Press, New York.
30. Lwoff A. (1953) Lysogeny. *Bacteriol Rev* **17**: 269–337.
31. Scott V.N. and Duncan C.L. (1978) Cryptic plasmids in *Clostridium botulinum* and *C. botulinum*-like organisms. *FEMS Microbiol Lett* **4**: 55–58.
32. Strom M.S., Eklund M.W. and Poysky F.T. (1984) Plasmids in *Clostridium botulinum* and related *Clostridium* species. *Appl Environ Microbiol* **48**: 956–963.
33. Ferriera J.L., Hamdy M.K., Zapatka F.A. and McCay S.G. (1987) Plasmids of *Clostridium botulinum* type A, B, E, F, and other clostridial species. *J Food Safety* **8**: 267–277.
34. Weickert M.J., Chambliss G.H. and Sugiyama H. (1986) Production of toxin by *Clostridium botulinum* type A strains cured of plasmids. *Appl Environ Microbiol* **51**: 52–56.
35. Binz T., Kurazona H., Wille M., Fevart J., Wernars K. and Niemann H. (1990) The complete sequence of botulinum neurotoxin A and comparison with other clostridial neurotoxins. *J Biol Chem* **265**: 9153–9158.
36. Thompson E.E., Brehm J.K., Oultram J.D., Swinfield T.-J., Shone C.C., Atkinson T., Melling J. and Minton N.P. (1990) The complete amino acid sequence of the *Clostridium botulinum* type A neurotoxin deduced by nucleotide sequence analysis of the coding gene. *Eur J Biochem* **189**: 73–81.
37. Lin W.-J. and Johnson E.A. (1991) Transposon Tn*916* mutagenesis in *Clostridium botulinum*. *Appl Environ Microbiol* **57**: 2946–2950.
38. Lin W.-J. and Johnson E.A. (1995) Genome analysis of *Clostridium botulinum* type A by pulsed-field gel electrophoresis. *Appl Environ Microbiol* **61**: 4441–4447.
39. Lin W.-J. (1992) The development of genetic methods to study group I *Clostridium botulinum*. PhD Thesis, University of Wisconsin, Madison.
40. Eklund M.W., Poysky L., Mseiteif M. and Strom M.S. (1988) Evidence for plasmid-mediated toxin and bacteriocin production in *Clostridium botulinum* type G. *Appl Environ Microbiol* **54**: 1405–1408.
41. Zhou Y., Sugiyama H., Nakano H. and Johnson E.A. (1995) The genes for the *Clostridium botulinum* type G toxin complex are on a plasmid. *Infect Immun* **63**: 2087–2091.
42. Hauser D., Gibert M., Marvaud J.C., Eklund M.W. and Popoff M.R. (1995) Botulinal neurotoxin C1 complex genes, clostridial neurotoxin homology and genetic transfer in *Clostridium botulinum*. *Toxicon* **33**: 515–526.
43. Hauser D,. Eklund M.W., Boquet P. and Popoff M.R. (1994) Organization of the botulinum neurotoxin C1 gene and its associated non-toxic protein genes in *Clostridium botulinum* C 468. *Mol Gen Genet* **243**: 631–640.
44. Hauser D., Gibert M., Eklund M.W., Boquet P. and Popoff M.R. (1993) Comparative analysis of C3 and botulinal neurotoxin genes and their environment in *Clostridium botulinum* type C and D. *J Bacteriol* **175**: 7260–7268.
45. Murphy E. (1983) Transposable elements in Gram-positive bacteria. In Berg D.E. and Howe M.M. (eds), *Mobile DNA*, pp. 269–288. American Society for Microbiology, Washington, D.C.
46. Hatheway C.L. (1995) Botulism: the present status of the disease. In Montecucco C. (ed.), *Clostridial Neurotoxins*, pp. 55-75. Springer, Berlin.

47. Hauser D., Gibert P., Boquet P. and Popoff M.R. (1992) Plasmid localization of a type E botulinal neurotoxin gene homologue in toxigenic *Clostridium butyricum* strains, and the absence of this gene in non-toxigenic *C. butyricum* strains. *FEMS Microbiol Lett* **99**: 251–256.
48. Zhou Y., Sugiyama H. and Johnson E.A. (1993) Transfer of neurotoxigenicity from *Clostridium butyricum* to a nontoxigenic *Clostridium botulinum* type E-like strain. *Appl Environ Microbiol* **59**: 3825–3831.
49. Arnon S.S. (1995) Botulism as an intestinal toxemia. In Blaser M.J., Smith P.D., Ravdin J.I., Greenberg H.B. and Guerrant R.L. (eds) *Infections of the Gastroinestinal Tract*, pp. 257–271. Raven Press, New York.
50. Paisley J.W., Lauer B.A and Arnon S.S. (1995) A second case of type F botulism caused by *Clostridium baratii. The Pediatr Inf Dis J* **14**: 912–914.
51. Fildes P. (1925) Tetanus. I. Isolation, morphology, and cultural reactions of *B. tetani. Brit J Exp Path* **6**: 62–70.
52. Mueller J.H. and Miller P.A. (1945) Production of tetanal toxin. *J Immunol* **47**: 15–22.
53. Hara T., Matsuda M. and Yoneda M. (1977) Isolation and some properties of nontoxigenic derivatives of a strain of *Clostridium tetani. Biken J* **20**: 105–115.
54. Prescott L.M. and Altenberg R.A. (1967) Inducible lysis in *Clostridium tetani. J Bacteriol* **93**: 1220–1226.
55. Prescot L.M. and Altenberg R.A. (1976) Detection of bacteriophages from two strains of *Clostridium tetani. J. Virol* **1**: 1085–1086.
56. Laird W.J., Aaronson W. Silver R.P., Habig W.H. and Hardegree M.C. (1980) Plasmid-associated toxigenicity in *Clostridium tetani. J Inf Dis* **142**: 623.
57. Finn C.W., Jr, Silver R.P., Habig W.H., Hardegree M.C., Zon G. and Garon C.F. (1984) The structural gene for tetanus neurotoxin is on a plasmid. *Science* **224**: 881–884.
58. Eisel U., Jarausch W., Goretzki K., Henschen A., Engels J., Weller U., Habermann E. and Niemann H. (1986) Tetanus toxin: primary structure, expression in *E. coli*, and homology with botulinum toxins. *EMBO J* **5**: 2495–2502.
59. Mueller J.H. and Miller P.A. (1956) Essential role of histidine peptides in tetanus toxin production. *J Biol Chem* **223**: 175–194.
60. Tsunashima I., Sato K., Shoji K., Yoneda M and Amano T. (1964) Excess supplementation of certain amino acids and its inhibitory effect on toxin production by *Clostridium tetani. Biken J* **7**: 161–163.
61. Schantz E.J. and Johnson E.A. (1992) Properties and use of botulinum toxin and other microbial toxins in medicine. *Microbiol Rev* **56**: 80–99.
62. Songer J.G. (1996) Clostridial diseases of domestic animals. *Clin Microbiol Rev* **9**: 216–234.
63. Rokos E.A., Rood J.I. and Duncan C.L. (1978) Multiple plasmids in different toxigenic types of *Clostridium perfringens. FEMS Microbiol Lett* **4**: 323–326.
64. Canard B., Saint-Joanis B. and Cole S.T. (1992) Genomic diversity and organization of virulence genes in the pathogenic anaerobe *Clostridium perfringens. Mol Microbiol* **6**: 1421–1429.
65. Anderson P. and Roth J. (1981) Spontaneous tandem genetic duplications in *Salmonella typhimurium* arise by unequal recombination between rRNA (*rrn*) cistrons. *Proc Natl Acad Sci USA* **78**: 3113–3117.
66. Cornillot E., Saint-Joanis B., Daube G., Katayama S-I., Granum P.E., Canard B. and Cole S.T. (1995) The enterotoxin gene (*cpe*) of *Clostridium perfringens* can be chromosomal or plasmid-borne. *Mol Microbiol* **15**: 639–647.
67. Brynestad S., Iwanejko L.A., Stewart G.S.A.B. and Granum P.E. (1994) A complex array of Hpr consensus DNA recognition sequences proximal to the enterotoxin gene in *Clostridium perfringens. Microbiol* **140**: 97–104.
68. Young M. and Cole S.T (1993) *Clostridium.* In Sonenshein A.L., Hoch J.A. and Losick R. (eds),

Bacillus subtilis and Other Gram-positive Bacteria: Biochemistry, Physiology and Molecular Genetics, pp. 35–52. American Society for Microbiology, Washington, D.C.
69. Daube G., Simon P., and Kaeckenbeeck A. (1993) IS1151, an IS-element of *Clostridium perfringens. Nucl Acids Res* **21**: 352.
70. Jin F., Matushita O., Katayama S.-I., Jin S., Matushita C, Minami J., and Okabe A. (1996) Purification, characterization, and primary structure of *Clostridium perfringens* lambda-toxin, a thermolysin-like metalloprotease. *Infect Immun* **64**: 230–237.
71. Blaschek H.P., and Solberg M. (1981) Isolation of a plasmid responsible for caseinase activity in *Clostridium perfringens* ATCC 3626B. *J Bacteriol* **147**: 262–266.

Chapter 4

Structure, Organization and Evolution of the Genome of *Clostridium perfringens*

STEWART T. COLE AND BRUNO CANARD
Unité de Génétique Moléculaire Bactérienne, Institut Pasteur, 28 Rue du Docteur Roux, 75724 Paris Cedex 15, France

I. INTRODUCTION

During the last decade our understanding of the structure and organization of bacterial chromosomes has advanced to an unparalleled extent thanks to the introduction of powerful new techniques for analyzing and physically mapping large DNA molecules. The advent of pulsed-field gel electrophoresis (PFGE) has freed the study of bacterial genomes from the constraints of recombination and linkage analysis by enabling the chromosomes of bacteria for which no genetic systems exist to be studied *in situ*. At the time of writing, detailed physical and gene maps of about 100 bacterial species have been reported and a vast body of information about the size, number and topology of prokaryotic chromosomes has accumulated.[1,2] Strange though it may seem, *Clostridium perfringens* was the first Gram-positive bacterium for which a complete genome map was constructed in this way and this map was described in 1989,[3] a few years prior to publication of the physical map of the chromosome of *Bacillus subtilis* 168,[4] the prototype of the endospore-forming bacteria. Initial work focused on the type A

The Clostridia: Molecular Biology and Pathogenesis
ISBN 0-12-595020-9

strain CPN50 but later studies addressed the issues of genome structure and evolution in the other major biotypes, B–E.[5,6]

II. THE GENOME MAP OF CPN50 – THE PARADIGM TYPE A STRAIN

One of the first strains of *C. perfringens* to attract the interest of bacteriologists was BP6K-N5 or ATCC10543[7] which produces the bacteriocin BCN5[8] that is encoded by the plasmid pIP404.[9] Plasmid pIP404 gave rise to a promising shuttle-vector system[10,11] that was later adopted as the basis of the popular *C. perfringens* vector, pJIR418.[12] BP6K-N5 was used extensively in the 1950s and was subsequently renamed CPN50.[9] As the minimal growth requirements of CPN50 were known[13] and its derivatives such as CPN52, a strain cured of pIP404, could be transformed by electroporation it represented the obvious choice for intensive genome mapping studies.

Initial studies showed that the A+T-rich genome of CPN50 could be cleaved into a limited number of fragments that were readily resolved by PFGE when a set of restriction enzymes with G+C-rich recognition sequences was used. This set included *Apa*I (GGGCCC), *Fsp*I (TGCGCA), *Mlu*I (ACGCGT), *Nru*I (TCGCGA) *Sac*II (CGGCGG) and *Sma*I (CCCGGG) which generated from five (*Mlu*I) to 16 (*Fsp*I) fragments ranging in size from 0.4 to 2280 kb. The physical map of the chromosome was obtained by a combination of PFGE techniques, including the analysis of fragments produced by single, double and partial digests, and by linking clone analysis in which probes spanning rare-cutter sites were used to demonstrate the contiguity of adjacent DNA fragments.[3] The resultant map, on which 105 restriction sites had been positioned, revealed the presence of a single circular chromosome of around 3.6 Mb. Subsequently, sites recognized by *Sfi*I and the intron-encoded endonuclease I-*Ceu*I were localized on an updated version.[14]

Although a number of the restriction sites, including those for I-*Ceu*I, occur in clusters (R) in the 10 *rrn* operons,[14,15] the majority are randomly distributed and delimit 68 mapping intervals, varying in size from 1 to 200 kb, in which cloned genes can be positioned by hybridization. Our current mapping strategy consists of an approximate localization using blots of I-*Ceu*I or *Mlu*I digests[16] followed by hybridization analysis of single and double digests especially with the informative, more frequently cutting enzymes, *Apa*I and *Fsp*I. The locations of almost 100 known genes have now been determined and the current map is presented in Figure 4.1. Rapid expansion of the map was achieved by a scanning strategy in which probes for sequence tagged sites, obtained by shotgun cloning and sequencing, were positioned.[14] As most of these corresponded to known genes this greatly improved the information content.

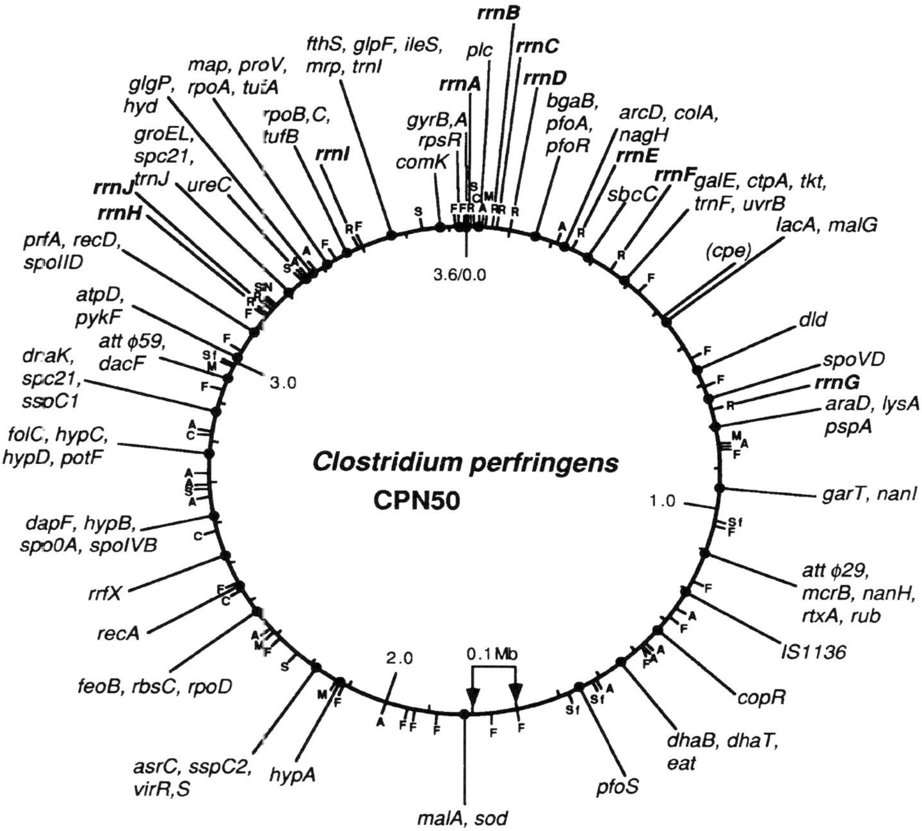

Figure 4.1 The current physical and gene map of the 3.6 Mb chromosome of *Clostridium perfringens* CPN50 showing the positions of the various cleavage sites and the *rrn* operons. Restriction enzyme sites; A, *Apa*I; C, *Sac*II (*Ksp*I); F, *Fsp*I (*Avi*II); M, *Mlu*I: N, *Nru*I; R, 2*Sac*II-*Sma*I-*Nru*I-2*Sma*I-*Nru*I-I-*Ceu*I; S, *Sma*I; Sf, *Sfi*I. The positions of genes are indicated by the circles situated in the center of the mapping intervals. When several genes map to the same site, their order is arbitrary. For details of genes and their functions see references 3 and 14. Reproduced with permission of the American Society for Microbiology from Katayama *et al.* (1995).[14]

III. FEATURES OF THE GENE MAP OF CPN50

The 10 *rrn* operons are organized around the origin of chromosomal replication, *oriC* (defined by *gyrBA*), at position 0, and are confined to a segment representing roughly one third of the chromosome, an arrangement reminiscent of that seen in *B. subtilis*.[17] Seven of the *rrn* operons are transcribed in a clockwise manner while the remaining three are transcribed in a counterclockwise direction (Figure 4.1). The known virulence genes of CPN50 are also distributed in a non-random manner as the *plc*, *pfoA*, *colA* and *nagH* genes encoding the α, θ, κ and μ toxins, respectively, are all localized within 250 kb of *oriC* while the global toxin regulatory locus *virRS*[18,19] that controls, among others, *plc*, *pfoA*, and *colA*, is close to the putative replication terminus. Other known genes, or candidates likely to be involved in the pathogenesis of *C. perfringens*, are also situated in this sector of the chromosome (Figure 4.1): notable among these are the sialidase genes, *nanH* and *nanI*, situated at *c.* 0.8 and 1 Mb from *oriC*, respectively, and *pspA* at 1.2 Mb, which encodes a putative surface protein homologous to the virulence factor PspA of *Streptococcus pneumoniae*.[14,20]

Comparison of the genome maps of three members of the endospore-forming Bacillaceae family, *B. subtilis*, *C. perfringens* and *C. beijerinckii* revealed that some heat shock (*groEL*, *dnaK*), sporulation (*spoVD*, *spo0A*, *spoIVB*, *spoIID*) and DNA damage response (*recA*, *uvrB*) genes occur in the same order and occupy roughly similar positions.[14] This common organization may indicate that all three were derived from a common spore-forming ancestor but that their genomes have subsequently diversified.

IV. GENOME MAPPING WITH OTHER BIOTYPES OF *CLOSTRIDIUM PERFRINGENS*

The tools developed in the course of mapping the genome of CPN50 proved useful for examining the chromosomes of strains belonging to other *C. perfringens* biotypes. In the first study aimed at assessing the extent of genomic diversity and localizing more genes encoding virulence determinants, detailed maps of the chromosomes of seven additional strains were established using linking clones, known genes and a battery of anonymous DNA fragments as probes.[21] This survey showed that in all cases a single, well-conserved circular chromosome of *c.* 3.6 Mb, carrying 10 *rrn* operons and a full complement of known virulence genes, was present (Table 4.1). Convincing proof that the genes for some of the typing toxins and other virulence factors were plasmid-borne has been obtained[21] and this is discussed below. In a subsequent study, I-*Ceu*I was employed to rapidly assess chromosomal variability, particularly among enterotoxin-producing strains and lower resolution maps were deduced by analogy with that of CPN50 and the consensus map.[16]

By means of pairwise map comparisons, and the identification of conserved restriction sites and genetic markers, a consensus map of the *C. perfringens* chromosome that was of use for phylogeny was deduced (Figure 4.2). Although global conservation of the chromosomal structure was found, a number of strains showed local variation from the consensus map and 10 variable regions differing in length (Figure 4.2, **a–j**) were identified. Interestingly, several of these variable

Table 4.1. *Clostridium perfringens* strains mapped

Strain	Biotype	Origin	Year of isolation	Genome map	I-*Ceu*I map	Size	Plasmids	Genes on plasmid
CPN50	A	Human, USA	1940s	+	+	3.58	+	*bcn*
NCTC6785	A	Human, USA	1944	+	+	3.55	+	
A99	A	Human, Germany	1986	+	+	3.5	?	
NCTC8798	A	Human, England	1953	+	+	3.07	+	
8-6	A	Spontaneous mutant of NCTC8798	1972	+	+	3.25	+	
NCTC9851	A	Human, England	1956	−	+	3.33[a]	?	
NCTC10239	A	Human, England	1961	−	+	3.05[a]	?	*cpe*
F3686/80	A	Human, England	1980	−	+	3.42[a]	+	*cpe*
44071-CO5	A	Goats, Belgium	1991	−	+		+	*cpe*
88L662MF	A	Pigs, Belgium	1988	−	+	3.61[a]	+	*cpe*
91E01793F	A	Cattle, Belgium	1991	−	+	3.60[a]	+	*cpe*
G23	B	Calves, France	1950s	+	+	3.5	+	
CN3922	B	Pigs, Belgium	1989	−	+		+	*btx?*
CWC235	C	Pigs, Denmark	?	−	−		+	*btx*
945P	D	Sheep, Belgium	1989	−	+		+	*cpe, etx*
NCTC2062	D	Sheep, England	?	+	+	3.65	+	*lam, etx*
ATCC27324	E	Rabbits, USA	1960s	+	+	3.65	+	*itxAB*

[a] Based on summing sizes of *Apa*I fragments and not on complete map values.

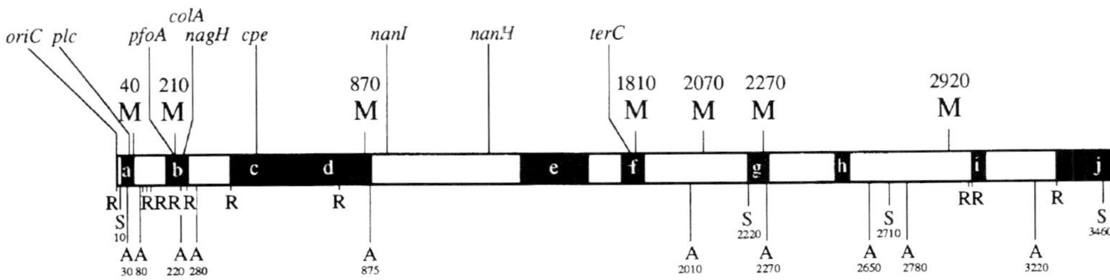

Figure 4.2 Consensus physical map of the *Clostridium perfringens* chromosome. The map was deduced from comparisons of the individual genome maps of eight strains (Tables 4.1 and 4.2) and shows the 10 *rrn* operons (denoted by the rare site cluster, R), 21 restriction sites plus key virulence genes and other markers. The order of the *rrn* operons is *rrnA–G*, *rrnJ*, *rrnI*, *rrnH* from left to right. The variable regions which display length polymorphisms, a–j, are indicated by black boxes.

regions were shown to be associated with the presence of genes involved in pathogenesis, such as *cpe*, as will be discussed in a later section. Strains such as NCTC8798, in which *cpe* was localized to variable region **c**, were found to deviate more extensively from the consensus map and to have the smallest chromosomes (Table 4.1).

V. GENOME MAPS AS PHYLOGENETIC INDEXES

The availability of detailed genome maps allows the extent of global relatedness between strains to be assessed by taking into consideration two evolutionary parameters: the divergence due to nucleotide variation, **d**, and to recombinational events like insertions and deletions, **g**. One means of monitoring **d** is to score well-defined restriction sites on the consensus map (Figure 4.2) as indicators of the sites of base variation leading to polymorphisms,[22] as shown in Table 4.2. The presence or absence of these sites is almost certainly the result of base pair substitutions and this has been proved experimentally in the case of the *Apa*I site at map coordinate 2270 kb. This restriction site is located in a constant genomic region[21] in all eight isolates but was lost by strain G23 as the result of a G–A transition, corresponding to the wobble position of a Gly codon, in the *Apa*I recognition sequence (S.T. Cole, unpublished data). Phylogenetic trees based on **d** values can then be constructed using the 'neighbor joining' method[23] and an example obtained with the best-mapped *C. perfringens* strains is shown in Figure 4.3 where it can be seen that the enterotoxin producer 8-6 is only distantly related to the others.

Pairwise map comparisons also identified regions **a–j** (Figure 4.2) that appear to differ in size as a result of recombinational events and, by expressing their

Table 4.2. Conservation of rare restriction sites in various *Clostridium perfringens* genomes

	*Mlu*I (position)[a]							*Sma*I (position)				*Apa*I (position)										
Strain	40	210	870	1810	2070	2270	2920	10	2220	2710	3460	30	80	220	280	875	2010	2270	2650	2780	3220	*n*
CPN50	1	0	1	0	1	1	1	1	1	1	1	1	0	1	0	1	0	1	1	1	1	16
8-6	1	0	0	1	1	1	1	1	1	1	0	1	0	1	1	1	0	1	0	0	1	14
NCTC6785	1	0	1	1	1	1	1	1	1	1	0	1	0	1	0	1	1	1	1	0	0	15
A99	1	1	1	0	1	1	1	1	1	1	0	1	0	1	0	1	0	1	1	0	1	15
G23	1	0	1	1	1	1	1	1	0	1	0	1	0	1	1	1	1	0	1	0	1	15
NCTC2062	1	1	1	1	1	1	1	1	1	1	0	1	1	1	0	1	0	1	1	0	1	17
ATCC27234	1	0	1	1	1	1	1	1	1	1	1	1	1	1	0	1	0	1	1	1	0	17

Sites whose presence (1) or absence (0) can be attributed to base substitutions and not to insertions or deletions were scored based on consensus map assignments.[6]

[a] Position refers to location in kilobases from *oriC* (Figure 4.2).

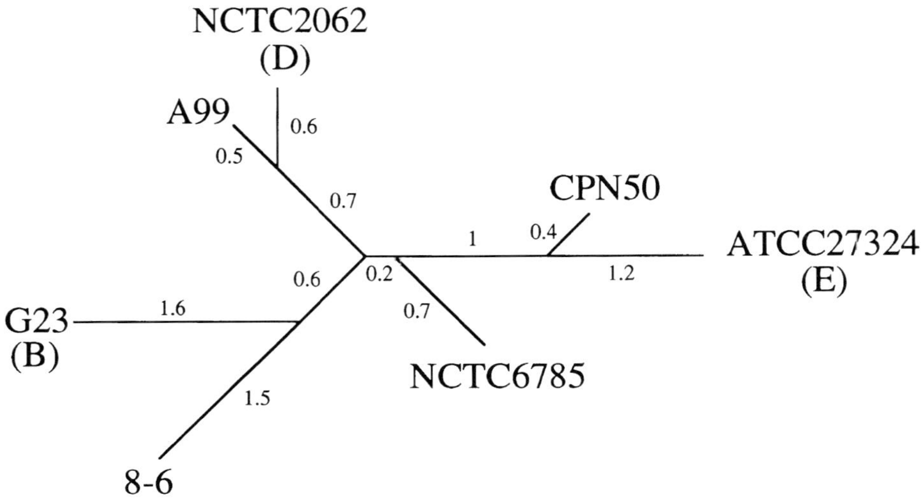

Figure 4.3 Phylogenetic tree constructed using nucleotide divergence **d**, and the neighbor joining method.[23] Relative branch lengths are indicated.

sum as a function of the genome size, as indicated in the legend to Table 4.3, **g** values can be expressed as percentage divergence. Again, it is apparent that strain 8-6 shows the highest phylogenetic distances when its **g** values are compared with those of the other isolates (Table 4.3) although its **d** values, i.e. at the nucleotide level, are more consensual. Indeed, **d** and **g** do not give the same values nor do they give the same relatedness patterns when used to build their corresponding phylogenetic tree (data not shown). This is an important finding for it is commonly believed that comparison of the PFGE restriction profiles is an accurate reflection of phylogenetic relatedness. Clearly, the situation is more complex and more reliable results can be obtained from map-based comparisons.

These combined studies indicate quite clearly that the classification of *C. perfringens* isolates into five biotypes by toxin typing is not a reflection of the genomic organization of the various strains. Indeed, as illustrated by strain 8-6, there are as many differences in map criteria between members of toxin-type A (Table 4.3), as there are between types A and D, or A and E. One can consider the *C. perfringens* strains that have been studied, and probably those that are circulating in nature as well, to be a relatively homogeneous group, showing near-constant chromosome organization and gene order. Many of the phenotypic differences which are observed can be attributed to plasmid-borne genes.

Table 4.3. Genomic parameters* of *Clostridium perfringens* strains

		d						
		CPN50	8-6	NCTC6785	A99	G23	NCTC2062	ATCC27234
g	CPN50		3.7 ± 1.6	2.9 ± 1.4	1.7 ± 1.0	4.3 ± 1.8	2.75 ± 1.3	1.6 ± 0.9
	8-6	11.7		3.2 ± 1.5	3.2 ± 1.5	3.2 ± 1.5	2.9 ± 1.4	4.3 ± 1.8
	NCTC6785	3.7	11.6		2.4 ± 1.3	2.4 ± 1.3	2.2 ± 1.2	2.2 ± 1.2
	A99	2	9.9	1.4		3.7 ±1.6	1.1 ± 0.8	3.5 ± 1.6
	G23	0.8	11.1	1.4	1.15		3.5 ± 1.6	4.8 ± 1.9
	NCTC2062	8.2	14.4	2.8	3.6	3.3		4.5 ± 1.8
	ATCC27234	2.5	14.4	2.8	4.5	3.3	10.7	

* The **d** values presented in the upper part were calculated by the neighbor joining method[27,30] using the following formula with values for CPN50 (x) or 8-6 (y): $m_x = 16$, $m_y = 15$, $m_{yy} = 12$, then $m_{av} = 15.5$ and $S = m_{yy}/m_{av} = 0.774$.
d = $(-\log_e S)/r$ where $r = 6$ for enzymes recognizing six base sites.
d = 0.0426.
Variance is given by:

$$V(\mathbf{d}) = \frac{(2 - S)(1 - S)}{2r2m_{av}S} = (0.018)^2$$

To illustrate how **g** was calculated for strain 8-6 relative to CPN50 the percentage differences in DNA content were as follows:

Zones	**a**	**b**	**c**	**d**	**e**	**f**	**g**	**h**	**i**	**j**
kb	0	−30 +10	−25	0	−280	0	−20	0	0	0

Genome sizes of CPN50 and 8-6 are $m_x = 3.58$ and $m_y = 3.25$ Mb respectively, and the shared DNA m_{xy} is 3.215 Mb deduced from the above Table. The divergence **g** can then be written as:[27,30]

$$\mathbf{g} = -2 \log_e \frac{m_{xy}}{(m_x m_y)^{0.5}} = 11.7\%$$

VI. PLASMIDS AND PFGE

The endonuclease I-*Ceu*I is particularly useful for distinguishing between genes localized on the chromosome and those situated on plasmids as its cleavage sites are confined to the *rrn* operons which are all chromosomal.[14,16,24] Other rare-cutting enzymes can also be used but some, such as *Apa*I, occasionally cleave plasmids.[21] In most, but not all, strains, plasmid DNA is readily apparent on PFGE-analysis of undigested DNA. Comparison of the hybridization profiles obtained with undigested and I-*Ceu*I-treated DNA thus enables one to distinguish between a plasmid and chromosomal localization as hybridization signals are obtained with chromosomal genes only after the chromosome has been digested thus enabling it to enter the gel. Plasmids <200 kb in size migrate freely into the gel but generally give complex hybridization patterns of at least three bands and this is illustrated in Figure 4.4. The fastest-moving species corresponds to the linearized form that is produced spontaneously during extraction of DNA from agarose-embedded bacteria whereas the other two probably correspond to open and closed circular forms of the plasmid. The apparent sizes of the circular forms vary with changes in the PFGE pulse times and ramping whereas the linear form behaves in a constant manner. Small plasmids are rarely detected by PFGE as they diffuse out of agarose plugs during preparation and treatment and the concentration of plasmids of <50 kb decreases during storage of the samples.

VII. VIRULENCE GENES LOCATED ON PLASMIDS

One of the principal findings of genome mapping studies with strains belonging to different biotypes of *C. perfringens* was the frequent occurrence of extra-chromosomal elements which carried the genes for toxins or other potential virulence factors. It has now been established that genes for three of the four typing determinants, the β-, ε- and ι-toxins are borne by plasmids, as are the genes for the λ-toxin, a protease, enterotoxin and urease. Details are presented in Table 4.2. It is not clear if these are genuine plasmids in all cases or whether some of them may correspond to temperate phages that can exist as episomes. The association of toxin genes with extrachromosomal elements in some pathogenic clostridia is well-known; see Chapter 3.[25] The genes for the botulinal neurotoxin are associated with phages in *C. botulinum* types C and D, and plasmids in type G strains. There is also strong evidence for the production of α-toxin in *C. novyi* resulting from lysogenization with phages[25] while the gene for the tetanus toxin is known to reside on a plasmid in *C. tetani*.[26]

As outlined above, based on genome-wide comparisons, *C. perfringens* isolates

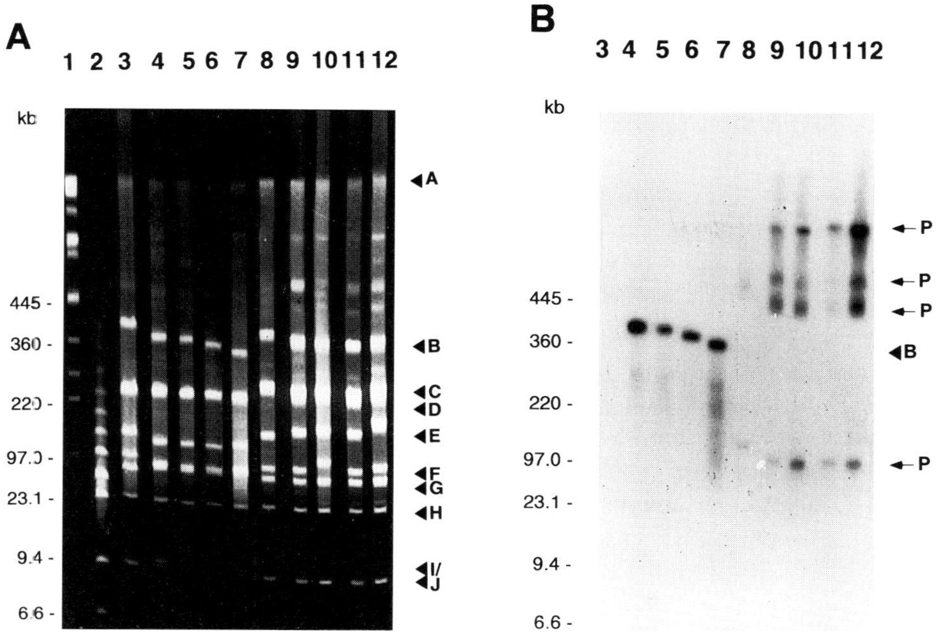

Figure 4.4 Hybridization analysis of I-*Ceu*I fragments from CPE-producing *Clostridium perfringens* strains separated by PFGE. A. I-*Ceu*I fragments from CPN50 and nine CPE-producing *C. perfringens* strains. Lane 1, *Saccharomyces cerevisiae* chromosomes; lane 2, λ concatemers and λ-*Hin*dIII markers; lane 3, CPN50 (CPE$^-$); lane 4, NCTC8798; lane 5, 8-6; lane 6, NCTC9851; lane 7, NCTC10239; lane 8, 945P; lane 9, F3686/80, lane 10, 44071-CO5; lane 11, 88L662MF; lane 12, 91E01793F. B. Southern blot of the gel shown in A hybridized with *cpe* gene probe. Note the presence of *cpe* on I-*Ceu*I fragment B from the chromosome (lanes 4–7), and on plasmids, P,[24] giving multiple bands (lanes 8–12).

constitute a relatively homogenous group and the maps of virtually all strains show extensive relatedness irrespective of the biotype. The constant chromosomal location of the α-toxin gene, *plc*, near the origin of replication[3,21] and the localization of the genes for the β-, ε-, and ι-toxins on plasmids provides a rational explanation for the biotype drift that is sometimes observed.[27] The unequal distribution of these plasmids, and their ability to be lost or acquired, would account for changes in biotype. The variation seen in the sizes of the various plasmids (Table 4.2) raises the possibility that they may carry additional genes involved in pathogenesis, such as the δ-toxin gene, or possibly genes involved in host range.

Evidence has been obtained for an association between a copy of the insertion sequence IS*1151*,[28] and the genes for enterotoxin,[24] and the β-, ε- and ι-toxins.[16]

The β-toxin gene appears to be highly unstable as strains readily lose activity[16,29] and this may be due to the influence of IS*1151*.

VIII GENOME MAPPING OF ENTEROTOXIN-PRODUCING *CLOSTRIDIUM PERFRINGENS* STRAINS

Not only do the maps of the well-characterized enterotoxin producer, NCTC8798, and its defective spore-producing mutant 8-6 deviate most extensively from the consensus *C. perfringens* map (Figure 4.2; Tables 4.2 and 4.3), they also differ from each other. The experimental evidence presented in Figure 4.5 indicates that the chromosome of 8-6 has acquired a 180 kb segment carrying a novel *Apa*I site in region **e**.[24] It is believed to have resulted from insertion of a plasmid into the chromosome and this may account for the production of defective spores. Comparison of the genome organization of a panel of strains carrying the enterotoxin gene revealed that in some cases *cpe* was on the chromosome in region **c** whereas in the majority it was situated on plasmids, often in association with other virulence factor genes such as *etx*.[24] When *cpe* was plasmid-borne, region **c** of the chromosome was organized similarly to that of CPN50.[21] This is consistent with the hypothesis outlined above, that virulence genes may be acquired by conjugal transfer of plasmids,

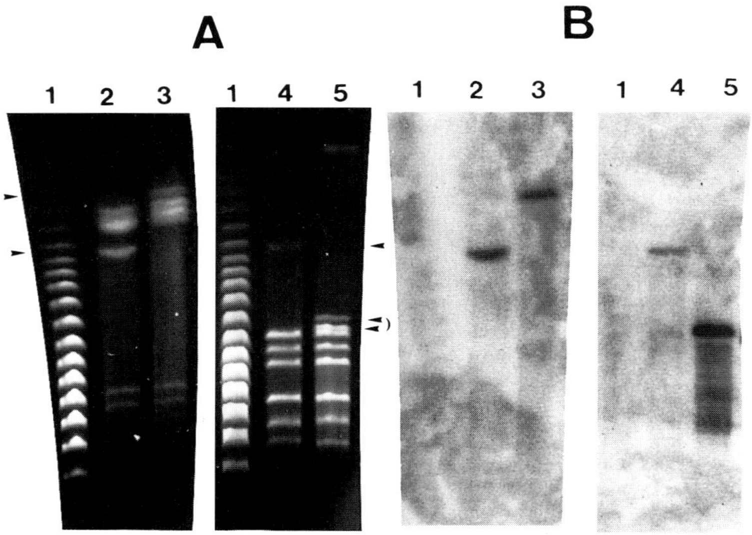

Figure 4.5 A. PFGE analysis of strains NCTC8798 (lanes 2 and 4) and 8-6 (lanes 3 and 5). Lane 1, λ DNA size marker; lanes 2,3, *Mlu*I digests; lanes 4,5; *Apa*I digests. B. Hybridization experiment using the *nanH* probe and the gel from A.

and with recombinational events being mediated by IS elements, as suggested by the following observations.

When the *cpe* gene is chromosomal it is located on a 5 kb *Nru*I fragment[24] and associated with two open reading frames, orf2, which is part of a repetitive element that may correspond to an insertion sequence,[30] and orf3.[31] The corresponding strains all originated from outbreaks of human food-poisoning. In contrast, when the gene is located on a plasmid it is usually linked to orf3, and often to IS*1151*, but no 5 kb *Nru*I fragment occurs. Such strains are invariably of veterinary origin. There is good reason to believe that strains of *C. perfringens* with chromosomal *cpe* genes may be quite distinct from most of the isolates as they generally have smaller chromosomes, they lack the *pfoA* and *nagH* genes, which results in variable region **b** being around 20 kb smaller than usual, and they have a distinctive I-*Ceu*I profile.[16]

IX. ACKNOWLEDGMENTS

Special thanks to Emmanuel Cornillot, Bruno Dupuy, Seiichi Katayama, and Brigitte Saint-Joanis for their enthusiastic practical and intellectual contributions to this work and to the Legs van Straelen for financial support.

X. REFERENCES

1. Cole S.T. and Saint-Girons I. (1994) Bacterial genomics. *FEMS Microbiol Rev* **14**: 139–160.
2. Fonstein M. and Haselkorn R. (1995) Physical mapping of bacterial genomes. *J Bacteriol* **177**: 3361–3369.
3. Canard B. and Cole S.T. (1989) Genome organization of the anaerobic pathogen *Clostridium perfringens. Proc Natl Acad Sci USA* **86**: 6676–6680.
4. Itaya M. and Tanaka T. (1991) Complete physical map of the *Bacillus subtilis* 168 chromosome constructed by a gene-directed mutagenesis method. *J Mol Biol* **220**: 631–648.
5. Hatheway C.L. (1990) Toxigenic Clostridia. *Clin Microbiol Rev* **3**: 66–98.
6. Rood J.I. and Cole S.T. (1991) Molecular genetics and pathogenesis of *Clostridium perfringens. Microbiol Rev* **55**: 621–648.
7. Boyd M.J., Logan M.A. and Tytell A.A. (1948) The growth requirements of *Clostridium perfringens* BP6K. *J Biol Chem* **174**: 1013–1025.
8. Garnier T. and Cole S.T. (1986) Characterization of a bacteriocinogenic plasmid from *Clostridium perfringens* and molecular genetic analysis of the bacteriocin-coding gene. *J Bacteriol* **168**: 1189–1196.
9. Bréfort G., Mago M., Ionesco H. and Sebald M. (1977) Characterization and transferability of *Clostridium perfringens* plasmids. *Plasmid* **1**: 52–66.
10. Garnier T. and Cole S.T. (1988) Identification and molecular genetic analysis of replication functions of the bacteriocinogenic plasmid, pIP404, from *Clostridium perfringens. Plasmid* **19**: 151–160.

11. Garnier T. and Cole S.T. (1988) Complete nucleotide sequence and genetic organization of the bacteriocinogenic plasmid pIP404 from *Clostridium perfringens. Plasmid* **19**: 134–150.
12. Sloan J., Warner T.A., Scott P.T., Bannam T.L., Berryman D.I. and Rood J.I. (1992) Construction of a sequenced *Clostridium perfringens–Escherichia coli* shuttle plasmid. *Plasmid* **27**: 207–219.
13. Sebald M. and Costilow R.N. (1975) Minimal growth requirements for *Clostridium perfringens* and isolation of auxotrophic mutants. *Appl Microbiol* **29**: 1–6.
14. Katayama S,. Dupuy B., Garnier T. and Cole S.T. (1995) Rapid expansion of the physical and genetic map of the chromosome of *Clostridium perfringens* CPN50. *J Bacteriol* **177**: 5680–5685.
15. Garnier T., Canard B. and Cole S.T. (1991) Cloning, mapping and molecular characterization of the ribosomal RNA operons of *Clostridium perfringens. J. Bacteriol* **173**: 5431–5438.
16. Katayama S., Dupuy B., Daube G., China B. and Cole S.T. (1996) Genome mapping of *Clostridium perfringens* strains with I-*Ceu*I shows many virulence genes to be plasmid-borne. *Mol Gen Genet* **251**: 720–726.
17. Anagnostopoulos, C., Piggot P.J. and Hoch J.A. (1993) The genetic map of *Bacillus subtilis*. In Sonenshein A.L., Hoch J.A. and Losick R. (eds), Bacillus subtilis *and Other Gram-positive Bacteria*, pp. 425–461. American Society for Microbiology, Washington, D.C.
18. Shimizu T., Ba-Thein W., Tamaki M. and Hayashi H. (1994) The *virR* gene, a member of a class of two-component response regulators, regulates the production of perfringolysin O, collagenase, and hemagglutinin in *Clostridium perfringens. J Bacteriol* **176**: 1616–1623.
19. Lyristis M., Bryant A.E., Sloan J., Awad M.M., Nisbet I.T., Stevens D.L. and Rood J.I. (1994) Identification and molecular analysis of a locus that regulates extracellular toxin production in *Clostridium perfringens. Mol Microbiol* **12**: 761–777.
20. Crain M.J., Waltman I., W.D., Turner J.S., Yother J., Tolkington D.K., McDaniel L.S., Gray B.M. and Briles D.E. (1990) Pneumococcal surface protein A (PspA) is serologically variable and is expressed by all clinically important capsular serotypes of *Streptococcus pneumoniae. Infect Immun* **58**: 3293–3299.
21. Canard B., Saint-Joanis B. and Cole S.T. (1992) Genomic diversity and organisation of virulence genes in the pathogenic anaerobe *Clostridium perfringens. Mol Microbiol* **6**: 1421–1429.
22. Nei M. (1987) *Molecular and Evolutionary Genetics*. Columbia University Press, New York.
23. Saitou N. and Nei M. (1987) A neighbour joining method: a new method for reconstructing evolutionary trees. *Mol Biol Evol* **44**: 406–425.
24. Cornillot E., Saint-Joanis B., Canard B., Granum P.-E., Daube G. and Cole S.T. (1995) The enterotoxin gene (*cpe*) of *Clostridium perfringens* can be chromosomal or plasmid-borne. *Mol Microbiol* **15**: 639–647.
25. Eklund M.W. (1992) The role of bacteriophages and plasmids in the production of toxins and other biologically active substances by *Clostridium botulinum* and *Clostridium novyi*. In M. Sebald (ed.), *Genetics and Molecular Biology of Anaerobic Bacteria*, pp. 179–194. Springer-Verlag, New York.
26. Finn C.W., Silver R.P., Habig W.H. and Hardegree M.C. (1984) The structural gene for tetanus toxin is on a plasmid. *Science* **224**: 881–884.
27. McDonel J.L. (1986) Toxins of *Clostridium perfringens* types A, B, C, D and E. In F. Dorner and J. Drews (eds), *Pharmacology of Bacterial Toxins*, pp. 477–517. Pergamon Press, Oxford.
28. Daube G., Simon P. and Kaeckenbeeck A. (1993) IS1151, an IS-like element of *Clostridium perfringens. Nucl Acid Res* **21**: 352.
29. Duncan C.L., Rokos E.A., Christenson C.M. and Rood J.I. (1978) Multiple plasmids in different toxigenic strains of *Clostridium perfringens*: possible control of beta toxin production. In D. Schlessinger (ed.), *Microbiology 1978*, pp. 246–248. American Society for Microbiology, Washington, D.C.
30. Granum P.E. and Stewart G.S.A. (1993) Molecular biology of *Clostridium perfringens* enterotoxin.

In M. Sebald (ed.), *Genetics and Molecular Biology of Anaerobic Bacteria*, pp. 235–247. Springer Verlag, New York.

31. Brynestad S., Iwanejko L.A., Stewart G.S.A.B. and Granum P.-E. (1994) A complex array of Hpr consensus DNA recognition sequences proximal to the enterotoxin gene in *Clostridium perfringens* type A. *Microbiology* **140**: 97–104.

Chapter 5

Genetic Analysis in *Clostridium perfringens*

JULIAN I. ROOD
Department of Microbiology, Monash University, Clayton 3168, Australia

I. INTRODUCTION

In the past ten years dramatic progress has been made in the development of genetic methods for the analysis of *Clostridium perfringens*, to the extent that *C. perfringens* is now the pathogenic clostridium most amenable to detailed genetic analysis. Reliable and reproducible methods for the transformation of *C. perfringens* cells with recombinant plasmids have been developed and several well-characterized shuttle vectors have been constructed. Methods for transposon mutagenesis and homologous recombination also have been elucidated and used for the functional analysis of *C. perfringens* genes.

II. TRANSFORMATION METHODS

The first method developed for the transformation of *C. perfringens* cells involved the transformation of L-phase variants generated by growth in the presence of penicillin.[1] This method could be used to introduce plasmids into L-forms of *C. perfringens*[2] and was used to construct some shuttle vectors.[3,4] However, the inability to either regenerate vegetative cells from the L-forms or to transform vegetative cells meant that these methods were of limited value.

The Clostridia: Molecular Biology and Pathogenesis
ISBN 0-12-595020-9

With the advent of electroporation technology, significant advances were made in the development of transformation methods for *C. perfringens*. Allen and Blaschek[5] were able to obtain pAMβ1- and pHR106-derived transformants of vegetative cells of strain 3624A by electroporation, although the frequency was still fairly low (1.2×10^3 transformants/μg DNA). Subsequent studies led to improved transformation efficiencies.[6,7] Scott and Rood[8] obtained relatively high transformation frequencies (3×10^5 transformants/μg DNA) using strain 13 as the transformation recipient, presumably because strain 13 lacks a restriction and modification system.[8,9] Other workers[7] tested eight *C. perfringens* strains for their ability to be transformed with the shuttle vector pAK201 and showed that strain 13 yielded by far the highest transformation frequency (9.6×10^6 transformants/μg DNA). As a result, strain 13 and its derivatives are now the most widely used strains for genetic studies on *C. perfringens*.[10–13] It should be noted that some *C. perfringens* strains cannot be transformed, even by electroporation[7] (P. Scott and J. Rood, unpublished results). Therefore, it is recommended that when commencing studies aimed at the analysis or manipulation of particular *C. perfringens* genes, an appropriate transformable strain be carefully selected.

III. SHUTTLE VECTORS

The development of transformation methods for *C. perfringens* rapidly led to the construction of *C. perfringens–Escherichia coli* shuttle vectors such as pJU16,[3] pHR106[4] and pAK201.[6] However, although these and other shuttle plasmids (Table 5.1) proved to be invaluable for the development and refinement of the transformation methods they have been virtually ignored as cloning or shuttle vectors.

The most widely used *C. perfringens–E. coli* shuttle vectors are derivatives of pJIR418.[14–16] This plasmid was constructed so that it contains the origins of replication of the *E. coli* plasmid pUC18 and the *C. perfringens* plasmid pIP404, the multiple cloning site and *lacZ′* gene from pUC18, and the *C. perfringens* *ermBP* and *catP* genes.[14] The last two genes confer erythromycin and chloramphenicol resistance, respectively, in both *E. coli* and *C. perfringens*. Because this plasmid does not confer ampicillin resistance, for ethical reasons it is ideal for use in *C. perfringens*, an organism in which penicillin resistance has not been observed. Recombinants can be detected by standard X-gal screening in *E.coli* or by insertional inactivation of the *ermBP* or *catP* genes in either *E. coli* or *C. perfringens*. Furthermore, the complete sequence of pJIR418 is known,[14] facilitating the analysis of recombinant derivatives. Two derivatives of pJIR418, pJIR750 and pJIR751, have been constructed by deletion of the *ermBP* and

Table 5.1. Properties of *Clostridium perfringens–Escherichia coli* shuttle vectors[a]

Shuttle plasmid	Size (kb)	*C. perfringens*		*E. coli*		Reference
		Replicon	Selection	Replicon	Selection	
pJU12	11.6	pJU121	Tc	pBR322	ApTc	3
pJU13	12.2	pJU122	Tc	pBR322	ApTc	3
pJU16	12.2	pJU122	Tc	pBR322	ApTc	3
pHR106	7.9	pJU122	Cm	pSL100	ApCm	4
pAK201	8.0	pHB101	Cm	pBR322	Cm	6
pSB92A2	7.9	pCP1	Cm	pHG165	ApCm	21
pTG67	6.6	pIP404	Cm	pUC18	ApCm	22
pJIR418	7.4	pIP404	EmCm	pUC18	EmCmXG	14
pJIR750	6.6	pIP404	Cm	pUC18	CmXG	15
pJIR751	6.0	pIP404	Em	pUC18	EmXG	15
pPSV	6.9	pIP404	Em	pUC18	EmXG	16
pTCATT	6.4	pIP404	Em	pUC18	Em	18

[a] Tc, Ap, Cm, Em represents resistance to tetracycline, ampicillin, chloramphenicol and erythromycin, respectively. XG represents screening for β-galactosidase production on X-Gal medium. Based on Table 5 of Rood and Cole.[9]

catP genes, respectively. These plasmids were specifically designed for the analysis of antibiotic resistance determinants and transposons but can also be used as general shuttle plasmids since the complete sequence of both plasmids is also known.[15]

Other workers have constructed both promoter probe and expression vectors.[16–19] The promoter probe plasmid pPSV[16] was constructed by deletion of the *catP* gene from pJIR418 and the subsequent addition of a promoterless *catP* gene downstream of the pUC18-derived multiple cloning site. Other studies have also used *catP* as a reporter system in *C. perfringens*.[17] In these experiments a promoter probe shuttle vector, pTCATT, was constructed and polymerase chain reaction (PCR) used to fuse the promoterless *catP* gene to the *plc* promoter. This vector contains upstream and downstream transcriptional terminators but requires the PCR-mediated regeneration of the 5′ terminus of the *catP* gene during the construction of promoter fusions. Phillips-Jones[18] constructed a plasmid which had the *luxAB* genes from *Vibrio fischeri* under the control of the *plc* promoter and showed that luciferase activity and bioluminescence were obtained in *C. perfringens*. This system appears to have some potential as a reporter system, although its use may be limited by the need to expose the cells to oxygen to obtain bioluminescence. Melville *et al.*[19] constructed reporter plasmids using the *E. coli gusA* gene and the shuttle vector pJIR750. In these studies, β-glucuronidase production was used successfully to monitor the sporulation-specific regulation of the *C. perfringens cpe* gene. Provided that most transformable *C. perfringens* strains are shown to have low endogenous β-glucuronidase levels, this system may have widespread applicability. In addition, constructing a transcriptional fusion with the *cpe* promoter offers the potential to obtain regulated sporulation-specific expression of cloned genes. Note that this scheme represents the only tightly regulated expression system currently available in *C. perfringens*.

IV. TRANSPOSON MUTAGENESIS

In other bacteria, transposon mutagenesis has been widely used for the genetic analysis of both chromosomal and plasmid-determined genes. The only transposons reported from *C. perfringens* are the chloramphenicol resistance transposons Tn*4451* and Tn*4452*, however, because these transposons have not been shown to transpose in *C. perfringens* they are not suitable for use as mutagenic agents and, until relatively recently, transposon mutagenesis techniques were not available in *C. perfringens*. Allen and Blaschek[9] developed a transposon mutagenesis method for *C. perfringens* by using the conjugative tetracycline resistance transposon Tn*916* from *Enterococcus faecalis*. *Clostridium perfringens* cells were transformed with a suicide plasmid carrying Tn*916* and transformants were selected on the basis of their ability to grow on medium containing tetracycline. Hybridization analysis showed that Tn*916* had transposed to different sites on the chromosome, although, as often occurs with Tn*916* insertions, multiple insertion events were common.

In my laboratory, Tn*916*-mutagenesis has been used to isolate a pleiotrophic mutant altered in its ability to produce several extracellular toxins.[10] Inverse PCR technology was used to clone the region flanking the single Tn*916* insertion and then to clone the wild-type gene. The mutant was shown to have arisen from the insertion of Tn*916* into the *virS* gene, which encodes the sensor kinase component of a single transduction system that regulates extracellular toxin production in *C. perfringens* (see Chapter 26). This study was the first published report of the isolation of defined *C. perfringens* mutants by transposon mutagenesis.[10] As such, it represented a significant development in *C. perfringens* genetics since genes identified by mutagenesis with sequenced transposons such as Tn*916* can readily be identified and cloned. Unfortunately, subsequent studies[23] have shown that although other mutants can be isolated by Tn*916*-mutagenesis, the high frequency of double insertion events complicates the analysis of the resultant mutants and makes the methodology less than ideal. Therefore, there is still a need for more efficient transposon mutagenesis methods that are more likely to yield single transposon insertion events at a high frequency.

V. HOMOLOGOUS RECOMBINATION

Modern methods of analysing gene function in bacteria often involve reverse genetics; a process whereby a gene is inactivated by transposon mutagenesis or homologous recombination, complemented by a cloned gene, and then a series of isogenic wild-type, mutant and complemented strains tested for whatever biological function is being examined. In this way it is possible to specifically test the role of individual structural or regulatory genes in quite complex biological reactions, such as bacterial virulence. With the advent of electroporation methods for *C. perfringens*, it has now become possible to use homologous recombination or allelic exchange specifically to inactivate chromosomal or plasmid-determined genes.

Homologous recombination may involve either single or double crossover events. To insertionally inactivate a chromosomal gene by a single crossover event an internal fragment of the gene to be inactivated is cloned into a suicide vector, that is, a vector that is unable to replicate in *C. perfringens*. An antibiotic resistance gene that is expressed in *C. perfringens* is added and the resultant plasmid used to transform *C. perfringens* to antibiotic resistance. The only way in which resistant transformants can be obtained is if the plasmid has been integrated into the chromosome by homologous recombination with the chromosomal gene. The advantages of this method are that the recombination efficiency is greater because only a single crossover event is required. The disadvantages are first that it requires an internal gene fragment, limiting the recombination efficiency, and second, that wild-type genes can be regenerated from the mutants by homologous recombination between the two gene copies on the chromosome. This may be a problem for *in vivo* studies. Single crossover events have been used to integrate suicide plasmids into the chromosomal *C. perfringens plc*[13] and *virR*[12] genes which has enabled the analysis of the functional role of these genes.

Allelic exchange also involves the use of homologous recombination to mutate specifically chromosomal or plasmid genes. In this process a cloned derivative of the gene to be mutated is insertionally inactivated by cloning an antibiotic resistance gene into an internal restriction site. The resultant gene is cloned onto a suicide vector and used to transform *C. perfringens*. Single crossover events will lead to the integration of the plasmid but not to the inactivation of the gene. However, mutants can be obtained from double crossover events involving gene regions located on either side of the resistance gene. In this situation the wild-type gene is replaced by the inactivated allele located on the suicide plasmid, although the suicide plasmid itself is not integrated and is lost. Because two crossover events are required allelic exchange is a less efficient process. However, it produces inherently more stable mutants because there is no possibility that a wild-type gene can be regenerated by further intragenic recombination events, as for single crossovers.

In my laboratory allelic exchange has been successfully used to construct defined mutants in *C. perfringens*.[11,20] Our experience is that the efficiency by which mutants are obtained by allelic exchange is very dependent upon the size of the homologous regions located on either side of the inactivated gene. We recommend that there be at least 1 kb of DNA on either side of the resistance gene, with larger regions preferred if possible. This requirement, presumably, is a reflection of either the relatively inefficient *C. perfringens* transformation process, or the substrate requirements of the RecA-mediated recombination machinery, or both. We have constructed both *plc* and *pfoA* mutants of strain 13 by allelic exchange. Analysis of an isogenic series of these mutants in a mouse myonecrosis model has enabled the functional significance of the toxins encoded by the genes to be determined and has shown that the *plc* gene encodes a toxin that is essential for the pathogenesis of gas gangrene.[11] Similarly, allelic exchange has been used to introduce an inactivated *tnpX* gene onto the copy of Tn*4451* located on the conjugative plasmid pIP401.[20] Genetic analysis of the resultant derivatives has shown that this gene encodes a site-specific recombinase that is essential for conjugative Tn*4451* excision in *C. perfringens* (see Chapter 6).

VI. CONCLUSIONS

The tools available for genetic analysis in *C. perfringens* have been developed to the stage that they are equivalent to, if not better than, many other Gram-positive bacteria. Transformation methods have been elucidated, well-defined shuttle vectors constructed, and both transposon mutagenesis and homologous recombination used to construct defined chromosomal mutations of genes involved in extracellular toxin production. *C. perfringens* clearly has the most well-developed genetic system of the pathogenic clostridia and as such should be considered as the paradigm bacterium for the functional genetic analysis of the pathogenesis of toxaemic clostridial diseases.

VII. ACKNOWLEDGMENTS

The studies carried out in my laboratory were generously supported by the Australian Research Council, the Australian National Health and Medical Research Council and CSL Ltd.

VIII. REFERENCES

1. Heefner D.L., Squires C.H., Evans R.J., Kopp B.J. and Yarus M.J. (1984) Transformation of *Clostridium perfringens. J Bacteriol* **159**: 460–464.
2. Mahony D.E., Mader J.A and Dubel J.R. (1988) Transformation of *Clostridium perfringens* L forms with shuttle plasmid DNA. *Appl Environ Microbiol* **54**: 264–267.
3. Squires C.H., Heefner D.L., Evans R.J., Kopp B.J. and Yarus M.J. (1984) Shuttle plasmids for *Escherichia coli* and *Clostridium perfringens. J Bacteriol* **159**: 465–471.
4. Roberts I., Holmes W.M. and Hylemon P.B. (1988) Development of a new shuttle plasmid system for *Escherichia coli* and *Clostridium perfringens. Appl Environ Microbiol* **54**: 268–270.
5. Allen S.P. and Blaschek H.P. (1988) Electroporation-induced transformation of intact cells of *Clostridium perfringens. Appl Environ Microbiol* **54**: 2322–2324.
6. Kim A.Y. and Blaschek H.P. (1989) Construction of an *Escherichia coli–Clostridium perfringens* shuttle vector and plasmid transformation of *Clostridium perfringens. Appl Environ Microbiol* **55**: 360–365.
7. Allen S.P. and Blaschek H.P. (1990) Factors involved in the electroporation-induced transformation of *Clostridium perfringens. FEMS Microbiol Lett* **70**: 217–220.
8. Scott P.T. and Rood J.I. (1989) Electroporation-mediated transformation of lysostaphin-treated *Clostridium perfringens Gene* **82**: 327–333.
9. Rood J.I. and Cole S.T. (1991) Molecular genetics and pathogenesis of *Clostridium perfringens. Microbiol Rev* **55**: 621–648.
10. Lyristis M., Bryant A.E., Sloan J., Awad M.M., Nisbet I.T., Stevens D.L. and Rood J.I. (1994) Identification and molecular analysis of a locus that regulates extracellular toxin production in *Clostridium perfringens. Mol Microbiol* **12**: 761–777.
11. Awad M.M., Bryant A.E., Stevens D.L. and Rood J.I. (1995) Virulence studies on chromosomal α-toxin and θ-toxin mutants constructed by allelic exchange provide genetic evidence for the essential role of α-toxin in *Clostridium perfringens*-mediated gas gangrene. *Mol Microbiol* **15**: 191–202.
12. Shimizu T., Ba-Thein W., Tamaki M. and Hayashi H. (1994) The *virR* gene, a member of a class of two-component response regulators, regulates the production of perfringolysin O, collagenase, and hemagglutinin in *Clostridium perfringens. J Bacteriol* **176**: 1616–1623.
13. Ninomiya M., Matsushita O., Minami J., Sakamoto H., Nakano M. and Okabe A. (1994) Role of alpha-toxin in *Clostridium perfringens* infection determined by using recombinants of *C. perfringens* and *Bacillus subtilis. Infect Immun* **62**: 5032–5039.
14. Sloan J., Warner T.A., Scott P.T., Bannam T.L., Berryman D.I. and Rood J.I. (1992) Construction of a sequenced *Clostridium perfringens–Escherichia coli* shuttle plasmid. *Plasmid* **27**: 207–219.
15. Bannam T.L. and Rood J.I. (1993) *Clostridium perfringens–Escherichia coli* shuttle vectors that carry single antibiotic resistance determinants. *Plasmid* **29**: 223–235.
16. Matsushita C., Matsushita O., Koyama M. and Okabe A. (1994) A *Clostridium perfringens* vector for the selection of promoters. *Plasmid* **31**: 317–319.
17. Bullifent H.L., Moir A. and Titball R.W. (1995) The construction of a reporter system and use for the investigation of *Clostridium perfringens* gene expression. *FEMS Microbiol Lett* **131**: 99–105.
18. Phillips-Jones M.K. (1993) Bioluminescence (*lux*) expression in the anaerobe *Clostridium perfringens. FEMS Microbiol Lett* **106**: 265–270.
19. Melville S.B., Labbe R. and Sonenshein A.L. (1994) Expression from the *Clostridium perfringens cpe* promoter in *C. perfringens* and *Bacillus subtilis. Infect Immun* **62**: 5550–5558.
20. Bannam T.L., Crellin P.K. and Rood J.I. (1995) Molecular genetics of the chloramphenicol resistance transposon Tn*4451* from *Clostridium perfringens*: the TnpX site-specific recombinase excises a circular transposon molecule. *Mol Microbiol* **16**: 535–551.

21. Phillips-Jones M.K. (1990) Plasmid transformation of *Clostridium perfringens* by electroporation methods. *FEMS Microbiol Lett* **66**: 221–226.
22. Garnier T. and Cole S.T. (1988) Identification and molecular genetic analysis of replication functions of the bacteriocinogenic plasmid pIP404 from *Clostridium perfringens. Plasmid* **19**: 151–160.
23. Awad M.M. and Rood J.I. (1997) Isolation of α-toxin, θ-toxin and κ-toxin mutant of *Clostridium perfringens* by Tn*916* mutagenesis. *Microbial Pathogen.* (In Press).

Chapter 6

Transposable Genetic Elements and Antibiotic Resistance Determinants from *Clostridium perfringens* and *Clostridium difficile*

DENA LYRAS AND JULIAN I. ROOD

Department of Microbiology, Monash University, Clayton 3168, Australia

I. INTRODUCTION

Antibiotic resistance determinants and transposons from both *Clostridium perfringens* and *Clostridium difficile* have been reviewed in recent years.[1–3] However, since these reviews were published there have been many advances in knowledge in this area. This chapter will focus on the most recent findings, as well as summarizing the earlier work. Where possible, our emphasis will be on comparisons between the *C. perfringens* and *C. difficile* antibiotic resistance determinants and transposons.

II. ANTIBIOTIC RESISTANCE DETERMINANTS

A. Tetracycline resistance

The tetracycline antibiotics exert their activity by binding to the 30S ribosomal subunit, thereby inhibiting the binding of aminoacyl-tRNA molecules and

The Clostridia: Molecular Biology and Pathogenesis
ISBN 0-12-595020-9

disrupting protein synthesis.[4] Bacterial resistance to the tetracyclines is mediated by two major mechanisms involving the synthesis of either a transmembrane protein which actively transports tetracycline out of the cell,[5] or a cytoplasmic protein which modifies the ribosomes and protects them from the action of the antibiotic. The latter process is mediated by a related group of determinants of the Tet M or Tet O class which encode large proteins that have similarity to the elongation factors Tu and G.[6–8]

1. The Tet P determinant from Clostridium perfringens

Tetracycline resistance is probably the most common antibiotic resistance phenotype found in *C. perfringens*.[2,9] Most tetracycline resistant strains cannot transfer their resistance genes, however, conjugative tetracycline resistance plasmids have been detected and are relatively common.[9–14] In all the conjugative strains examined, transfer is mediated by a plasmid that is either indistinguishable from, or closely related to, the prototype conjugative *C. perfringens* R-plasmid, pCW3.[15] In nonconjugative isolates tetracycline resistance is generally constitutively expressed whereas in conjugative isolates it is inducible.[9] The molecular basis for this difference is not known.

The best characterized tetracycline resistance determinant from *C. perfringens*, Tet P, was isolated from pCW3. This determinant is found on all known tetracycline resistance plasmids from *C. perfringens*.[15–18] Hybridization studies showed that Tet P did not hybridize to a variety of tetracycline resistance genes from both Gram-positive and Gram-negative bacteria, indicating that this determinant was unique and belonged to a separate hybridization class.[17] Early studies on the Tet P determinant involved the subcloning and localization of the resistance gene(s).[16,17] Subsequent studies[19] involved the complete nucleotide sequence analysis of the Tet P determinant. The results showed that it encodes two functional tetracycline resistance genes designated *tetA*(P) and *tetB*(P) (Figure 6.1). These genes overlap by 17 bp, which suggests that the expression of the encoded proteins may be translationally coupled.[19] Potential promoters could only be identified upstream of *tetA*(P), indicating that the *tetA*(P) and *tetB*(P) genes may be linked in an operon. This represents a novel genetic arrangement for tetracycline resistance determinants because there are no other reported examples where two distinct tetracycline resistant genes are juxtaposed.[19]

The *tetA*(P) gene is 1260 bp long and encodes a putative 46 kDa protein, TetA(P), that is predicted to have 12 transmembrane domains. The TetA(P) protein was shown to mediate the active efflux of tetracycline from resistant *Escherichia coli* cells. Although some amino acid sequence similarity was observed between the putative TetA(P) protein and other tetracycline efflux

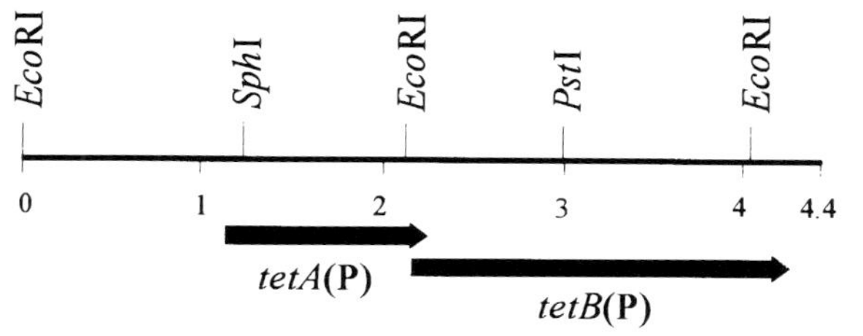

Figure 6.1 Genetic organization of the Tet P determinant. Sizes are in kb. Modified from Sloan *et al.*[19]

proteins, hydrophobicity analysis indicated that TetA(P) represented a different type of efflux protein.[19] The prototype tetracycline efflux proteins all have two related six-transmembrane domains which are separated by a large central cytoplasmic domain, whereas TetA(P) appears to have two major hydrophilic domains that are not centrally located, even though it appears to have the twelve-transmembrane domains typical of such proteins.[19] Site-directed mutagenesis has been used to confirm that amino acids which were predicted, from the proposed transmembrane model, to be involved in tetracycline resistance mediated by *tetA*(P), namely Glu-52, Glu-59 and Glu-89,[19] are essential for the active efflux of tetracycline from the cell (R.M. Kennan and J.I. Rood, unpublished results).

The *tetB*(P) gene is 1956 bp long and encodes a putative 72.6 kDa hydrophilic protein that has significant amino acid sequence identity to Tet M-like cytoplasmic tetracycline resistance proteins (37–39%) and which therefore probably functions in a similar manner.[19] However, the *tetB*(P) gene and the *tet*(M) gene from the *Enterococcus faecalis* transposon Tn*916* do not hybridize with each other.[20] In addition to conferring tetracycline resistance, *tetB*(P) was shown to confer minocycline resistance, as do the *tet*(M)-like genes, whereas *tetA*(P) confers only tetracycline resistance.[19]

These findings contrast with those of other workers who suggested that the Tet P region from pJIR39, which carries all of *tetA*(P) but only part of *tetB*(P),[16,19] encodes two proteins of 10 kDa and 20 kDa.[21] This result disagrees with the protein sizes predicted from the nucleotide sequence analysis of Tet P.[19] Saksena and Truffaut[21] concluded that the Tet P protein(s) act at the level of protein synthesis which, although highly likely for TetB(P), was based on the analysis of plasmids such as pJIR39 which carry only *tetA*(P). The *tetA*(P) gene has been shown clearly to mediate tetracycline resistance by an active efflux mechanism.[19]

To determine the distribution of the *tetA*(P) and *tetB*(P) genes in *C. perfringens*, a fragment from within each of these genes was used as a probe in hybridization studies. Early studies, which used only an internal *tetA*(P) probe, showed that this gene was present in eight nonconjugative *C. perfringens* isolates and a strain of *Clostridium paraputrificum*, but was not found in a *Clostridium sporogenes* isolate or in six *C. difficile* isolates.[17] More recent studies,[20] using both *tetA*(P) and *tetB*(P)-specific probes, confirmed the results from the earlier work. Hybridization and polymerase chain reaction (PCR) analysis of 81 tetracycline resistant isolates of *C. perfringens* showed that they all carried the *tetA*(P) gene. Most of these isolates (93%) carried a second tetracycline resistance gene, with 53% carrying *tetB*(P) and, unexpectedly, 40% carrying a *tet*(M)-like gene. These studies therefore showed that some *C. perfringens* isolates carried a *tet*(M)-like gene. Despite the wide distribution of the *tetB*(P) and *tet*(M) genes, no isolate which carried both of these determinants was detected. In all of the isolates that carried both *tetA*(P) and *tetB*(P) these genes were found to overlap as in pCW3. In contrast, there was no similar overlap between the *tetA*(P) and *tet*(M) genes in isolates carrying these determinants. In *C. perfringens* isolates that did not carry *tetB*(P), nucleotide sequencing revealed that the *tetA*(P) gene had diverged at its 3′ end in comparison to the *tetA*(P) gene from pCW3. The divergent gene, designated *tetA408*(P), was 37 bp shorter than *tetA*(P), and therefore encoded a protein consisting of 408 amino acids, 12 amino acids fewer than that encoded by *tetA*(P). There were also four conservative amino acid changes within the last transmembrane domain of the protein. Except for the above, the transmembrane domains of the protein were not significantly altered and therefore it is not surprising that these changes have not affected the efflux phenotype.[20]

Preliminary studies have also indicated that in isolates which carry both the *tetA408*(P) and *tet*(M)-like genes, resistance is not plasmid associated, whereas the *tetA*(P)/*tetB*(P) genes are found on plasmids.[20] In addition, hybridization experiments carried out using the left and right ends of Tn*916* (excluding the *tet*(M) gene) as probes indicate that the *C. perfringens tet*(M)-like determinant may have been associated with a conjugative transposon since cross-hybridization was observed (D. Lyras and J.I. Rood, unpublished data). This result was not unexpected given that *tet*(M) genes are often associated with conjugative transposons such as Tn*916*. Conjugative transfer of the *C. perfringens tet*(M)-like determinant has not, however, been observed (J.I. Rood, unpublished results).[22] It has been postulated that *tetB*(P) evolved from the transfer into *C. perfringens* of a conjugative transposon carrying a *tet*(M) gene from a Gram-positive organism, and that this gene has subsequently diverged. However, the *tetB*(P) and *tet*(M) genes are different and so this event must have occurred in the distant evolutionary past.[19] The discovery that a Tet M-like determinant, which is possibly transposon associated, is found in *C. perfringens* supports this hypothesis. Nucleotide sequencing of the *C. perfringens* Tet M-like gene and its flanking DNA and comparison to the *tetB*(P) region may reveal whether the two genes are evolutionarily related.

2. *Tetracycline resistance determinants from* Clostridium difficile

Studies on the *C. difficile* tetracycline resistance determinants are not as advanced as those on the *C. perfringens* Tet P determinant. Early reports indicated that the tetracycline resistance determinant from strain 630 was transferable at a low level and was constitutively expressed.[23,24] Subsequent experiments have shown that the tetracycline resistance gene is located on a conjugative chromosomal transposon similar to those found in the enterococci and streptococci.[25,26] This conjugative transposon has been designated Tn*5397*.[27]

Molecular analysis of the tetracycline resistance determinant from the conjugative *C. difficile* strains 630 and 662 showed that this determinant was of the Tet M type.[25] Subsequent studies involved the cloning of a 3.4 kb *Sau*3A strain 630-derived fragment which conferred tetracycline resistance in *E. coli*.[26] A restriction map was constructed and Tn*1000* mutagenesis showed that the tetracycline resistance determinant covered a region of more than 1.9 kb. Also, *in vitro* transcription and translation revealed that this region encoded a protein of 68 kDa, which is similar in size to the Tet M and Tet O proteins.[26] However, since the *tet*(M)-like gene from *C. difficile* has not been sequenced, its precise relationship to *tet*(M) and *tet*(O) is yet to be determined.

B. Chloramphenicol resistance

Resistance to chloramphenicol is most commonly mediated by the enzymatic modification and inactivation of the antimicrobial agent, as a result of the action of chloramphenicol acetyltransferase (CAT). This enzyme catalyses the transfer of an acetyl group from acetyl coenzyme A to the primary hydroxyl group of chloramphenicol.[28] The acetylated chloramphenicol derivatives do not bind to bacterial ribosomes and consequently do not exhibit antimicrobial activity. Many CAT enzymes have been characterized from a diverse range of bacterial species. These enzymes are all trimers composed of identical subunits of 20–26 kDa.[29,30] The nucleotide sequences of the genes which encode the monomers often have little similarity but the resultant enzymes are functionally very similar and their amino acid sequences are highly conserved, especially in the region which surrounds the histidyl residue that forms part of the catalytic site.[31,32]

Chloramphenicol resistance in *C. perfringens* is not as common as tetracycline or erythromycin resistance[14,33] and has been shown to be mediated by the production of CAT enzymes.[13,34,35] The strains CP590 and CP600 have been shown to carry the conjugative plasmids pIP401 and pJIR25, respectively, which encode both tetracycline and chloramphenicol resistance.[10] Other plasmids of this type, such as pJIR27, have also been found in other strains.[22,36] Further studies have shown that the determinants from these plasmids are located on transposons.[37]

The chloramphenicol resistance determinants from pIP401 and pJIR27 have been cloned to form the recombinant plasmids pJIR45 and pJIR97, respectively. Both plasmids confer chloramphenicol resistance in *E. coli*.[22,37] Subcloning and deletion analysis of pJIR45 localized the resistance gene to within a 1.5 kb region and hybridization studies showed that five of the six chloramphenicol resistant isolates that were tested carried homologous resistance determinants which were located on large plasmids.[22,36] This determinant did not hybridize to other known *cat* genes[36] and was therefore designated *catP*. In contrast to other *cat* genes from Gram-positive bacteria, the expression of *catP* was constitutive.[35] Nucleotide sequencing has shown that the *catP* gene is 621 bp long and encodes a 24.4 kDa protein, which is similar in size to other CAT determinants.[38,39]

Comparative analysis of the nucleotide sequence of the *catP* gene, and the amino acid sequence of the deduced protein, showed that although *catP* does not have significant DNA sequence similarity to other *cat* genes, except for *catD*, there is significant similarity at the amino acid sequence level. The common regions include the known active site, which is the carboxy-terminal region of the protein. The *catP* gene product has 42–45% amino acid sequence similarity with CAT proteins from other genera[40] and is more closely related to CAT monomers from *Vibrio anguillarum* and *Campylobacter coli* than to monomers from other bacteria.[41,42] As is found with the *V. anguillarum* and *C. coli* CAT monomers, the CATP monomer has a four amino-acid deletion in comparison to other CAT determinants.[41]

A chloramphenicol resistance determinant which does not hybridize to *catP* has been detected in *C. perfringens* strain CW92.[36] Hybridization studies showed that this *catQ* gene does not hybridize with *cat* genes from a variety of different bacterial species and therefore represents a distinct hybridization class.[36] Nucleotide sequencing of this determinant revealed an open reading frame of 657 bp encoding a putative 26 kDa protein.[41] Expression studies have indicated that *catQ* is constitutively expressed in both *C. perfringens* and *E. coli* (D. Lyras and J.I. Rood, unpublished results). In contrast to the *catP* determinant, which is transposon-associated and located on a plasmid, *catQ* was found to be chromosomally encoded. There is no evidence to suggest that it is associated with a transposon.[36] The CATQ monomer has maximal sequence similarity (72%) with the CATB monomer,[40] encoded by the *catB* gene from *C. butyricum*.[43] The CATQ monomer is as closely related to the *Staphylococcus aureus* CAT monomers (47–48% similarity) as it is to CATP (54%), and does not contain the four amino acid deletion present in the other clostridial genes.[40,41] Thus, this determinant appears to have evolved independently of *catP*, although both genes must have a common primordial gene.[41]

The chloramphenicol resistance gene, *catD*, from *C. difficile* has also been cloned.[44] Hybridization studies using both *catP*- and *catD*-specific probes indicated that these genes were related.[36] Nucleotide sequence comparisons have shown that the CATD and CATP monomers are almost identical – they have 97% amino acid sequence identity.[42] The CATD monomer also had a deletion of four amino acids in comparison to the other CAT determinants. It is concluded that the *catP* and *catD* genes are closely related and have evolved fairly recently from a common ancestor.[42,45] In contrast to *catP*, the *catD* gene appears to be present in multiple copies on the *C. difficile* chromosome.[44]

C. Erythromycin resistance

Resistance to erythromycin and macrolide–lincosamide–streptogramin B (MLS) antibiotics has been found in a number of Gram-positive and Gram-negative genera.[46,47] Although the MLS antibiotics are structurally and chemically distinct, they inhibit bacterial protein synthesis by similar mechanisms. Resistance is often mediated by a 23S rRNA methylase encoded by a gene designated *erm* (*e*rythromycin *r*esistance *m*ethylase) and is the result of dimethylation of 23S rRNA which reduces the affinity between the antibiotic and the target ribosome.[48,49] To date, eight Erm hybridization classes (ErmA–ErmG and ErmQ) have been described from a range of bacterial genera.[46,50] Although the hybridization classes are distinct, sequence analysis has shown that the nucleotide and deduced amino acid sequences have from 17 to 70% identity.[46,50,51]

Erythromycin resistance in *C. perfringens* was first reported in 1975.[52] This study identified the *C. perfringens* strain CP590 which not only carried the conjugative tetracycline and chloramphenicol resistance plasmid pIP401 but was also resistant to erythromycin, clindamycin and lincomycin. Further studies indicated that the MLS resistance determinant was carried on a 63 kb nonconjugative plasmid, pIP402.[10] Cloning and molecular analysis of this determinant indicated that it belonged to the ErmB–ErmAM hybridization class and it was therefore designated *ermBP*.[53] Hybridization studies showed that five out of 40 erythromycin-resistant *C. perfringens* strains hybridized with the *ermBP* probe,[53] indicating that this determinant was not widespread in *C. perfringens*.

Sequence analysis of the 738 bp *ermBP* gene has shown that it is identical to the *erm* gene from the promiscuous *Enterococcus faecalis* plasmic pAMβ1 and that it has at least 98% nucleotide sequence identity with other *ermB*/*ermAM* genes such as those from pIP501, pAM77, Tn*917* and the *ermBC* gene from *E. coli*.[54] In contrast to many of the other *ermB*/*ermAM* genes, *ermBP* is not preceded by a leader peptide sequence and is therefore constitutively expressed,[54] as is the *erm* gene from pAMβ1.[55,56] Comparison of these gene regions shows that the identical *erm* leader sequence region, which is absent in pAMβ1, is missing in the equivalent *ermBP* region.[54] In addition, an open reading frame (ORF) of unknown function, which is present downstream of the Tn*917*, pAM77 and pIP501 *erm* genes, and is designated ORF3, is also found 4 bp downstream of *ermBP*.[54]

The *ermBP* determinant is located between two almost identical directly repeated sequences (Figure 6.2), DR1 and DR2 (1341 and 1340 bp, respectively).[54] These repeats appear to encode a protein of 298 (or 284) amino acids which has similarity to chromosomal and plasmid partitioning proteins and which therefore may be associated with partitioning of the parent plasmid pIP402 in *C. perfringens*. In both DR1 and DR2, there are 47 bp palindromic sequences, designated as *palA* sequences, located upstream of the ORF298 start codons, and 48 bp sequences designated *palB*, located downstream of ORF298 (Figure 6.2). The *palA* and *palB* sequences are almost identical. Comparative analysis has shown both the pAMβ1 (*E. faecalis*) and pIP501 (*Strep. agalactiae*) *erm* gene regions have DR2 but have a 975 bp internal deletion in DR1 (Figure 6.2). Other *ermB*/*ermAM* determinants had small portions of DR1 but none have complete copies.[54] The deletion endpoints in pAMβ1 and pIP501 are located within the *palA* and *palB* sequences, but they appear to have arisen from separate deletion events because the exact points of divergence are different. It has been suggested that recombination between the *palA* and *palB* homologs in the progenitor *erm* determinants resulted in the deletions observed in pAMβ1 and pIP501.[54]

The extent of the similarity between the *erm* determinants suggests that they are derived from a common progenitor. However, it appears that the sequences flanking the *erm* genes have diverged extensively even though the *erm* genes are highly conserved. The major conclusion from this analysis is that the *ermBP* determinant is probably derived from an ancestral enterococcal or streptococcal determinant but that it represents a more primordial form of the *ermB*/*ermAM* determinant than the determinants from pAMβ1 and pIP501.[54] This conclusion is supported by the fact that many of these determinants are located on conjugative plasmids or transposons which are easily disseminated. Note that not all erythromycin-resistant *C. perfringens* isolates carry *ermBP* genes that are flanked by the DR1 and DR2 sequences.[54] These isolates may have been derived from the chromosomal integration of a plasmid-determined *ermBP* gene.

Initial studies of erythromycin resistance in *C.difficile* demonstrated that, in contrast to the situation in *C. perfringens*,[9,10,14] the conjugative transfer of MLS resistance in *C. difficile* was common. Transfer appeared to involve a chromosomal determinant,[24] subsequently designated Tn*5398*.[57,58] Interestingly, Tn*5398* was identified in strain 630 which also carries the tetracycline resistance transposon Tn*5397*.

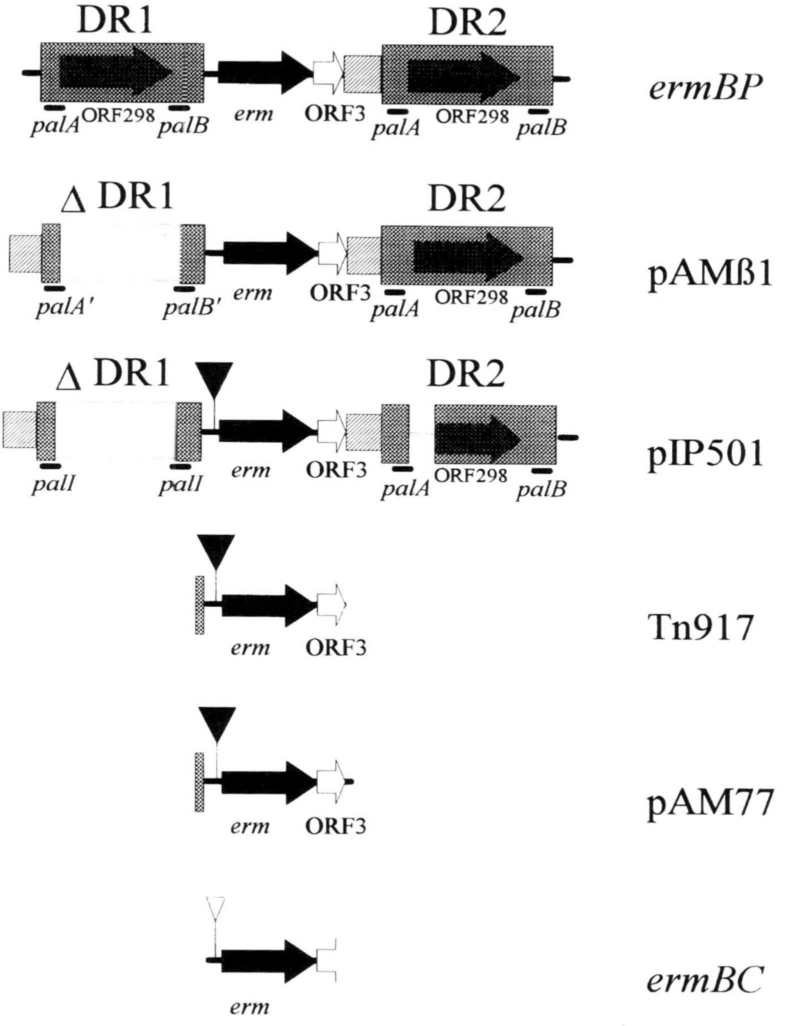

Figure 6.2 Comparison of the genetic organization of the *ermB–ermAM* gene regions. Regions of nucleotide sequence similarity are indicated by the same shading. The approximate extents of the deletions in the pAMβ1 and pIP501 DR1-like sequences (ΔDR1) are indicated by the dotted lines. The filled arrows indicate the individual ORFs and their direction of transcription. The approximate locations of the palindromic sequences (*palA*, *palB* and *palI*) but not their sizes are indicated by the bold lines below the filled boxes. The *palA'* and *palB'* sequences in pAMβ1 and the *palI* sequences in pIP501 represent the portions of the *ermBP*-derived *palA* and *palB* homologs that are present at the ends of the deletion in these DR1-like sequences. Functional and nonfunctional leader peptide sequences are indicated by filled and open inverted triangles, respectively. Modified from Berryman and Rood.[54]

The erythromycin resistance determinant from *C. difficile* strain 630 also belongs to the *ermB/ermAM* class.[57] Subsequent hybridization studies using the *ermBP* gene as a probe, and other *C. difficile* isolates, confirmed these results.[53] The *C. difficile* gene was therefore designated *ermBZ*.[2] More recent studies have shown that ORF3 is present downstream of *ermBZ*, as with *ermBP* and other *erm* genes, and that *C. difficile* strains do not seem to have complete DR1 or DR2 sequences.[54] However, preliminary data suggests that there may be a derivative of DR2 located downstream of *ermBZ* (K.A. Farrow, D. Lyras and J.I. Rood, unpublished results).

Other workers have carried out hybridization studies on an extensive range of *C. difficile* isolates.[59] It was reported that *C. difficile* strains could be detected which carry resistance genes that hybridized with either the *C. perfringens tetP* or *catQ* genes, or the *ermFS* gene from *Bacteroides fragilis*. Since these findings did not involve the cloning and sequence analysis of these determinants they must at this time be considered preliminary and unconfirmed.

Examination of the *C. perfringens* strains that did not hybridize with *ermBP*, using probes representing each of the Erm hybridization classes, showed that none of these isolates hybridized with any of the recognized *erm* gene classes.[53] Further studies on one such isolate, JIR100, led to the cloning and sequencing of the 771 bp *ermQ* gene, which represented a distinct hybridization class, Erm Q.[50] Comparison of the nucleotide sequence with equivalent sequences from each of the Erm classes showed that *ermQ* had between 31% and 57% DNA sequence identity to these *erm* genes. Alignment of the putative sequence of the 257 amino acid ErmQ protein with that of other Erm proteins showed that it had 20–43% identity, confirming that ErmQ represents a distinct class of MLS resistance determinant.[50] The ErmQ determinant was found to be widespread in *C. perfringens* since 30 out of 38 erythromycin resistant isolates hybridized to an *ermQ*-specific probe. It therefore represents the most common erythromycin resistance determinant in *C. perfringens*, which may reflect differences in the mechanisms by which *ermQ* and *ermBP* are disseminated.[50]

D. Summary

As a result of extensive cloning, hybridization and DNA sequence analysis, our knowledge of antibiotic resistance determinants from *C. perfringens* and *C. difficile* has grown considerably in recent years. The latest studies on tetracycline resistance in *C. perfringens*[20] have proven to be interesting as they have shown that, for reasons which are not yet known, virtually all tetracycline resistant isolates carry two resistance genes, which is most unusual. In addition, the finding that a *tet*(M)-like determinant exists in *C. perfringens* is significant because it implies that exchange of genetic information must be possible with other clostridia or with other genera. This conclusion could not be reached from the analysis of Tet P, since both *tetA*(P) and *tetB*(P) appear, at least from the data obtained to date, to be almost unique to *C. perfringens*. *C. difficile* also carries a *tet*(M)-like gene and it would be interesting to determine how similar this gene is to the equivalent gene from *C. perfringens*.

Studies on the chloramphenicol resistance determinants clearly imply that exchange between *C. perfringens* and *C. difficile* is possible because of the near identity between the *catP* and *catD* genes. The probability of gene transfer between these species is further supported by comparative analysis of erythromycin resistance genes, since the *ermBP* and *ermBZ* determinants also are very similar. Whether this exchange occurs directly between *C. perfringens* and *C. difficile* cannot, however, be determined from these data. Furthermore, given the similarity between the *erm* and *tet*(M)-like genes from *C. perfringens* and *C. difficile* and those from *E. faecalis*, it is possible that genetic exchange occurs between these organisms. In contrast, the *catQ* and *ermQ* determinants appear to be restricted to *C. perfringens*, which indicates that these genes are not on elements that are readily exchanged with other species or genera.

III. TRANSPOSONS AND INSERTION SEQUENCES

A. The chloramphenicol resistance transposon Tn*4451* from *Clostridium perfringens*

The *C. perfringens catP* gene is located on a transposable genetic element. Conjugation experiments revealed that transfer of the plasmids pIP401, pJIR25, or pJIR27 from their host strains was often associated with the concomitant loss of the chloramphenicol resistance determinant and approximately 6 kb of DNA.[10,22] Subsequent studies involved the mapping of this 6 kb segment of DNA on the parent plasmids,[22,60] the cloning of the chloramphenicol resistance determinants from pIP401 and pJIR27[22,37] and analysis of the respective recombinant plasmids pJIR45 and pJIR97. Both plasmids are unstable in *recA* strains of *E. coli* with deletion derivatives, which do not confer chloramphenicol resistance, being isolated spontaneously at a high frequency.[37] These deletents lack the identical region that is lost at high frequency on conjugation in *C. perfringens*. Restriction and sequence analysis showed that the identical region was always deleted and that the excision process was a precise event in both *E. coli* and *C. perfringens*.[61] These data, together with the fact that both ends of the 6 kb region are required for spontaneous excision,[37] suggested that the chloramphenicol resistance locus was carried on a transposable genetic element. This

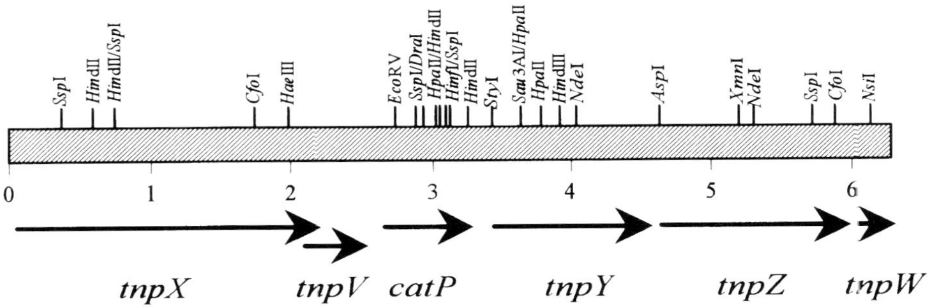

Figure 6.3 Genetic organization of Tn*4451*. Sizes are in kb. Modified from Bannam *et al.*[39]

conclusion was confirmed by the demonstration that these elements could transpose in *E. coli*.[37] Note that transposition has not been demonstrated in *C. perfringens*. Recently, the complete nucleotide sequence of the transposon from pIP401, designated Tn*4451*, has been determined. The 6338 bp transposon encodes six genes, as shown in Figure 6.3.[39] The equivalent transposon from pJIR27, Tn*4452*, has not been sequenced but is closely related to Tn*4451*.[37]

Genetic analysis indicated that the *tnpX* gene was required for the spontaneous excision of the transposon in both *E. coli* and *C. perfringens* since a derivative of Tn*4451* with an internal deletion in *tnpX* was stable in both organisms.[39] The provision of *tnpX in trans* restored the ability of the transposon to excise, therefore it was concluded that this gene encoded a *trans*-acting site-specific recombinase. The TnpX protein was also found to be required for the conjugative excision of Tn*4451* in *C. perfringens*. Furthermore, hybridization and PCR studies showed that TnpX-mediated transposon excision resulted in the formation of a circular form of the transposon.[39] The TnpX recombinase appears to be unique because it potentially contains motifs from two independent site-specific recombinase families, specifically the resolvase/invertase and integrase families. Sequence analysis indicated that the resolvase/invertase domain of TnpX was probably involved in the excision process by catalyzing the formation of a 2 bp staggered nick on either side of the GA dinucleotide located at the ends of Tn*4451* and at the junction of the circular form.[39]

The TnpX-mediated formation of a circular form of Tn*4451* mirrors the behavior of the conjugative transposons Tn*916* and Tn*1545*, where the first step in transposition involves the formation of a non-replicating circular intermediate.[62,63] It has therefore been proposed that the formation of the circular form represents the first step in the transposition of Tn*4451*.[39] Excision of Tn*916* and Tn*1545* involves the Int-Tn and Xis-Tn proteins, the latter acting as an enhancer of excision.[64] It is possible that TnpX acts as a functional homolog of Int-Tn. The TnpV protein has limited similarity to the Xis-Tn protein but

TnpV does not appear to be required for the excision of Tn*4451*,[39] and therefore probably does not act in a similar manner to Xis-Tn.

Tn*4451* appears to encode a second potential site-specific recombinase, TnpZ. This protein has amino acid sequence similarity to plasmid-encoded Mob/Pre proteins, which are involved in plasmid mobilization and multimer formation.[65] Located upstream of *tnpZ* is a region with similarity to the site of interaction of these mobilization proteins, the RS_A site.[39] Interaction between Mob/Pre proteins and an RS_A site is thought to result in plasmid nicking, which is the first step in DNA mobilization.[65] This nicking reaction is also thought to be the first step of plasmid recombination in the formation of multimers.[66] It was therefore proposed that TnpZ is a site-specific recombinase that acts at the upstream RS_A-like site.[39] Furthermore, if TnpZ has a similar function to the Mob/Pre proteins, it is possible that Tn*4451* may be a mobilizable transposon. The presence of mobilizable transposons can enable nonconjugative plasmids to be mobilized in the presence of other transfer factors such as conjugative transposons or plasmids.[67,68] This raises the interesting possibility that TnpZ potentially plays a role in the transmission, and therefore wide-spread dissemination, of other resistance genes.

The *catD* gene from *C. difficile* is closely related to the Tn*4451*-encoded *catP* gene but is located on the chromosome. There is no evidence that *catD* can be transferred by conjugation, either within or between species.[24,69] Preliminary PCR analysis suggests that the *catD* gene may be located on a transposon similar to Tn*4451*, which we have designated Tn*4453* (D. Lyras, C. Storie, A.S. Huggins and J.I. Rood, unpublished results). This observation may explain why there are multiple copies of the *catD* gene on the *C. difficile* chromosome.[44]

B. The conjugative erythromycin resistance transposon Tn*5398* from *Clostridium difficile*

The *ermBZ* determinant from *C. difficile* strain 630 is transferable at a low frequency to the genetically marked *C. difficile* recipient strain CD37.[24] The determinant can be transferred to *S. aureus*[57] and *Bacillus subtilis*[58] in the absence of any detectable plasmid DNA transfer. Transconjugants derived from the latter species can act as donors in matings using strain CD37 as the recipient.[58] Based on these data it was concluded that *ermBZ* was located on a conjugative transposon, Tn*5398*, that is chromosomally determined in strain 630.[58] Hybridization analysis revealed that Tn*5398* integrated in a site-specific manner in *C. difficile* but randomly in *B. subtilis*. This element therefore shares the same properties as Tn*916* and Tn*5397* in transposing to the *B. subtilis* chromosome at multiple sites but being site-specific in *C. difficile*.[26,70] It is interesting that Tn*5398* appears to carry a region that is homologous to part of the *C*-terminal repeat region of the *C. difficile* toxin A gene.[58] The significance of this finding is not known. Very little is known about the structure or transposition mechanism of Tn*5398*. The complete transposon has not been cloned or analyzed at the molecular level and even its size remains unknown.

C. The conjugative tetracycline resistance transposon Tn*5397* from *Clostridium difficile*

The *C. difficile* isolate that carries Tn*5398*, strain 630, also carries a second conjugative transposon, Tn*5397*, which encodes a *tet*(M)-like tetracycline resistance determinant and which is transferred independently of Tn*5398*.[1,23–25,27,69] Hybridization studies showed that at least six out of seven internal or junction fragments of Tn*916*, representing 95% of its length, hybridized with Tn*5397*.[25] These findings were confirmed by Mullany *et al.*[26] who cloned the *tet*(M) gene and some of the flanking DNA from strain 630 and showed that the similarity between Tn*5397* and Tn*916* was significant near the *tet*(M) genes, where several corresponding restriction sites were mapped. Further restriction mapping and hybridization studies have indicated that significant differences exist between these transposons. In particular, the ends of the two elements are different because oligonucleotides complementary to the ends of Tn*916* do not hybridize with the *C. difficile* element.[26] It is concluded that the *C. difficile* tetracycline resistance transposon Tn*5397* is a conjugative transposon that is related to, but distinct from, Tn*916*.[26,70] Recent studies have indicated that Tn*5397* appears to be a modular conjugative transposon with some regions that are almost identical to the Tn*916* family and other regions, including those involved in integration and excision, being completely different (P. Mullany, personal communication).

Like Tn*5398*, Tn*5397* can be transferred to *B. subtilis* and then back to *C. difficile*.[26] The determinant integrates into the *C. difficile* chromosome at two specific sites whereas it integrates into *B. subtilis* at various sites.[26] Other work has suggested that there is also a hot spot for the integration of Tn*916*-like transposons in the *C. difficile* chromosome,[70] as there is for the erythromycin resistance transposon Tn*5398*.[58]

The 22 kb transposon has been cloned and a recombinant plasmid carrying Tn*5397* has been used to transform *B. subtilis* to tetracycline resistance. The resultant transformants could transfer Tn*5397* to *C. difficile* by filter mating. Nucleotide sequence analysis of the region upstream of the *tet*(M) gene has revealed a reverse transcriptase gene that is transcribed in the same direction as *tet*(M). This reverse transcriptase gene was found to be homologous to the reverse transcriptase genes from Group II introns, which appears to represent the first report of such a gene from within a mobile genetic element.[27] The reverse transcriptase gene is contained within a nucleotide sequence that can be folded into a structure typical of a group II intron when transcribed. Furthermore, sites typical of 5′ and 3′ splice sites have been identified. The group II intron interrupts the p60 homolog of Tn*5397*. Tn*916* also contains a p60 homolog that is almost identical to that found in Tn*5397*. However, the Tn*916* gene is not interrupted by a group II intron (P. Mullany, personal communication).

It is possible that there is more than one type of conjugative tetracycline resistance transposon in *C. difficile*. Hybridization studies in our laboratory have shown that DNA isolated from nine *C. difficile* isolates hybridizes with both the left and right sides of Tn*916*, and also with the *tet*(M) gene, but only six of these isolates hybridize with a probe internal to the reverse transcriptase gene from Tn*5397* (D. Lyras, P. Mullany and J.I. Rood, unpublished results). Further comparative studies on the tetracycline resistance elements from *C. difficile* are required to determine the relationship between these elements and Tn*916*.

Finally, it is possible that the *C. perfringens* Tet M-like determinant previously discussed may also be associated with a conjugative transposon-like element because hybridization also is observed between DNA from isolates carrying this determinant and probes consisting of the left and right ends of Tn*916*.[20] Further studies are being carried out to define and characterize this putative element and also to determine if it is related to the *C. difficile* tetracycline resistance elements.

D. The *Clostridium perfringens* IS-like element, IS*1151*

Only one insertion sequence, IS*1151*, from *C. perfringens* has been described. This 1696 bp element was identified 96 bp upstream of the plasmid-encoded[71] epsilon-toxin structural gene, *etx*, from a type D strain.[72] IS*1151* was found to contain two terminal 23 bp regions which constitute a perfect inverted repeat. Database searches revealed sequence similarity to the IS*231* family of insertion elements from *Bacillus thuringiensis*, especially in the inverted repeat region, and also to IS*4*, IS*186* and IS*421* from *E. coli*.[72] The sequence of the 474 amino acid protein deduced from the open reading frame found within IS*1151* was found to resemble the putative transposases encoded by the related insertion sequences.[72] The putative IS*1151*-encoded transposase appeared to be most closely related to transposases of the IS*231* family, having four of the five proposed conserved domains of the IS*231* transposase.[73] It was therefore postulated that IS*1151* may be derived from IS*231*.[72]

It has been suggested that, like IS*231* in *B. thuringiensis*, IS*1151* may be involved in virulence gene transfer in *C. perfringens*. This hypothesis is based on the finding that an IS*1151*-like element was located near the epsilon toxin structural gene in all of the type B and type D isolates tested. Homologous sequences were also found in some strains that did not produce epsilon toxin.[72]

Recent studies have indicated that in some *C. perfringens* strains IS*1151* also may be linked to the enterotoxin gene, *cpe*.[74] The *cpe* gene may be chromosomal or plasmid determined, and in one strain it was found to be located on the same plasmid as the *etx* gene. The analysis of several enterotoxin producing strains indicates that IS*1151* is found in some, but not all, enterotoxigenic isolates of animal origin. Importantly, IS*1151* is confined to isolates which have plasmid-born *cpe* genes.[74] In one such isolate, nucleotide sequencing has shown that IS*1151* is situated approximately 260 bp downstream of the *cpe* gene whereas in another isolate it was located between the *etx* and *cpe* genes.[74] Interestingly, the *cpe* gene may also be associated with another repetitive sequence in addition to IS*1151*. When *cpe* is located on the chromosome it is preceded by a repeated sequence, a *Hin*dIII repeat, and ORF3, which has similarity to a gene from IS*200* from *Salmonella enterica* ser. typhimurium.[74] The *cpe* gene was linked to ORF3 in virtually all of the strains examined. These observations have led Cornillot *et al.*[74] to suggest that insertion sequences, such as IS*1151*, or other transposable elements, may be involved in the mobilization or transfer of virulence genes in *C. perfringens*.

E. Future perspectives

The only transposons which have been detected to date in *C. perfringens* are the closely related Tn*4451* and Tn*4452*. In contrast three different transposons have been detected in *C. difficile*, that is Tn*4453*, Tn*5397*, and Tn*5398*. Apart from Tn*4451*, little is known about these transposons. Further work is therefore necessary to further characterize all of these elements, especially those carried by *C. difficile*. Elucidation of their mode of action will allow comparison with other well-characterized transposons and conjugative elements, and therefore will lead to a greater understanding of how these transposons are disseminated and of the evolutionary relationships between the clostridial transposons and those from other bacteria.

The proposal that insertion elements such as IS*1151* may be involved in virulence gene transfer in *C. perfringens* is intriguing and clearly warrants further investigation. In particular, it is important that direct experimental evidence for the insertion sequence-mediated transfer of either the *etx* or *cpe* genes is obtained. The genetic tools now available in *C. perfringens* should enable such studies to be carried out.

IV. ACKNOWLEDGMENT

The work carried out in this laboratory was generously supported by grants from the Australian National Health and Medical Research Council.

V. REFERENCES

1. Hächler H. and Kayser F.H. (1993) Genetics and molecular biology of antibiotic resistance in *Clostridium difficile*: general and specific overview. In Sebald M. (ed.), *Genetics and Molecular Biology of Anaerobic Bacteria*, pp. 156–173. Springer, New York.
2. Rood J.I. and Cole S.T. (1991) Molecular genetics and pathogenesis of *Clostridium perfringens*. *Microbiol Rev* **55**: 621–648.
3. Rood J.I. (1993) Antibiotic resistance determinants of *Clostridium perfringens*. In Sebald M. (ed.), *Genetics and Molecular Biology of Anaerobic Bacteria*, pp. 141–155. Springer, New York.
4. Levy S.B. (1984) Resistance to the tetracyclines. In Bryan L.E. (ed.), *Antimicrobial Drug Resistance*, pp. 191–240. Academic Press, Orlando, FL.
5. Levy S.B. (1992) Active efflux mechanisms for antimicrobial resistance. *Antimicrob Agents Chemother* **36**: 695–703.
6. Burdett V. (1991) Purification and characterization of Tet(M), a protein that renders ribosomes resistant to tetracycline. *J Biol Chem* **266**: 2872–2877.
7. Manavathu E.K., Fernandez C.L., Cooperman B.S. and Taylor D.E. (1990) Molecular studies on the mechanism of tetracycline resistance mediated by Tet(O). *Antimicrob Agents Chemother* **34**: 71–77.
8. Taylor D.E. and Chau A. (1996) Tetracycline resistance mediated by ribosomal protection. *Antimicrob Agents Chemother* **40**: 1–5.
9. Rood J.I. (1983) Transferable tetracycline resistance in *Clostridium perfringens* strains of porcine origin. *Can J Microbiol* **29**: 1241–1246.
10. Brefort G., Magot M., Ionesco H. and Sebald M. (1977) Characterization and transferability of *Clostridium perfringens* plasmids. *Plasmid* **1**: 52–66.
11. Miyoshi Y. (1984) Transferability of tetracycline resistance to *Clostridium perfringens* isolated from human feces. *Chemotherapy* **30**: 170–174.
12. Miyoshi Y. and Higa A. (1984) Interrelationships between drug resistance and bacteriocinogeny of *C. perfringens*. *Microbiol Immunol* **28**: 281–289.
13. Rood J.I., Scott V.N. and Duncan C.L. (1978) Identification of a transferable resistance plasmid (pCW3) from *Clostridium perfringens*. *Plasmid* **1**: 563–570.
14. Rood J.I., Buddle J.R., Wales A.J. and Sidhu R. (1985) The occurrence of antibiotic resistance in *Clostridium perfringens* from pigs. *Aust Vet J* **62**: 276–279.
15. Abraham L.J. and Rood J.I. (1985) Molecular analysis of transferable tetracycline resistance plasmids from *Clostridium perfringens*. *J Bacteriol* **161**: 636–640.
16. Abraham L.J. and Rood J.I. (1985) Cloning and analysis of the *Clostridium perfringens* tetracycline resistance plasmid, pCW3. *Plasmid* **13**: 155–162.
17. Abraham L.J., Berryman D.I. and Rood J.I. (1988) Hybridization analysis of the class P tetracycline resistance determinant from the *Clostridium perfringens* R-plasmid, pCW3. *Plasmid* **19**: 113–120.
18. Saksena N.K. and Truffaut N. (1992) Cloning of tetracycline-resistance genes from various strains of *Clostridium perfringens* and expression in *Escherichia coli*. *Can J Microbiol* **38**: 215–221.
19. Sloan J., McMurry L.M., Lyras D., Levy S.B. and Rood J.I. (1994) The *Clostridium perfringens* Tet P determinant comprises two overlapping genes: *tetA*(P), which mediates active tetracycline efflux,

and *tetB*(P), which is related to the ribosomal protection family of tetracycline-resistance determinants. *Mol Microbiol* **11**: 403–415.

20. Lyras D. and Rood J.I. (1996) Genetic organisation and distribution of tetracycline resistance determinants from *Clostridium perfringens. Antimicrob Agents Chemother* **40**: 2500–2504.
21. Saksena N.K. and Truffaut N. (1993) Identification of the product of *tetP* gene: a possible mechanistic basis for tetracycline resistance in *Clostridium perfringens. Folia Microbiol* **38**: 15–21.
22. Abraham L.J., Wales A.J. and Rood J.I. (1985) Worldwide distribution of the conjugative *Clostridium perfringens* tetracycline resistance plasmid, pCW3. *Plasmid* **14**: 37–46.
23. Smith C.J., Markowitz S.M. and Macrina F.L. (1981) Transferable tetracycline resistance in *Clostridium difficile. Antimicrob Agents Chemother* **19**: 997–1003.
24. Wüst J. and Hardegger U. (1983) Transferable resistance to clindamycin, erythromycin, and tetracycline in *Clostridium difficile. Antimicrob Agents Chemother* **23**: 784–786.
25. Hächler H., Kayser F.H. and Berger-Bächi B. (1987) Homology of a transferable tetracycline resistance determinant of *Clostridium difficile* with *Streptococcus* (*Enterococcus*) *faecalis* transposon Tn*916*. *Antimicrob Agents Chemother* **31**: 1033–1038.
26. Mullany P., Wilks M., Lamb I., Clayton C., Wren B. and Tabaqchali S. (1990) Genetic analysis of a tetracycline resistance element from *Clostridium difficile* and its conjugal transfer to and from *Bacillus subtilis. J Gen Microbiol* **136**: 1343–1349.
27. Mullany P., Wilks M., Pallen M., Allen E. and Tabaqchali S. (1995) Cloning and genetic analysis of a conjugal broad host range Group II intron from *Clostridium difficile. ASM First International Conference on the Molecular Genetics and Pathogenesis of the Clostridia*, p. 10. American Society for Microbiology, Washington, D.C.
28. Shaw W.V. (1967) The enzymatic acetylation of chloramphenicol by extracts of R factor-resistant *Escherichia coli. J Biol Chem* **242**: 687–693.
29. Harding S.E., Rowe A.J. and Shaw W.V. (1987) The molecular mass and trimeric nature of chloramphenicol acetyltransferase. *Biochem Soc Trans* **15**: 513.
30. Leslie A.G.W., Moody P.C.E. and Shaw W.V. (1988) The structure of chloramphenicol acetyltransferase at 1.75 Å resolution. *Proc Natl Acad Sci USA* **85**: 4133–4137.
31. Lewendon A., Murray I.A., Kleanthous C., Cullis P.M. and Shaw W.V. (1988) Substitutions in the active site of chloramphenicol acetyltransferase: role of a conserved aspartate. *Biochemistry* **27**: 7385–7390.
32. Shaw W.V. and Leslie A.G.W. (1989) Chloramphenicol acetyltransferases. In Bryan L.E. (ed.) *Microbial Resistance to Drugs*, pp. 313–324. Springer-Verlag K.G., Berlin.
33. Rood J.I., Maher E.A., Somers E.B., Campos E. and Duncan C.L. (1978) Isolation and characterization of multiply antibiotic-resistant *Clostridium perfringens* strains from porcine feces. *Antimicrob Agents Chemother* **13**: 871–880.
34. Zaidenzaig Y., Fitton J.E., Packman L.C. and Shaw W.V. (1979) Characterization and comparison of chloramphenicol acetyltransferase variants. *Eur J Biochem* **100**: 609–618.
35. Steffen C. and Matzura H. (1989) Nucleotide sequence analysis and expression studies of a chloramphenicol-acetyltransferase-coding gene from *Clostridium perfringens. Gene* **75**: 349–354.
36. Rood J.I., Jefferson S., Bannam T.L., Wilkie J.M., Mullany P. and Wren B.W. (1989) Hybridization analysis of three chloramphenicol resistance determinants from *Clostridium perfringens* and *Clostridium difficile. Antimicrob Agents Chemother* **33**: 1569–1574.
37. Abraham L.J. and Rood J.I. (1987) Identification of Tn*4451* and Tn*4452*: Two chloramphenicol resistance transposons from *Clostridium perfringens. J Bacteriol* **169**: 1579–1584.
38. Sloan J., Warner T.A., Scott P.T., Bannam T.L., Berryman D.I. and Rood J.I. (1992) Construction of a sequenced *Clostridium perfringens–Escherichia coli* shuttle plasmid. *Plasmid* **27**: 207–219.
39. Bannam T.L., Crellin P.K. and Rood J.I. (1995) Molecular genetics of the chloramphenicol-resistance transposon Tn*4451* from *Clostridium perfringens*: the TnpX site-specific recombinase excises a circular transposon molecule. *Mol Microbiol* **16**: 535–551.

40. Huggins A.S., Bannam T.L. and Rood J.I. (1992) Comparative sequence analysis of the *catB* gene from *Clostridium butyricum*. *Antimicrob Agents Chemother* **36**: 2548–2551.

41. Bannam T.L. and Rood J.I. (1991) The relationship between the *Clostridium perfringens catQ* gene product and chloramphenicol acetyl-transferases from other bacteria. *Antimicrob Agents Chemother* **35**: 471–476.

42. Bannam T.L. (1994) Molecular genetics of the transposon Tn*4451* from *Clostridium perfringens*. PhD Thesis. Monash University, Melbourne, Australia.

43. Dubbert W., Luczak H. and Staudenbauer W.L. (1988) Cloning of two chloramphenicol acetyltransferase genes from *Clostridium butyricum* and their expression in *Escherichia coli* and *Bacillus subtilis*. *Mol Gen Genet* **214**: 328–332.

44. Wren B.W., Mullany P., Clayton C. and Tabaqchali S. (1988) Molecular cloning and genetic analysis of a chloramphenicol acetyltransferase determinant from *Clostridium difficile*. *Antimicrob Agents Chemother* **32**: 1213–1217.

45. Wren B.W., Mullany P., Clayton C. and Tabaqchali S. (1989) Nucleotide sequence of a chloramphenicol acetyl transferase gene from *Clostridium difficile*. *Nucl Acids Res* **17**: 4877.

46. Leclercq R. and Courvalin P. (1991) Bacterial resistance to macrolide, lincosamide, and streptogramin antibiotics by target modification. *Antimicrob Agents Chemother* **35**: 1267–1272.

47. Leclercq R. and Courvalin P. (1991) Intrinsic and unusual resistance to macrolide, lincosamide, and streptogramin antibiotics in bacteria. *Antimicrob Agents Chemother* **35**: 1273–1276.

48. Lai C.J. and Weisblum B. (1971) Altered methylation of ribosomal RNA in an erythromycin-resistant strain of *Staphylococcus aureus*. *Proc Natl Acad Sci USA* **68**: 856–860.

49. Lai C.J., Weisblum B., Fahnestock S.R. and Nomura M. (1973) Alteration of 23S ribosomal RNA and erythromycin induced resistance to lincomycin and spiramycin in *Staphylococcus aureus*. *J Mol Biol* **74**: 67–72.

50. Berryman D.I., Lyristis M. and Rood J.I. (1994) Cloning and sequence analysis of *ermQ*, the predominant macrolide-lincosamide-streptogramin B resistance gene in *Clostridium perfringens*. *Antimicrob Agents Chemother* **38**: 1041–1046.

51. Dubnau D. and Monod M. (1986) The regulation and evolution of MLS resistance, *Banbury Rep* **24**: 369–387.

52. Sebald M., Bouanchaud D. and Bieth G. (1975) Nature plasmidique de la resistance à plusiers antibiotiques chez *C. perfringens* type A, souche 659. *Compt Rend Acad Sci Paris Serie D* **280**: 2401–2404.

53. Berryman D.I. and Rood J.I. (1989) Cloning and hybridization analysis of *ermP*, a macrolide-lincosamide-streptogramin B resistance determinant from *Clostridium perfringens*. *Antimicrob Agents Chemother* **33**: 1346–1353.

54. Berryman D.I. and Rood J.I. (1995) The closely related *ermB/AM* genes from *Clostridium perfringens, Enterococcus faecalis* (pAMβ1), and *Streptococcus agalactiae* (pIP501), are flanked by variants of a directly repeated sequence. *Antimicrob Agents Chemother* **39**: 1830–1834.

55. Brehm J., Salmond G. and Minton N. (1987) Sequence of the adenine methylase gene of the *Streptococcus faecalis* plasmid pAMβ1. *Nucl Acids Res* **15**: 3177.

56. Martin B., Alloing G., Méjean V. and Claverys J.-P. (1987) Constitutive expression of erythromycin resistance mediated by the *ermAM* determinant of plasmid pAMβ1 results from deletion of 5′ leader peptide sequences. *Plasmid* **18**: 250–253.

57. Hächler H., Berger-Bächi B. and Kayser F.H. (1987) Genetic characterization of a *Clostridium difficile* erythromycin-clindamycin resistance determinant that is transferable to *Staphylococcus aureus*. *Antimicrob Agents Chemother* **31**: 1039–1045.

58. Mullany P., Wilks M. and Tabaqchali S. (1995) Transfer of macrolide-lincosamide-streptogramin B (MLS) resistance in *Clostridium difficile* is linked to a gene homologous with toxin A and is mediated by a conjugative transposon, Tn*5398*. *J Antimicrob Chemother* **35**: 305–315.

59. Roberts M.C., McFarland L.V., Mullany P. and Mulligan M.E. (1994) Characterization of the genetic basis of antibiotic resistance in *Clostridium difficile*. *J Antimicrob Chemother* **33**: 419–429.
60. Magot M. (1984) Physical characterization of the *Clostridium perfringens* tetracycline-chloramphenicol resistance plasmid pIP401. *Ann Microbiol (Inst Pasteur)* **135B**: 269–282.
61. Abraham L.J. and Rood J.I. (1988) The *Clostridium perfringens* chloramphenicol resistance transposon Tn*4451* excises precisely in *Escherichia coli*. *Plasmid* **19**: 164–168.
62. Scott J.R. (1993) Conjugative transposons. In Sonenshein A.L., Hoch J.A. and Losick R. (eds), Bacillus subtilis *and Other Gram-positive Bacteria*, pp. 597–614. American Society for Microbiology, Washington, D.C.
63. Clewell D.B. and Flannagan S.E. (1993) The conjugative transposons of Gram-positive bacteria. In Clewell D.B. (ed.), *Bacterial Conjugation*, pp. 369–393. Plenum Press, New York.
64. Poyart-Salmeron C., Trieu-Cuot P., Carlier C. and Courvalin P. (1989) Molecular characterization of two proteins involved in the excision of the conjugative transposon Tn*1545*: homologies with other site-specific recombinases. *EMBO J* **8**: 2425–2433.
65. Waters V.L. and Guiney D.G. (1993) Processes at the nick region link conjugation, T-DNA transfer and rolling circle replication. *Mol Microbiol* **9**: 1123–1130.
66. Gennaro M.L., Kornblum J. and Novick R.P. (1987) A site-specific recombination function in *Staphylococcus aureus* plasmids. *J Bacteriol* **169**: 2601–2610.
67. Hecht D.W. and Malamy M.H. (1989) Tn*4399*, a conjugal mobilizing transposon of *Bacteroides fragilis*. *J Bacteriol* **171**: 3603–3608.
68. Smith C.J. and Parker A.C. (1993) Identification of a circular intermediate in the transfer and transposition of Tn*4555*, a mobilizable transposon from *Bacteroides* spp. *J Bacteriol* **175**: 2682–2691.
69. Ionesco H. (1980) Transfert de la résistance à la tétracycline chez *Clostridium difficile*. *Ann Inst Pasteur Microbiol* **131A**: 171–179.
70. Mullany P., Wilks M. and Tabaqchali S. (1991) Transfer of Tn*916* and Tn*916*ΔE into *Clostridium difficile*: demonstration of a hot-spot for these elements in the *C. difficile* genome. *FEMS Microbiol Lett* **79**: 191–194.
71. Canard B., Saint-Joanis B. and Cole S.T. (1992) Genomic diversity and organization of virulence genes in the pathogenic anaerobe *Clostridium perfringens*. *Mol Microbiol* **6**: 1421–1429.
72. Daube G., Simon P. and Kaeckenbeeck A. (1993) IS*1151*, an IS-like element of *Clostridium perfringens*. *Nucl Acids Res* **21**: 352.
73. Mahillon J., Rezsöhazy R., Hallet B. and Delcour J. (1994) IS*231* and other *Bacillus thuringiensis* transposable elements: a review. *Genetica* **93**: 13–26.
74. Cornillot E., Saint-Joanis B., Daube G., Katayama S., Granum P.E., Canard B. and Cole S.T. (1995) The enteroxin gene (*cpe*) of *Clostridium perfringens* can be chromosomal or plasmid-borne. *Mol Microbiol* **15**: 639–647.

C: Clostridia in Clinical Practice

Chapter 7

Human Tetanus and Human Botulism

STEPHEN S. ARNON

Infant Botulism Prevention Program, California Department of Health Services, 2151 Berkeley Way, Berkeley, CA 94704–1011, USA

I. INTRODUCTION

Generalized tetanus is one of the most dramatic, devastating and globally prevalent diseases of humans and other vertebrate animals. Tetanus is also one of the most preventable of diseases, through immunization with tetanus toxoid. Botulism, in contrast, is rare and therefore often mistaken for common illnesses. Botulism is also an important veterinary disease. Both diseases originate from one of eight clostridial neurotoxins whose remarkable similarities of structure and mechanism indicate a common ancestral gene, yet the principal manifestations of the two diseases – spastic paralysis for tetanus (Figure 7.1) and flaccid paralysis for botulism (Figure 7.2) – paradoxically represent polar opposites of clinical immobility.

By the middle of the twentieth century, the differences in the clinical features of botulism and tetanus were known to result from the different sites of action of their respective toxins.[1] All seven serotypes (A–G) of botulinum toxin block acetylcholine release at the neuromuscular junction, thereby preventing muscle contraction (flaccid paralysis), whereas tetanus toxin blocks release of the neurotransmitters glycine and gamma-amino-butyric acid (GABA) from certain

The Clostridia: Molecular Biology and Pathogenesis
ISBN 0–12–595020–9

Figure 7.1 Sir Charles Bell's portrait (*c.* 1821) of a soldier wounded in the Peninsular War in Spain with generalized tetanus and evident *opisthotonos* and *risus sardonicus*. Original in the Royal College of Surgeons of Edinburgh. Reproduced with the permission of the President and Fellows of the Royal College of Surgeons of Edinburgh. (See colour plate.)

Figure 7.2 Mother and 7-week-old patient with infant botulism. Note expressionless face, ptosis, inability to hold head erect, loss of extremity and truncal muscle tone. (See colour plate.)

spinal cord and brainstem neurons, thereby preventing muscle relaxation. Both toxins initially bind at the neuromuscular junction, from which point they are internalized into the motoneuron cell. Tetanus toxin is then transported via retrogade axoplasmic flow to the spinal cord or brainstem, whereas botulinum toxin remains and acts immediately proximal to the terminal motoneuron membrane. The seven serotypes of botulinum toxin are distinguished by the inability of neutralizing antibody raised against one toxin type to neutralize any of the other toxin types.

Botulinum and tetanus toxins share the dubious distinction of being the most poisonous substances known; the potency of tetanus toxin (human lethal dose estimated at about 1 ng/kg) is exceeded only by the potency of botulinum toxin (human lethal dose estimated at about $\leqslant$0.1 ng/kg).[2] These estimates are based on systemic exposure to toxin in experimental animals; botulinum toxin is equally potent and effective whether administered by intravenous injection, by intraperitoneal injection, or by aerosol inhalation.[3] A dilute solution of purified botulinum toxin is colorless, odorless and, as far as is known, tasteless.

The potency of these two toxins defies ordinary experience. Assuming a global population of approximately 5.6 billion people (the current US Census Bureau estimate) who on average weigh 70 kg each, then just 392 g (less than 1 lb) of tetanus toxin and 39.2 g (slightly more than 1 ounce) of botulinum toxin would be sufficient to eradicate humankind. In comparison with more widely known poisons, botulinum toxin is approximately 800 million times more potent than cyanide, 400 million times more potent that arsenic, 300 million times more potent than rattlesnake venom, and 100 million times more potent than cobra venom. The comparable potencies for tetanus toxin are those of botulinum toxin reduced by a factor of 10.

With the recent illuminating discoveries of the molecular mechanism and sites of action of tetanus toxin and botulinum toxin, the biochemical bases of these ancient diseases are now almost entirely understood (see Chapter 18). Both botulinum and tetanus toxins exist as simple bi-chain proteins, whose larger component (a 'heavy' chain of 100 000 daltons) contains the cellular binding site and whose smaller ('light' chain of 50 000 daltons) component is a Zn^{2+}-containing endoprotease. The enzymatic light chain cleaves one of three components of the 'docking' or 'fusion' complex by which the synaptic vesicle fuses with the terminal cell membrane and releases its neurotransmitter.[4,5]

This chapter describes clinical and epidemiological features of human tetanus and human botulism, the paralytic diseases caused by these most poisonous poisons. Brief mention will also be made of botulinum toxin's use as a therapeutic agent to produce muscle relaxation in a variety of clinical situations, a profoundly useful and relatively recent development that illustrates the humanitarian benefit now available from this ancient toxin. In contrast, at least one malevolent government is known to have recently mass-produced approximately 19 000 l of botulinum toxin, of which approximately 10 000 l was incorporated into artillery shells and SCUD missiles. Many of these armaments may still remain in existence.[6]

II. TETANUS

A. The disease, the organism, the toxin

Tetanus, known familiarly as lockjaw, is an acute, spastic paralytic illness caused by the neurotoxin produced by *Clostridium tetani*. In past years this neurotoxin was referred to as tetanospasmin, but it is now most commonly known as tetanus toxin. The light chain of tetanus toxin is a Zn^{2+}-containing endoprotease whose substrate is synaptobrevin, one of the constituent proteins of the docking complex that enables a synaptic vesicle to fuse with the terminal cell membrane.[7] The heavy chain of the toxin contains the binding domain by which it attaches specifically to the motoneuron.

Clostridium tetani is a motile, Gram-positive, spore-forming obligate anaerobe whose natural habitat worldwide is soil, dust and the intestinal tracts of various animals. It forms spores terminally, producing a microscopic 'drumstick' or 'tennis racket' appearance. Tetanus spores can survive boiling but not autoclaving. The vegetative cells of *C. tetani* are killed by antibiotics, heat and standard disinfectants. *C. tetani* is not an invasive organism. After introduction of *C. tetani* spores and their subsequent germination, the toxin-producing vegetative cells remain at the site of injury, which may or may not display local inflammatory changes or a mixed infectious flora.

B. Pathogenesis

Illness occurs after introduced tetanus spores germinate, multiply and produce tetanus toxin in the low oxidation–reduction potential (E_h) of an infected injury site. A plasmid carries the toxin gene;[8] the toxin is released following the death and lysis of the vegetative bacterial cell. Tetanus toxin binds at the neuromuscular junction and is then endocytosed by the motor nerve, after which it undergoes retrograde axonal transport to the motoneurons in the spinal cord and brain stem at a rate of 3–10 mm/h.[9] After reaching the spinal cord the toxin exits the motoneuron and next enters the adjacent, spinal and brainstem inhibitory interneurons. Tetanus toxin then prevents these interneurons from releasing their inhibitory neurotransmitters. In this manner, tetanus toxin blocks the relaxation of opposing muscles that is the basis of normal, coordinated movement between antagonistic muscle groups; in consequence, affected muscles sustain maximal contraction. The autonomic nervous system is also rendered unstable in tetanus.[10]

C. Epidemiology

Tetanus occurs worldwide and is considered endemic in 90 developing countries, but its incidence varies considerably among them. The most common form of the disease, neonatal (umbilical) tetanus, kills at least 500 000 infants each year simply because their mothers were not immunized. Over 70% of these deaths occur in just 10 tropical Asian and African countries.[11] Various customs, such as using dung, dirt or local butter (ghee) to pack the umbilical stump, often are responsible for infant infection.[12,13] In addition, an estimated 15 000–30 000 unimmunized women worldwide die each year from maternal tetanus, the result of post-partum, post-abortal or post-surgical wound infection with *C. tetani*.[14] In the United States approximately 50 cases of tetanus are reported each year, mostly in persons aged 60 years or older, but cases in toddlers and neonates also occur.[15]

Most non-neonatal cases of tetanus are associated with a traumatic injury, often a penetrating wound inflicted by a dirty object such as a nail, splinter, fragment of glass or unsterile injection, but some cases may have no history or evidence of trauma. Tetanus following illicit drug injection is becoming more common,[16] while uncommon occurrences include animal bites, abscesses (including dental abscesses), ear piercing, chronic skin ulceration, burns, compound fractures, frostbite, gangrene, intestinal surgery, ritual scarification and female circumcision.[17,18] The disease has also occurred after the use of contaminated suture material or after intramuscular injection of medicines, most notably quinine (which may result in local muscle necrosis) for chloroquine-resistant falciparum malaria.[19]

D. Clinical aspects

Tetanus may be either localized or generalized, the latter being more common. The incubation period typically is 2–14 days, but may be as long as several months after the injury. In *generalized tetanus*, trismus (masseter muscle spasm, or lockjaw) is the presenting symptom in about half the cases. Headache, restlessness and irritability are early symptoms, often followed by stiffness, difficulty chewing, dysphagia and neck muscle spasm. The so-called sardonic smile of tetanus (risus sardonicus) results from the intractable spasm of facial and buccal muscles. When the paralysis extends to abdominal, lumbar, hip and thigh muscles, the patient may assume an arched posture, opisthotonos, in which only the back of the head and the heels touch ground. Opisthotonos is an end-state position that results from the unrelenting total contraction of opposing muscle groups, all of which display the typical 'board-like' rigidity of tetanus (Figure 7.1). Laryngeal and respiratory muscle spasm can lead to airway obstruction and asphyxiation.

Because tetanus toxin does not affect sensory nerves or cortical function, the patient unfortunately remains conscious, in extreme pain and in anxious anticipation of the next tetanic seizure. These seizures are characterized by sudden, severe tonic contractions of the muscles, with fist clenching, flexion and adduction of the arms and hyperextension of the legs. Without treatment, the seizures range from a few seconds to a few minutes with intervening respite periods, but as the illness progresses the spasms become sustained and exhausting. The smallest disturbance by sight, sound or touch may trigger a tetanic spasm. Dysuria and urinary retention result from bladder sphincter spasm; forced defecation may occur. Fever, with temperatures occasionally as high as 40°C, is common because of the substantial metabolic energy consumed by spastic muscles. Notable autonomic effects include tachycardia, arrhythmias, labile hypertension, diaphoresis, and cutaneous vasoconstriction. The tetanic paralysis usually becomes more severe in the first week after onset, stabilizes in the second week, and ameliorates gradually over the ensuing 1–4 weeks.

Neonatal tetanus (tetanus neonatorum), the infantile form of generalized tetanus, typically manifests within 3–12 days of birth as progressive difficulty in feeding (i.e., sucking and swallowing), with associated hunger and crying. Paralysis or diminished movement, stiffness to the touch and spasms, with or without opisthotonos, characterize the disease. The umbilical stump, the usual portal of entry of the spores, may hold remnants of dirt, dung, clotted blood or serum, or it may appear relatively benign.

Localized tetanus results in painful spasms of the muscles adjacent to the wound site and may precede generalized tetanus. *Cephalic tetanus* is a rare form of localized tetanus involving the bulbar musculature that occurs with wounds or foreign bodies in the head, nostrils or face. It also occurs in association with chronic otitis media. Cephalic tetanus is characterized by retracted eyelids, deviated gaze, trismus, risus sardonicus and spastic paralysis of tongue and pharyngeal musculature.

The image of tetanus is one of the most dramatic in medicine, and the diagnosis is most often made clinically. The typical setting is an unimmunized person who either gave birth, was injured or was born within the preceding 2 weeks and who presents with trismus, other rigid muscles and a clear sensorium. Only intoxication with strychnine produces a clinical picture that may be truly confused with tetanus.

Routine laboratory studies are remarkable for their normality. A peripheral leukocytosis may result from a secondary bacterial infection of the wound or may be stress-induced from the sustained tetanic spasms. Cerebro-spinal fluid (CSF) cellular and chemistry indices are normal, although the intense muscle contractions may raise CSF pressure. Neither the electroencephalogram (EEG) nor the electromyogram (EMG) show a characteristic pattern. *C. tetani* is not always visible on Gram stain of wound material, and it is isolated in only about one-third of cases.

Management of tetanus requires eradication of *C. tetani* and the wound environment conducive to its anaerobic multiplication, neutralization of all accessible tetanus toxin, control of seizures and respiration, palliation and provision of meticulous supportive care, and finally, prevention of recurrences.[20]

Surgical wound excision and debridement is often needed to remove the foreign body or devitalized tissue that created anaerobic growth conditions. Surgery should be performed promptly, but preceded by the administration of human tetanus immune globulin (TIG) and antibiotics. Sedation is needed unless general anesthesia is used. Excision of the umbilical stump in neonatal tetanus is no longer recommended.

Once tetanus toxin has begun its axonal ascent to the spinal cord or brainstem, it can no longer be neutralized by TIG. Accordingly, TIG is given as soon as possible to neutralize toxin that diffuses from the wound into the circulation before the toxin can bind to distant neuromuscular sites. An optimal dose of TIG has not been determined, and expert opinion varies. A single intramuscular injection of 500 U of human TIG is sufficient to neutralize systemic tetanus toxin, but doses as high as 3000–6000 have been recommended. Infiltration of TIG into the wound is now considered unnecessary. If TIG is unavailable, use of human intravenous immunoglobulin (IGIV), which contains 4–90 U/ml of TIG, or of equine- or bovine-derived tetanus antitoxin (TAT) may be necessary.[21,22] Intrathecal TIG, given to neutralize tetanus toxin in the spinal cord, has not on meta-analysis proved effective.[23]

Metronidazole has displaced penicillin G as the first-choice antibiotic because of its effective clostridiocidal action, its lack of GABA antagonism and its diffusability – an important consideration because blood flow to injured tissue may be compromised. The dose is 500 mg every 6 h or 1000 mg every 12 h intravenously for 10–14 days.[9] If metronidazole is not available, erythromycin and tetracycline (in patients 9 or more years old) are alternatives for penicillin-allergic patients.[20]

All patients with generalized tetanus need muscle relaxants. Remarkably, botulinum toxin type A was recently and successfully applied for this purpose in a case of cephalic tetanus.[24] Because of their more limited durations of action (4–5 weeks versus 3–4 months), botulinum toxin types B and F should especially be considered for this treatment.[25] As standard therapy, diazepam provides both relaxation and seizure control; an initial dose of 0.1–0.2 mg/kg is given intravenously and every 3–6 h as needed to control the tetanic spasms, after which the diazepam is sustained for 2–6 weeks before its tapered withdrawal. Magnesium sulfate, other benzodiazepines, chlorpromazine, dantrolene and baclofen have also been used. Intrathecal baclofen produces such complete muscle relaxation that apnea often ensues; like most other agents listed, baclofen should be used only by experienced persons in an intensive care unit setting. The best survival rates in generalized tetanus have been achieved with neuromuscular blocking agents like vecuronium and pancuronium, which produce a general flaccid paralysis then managed by mechanical ventilation. Autonomic instability has been regulated with standard alpha- or beta- (or both) blocking agents; morphine has also proved useful.

Meticulous supportive care of the patient in a quiet, dark secluded setting is highly desirable. Because tetanic spasms may be triggered by minor sensory stimuli, the patient should be sedated and protected from all unnecessary sounds, sights and touch. All therapeutic and other manipulations must be carefully scheduled and coordinated. Endotracheal intubation may not be required, but it should be done to prevent aspiration of secretions before laryngospasm develops. A tracheotomy kit should be at hand for unintubated patients. However, endotracheal intubation and suctioning easily provoke reflex tetanic seizures and spams, and so early tracheostomy deserves consideration for severe cases not managed by pharmacologically induced flaccid paralysis. Cardiorespiratory monitoring, frequent suctioning, and maintenance of the substantial fluid, electrolyte and caloric needs are fundamental. Careful nursing attention to mouth, skin, bladder and bowel function is needed to avoid ulceration, infection and obstipation. Prophylactic subcutaneous heparin use is advised.

The seizures and the severe, sustained rigid paralysis of tetanus predispose the patient to many complications. Aspiration of secretions and pneumonia may have occurred before the first medical attention is received. Maintenance of airway patency often mandates endotracheal intubation and mechanical ventilation with their attendant hazards, including pneumothorax and mediastinal emphysema. The seizures may result in lacerations of the mouth or tongue, in intramuscular hematomas or rhabdomyolysis with myoglobinuria and renal failure, or in long-bone or spinal fractures. Venous thrombosis, pulmonary embolism, gastric ulceration with or without hemorrhage, paralytic ileus, and decubitus ulceration are constant hazards. Excessive use of muscle relaxants, although an integral part of care, may produce iatrogenic apnea. Cardiac arrhythmias, including asystole, unstable blood pressure and labile temperature regulation reflect disordered autonomic nervous system control that may be aggravated by inattention to the maintenance of intravascular volume needs.

Recovery from tetanus occurs through regeneration of synapses within the spinal cord and brainstem and thereby, the restoration of muscle relaxation. However, because illness with tetanus does not induce formation of toxin neutralizing antibodies, initiating active immunization with tetanus toxoid at discharge, with arrangements for completion of the primary series, is mandatory.

The most important factor influencing outcome is the quality of supportive care. Mortality is highest in the very young and the very old. A favorable prognosis is associated with a long incubation period, with the absence of fever and with localized disease. An unfavorable prognosis is associated with an interval of 1 week or less between the injury and the onset of trismus or with 3 days or less between trismus and onset of generalized tetanic spasms. Sequelae of hypoxic brain injury, especially in infants, include cerebral palsy, diminished mental abilities and behavioral difficulties. Most fatalities occur within the first week of illness. Reported case fatality rates for generalized tetanus range between 5% and 35% and for neonatal tetanus extend from less than 10% with intensive care unit management to more than 75% without it. Cephalic tetanus has an especially poor prognosis because of breathing and feeding difficulties.

Tetanus is an entirely preventable disease; a serum antibody level of ≥0.01 U/ml, as determined by a suitable *in vivo* neutralization assay, is considered protective.[26] Active immunization should begin in early infancy with combined diphtheria toxoid–tetanus toxoid–pertussis (DTP) vaccine at 2, 4 and 6 months of age, with a booster at 4–6 years of age and at 10-year intervals thereafter throughout adult life with tetanus–diphtheria (Td) toxoids. Immunization of women of childbearing age with tetanus toxoid will prevent neonatal tetanus; a single dose of toxoid that contains 250 Lf (limits of flocculation) units may be safely given in the third trimester of pregnancy and is capable of providing enough transplacental antibody to protect the child for at least 4 months.[27] For unimmunized persons 7 or more years of age, the primary immunization series consists of three doses of Td toxoid given intramuscularly, the second 4–6 weeks after the first and the third 6–12 months after the second.

Tetanus prevention measures following trauma consist of inducing active immunity to tetanus toxin and of passively providing neutralizing antitoxin antibody. Tetanus prophylaxis is an essential part of all wound management, but the specific measures taken depend on the nature of the injury and the immunization status of the patient. Tetanus toxoid should be given after a dog or other animal bite, even though *C. tetani* is infrequently found in canine mouth flora, if five or more years have elapsed since the last booster. Human TIG is not given for clean and minor wounds, regardless of prior immunization status; it is also not given to patients with 'dirty' or other traumatic injuries who have received a full primary immunization series and a tetanus toxoid booster within the last five years. In other circumstances (e.g., patients with an unknown or incomplete immunization history, with crush, puncture or projectile wounds or wounds contaminated with saliva, soil or feces, avulsion injuries, compound fractures, frostbite, etc.), 250 U of TIG should be given intramuscularly, and increased to 500 U for highly tetanus prone wounds (i.e., undebridable, with substantial bacterial contamination, or more than 24 h old). If TIG is unavailable, then use of human IGIV may be considered.[22]

The wound itself should have immediate, thorough surgical cleansing and debridement to remove foreign bodies and any necrotic tissue in which anaerobic conditions might develop. Tetanus toxoid should be given to stimulate active immunity and may be administered concurrently with TIG (or TAT) if given in separate syringes at separate sites. A tetanus toxoid booster (preferably Td) is given to all persons with any wound if their tetanus immunization status is unknown or incomplete. Also, a booster is given to injured persons who have completed their primary immunization series if (1) the wound is clean and minor but 10 years or more have passed since the last booster, or (2) the wound is more than clean and minor, and 5 years or more have passed since the last booster. With delayed wound care, active immunization should be started at once. Although fluid tetanus toxoid produces a more rapid immune response than the adsorbed or precipitated toxoids, the adsorbed toxoid results in a more durable titer. The constant threat of tetanus, both in developed and in less developed countries, is highlighted by the recent finding that one-fifth of children aged 10–16 years in the United States did not possess a protective level of tetanus antibody.[15]

III. BOTULISM

A. The disease, the organism, the toxins

Botulism is the acute, flaccid paralysis caused by the neurotoxins of *Clostridium botulinum*, or rarely, by an equivalent neurotoxin produced by unique strains of *Clostridium butyricum* or *Clostridium baratii*. Three forms of human botulism are known: infant botulism (the most common in the United States), foodborne (classical) botulism, and wound botulism. Each form of human botulism has a different pathogenesis and epidemiology.

As already mentioned, botulinum toxin is the most poisonous substance known.[2] Its potency derives from its ability to block neuromuscular transmission and cause death through paralysis of airway and respiratory musculature. Seven antigenic toxin types, arbitrarily assigned the letters A–G, can be distinguished by the inability of protective (neutralizing) antibody raised against one toxin type to protect against poisoning by a different toxin type. The seven toxin types are often used as convenient clinical and epidemiological markers. Neurotoxigenic *C. butyricum* strains produce a type E-like toxin, while neurotoxigenic *C. baratii* strains produce a type F-like toxin. Toxin types A, B, E, and F are well-established causes of human botulism, while types C and D cause illness in other animals. Type G was discovered in an Argentinean cornfield in 1970 and has not been established as a cause of either human or animal disease.[28]

The phenomenal potency of botulinum toxin was finally explained by the recent discovery that its seven light chains are Zn^{2+}-containing endopeptidases, whose substrates are one of three protein components of the docking complex by which synaptic vesicles fuse with the terminal cell membrane and release acetylcholine into the synaptic cleft. Of these three substrates, synaptobrevin is cleaved by tetanus toxin and by botulinum toxin types B, D, F, G; SNAP-25 is cleaved by botulinum toxins types A, C and E, and syntaxin is cleaved by botulinum toxin type C (see Chapter 18).[29]

Clostridium botulinum is a Gram-positive, spore-forming obligate anaerobe whose natural habitat worldwide is soil, dust and marine sediments, and consequently, it is frequently found in a wide variety of fresh and cooked agricultural products. Spores of some *C. botulinum* strains can survive boiling for several hours, which enables the organism occasionally to outlast human efforts at food preservation. In contrast, botulinum toxin is heat-labile and is easily destroyed by heating at 80°C or above for 5–10 min. Little is known about the ecology of neurotoxigenic strains of *C. butyricum* and *C. baratii*.

B. Pathogenesis

All three forms of human botulism result from a final common disease pathway: botulinum toxin is carried by the bloodstream to peripheral cholinergic synapses, where it binds irreversibly, blocking acetylcholine release and causing impaired autonomic and neuromuscular transmission. *Infant (intestinal toxemia) botulism* is an infectious disease that results from ingesting the spores of any of the three botulinum-toxin-producing clostridial strains, with subsequent spore germination, multiplication and production of botulinum toxin in the large intestine.[30] *Foodborne botulism* is an intoxication that results when preformed botulinum toxin contained in an improperly preserved food is swallowed.[31] *Wound botulism* results from spore germination and colonization of traumatized tissue by *C. botulinum*; it is the analog of tetanus, but much rarer.[32,33]

Sprouting of new unmyelinated motoneuron terminals occurs as a healing process in botulism. When these new 'twigs' locate noncontracting muscle fibers and re-innervate them by inducing formation of a new motor end-plate, movement resumes. Clinically, and in experimental animals, this process takes, on average, about 4 weeks. No diagnostic technique is available to identify botulinum toxin bound at the neuromuscular junction.

C. Epidemiology

Each form of human botulism has a different epidemiology. Infant botulism has been reported from all inhabited continents except Africa. Notably, the infant is the only family member who is ill. To date, almost all recognized cases have been hospitalized, with epidemiological generalizations derived from them. The most striking epidemiological feature of infant botulism is its age distribution, in which 95% of cases are aged between 3 weeks and 6 months, with a broad peak between 2 and 4 months old. This pattern is matched only by one other condition, the sudden infant death syndrome. However, cases as young as 6 days or as old as 363 days at onset have been reported. The male:female ratio of cases is essentially 1:1, and cases have occurred in all major racial and ethnic groups.

Infant botulism is an uncommon and often unrecognized illness. In the United States about 75–100 cases are diagnosed annually, and 1290 cases were reported for the years 1976–94. Almost half of US cases have been reported from the State of California. Reflecting the known asymmetry of the soil distribution of toxin types of *C. botulinum*, most cases west of the Mississippi River have been caused by type A strains, while most cases east of it have been caused by type B strains. One case each in the states of New Mexico and Oregon resulted from *C. baratii* and its type F-like toxin, while two cases in Rome, Italy, resulted from *C. butyricum* and type E-like toxin. Identified risk factors for illness include the ingestion of honey and a slow intestinal transit time (less than one stool per day). Breast-feeding appears to provide protection against fulminant, sudden death from infant botulism.[34] Under exceptional circumstances of altered intestinal anatomy, physiology and microflora, older children and adults may contract infant botulism (summarized by Arnon[30]).

Foodborne botulism results from the ingestion of a food in which *C. botulinum* has multiplied and produced its toxin. Recent outbreaks in North America associated with restaurants in which such unlikely foods as baked potatoes, sautéd onions, and chopped garlic were implicated have revised the traditional view of foodborne botulism as resulting mainly from home-canned foods. Other recent US outbreaks have been caused by commercial foods sealed in plastic pouches that relied solely on refrigeration to prevent outgrowth of *C. botulinum* spores. Non-canned foods, either sold commercially or served in restaurants, have been responsible for recent, unusual foodborne botulism episodes in the United States, Europe and the Middle East. The foods implicated included hazelnut flavoring added to yogurt,[35] sautéd onions in 'patty melt' sandwiches,[36] chopped garlic-in-oil,[37] and both freshwater and saltwater dried, uneviscerated fish.[38,39] Foodborne botulism outbreaks have resulted from commercial preserved peanuts in Asia[40] and from white ants (termites) stored in a polyethylene bag in Africa.[41]

A recent trend toward a single case per outbreak or of cases presenting separately in different cities or hospitals has meant that the physician cannot rely on the temporal and geographical clustering of illness (often in a family) to suggest the diagnosis of foodborne botulism. Most foods that man has attempted to preserve have been implicated at one time or another in foodborne botulism, but the usual offenders in the United States are the 'low-acid' (pH 6.0 and above) home-canned foods such as jalapeño peppers, asparagus, olives and beans. The potential for foodborne botulism exists throughout the world, but outbreaks occur most commonly in the temperate zones where winters are long, rather than in the tropics, where preservation of fruits, vegetables and other foods is less common.[42]

In the last decade approximately 250 cases of foodborne botulism from about 15 outbreaks per year have occurred in the United States. Most of the continental US outbreaks resulted from either proteolytic type A or type B strains, which produce a strongly putrefactive odor in the food that some people, unfortunately, find necessary to verify by tasting. In contrast, in Alaska most foodborne outbreaks have resulted from nonproteolytic type E strains and Native American foods, such as salmon eggs and seal flippers, that do not exhibit signs of spoilage. A further hazard of type E strains is their ability to grow at temperatures as low as 5°C, the inside temperature of household refrigerators.

Wound botulism is an exceptionally rare disease, with about 200 cases reported worldwide. Most cases have occurred in young, physically active males who are presumably at greatest risk of traumatic injury, but wound botulism has also occurred from crush injuries in which no break in the skin was evident. About one-quarter of wound botulism cases have occurred in teenage or younger persons. In the last decade wound botulism has become increasingly common in drug abusers, resulting from self-injection or nasal inhalation of nonsterile products, not always with evident abscess formation.[33]

D. Clinical manifestations

Because botulinum toxin is hematogenously distributed and because relative blood flow and density of innervation are greatest in the bulbar musculature, all three forms of botulism manifest clinically as a symmetrical flaccid paralysis that first affects and descends from the muscles of the head, face, mouth and throat. It is not possible to have botulism without having bulbar palsies, yet in infants symptoms such as poor feeding, weak suck, feeble cry, drooling and even obstructive apnea are often not recognized as bulbar in origin. Patients with evolving illness may already have generalized weakness and loss of muscle tone in addition to cranial nerve palsies when first seen.

In older children with foodborne or wound botulism the onset of neurological symptoms follows a characteristic pattern of diplopia, blurred vision, ptosis, dry mouth, dysphagia, dysphonia and dysarthria, with deceased gag and corneal reflexes. Importantly, because the toxin acts only on motor nerves, paresthesias are not seen in botulism, except when a patient hyperventilates from anxiety. The sensorium remains clear, but this may be difficult to ascertain because of slurred speech.

Foodborne botulism begins with gastrointestinal symptoms of nausea, vomiting and diarrhea in about one-third of cases. These symptoms are thought to result from metabolic byproducts of growth of *C. botulinum* or from the presence of other toxic contaminants in the food, as gastrointestinal distress is not seen in wound botulism. However, constipation is common in foodborne botulism once flaccid paralysis become evident. Illness usually begins 18–36 h after ingestion of the contaminated food, but can range from as little as 2 h to as long as 8 days. The incubation period in wound botulism is 4–14 days. Fever may be present in wound botulism but is absent in foodborne botulism unless a secondary infection (e.g., pneumonia) is present. All three forms of botulism display a wide spectrum in their clinical severity, from the very mild with minimal ptosis, flattened facial expression, minor dysphagia and dysphonia to the fulminant, with rapid onset of extensive paralysis, respiratory distress and frank apnea. Fatigability with repetitive muscle activity is the clinical hallmark of botulism.

Infant botulism differs in the apparent initial symptoms of illness only because the infant cannot verbalize them. Almost invariably, the first indication of illness is constipation (defined as three or more days without defecation), although this sign is frequently overlooked by both parents and physicians. Parents usually notice lethargy, listlessness, poor appetite, weak cry and diminished spontaneous movement. Dysphagia may be evident as secretions drooling from the mouth. Gag, suck and corneal reflexes diminish as the paralysis advances. Oculomotor palsies may be evident only with sustained observation. Paradoxically, the pupillary light reflex may be unaffected until the child is severely paralyzed, or it may be initially sluggish. Loss of head control is typically a prominent sign as the weakness becomes generalized. Respiratory arrest may occur suddenly from airway occlusion from unswallowed secretions or from obstructive flaccid pharyngeal musculature. Occasionally, the diagnosis of infant botulism is suggested by a respiratory arrest that occurs after the infant has been curled into position for lumbar puncture.

In mild cases or in the early stages of illness, the physical signs of infant botulism may be subtle and easily missed. Eliciting cranial nerve palsies and fatigability of muscular function requires careful examination. It is common for patients to be seen by several physicians over successive days until the paralysis has become sufficiently evident to suggest the diagnosis. Ptosis may not be seen unless the head of the child is kept erect. Decreased anal sphincter tone may suggest a generalized neuromuscular disease.

The classic picture of botulism is the acute onset of a flaccid descending paralysis with clear sensorium, no fever and no paresthesias. The rarity of foodborne and wound botulism makes them easily confused with other diseases. Routine laboratory studies, including the cereobrospinal fluid, are normal in botulism unless dehydration or starvation ketosis are present. EMG may demonstrate a defect in neuromuscular transmission, and the typical finding in foodborne and wound botulism is facilitation (potentiation) of the elicited muscle action potential at high frequency (50 Hz) stimulation. In infant botulism a characteristic pattern, known by the acronym BSAP (brief, small, abundant motor-unit action potentials), is present only in clinically weak muscles.

The diagnosis of botulism is unequivocally established by demonstrating the presence of botulinum toxin in serum, or of *C. botulinum* toxin or organisms in wound material or feces. *C. botulinum* is not part of the normal resident intestinal flora of humans, and its presence in the setting of acute flaccid paralysis is diagnostic. Suspected botulism represents a medical and public health emergency that is immediately reportable in virtually all US health jurisdictions. An epidemiological diagnosis of foodborne botulism can be made when *C. botulinum* organisms and toxin are found in a food item incriminated by ingestion history.

Management of botulism rests on three principles: (1) fatigability with repetitive muscle activity is the clinical hallmark of the disease; (2) complications are best avoided by anticipating them; and (3) meticulous supportive care is necessary. The first principle applies mainly to feeding and breathing. Correct positioning is imperative to protect the airway and improve respiratory mechanics.

Endotracheal intubation should be done prophylactically to maintain airway patency and avoid aspiration. A rising pCO_2 signals alveolar hypoventilation and irreversible muscle fatigue. About one-half of hospitalized patients will need intubation. Tracheostomy is almost never required because with proper positioning patients with global flaccid paralysis can tolerate intubation for weeks or months without permanent sequelae.

Feeding should be done by nasogastric or nasojejunal tube until sufficient oropharyngeal strength and coordination has returned to enable an entire feeding by breast or bottle. Expressed breast milk is the most desirable food for the affected infant, in part because of its immunological components (sIgA, lactoferrin, leukocytes, etc.). Tube feeding also assists in the restoration of peristalsis, a nonspecific but probably essential part of eliminating *C. botulinum* from the intestinal flora. Intravenous feeding (hyperalimentation) is discouraged because of the potential for infection and the advantages of tube feeding.

Antibiotic therapy is not part of the treatment of uncomplicated infant or foodborne botulism because the toxin is primarily an intracellular molecule that is released into the intestinal lumen with vegetative bacterial cell death and lysis. Antibiotics should be reserved for the treatment of secondary infections, and in this setting a nonclostridiocidal antibiotic such as trimethoprim/sulfamethoxazole or nalidixic acid is preferred. Aminoglycoside antibiotics should be avoided because they may potentiate the blocking action of botulinum toxin at the neuromuscular junction. However, wound botulism requires aggressive treatment with antibiotics and antitoxin in a manner analogous to tetanus. The currently available botulinum antitoxin is a horse-serum-derived product that has side-effects of serum sickness, anaphylaxis and potential life-long sensitization to equine proteins; its use in children require care. A randomized trial of botulism immune globulin, a human-derived botulinum antitoxin for the treatment of infant botulism, is now underway in California.[43]

Avoidance of complications is best accomplished by noting past experience. Even so, some critically ill, paralyzed patients who must spend weeks or months on ventilators in intensive care units will inevitably develop complications. Because patients with botulism display little daily variation in their clinical condition while their nerve endings regenerate, physicians and nurses may become complacent and overlook the beginning of a complication. Infant botulism does not have a relapsing course, and suspected 'relapses' usually reflect premature hospital discharge or an undiscovered underlying complication, such as pneumonia.[44]

In the absence of complications, particularly hypoxic ones, the prognosis in infant botulism is for full and complete recovery. Once the regenerating nerve endings have induced formation of a new motor end-plate, neuromuscular transmission is restored. In the United States, the case:fatality ratio for hospitalized infant botulism is less than 1%, which is a tribute to the availability and high quality of intensive care. In other countries the outcome has not been as favorable. The case:fatality ratio in foodborne and wound botulism varies with age, with younger patients having the best prognosis. Some adults with botulism have reported chronic weakness and fatigue as sequelae.

Hospital stay in infant botulism averages approximately 1 month, but differs significantly by toxin type, with type B cases being hospitalized a mean of 3.7 weeks and type A cases 5.6 weeks. Infant botulism is also a costly illness. For all California patients in the years 1984–94, hospital costs averaged \$2700 per day in constant 1995 dollars (physician fees not included). Twenty-five percent of

Figure 7.1 (above) Sir Charles Bell's portrait of a soldier with generalized tetanus. Reproduced with the permission of the President and Fellows of the Royal College of Surgeons of Edinburgh. See page 96.

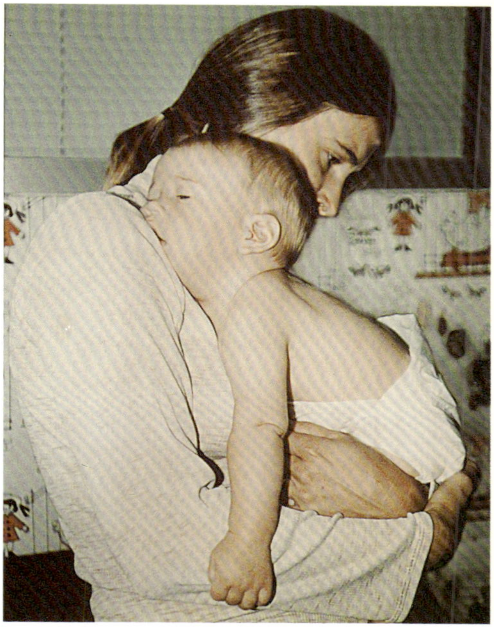

Figure 7.2 (right) Mother and 7-week-old patient with infant botulism. See page 96.

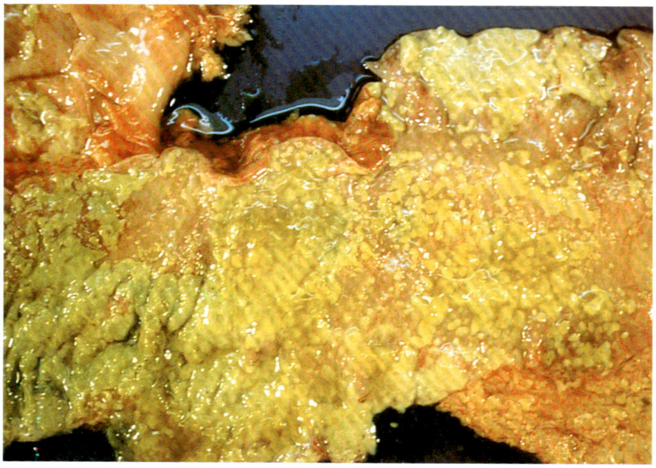

Figure 8.2 Plaques on the colonic mucosa of patient with pseudomembranous colitis. See page 122.

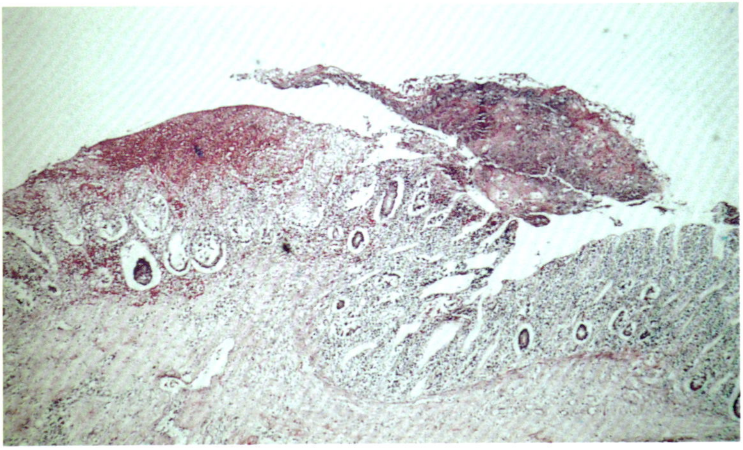

Figure 8.3 Microscopic appearance of pseudomembranous colitis. See page 123.

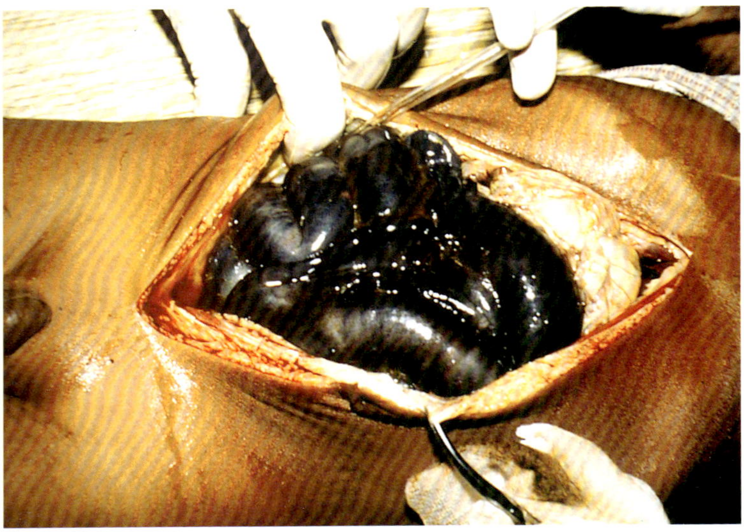

Figure 8.5 Extensive small bowel necrosis at autopsy of young Khmer boy with enteritis necroticans. See page 132.

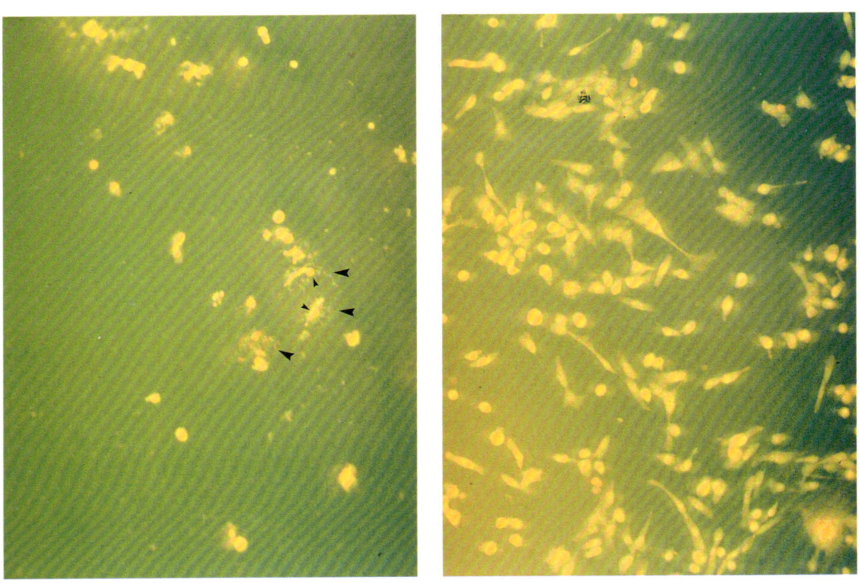

A B

Figure 25.1 MDCK cells (A) exposed to ε-toxin, stained and examined by fluorescence microscopy. Control cells are shown in (B). See page 442.

patients had hospital stays costing $100 000 or more; the most expensive case was hospitalized for 10 months at a cost in excess of $890 000.

Foodborne botulism is best prevented by adhering to safe methods of home-canning (altitude-adjusted pressure cooker and acidification), by avoiding suspicious foods and by heating all home-canned foods to 80°C for at least 5 min. Wound botulism is best prevented by thorough cleansing and surgical debridement of contaminated traumatic injuries with provision of appropriate antibiotics and by avoiding illicit drug use.

Circumstantial evidence suggests that most infant botulism patients first inhaled and then swallowed airborne clostridial spores; such cases are presently unpreventable. The one identified, avoidable source of botulinum spores for infants is honey, and all major pediatric and public health agencies in the United States have recommended that honey should not be fed to any child less than 1 year old. Corn syrups were once thought to be a possible source of botulinum spores, but recent evidence indicates otherwise. Breast-feeding appears to slow the onset of infant botulism and to diminish the risk of respiratory arrest in infants in whom the disease develops.[34] Immunization to prevent botulism can be accomplished through use of botulinum toxoid, a US Food and Drug Administration-designated Investigational New Drug (IND) used mainly to protect laboratory personnel against accidental exposure to botulinum toxin. Botulinum toxoid was also used to immunize Allied troops during the Gulf War of 1991 because intelligence reports had identified the production capability for botulinum toxin in Iraq.[45] The emergence of botulinum toxin as a therapeutic agent (see below) provides a compelling reason to restrict use of the toxoid to those who should have it for occupational safety, in that immunity to a particular toxin type deprives a person of ever benefiting from its therapeutic benefits.[25,47,48]

IV. THERAPEUTIC USES OF BOTULINUM TOXIN

However paradoxical it may seem, botulinum toxin has over the past decade risen to the status of 'wonder drug'.[47,48] It is used to treat a variety of diseases characterized by spasm or overactivity of a particular muscle or group of muscles. In many of these illnesses the muscular hyperactivity is the primary disorder (e.g., cervical dystonia), while in others, it is secondary to another disease (e.g., rigidity and tremor in Parkinson's disease). In most of these conditions intramuscular injection of botulinum toxin has become the treatment of choice and has replaced previous and much less satisfactory surgical or pharmacological alternatives. Clinically effective paralysis with type A botulinum toxin typically lasts three or four months, but it can be longer.[47,48]

The widespread applicability of botulinum toxin's ability to relax tense muscles originates in the work of two investigators. Alan Scott was an ophthalmologist who pioneered preclinical evaluation of botulinum toxin in monkeys and its subsequent clinical use in patients with strabismus (wandering eye) and blepharospasm;[49] Edward Schantz was a protein chemist who had crystallized botulinum toxin type A and provided it on request to Scott.[50] Botulinum toxin type A was licensed by the US Food and Drug Administration on December 1989 for the treatment of blepharospasm and strabismus, thus culminating more than a decade of collaborative effort by Scott and Schantz.

Once the effectiveness of botulinum toxin had become evident to clinicians, its application to other disorders or muscle contraction and movement quickly followed.[47,48] The list of disorders for which botulinum toxin is being evaluated continues to lengthen and now includes smooth muscle (gastrointestinal) disorders and also a variety of skeletal muscle conditions that include cerebral palsy and extremity spasms after injury to the central nervous system. Botulinum toxin has even been used cosmetically for the removal of frown lines and 'crow's feet' wrinkles on the face.[47]

At present, there are two commercial preparations of botulinum toxin type A in use, one made in the United States (BOTOX®, Allergan) and one made in England (Dysport®, Speywood Pharmaceuticals): The potency of the two toxin preparations are not identical; 1 ng of BOTOX® contain approximately 2.5 units of toxin, whereas 1 ng of Dysport® contains 40 units (one unit of toxin is defined as the mouse LD_{50} dose).[48]

The clinical use of botulinum toxin is not without its problems. Several side-effects have been noted. Unwanted paralysis of adjacent muscles (e.g., ptosis obscuring vision, dysphagia resulting in aspiration) is caused by the diffusion of toxin away from the injection site. This complication limits the amount of toxin that can be injected at a given treatment session. Other side-effects include a transient influenza-like syndrome suggestive of an allergic reaction and the development of neutralizing antibodies that render the patient resistant to further treatment with botulinum toxin type A.[46–48,51]

Clinical experience with botulinum toxin has accumulated rapidly, and international conferences on its use now occur with regularity.[51–53] Even so, the therapeutic application of botulinum toxin remains in its infancy. Much refinement and improvement in its usefulness can be expected in the years ahead. It is likely that additional botulinum toxin serotypes will come into widespread use (types B and F toxins have already been studied[25,47,50,51]), including type C toxin which, uniquely among the eight clostridial neurotoxins, has syntaxin as its substrate. Various combinations of toxins will likely be tried in order to prolong the duration of desired weakening or of frank paralysis (e.g., a mixture of toxins A, B, and C would cleave the three different protease substrates), and the problem of neutralizing antibody formation will also probably be solved.[43] Tetanus toxin's retrograde axonal transport may be harnessed as a means to deliver various medications to the spinal cord and perhaps beyond. Finally, molecular engineering of the toxins may improve their duration of action or permit delivery of the light chain to other cell types, where it could block unwanted secretory activity.[53]

The paradoxical emergence of the world's most poisonous substance as a wonder drug at the very time that at least one nation was preparing to use it as a weapon of mass destruction illustrates once again the constant potential for the perversion of scientific knowledge. This paradox re-emphasizes the need for humanitarian principles, as well as humanitarians, always to prevail in the use of such knowledge.

V. REFERENCES

1. Finegold S.M. (1977) *Anaerobic Bacteria in Human Disease.* Academic Press, New York.
2. Gill D.M. (1982) Bacterial toxins: table of lethal amounts. *Microbiol Rev* **46**: 86–94.
3. Franz D.R., Pitt L.M. Clayton M.A., Hanes M.A. and Rose K.J. (1993) Efficacy of prophylactic and therapeutic administration of antitoxin for inhalation botulism. In DasGupta B.R. (ed.), *Botulinum and Tetanus Neurotoxins: Neurotransmission and Biomedical Aspects*, pp. 473–476. Plenum, New York.
4. Montecucco C. and Schiavo G. (1993) Tetanus and botulism neurotoxins: a new group of zinc proteases. *Trends Biochem Sci* **18**: 324–327.
5. Niemann H., Blasi J. and Jahn R. (1994) Clostridial neurotoxins: new tools for dissecting exocytosis. *Trends Cell Biol* **4**: 179–185.
6. UNSCOM (1995) Iraq biological weapons and ballistic missile development programs continue. *Near East Rep* **39**: 145–148.
7. Schiavo G., Benfenati F., Poulain B., Rossetto O., Polverino de Laureto P., DasGupta B.R. and Montecucco C. (1992) Tetanus and botulinum-B neurotoxins block neurotransmitter release by proteolytic cleavage of synaptobrevin. *Nature* **359**: 832–834.
8. Eisel U., Jarausch W., Goretzki K., Henschen A., Engels J., Weller U., Hudel M., Haberman E. and Niemann H. (1986) Tetanus toxin: primary structure, expression in *E. coli*, and homology with botulinum toxins. *EMBO J* **5**: 2495–2502.
9. Sanford J.P. (1995) Tetanus – forgotten but not gone (editorial). *N Engl J Med* **332**: 812–813.
10. Wright D.K., Lalloo U.G., Nayiger S. and Govender P. (1989) Autonomic nervous system dysfunction in severe tetanus: current perspectives. *Crit Care Med* **17**: 371–375.
11. Expanded Program on Immunization (1993) The global elimination of tetanus: progress to date. *WHO Wk Epidemiol Rec* **68**: 277–284.
12. Gasse F. (1995) Neonatal tetanus: review of progress. *Intl J Gyn Obs* 1995; **50**: S67–S72.
13. Bennett J., Azhar N., Rahim F., Kamil S., Traverso H., Killgore G. *et al* (1995) Further observations on ghee as a risk factor for neonatal tetanus. *Int J Epidemiol* **24**: 643–647.
14. Fauveau V., Mamdani M., Steinglass R. and Koblinsky M. (1993) Maternal tetanus: magnitude, epidemiology and potential control measures. *Int J Gynecol Obstet* **40**: 3–12.
15. Gergen P.J., McQuillan G.M., Kiely M., Ezzati-Rice T.M., Sutter R.W. and Virella G. (1995) A population-based serologic survey of immunity to tetanus in the United States. *N Engl J Med* **332**: 761–766.
16. Sun K.O., Chan Y.W., Cheung R.T., So P.C., Yu Y.L. and Li P.C. (1994) Management of tetanus: review of 18 cases. *J Roy Soc Med* **87**: 135–137.
17. Luisto M. (1993) Unusual and iatrogenic sources of tetanus. *Ann Chir Gynaec Fenn* **82**: 25–29.

18. Luisto M. and Livanainen M. (1993) Tetanus of immunized children. *Devel Med Child Neurol* **35**: 346–358.
19. Yen L.M., Dao L.M., Day N.P., *et al.* (1994) Role of quinine in the high mortality of intramuscular injection tetanus. *Lancet* **344**: 786–787.
20. Bleck T.P. (1991) Tetanus: pathophysiology, management, and prophylaxis. *Dis Mon* **37**: 545–603.
21. Sesardic D., Wong M.Y., Gaines Das R.E. and Corbel M.J. (1993) The first international standard for antitetanus globulin, human: pharmaceutical evaluation and international collaborative study. *Biologicals* **21**: 67–75.
22. Lee D.C. and Lederman H.M. (1992) Anti-tetanus antibodies in intravenous gamma globulin: an alternative to tetanus immune globulin. *J Inf Dis* **166**: 642–645.
23. Abrutyn E and Berlin J.A. (1991) Intrathecal therapy in tetanus: a meta-analysis. *JAMA* **226**: 2262–2267.
24. Andrade L.A. and Brucki S.M. (1994) Botulinum toxin A for trismus in cephalic tetanus. *Arq Neuropsiquiatr* **52**: 410–413.
25. Greene P.E. and Fahn S. (1993) Use of botulinum toxin type F injections to treat torticollis in patients with immunity to botulinum toxin type A. *Mov Disord* **8**: 479–483.
26. Wassilak S.G.F., Orenstein W.A. and Sutter R.W. (1994) Tetanus toxoid. In Plotkin S.A. and Mortimer E.A. Jr (eds), *Vaccines*, 2nd edn, pp. 57–90. WB Saunders, Philadelphia, PA.
27. Dastur F.D., Awatramani V.P., Chitre S.K. and D'Sa J.A. (1993) A single dose vaccine to prevent neonatal tetanus. *J Assn Phys India* **41**: 97–99.
28. Suen J.G., Hatheway C.L., Steigerwalt A.G. and Brenner D.J. (1988) *Clostridium argentinense* sp.nov.: a genetically homogeneous group composed of all strains of *Clostridium botulinum* toxin type G and some non-toxigenic strains previously identified as *Clostridium subterminale* or *Clostridium hastiforme. Int J Syst Bacteriol* **38**: 375–381.
29. Schiavo G., Rossetto O., Tonello F. and Montecucco C. (1995) Intracellular targets and metalloprotease activity of tetanus and botulism neurotoxins. *Curr Top Microbiol Immunol* **195**: 257–274.
30. Arnon S.S. (1995) Botulism as an intestinal toxemia. In Blaser M.J., Smith P.D., Ravdin, J.I., Greenberg H.B. and Guerrant R.L. (eds) *Infections of the Gastrointestinal Tract*, pp. 257–271. Raven Press, New York.
31. Werner S.B. (1992) Food Poisoning. In Last J.M. and Wallace R.B. (eds) *Public Health and Preventive Medicine*, 13th edn, pp. 193–202. Appleton & Lange, San Mateo, CA.
32. Weber J.T., Goodpasture H.C., Alexander H., Werner S.B., Hatheway C.L. and Tauxe R.V. (1993) Wound botulism in a patient with a tooth abscess: case report and review. *Clin Inf Dis* **16**: 635–639.
33. Anonymous (1995) Wound botulism – California. *Morbid Mortal Wkly Rept* **48**: 889–892.
34. Arnon S.S., Damus K., Thompson B., Midura T.F. and Chin J. (1982) Protective role of human milk against sudden death from infant botulism. *J Pediatr* **100**: 568–573.
35. O'Mahony M., Mitchell E., Gilbert R.J., Hutchinson D.N., Begg N.T., Rodhouse J.C. and Morris J.E. (1990) An outbreak of foodborne botulism associated with contaminated hazelnut yoghurt. *Epidemiol Infect* **104**: 389–395.
36. MacDonald K.L., Spengler R.F., Hatheway C.L., Hargrett N.T. and Cohen M.L. (1985) Type A botulism from sauteed onions: clinical and epidemiological observations. *JAMA* **253**: 1275–1278.
37. St. Louis M.E., Peck S.H.S., Bowering D., Morgan G.B., Blatherwick J., Banerjee S., Kettyls G.D.M., Black W.A., Milling M.E., Hauschild A.H.W. and Tauxe R.V. (1988) Botulism from chopped garlic: delayed recognition of a major outbreak. *Ann Int Med* **108**: 363–368.
38. Slater P.E., Addiss D.G., Cohen A., Leventhal A., Chassis G., Zehavi H., Bashari A. and Costin C. (1989) Foodborne botulism: an international outbreak. *Int J Epidemiol* **18**: 693–696.
39. Weber J.T., Hibbs R.G. Jr, Darwish A., Mishu B., Corwin A.L., Rakha M., Hatheway C.L., El Sharkawy S., El-Rahim S.A., Al-Hamd M.F.S., Sarn J.E., Blake P.A. and Tauxe R.V. (1993) A massive outbreak of type E botulism associated with traditional salted fish in Cairo. *J Inf Dis* **167**: 451–454.

40. Chou J.H., Hwang P.H. and Malison M.D. (1988) An outbreak of type A foodborne botulism in Taiwan due to commercially preserved peanuts. *Int J Epidemiol* **17**: 899–902.
41. Nightingale K.W. and Ayim E.N. (1980) Outbreak of botulism in Kenya after ingestion of white ants. *Br Med J* **281**: 1682–1683.
42. Hauschild A.H.W. and Dodds K.L. (1993) *Clostridium Botulinum: Ecology and Control in Foods*. Marcel Dekker, New York.
43. Arnon S.S. (1993) Clinical trial of human Botulism Immune Globulin. In DasGupta B.R. (ed.), *Botulinum and Tetanus Neurotoxins: Neurotransmission and Biomedical Aspects*, pp. 477–482. Plenum, New York.
44. Glauser T.A., Maquire H.C. and Sladky J.T. (1990) Relapse of infant botulism. *Ann Neurol* **28**: 187–189.
45. Ember L. (1991) Chemical weapons: plan proposed to destroy Iraqi arms. *Chem Eng News* **69**: 6.
46. Greene P., Fahn S. and Diamond B. (1994) Development of resistance to botulinum toxin type A in patients with torticollis. *Mov Disord* **9**: 213–217.
47. Jankovic J. and Hallet M. (eds) (1994) *Therapy with Botulinum Toxin*. Marcel Dekker, New York.
48. Moore P. (ed.) (1995) *Handbook of Botulinum Toxin Treatment*. Blackwell Science, Oxford, 1995.
49. Scott A.B. (1994) Foreword. In Jankovic J. and Hallet M. (eds), *Therapy with Botulinum Toxin*, pp. vii–ix. Marcel Dekker, New York.
50. Schantz E.J. (1994) Historica perspective. In Jankovic J. and Hallet M. (eds), *Therapy with Botulin Toxin*, pp. xxiii–xxvi. Marcel Dekker, New York.
51. DasGupta B.R. (ed.) (1993) *Botulinum and Tetanus Neurotoxins: Neurotransmission and Biomedical Aspects*. Plenum, New York.
52. Brin M.F. (ed.) (1995) International conference on botulinum toxin: basic science and clinical therapeutics (abstracts). *Mov Disord* **10**: 361–404.
53. Foster K.A. et al. (1997) International Conference on Biomedical Aspects of Clostridial Neurotoxins. Oxford, July, 1996. Centre for Applied Microbiology and Research, Oxford.

Chapter 8

Enterotoxemic Infections

STUART JOHNSON AND DALE N. GERDING*
*Medicine Service, VA Lakeside Medical Center, 333 East Huron, Chicago, IL 60611 and *Department of Medicine, Northwestern University Medical School, Chicago, Illinois, USA*

I. INTRODUCTION

Clostridia cause distinct gastrointestinal disease syndromes, most of which can be attributed to the action of specific toxins on the intestinal mucosa. Clostridial toxins produced by *Clostridium difficile* and *Clostridium perfringens*, as distinct from the classical enterotoxin, cholera toxin, cause gross cellular morphological disruption but still result in net fluid secretion in the ileal loop assay for enterotoxin. This chapter deals with clostridial infections that cause intestinal disease in humans and in which diarrhea is a major or significant feature of the clinical syndrome. In addition to the following distinct disease syndromes, clostridia are often involved in polymicrobial infections of the gastrointestinal tract and present without specific features except for the occasional association of gas at the infected site. The production of hydrogen gas by clostridia and *C. perfringens*, in particular, probably accounts for radiographic and pathologic findings associated with emphysematous cholecystitis and pneumatosis cystoides intestinalis.[1,2] Although gas associated with the former syndrome portends a serious illness with an aggressive clinical course, pneumatosis cystoides intestinalis is usually a benign condition.

The Clostridia: Molecular Biology and Pathogenesis
ISBN 0-12-595020-9

II. *CLOSTRIDIUM DIFFICILE* DIARRHEA AND COLITIS

A. Background/epidemiology

Clostridium difficile is a unique enteric pathogen in that the affected hosts have nearly always been exposed to antimicrobials.[3] The role of antimicrobials in the pathogenesis of *C. difficile* disease appears to be in the alteration of the 'normal intestinal flora' that normally provides protection or resistance to infection with *C. difficile*.[4] Although prior antimicrobial therapy increases the risk of infection with other enteric pathogens such as *Salmonella*,[5] this association is nearly universal with *C. difficile* and has led to confusion about the pathogenesis of this disease. Specifically, *C. difficile* was thought to be part of the normal human intestinal flora, present in low numbers but proliferating under the influence of antimicrobial pressure on the other indigenous flora. In addition, most cases of *C. difficile* diarrhea and colitis are recognized in patients who are, or recently have been hospitalized.[6] Despite this association, recognition that *C. difficile* is a true nosocomial infection was slow. Pseudomembranous colitis, a distinct clinical manifestation of *C. difficile* infection, was originally considered a complication of antimicrobials (particularly clindamycin) by most hospital infection control practitioners and was not recognized as an infection transmitted between patients in the hospital.[7] Although *C. difficile* diarrhea has traditionally been discussed separately from other enteric pathogens and other nosocomial pathogens, it is now clear that *C. difficile*, like other enteric pathogens, is exogenously acquired and that hospitalized patients exposed to antimicrobials are uniquely susceptible.

Antimicrobial therapy was the first recognized risk factor for *C. difficile* disease and nearly all antibacterial agents have been implicated. Clindamycin, ampicillin, and cephalosporins have been most frequently associated, whereas parenteral aminoglycosides, vancomycin, and metronidazole have been infrequently implicated.[8,9] Although more cases of *C. difficile* diarrhea have probably resulted from ampicillin and cephalosporin use, the unique predisposition of patients treated with clindamycin has been repeatedly documented.[6,7,10] The mechanism of this unique propensity is unknown, but clindamycin has marked activity against anaerobic bacteria and colonic flora, and clindamycin resistance has been a marker for *C. difficile* strains implicated in two reported outbreaks.[10,11] In addition, *C. difficile* diarrhea may develop during administration of the antimicrobial agent or may not start until several days (or occasionally weeks) after the offending drug has been discontinued. One-third of the patients in the original report of clindamycin-associated pseudomembranous colitis developed symptoms after clindamycin was discontinued[7] which, in retrospect, most likely reflected the intermittent exposure of these hospitalized patients to *C. difficile*.

Hospitalization is the next major risk factor for *C. difficile* infection and reflects the opportunity for repeated exposure of susceptible patients to a major reservoir for this pathogen. Whereas fecal carriage of *C. difficile* in healthy adults is uncommon,[12] the rate of positive stool cultures for *C. difficile* in hospitalized adults is often 20% or greater for patients hospitalized more than 1 week.[13] One study that prospectively cultured all adult patients admitted to a hospital ward (weekly rectal swab cultures) over a 1-year period showed that the risk of *C. difficile* acquisition increased in direct proportion to length of hospital stay (Figure 8.1).[14] The rate of acquisition was 13% for those patients hospitalized for 1–2 weeks and increased to 50% for those hospitalized for more than 4 weeks. In contrast to adults, carriage of *C. difficile* in healthy neonates is very common, but the carriage rate decreases quickly during the first year of life.[12] The carriage rate in neonates at one hospital ranged from 2% to 52%, depending on the post-natal ward and the data suggested that *C. difficile* was nosocomially acquired in this setting rather than by acquisition from the mother's intestinal flora.[15] Interestingly, these children often have detectable toxin activity in their stools without symptoms and therapy is not indicated.[12,15] Among hospitalized adults, the incidence of *C. difficile* diarrhea varied from 0.3 to 78 cases per 1000

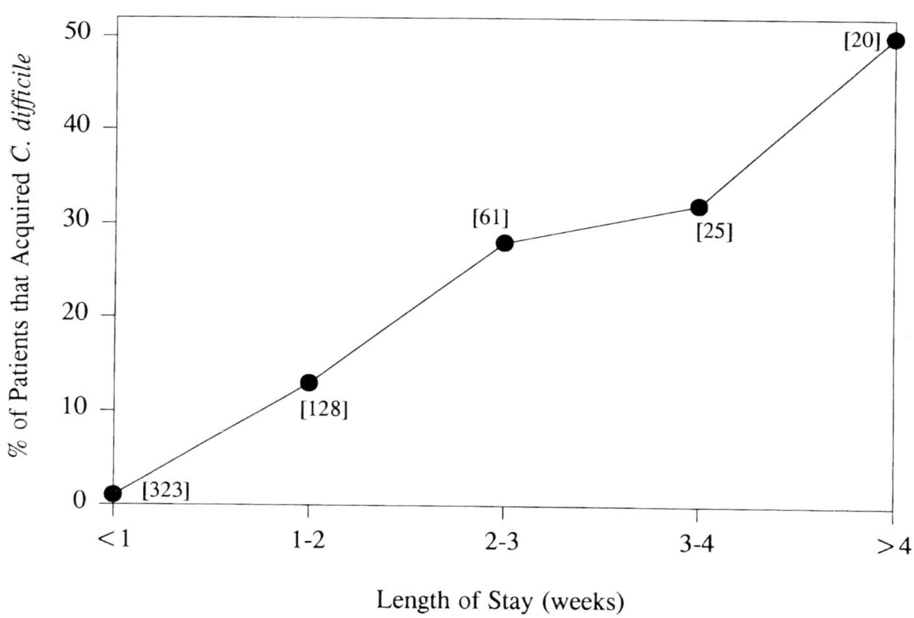

Figure 8.1 Rate of *Clostridium difficile* acquisition as a function of length of hospital stay in weeks. The number of patients at risk (initially culture-negative) is indicated in brackets. All patients admitted to one hospital ward over a 9-month period (of whom 557 were initially *C. difficile* culture-negative) were prospectively monitored by weekly rectal swab culture.[14]

admissions, depending on the hospital[16,17] and the incidence varied from year to year in the same institution.[18] Although other reservoirs are likely to exist for *C. difficile*, and community-acquired cases do occur, the incidence is low (less than one case per 10 000 antibiotic prescriptions in one study) and the majority of *C. difficile* diarrhea cases can be traced to hospitals or other chronic care institutions.[19]

Acquisition of *C. difficile* in the hospital setting can potentially occur directly from the environment or indirectly from health-care personnel. Although the environment is often heavily contaminated and fomite transmission has occasionally been reported, the main route of transmission is likely to be via the hands of personnel caring for multiple patients.[17,20] One prospective, controlled study showed a significant decline in the incidence of *C. difficile* diarrhea and a decline in the overall carriage rates after the adoption of intensive glove use for handling body substances.[20] Specific fomites that have been linked to spread of *C. difficile* include a contaminated commode chair,[21] a nursery bath,[15] and contaminated electronic rectal thermometers.[22]

Sensitive typing methods such as restriction endonuclease analysis (REA) have helped to elucidate several epidemiologic aspects of *C. difficile* infections, including the wide diversity of strains and the role of asymptomatic carriers. Over 200 unique types organized into 75 closely-related groups have been identified by *Hin*dIII restriction analysis of genomic DNA.[23] This wide strain diversity is also apparent within a particular hospital[24] and even during an outbreak caused by one unique strain, different strains are usually present in the background.[13] When surveillance is performed, asymptomatic infections often outnumber symptomatic infections.[13,17] Some of these strains are nontoxigenic and therefore unlikely to cause illness, but highly toxigenic strains including epidemic strains responsible for disease in other patients can also be carried asymptomatically.[13] The potential importance of asymptomatic carriers has been demonstrated in the prospective study of one hospital ward in which 19 distinct *Hin*dIII REA types were both introduced and acquired on that ward by different patients.[14] Nosocomial acquisition of a strain was preceded by a documented introduction of that strain to the ward by another asymptomatic carrier in 16 (84%) out of 19 instances, implicating asymptomatic carriers as the source of infection for other patients.

Although asymptomatic carriers may serve as a reservoir for *C. difficile* infections, there is little basis for local policies restricting transfer of patients with positive cultures to nursing homes from hospitals. Attempts at eradication of the organism in colonized patients are not practical, as treatment with metronidazole is not effective and only temporarily effective with vancomycin.[25] Carriage rates of *C. difficile* in nursing homes are generally somewhat lower than rates in acute care hospitals but higher than rates in outpatient populations.[26] The epidemiology of isolated cases and outbreaks of *C. difficile*-associated diarrhea in institutions is complex and involves patterns of antimicrobial use[10] and host factors, in addition to exposure to *C. difficile*, some strains of which are non-toxigenic.[13]

Accumulated data indicate that some *C. difficile* strains have increased virulence or increased capacity to cause epidemics and the ratio of symptomatic to asymptomatic infections, as detected by surveillance studies, may predict the propensity of a strain for epidemics.[27] Recently, large outbreaks of *C. difficile* diarrhea in three widely separated geographic locations of the US have been shown to be caused by the same strain.[28] This strain is highly resistant to clindamycin and exposure to this antimicrobial was a risk factor in all of the three outbreaks in which this data was collected.[28]

B. Clinical syndromes

The most prominent manifestation of infection with *C. difficile* is diarrhea. Although the pathologic lesions, when seen, are those of colitis, bloody stools are rarely present and some infected patients present with profuse watery diarrhea and signs of mild to moderate dehydration suggestive of a noninflammatory diarrheal syndrome. The reason for this overlap in features of noninflammatory and inflammatory (colitis) syndromes may be related to the nature of the *C. difficile* enterotoxin, toxin A. Unlike the classical enterotoxin, cholera toxin, toxin A causes gross cellular morphologic changes in addition to fluid secretion *in vivo* (see Chapter 21).[29–31] Although diarrhea may resolve spontaneously in 25% of patients after simple withdrawal of the offending antibiotic,[32] the symptoms more commonly become chronic and severe if the etiology is not recognized and specific therapy is not initiated.[33] In addition to diarrhea, abdominal pain, ileus, and fever are associated findings in 22%, 21%, and 28% of cases, respectively.[6] Leukocytosis is seen in 50% of cases[6] and occasionally leukemoid reactions have been noted.[33] An additional feature characteristic of *C. difficile* diarrhea is that, following successful therapy,[33] there is a high rate of clinical recurrence which may result from either relapse with the same strain or reinfection with a new strain.[34,35]

In addition to uncomplicated diarrhea, a wide variety of syndromes may follow acquisition of *C. difficile*, ranging from asymptomatic colonization to nonspecific colitis to pseudomembranous colitis to toxic megacolon. Asymptomatic colonization is the most frequent outcome and this lack of symptoms may depend on strain and host factors, but probably not antimicrobial exposure.[13] Subclinical infection with *C. difficile* is not detrimental to the host and contrary to earlier reports, asymptomatic colonization is not associated with protein-losing enteropathy.[36] These patients are not at increased risk of subsequent illness and attempts at eradication are not recommended as metronidazole is ineffective and vancomycin is only temporarily effective.[13,25]

Pseudomembranous colitis was the first recognized manifestation of *C. difficile* infection and was originally thought to be a side-effect of clindamycin therapy.[7] The pseudomembranous intestinal lesions associated with *C. difficile* have a characteristic gross and histologic appearance (Figures 8.2 and 8.3). Early in the disease course small (1–2 mm) raised, yellowish-white plaques are noted which may enlarge and coalesce.[37] These lesions, which are composed of fibrin, mucous, necrotic epithelial cells and leukocytes, are restricted to the colon and stop at the ileocecal junction. Although pseudomembranous colitis can be visualized by the sigmoidoscope in 90% of the cases, some patients have disease limited to the right colon and the presentation may mimic appendicitis or Crohn's disease.[38,39] Complications of severe colitis include toxic megacolon and colonic perforation. Toxic megacolon is an important syndrome to recognize as it is associated with a high mortality and may occur without diarrhea.[40] Toxic megacolon is a life-threatening condition that involves acute dilation of the colon (usually defined radiographically as >6 cm diameter) with systemic toxicity.[41] This complication has been noted in the setting of inflammatory bowel disease as well as in patients with infectious enteritis caused by other

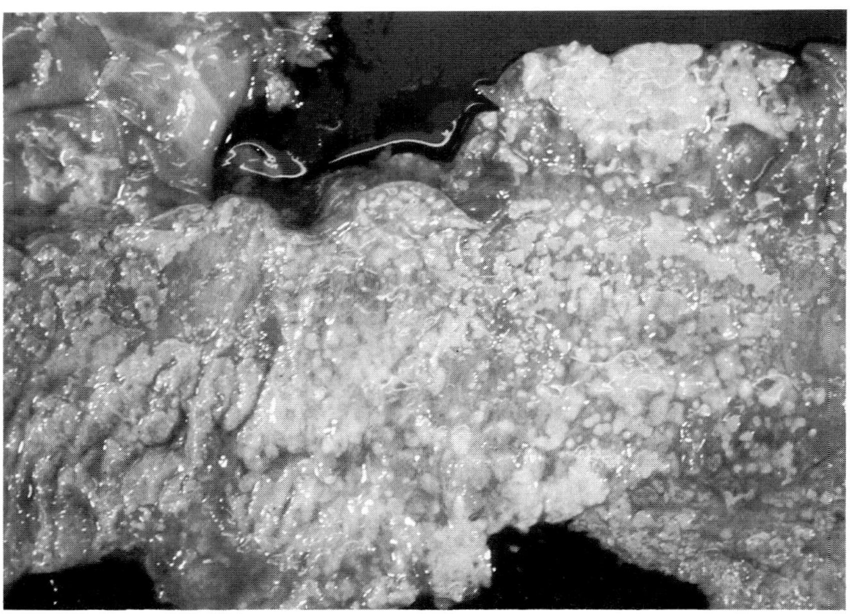

Figure 8.2 Raised, adherent yellow plaques which merge into confluent plaques on the colonic mucosa at autopsy of a patient with severe pseudomembranous colitis. (See colour plate.)

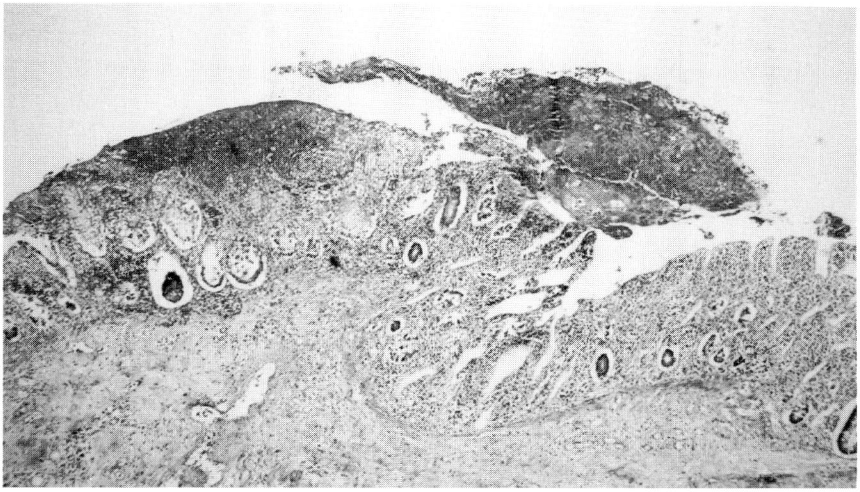

Figure 8.3 Microscopic appearance of pseudomembranous colitis with a typical mushroom-like lesion erupting from the mucosa. (See colour plate.)

pathogens, such as *Shigella* sp. In addition, this syndrome is often mistaken for other forms of acute abdomen and laparotomy performed without this consideration may be deterimental. Cases of toxic megacolon that are refractory to medical management may require colectomy.[39,41]

Although *C. difficile* is not an invasive pathogen and bacteremia is rarely documented, a variety of extraintestinal infections have been reported.[42] These rare manifestations include splenic abscess, bacteremia, wound infections, osteomyelitis, pleuritis, peritonitis, and urogenital tract infections.[42,43] Similar to other enteric infections a reactive arthritis has been noted in association with *C. difficile* infection.[44]

C. Diagnosis

Clostridium difficile diarrhea and colitis are toxin-mediated and detection of either toxin A or B in the stool of symptomatic patients is the most specific non-invasive test strategy for diagnosis of *C. difficile*-associated disease.[45] The traditional laboratory test for toxin involves inoculation of a tissue-cell culture with filtered supernatants of the stool specimen and neutralization of the cytopathic effect with specific antisera.[45] Many cell lines have been used including human foreskin fibroblasts, HEp-2 cells, and MRC-5 cells.[45] These assays detect primarily toxin B because of its relatively greater cytotoxicity on most cell lines compared with toxin A.[46] Although toxin A may be the more important toxin involved in the initial pathogenesis of the disease, virtually all pathogenic isolates produce both toxins, making this diagnostic distinction a moot point clinically.[47] Recently, several enzyme-linked immunoassays have been developed which detect either toxin A or toxin A and B.[48] These tests are replacing the traditional cytotoxin test in clinical laboratories because of their rapid turnaround time and the reluctance of many laboratories to maintain tissue culture cell lines. In general, these tests have similar specificities to cell culture assays but are less sensitive. Sensitivity rates reported from studies using both clinical and test-based criteria for *C. difficile* diarrhea range from 63% to 85% from ELISA and from 67% to 100% for cell culture assays.[49]

An alternative strategy to toxin testing for *C. difficile* disease is detection of the organism. In an experienced laboratory, culture for *C. difficile*, using selective media (cycloserine-cefoxitin-fructose-agar, CCFA) is the most sensitive test available and, in addition, it allows for epidemiologic typing analyses.[49] However, interpretation of culture results must first take into account the clinical setting, particularly in light of the high frequency of asymptomatic carriers in hospital settings where *C. difficile* diarrhea is noted.[13,17] Patients with diarrhea who have a stool culture positive for *C. difficile* but a negative toxin test should have toxicity of the isolate tested *in vitro*. Empirical therapy should be used if other clinical criteria support the diagnosis (e.g. persistent diarrhea following recent antimicrobial exposure). Failure to treat patients with negative stool toxin test results in one study was associated with increased morbidity and mortality.[50] A latex agglutination test is also available that was originally thought to detect Toxin A, but was subsequently shown to detect a different *C. difficile* protein, glutamate dehydrogenase.[50] The test is a rapid, relatively inexpensive alternative to culture, but lacks both sensitivity and specificity and should be considered a suboptimal test, particularly when used alone.[49] The use of polymerase chain reaction (PCR) to detect *C. difficile* in stool is still an experimental procedure, but could potentially replace culture as a diagnostic test.[52]

Detection of fecal leukocytes by direct microscopy or indirectly by lactoferrin assay is moderately predictive for the presence of *C. difficile* toxin but its specificity and sensitivity are not sufficient to be helpful in the diagnosis of *C. difficile* diarrhea.[53–55] In addition to the laboratory tests, endoscopy can be utilized to make a rapid, indirect diagnosis of *C. difficile* disease by the demonstration of pseudomembranous colitis with endoscopic visualization or by biopsy.[37] Endoscopy may be particularly helpful in evaluating a patient with an acute abdomen (necessitating surgical exploration) for whom a laparotomy is considered, but this procedure is insensitive for the diagnosis of *C. difficile* disease, in general, and detects pseudomembranes in only about 50% of patients with diarrhea who have stools positive for both cytotoxin and *C. difficile* organisms.[6]

D. Treatment

Response rates to specific treatment with oral metronidazole or vancomycin are typically better than 95% and these drugs constitute the most widely used therapies for *C. difficile* diarrhea.[49] It should be remembered that this disease is invariably precipitated by antimicrobial therapy and eventual recovery for some patients likely involves the re-establishment of a protective colonic bacterial flora after discontinuation of all antimicrobials. Up to 23% of patients may respond to simple withdrawal of the offending antimicrobial and this strategy should be considered for patients with mild symptoms.[32] Specific therapy should be administered orally and continued for 10 days. Antiperistaltic agents, such as phenoxylate-atrophine (Lomotil) should be avoided as these drugs may predispose patients to toxic megacolon when used alone.[49] Vancomycin was the first effective therapy for *C. difficile* diarrhea and fecal drug concentrations several $\log_{10}$ higher than the minimum inhibitory concentration (MIC) for *C. difficile* can be achieved because the drug is not absorbed.[56] There has been some concern about the efficacy of metronidazole because it is well absorbed and fecal levels are undetectable in patients without diarrhea.[25] In practice, however, metronidazole provides a highly effective therapy with similar response and relapse rates to those achieved with vancomycin and bactericidal concentrations of metronidazole are present in the diarrheal stool specimens of patients with symptomatic *C. difficile* infection.[32,57] More recently, concern over the spread of vancomycin resistance in enterococci (and potentially other pathogens) has led to a recommendation by the United States Hospital Infection Control Policy Advisory Committee (HICPAC) to discourage the use of oral vancomycin in the treatment of *C. difficile* diarrhea except for severe, potentially life-threatening illness.[58] Other agents that have been studied in randomized trials of *C. difficile* diarrhea include teicoplanin, bacitracin, and colestipol.[49] Teicoplanin treatment has resulted in similar response rates to those achieved with vancomycin, but bacitracin gives lower cure rates and colestipol is clearly inferior to any of the above-mentioned agents for the initial treatment of *C. difficile* diarrhea.[49,59–62]

Recurrence of *C. difficile* diarrhea is a problematic feature of this illness, occurring in 5–30% of successfully treated patients, and may be due to a relapse of the original infecting strain or acquisition of a new strain.[34,35,49] Regardless of the initial therapy (metronidazole or vancomycin) or the mechanism of recurrence, patients invariably respond again to specific therapy with the same agent that was used for treatment of the first episode. A small subset of patients will develop multiple recurrences and a number of empirical therapies have been tried for this frustrating complication, including combination treatment with vancomycin and rifampin,[63] vancomycin followed by cholestyramine,[64,65] treatment with tapering doses of vancomycin,[66] and no treatment with careful observation.[49] Non-antimicrobial attempts to treat this condition include administration of fecal bacterial enemas,[67] administration of a nontoxigenic strain of *C. difficile*,[68] and intravenous administration of gamma globulin to children with low serum antibodies to *C. difficile* toxin A.[69] Recently, a blinded, controlled trial showed that the addition of 4 weeks of orally administered *Saccharomyces boulardii* (a non-pathogenic yeast) to standard therapy reduced the rate of recurrence compared with standard therapy alone for patients with recurrent disease.[70]

Treatment of patients with severe *C. difficile* infections complicated by toxic megacolon or ileus is compromised by the inability to ensure therapeutic drug concentrations at the site of infection by the oral administration route. One empirical approach that has been used is to instill vancomycin by retention enema in addition to nasograstric tube delivery and intravenous administration of metronidazole.[18] Surgical intervention in the management of toxic megacolon may be required and subtotal colectomy with sparing of the rectal stump appears to be the preferred procedure, although efficacy is difficult to establish in this uncommon complication with associated high mortality.[39,41]

III. *CLOSTRIDIUM PERFRINGENS* TYPE A DIARRHEA

A. Background/epidemiology

Clostridium perfringens type A produces food poisoning as its most common manifestation of clinical illness. The symptoms are mediated by an enterotoxin produced during sporulation of the organism in the intestine following ingestion of contaminated food. First reported in England in 1943, and in the United States in 1945, it has been included in foodborne illness reporting since 1959 in the USA where it accounts for about 10% of foodborne outbreaks currently.[71,72] The most common source for food poisoning is a meat product, usually beef, that has not been properly stored or has been held at a temperature that allows growth of the organism. Inadequate cooking temperatures account for about one-third of outbreaks. Other foods implicated include Mexican food, turkey, chicken, pork, seafood, dairy products, and fruits and vegetables; however, meat and meat products account for over 75% of outbreaks.

High contamination of the ingested food by enterotoxin-producing *C. perfringens* (10^8 organisms) is thought to be necessary for disease to occur. The spores presumably survive cooking, then germinate and grow optimally at temperatures from 43–47°C when the food is inadequately reheated or cooled to prevent bacterial growth. The sites where most food poisonings occur include restaurants (56%), schools (11%), camps (5%), nursing homes (3%), and prisons (3%). Private homes account for only 3% of documented outbreaks, possibly because the disease is relatively mild and may not be detected if small numbers of individuals are involved.[72]

In addition to classical food poisoning, a more chronic form of diarrhea has been described in institutionalized persons, many of whom are elderly.[73–77] These patients often have blood in their stools, they may have a prior association with antimicrobials, and disease may occur in clusters, with the suggestion that the organism may be spread from person-to-person rather than by ingestion of a common food. Enterotoxigenic *C. perfringens* has been isolated from the stools of these patients in high counts compared with control patients without diarrhea.[73] Sporadic diarrhea in nonhospitalized patients has also been associated with enterotoxigenic *C. perfringens*.[74,75] This syndrome also affected predominately elderly patients who suffered protracted (⩾3 days) diarrhea.

B. Clinical syndromes

Most patients with *C. perfringens* type A food poisoning have diarrhea (92%) and abdominal pain (81%). Nausea (25%), vomiting (9%), and fever (24%) are much less common.[72] Illness is usually mild, lasting less than 24 h, but mortality has been reported in debilitated patients over age 60 in an outbreak in a psychiatric hospital.[72] The incubation period is 7–15 h following ingestion of the contaminated food. The mean attack rate for persons ingesting the implicated food is 57%, with a wide range from <1% to 100%, reflecting perhaps the wide variation in level of food contamination and amount of food ingested in various outbreaks. Most outbreaks occur in the spring or fall.

In contrast to food poisoning, symptoms associated with the infectious diarrhea syndrome caused by enterotoxigenic *C. perfringens* type A tend to be protracted and more severe. This syndrome has been characterized in 50 hospitalized patients with diarrhea from whom *C. perfringens* was isolated.[73] These isolates represent 1.8% of 2759 stool specimens submitted from multiple hospitals in the United Kingdom. The average age of cases was 76 years, 86% were more than 60 years old, and 72% were women. Mode and median duration of diarrhea was 7 days. Abdominal pain was present in 49% of patients, 35% had blood in their stools, and 62% had received antimicrobials before onset of diarrhea. *Clostridium perfringens* enterotoxin was detected in the stools of all of the patients. A total of 24 different serotypes of *C. perfringens* were isolated from 13 different hospitals. With one exception, serotype 41, there were no temporal or spatial clusters of the cases to suggest a common foodborne source.

C. Diagnosis

Diagnosis of *C. perfringens* food poisoning is difficult and this may account for missed outbreaks. Onset of diarrhea and abdominal pain symptoms from 7–15 h following ingestion of a suspect food should precipitate an investigation. *Clostridium perfringens* food poisoning should be suspected if (1) diarrheal stools contain 10^6 colony-forming units (cfu) or more *C. perfringens* spores per gram and (2) *C. perfringens* is detected at 10^5 cfu or more per gram of suspect food.[78] Because the finding of *C. perfringens* in stools is common, quantitation is important, and in one study patients with diarrhea had at least 10^7 cfu/g of stool.[73] Potentially, more specific diagnoses could be made by assaying stools for enterotoxin using latex agglutination, enzyme-linked immunosorbent assay (ELISA), or Vero cell cytotoxicity with antibody neutralization.[73,78–80] Although each of these tests has drawbacks, including high false-positive rates (latex agglutination) and lower sensitivity (Vero cell assay), ELISA holds the most promise but is not yet approved for clinical use. The diagnosis has been made in hospitalized patients with protracted diarrhea by use of quantitative stool culture for *C. perfringens* ($>10^6$ cfu/g stool), presence of enterotoxin in the stool, and absence of other diarrheal pathogens in the stool.[73] In one study, the presence of 10^4 or more *C. perfringens* and enterotoxin in the stool were used as diagnostic criteria in patients who had diarrhea.[76] However, other studies have found asymptomatic carriers with fecal counts of $>10^4$ cfu/g and the diagnosis of *C. perfringens*-associated diarrhea on the basis of spore counts is questionable.[77]

D. Treatment

Nearly all patients with acute *C. perfringens* type A food poisoning recover spontaneously and do not require specific treatment. For elderly institutionalized patients who have protracted diarrhea caused by *C. perfringens* type A, oral metronidazole 400 mg three times a day for 10 days was effective in stopping diarrhea and reducing *C. perfringens* and enterotoxin in the stool.[76] As with treatment of *C. difficile* diarrhea, relapses of diarrhea following successful treatment have been observed following treatment of *C. perfringens* type A diarrhea.[76]

IV. *CLOSTRIDIUM PERFRINGENS* TYPE C ENTERITIS

A. Background/epidemiology

Infection with *C. perfringens* type C has been linked to a unique syndrome of segmental necrotizing enteritis of the proximal small intestine and has been locally referred to as 'Darmbrand' (or 'fire-bowels') in Germany where outbreaks occurred in the post-World War II period and as 'pigbel' in Papua New Guinea where it was originally associated with ritual pork feasts.[81,82] Several lines of evidence have implicated the β-toxin produced by *C. perfringens* type C as the primary virulence factor in this disease, including serological studies in surviving patients,[82,83] reproduction of the lesions in guinea pigs,[84] and the clinical efficacy of a toxoid vaccine in an area in which the organism is endemic.[85,86] The epidemiology of this disease has been best described in the Highlands of Papua New Guinea where the disease is endemic and affects primarily children under the age of 13 years (see Chapter 12).[88]

Enteritis necroticans was originally recognized in Papua New Guinea when Highlanders would present with painful, swollen bellies several hours to a few days after large celebrations in which pigs are butchered and prepared under nonhygienic conditions.[88] Pork consumption is significantly associated with this disease,[89] but many cases occur without pork consumption[88,90] and other reservoirs of this infection likely exist. Serological studies indicate high rates (*c.* 75%) of antibody to β-toxin among healthy individuals living in areas with high rates of the disease.[82] *Clostridium perfringens* type C has not been identified by culture of stool specimens from healthy individuals in these regions,[91] perhaps because of difficulties in distinguishing these strains from other *C. perfringens* types which are identified only by toxin testing.[92] In contrast, fluorescent antibody staining of stool specimens suggests that 75% of people in highly endemic regions carry *C. perfringens* type C in low numbers.[93] Similar results were obtained with a fluorescent antibody assay for *C. difficile*, but the test was felt to be nonspecific when applied to a population of normal adults.[94] Now that the gene for β-toxin has been identified and sequenced[95] these studies could be corroborated with DNA-based assays; however, there are no such studies at present.

Helminthic co-infection is frequently noted in cases of enteritis necroticans but helminthic disease is also common in the general population at risk for enteritis necroticans, particularly children. *Ascaris* infection has been postulated to contribute to the pathogenesis of pigbel through elaboration of trypsin inhibitors which, in combination with dietary sources of trypsin inhibitors, circumvent host protease activity toward β-toxin (see Chapter 12).[88,96] *Strongyloides* infection was noted in 69% of cases in a Khmer refugee camp in Thailand compared with 14% of other children in the camp and could potentially interfere with the mucosal defenses through penetration of the mucosa by the adult worms.[83] The proximal small intestine is the preferential site of infection for both strongyloidiasis and enteritis necroticans and could potentially explain this feature of the latter.

In the early 1960s enteritis necroticans was the most common cause of acute abdominal emergencies requiring surgery and accounted for 23% of all deaths on the pediatric wards at the base hospital in Goroka, Papua New Guinea.[88,97] The incidence of disease in nonimmunized children aged 1–15 years in the Chimbu province of the eastern Highlands as determined during a vaccine trial in the late 1970s was >60 out of 10 000 over 24 months with six deaths.[85] The incidence was 7.9 out of 10 000 in children who received two doses of adsorbed β-toxoid and there were no deaths. Following introduction of this vaccine to the immunization protocol of the Highland health services in 1980, admission rates to referral hospitals for pigbel decreased by 80%.[86]

Outbreaks of necrotizing enteritis have been reported from countries in South and Southeast Asia, Africa, and in northern Germany in the late 1940s.[81,98–105] Careful bacteriological studies have not been performed in most of these studies and the etiology is not always known. However, the pathology of the lesions often resembles that of pigbel and *C. perfringens* type C has been implicated in several outbreaks and isolated cases outside of Papua New Guinea.[81,83,90,104,106] The outbreak in northern Germany occurred in malnourished people shortly after World War II and although clinical features of the disease differed from pigbel, *C. welchii* type F was recovered from many of the cases.[81] *Clostridium welchii* type F has been reclassified and is identical to *C. perfringens* type C, except that the former strains were reported to be heat-resistant.[107] *Clostridium perfringens* type C has been isolated from cases of necrotizing enteritis in Uganda[104] and patients with diarrhea in Bangladesh.[108]

Clostridium perfringens type C was also identified as the etiology of enteritis necroticans in a Khmer refugee camp in Thailand.[83] The clinical features of this disease also differed somewhat from pigbel (bloody diarrhea and fever were the predominant features) but histological (five cases), culture (two cases), and serological (five out of nine survivors versus zero out of ten controls) evidence supported the same etiological agent. Pork consumption was only noted in one in three of the cases and the reservoir of infection was never identified although some environmental factor was undoubtedly present at this camp as no outbreaks were identified at any of the other eight Khmer refugee camps along the Thai border. Necrotizing enteritis with similar features to the *C. perfringens* type C-associated enteritis in the Khmer camp have been reported from three other regions in Thailand and have been referred to as acute segmental ischemic enteritis based on histological features similar to ischemic bowel disease; however, no detailed microbiologic investigations have been performed.[98–100] In summary, outbreaks of necrotizing enteritis with predominant localization in the proximal small intestine and associated high mortality occur throughout the developing world, predominantly affecting children. Microbiological studies to investigate the role of *C. perfringens* type C and β-toxin in these outbreaks may now be easier to conduct using DNA-based assays.[90] Isolated cases have also been reported in developed, temperate climates,[90,106,109,110] but other cases may go unrecognized because of unfamiliarity with disease.

B. Clinical syndromes

The typical presentation of pigbel in Papua New Guinea is that of a patient who presents with severe abdominal pain 12 h to several days after consumption of pork or other protein meal.[111] Vomiting and diarrhea, often bloody, are associated with the pain in many cases, although fever is not typical in the initial phase.[111,112] The abdomen becomes distended and thickened loops of bowel are frequently palpable on examination. The abdominal X-ray pattern is that of small intestinal obstruction or ileus (Figure 8.4). At laparotomy or autopsy, patchy

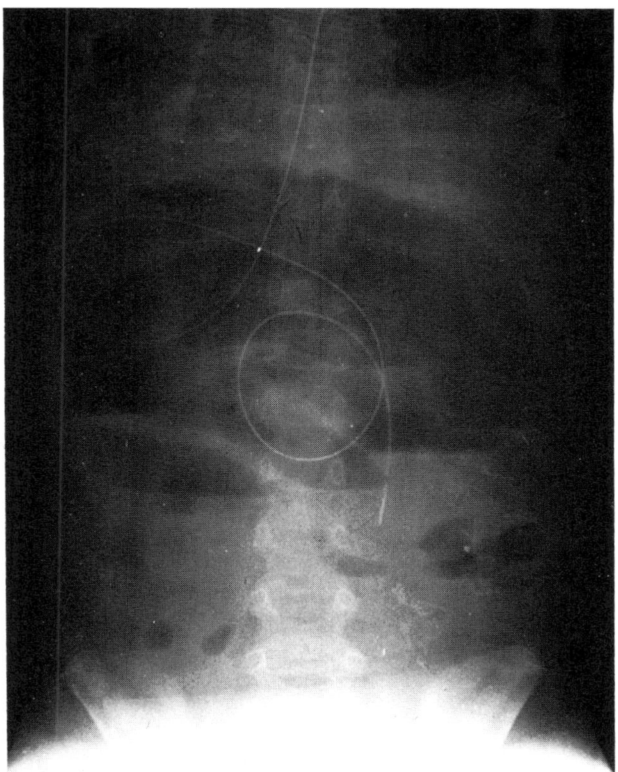

Figure 8.4 Abdominal roentgenogram of a young Khmer girl with enteritis necroticans.[83] Dilated loops of small intestine with air-fluid levels and minimal colonic shadowing indicate small bowel obstruction.

necrosis of the intestine can be seen on the serosal surface interspaced with normal appearing bowel. The affected segments can vary in length and can occur anywhere in the small intestine but the proximal jejunum is most frequently involved. In some cases the entire small bowel is involved (Figure 8.5). Histologically, full-thickness necrosis involving all layers from the mucosa to the serosa is evident with sharp demarcation from noninvolved segments with acute inflammatory cell infiltrates and endothelial proliferation of the venules in the adjacent submucosa.[112]

In addition to the typical acute surgical cases, other syndromes have been described including an acute toxic form in which individuals present with signs of toxemia and shock. These cases are rapidly fatal and usually never reach the hospital in Papua New Guinea. Differences in the clinical features of pigbel in Papua New Guinea and enteritis necroticans in the Khmer refugee camp may be partly explained by the confined population in the latter setting where cases may have been seen earlier in the course of their illness.[83] Bloody diarrhea and fever were the most prominent symptoms in the Khmer refugee camp outbreak and palpable loops of bowel were not found on examination, although in some cases signs of intestinal obstruction were noted radiographically (Figure 8.4).[83] In

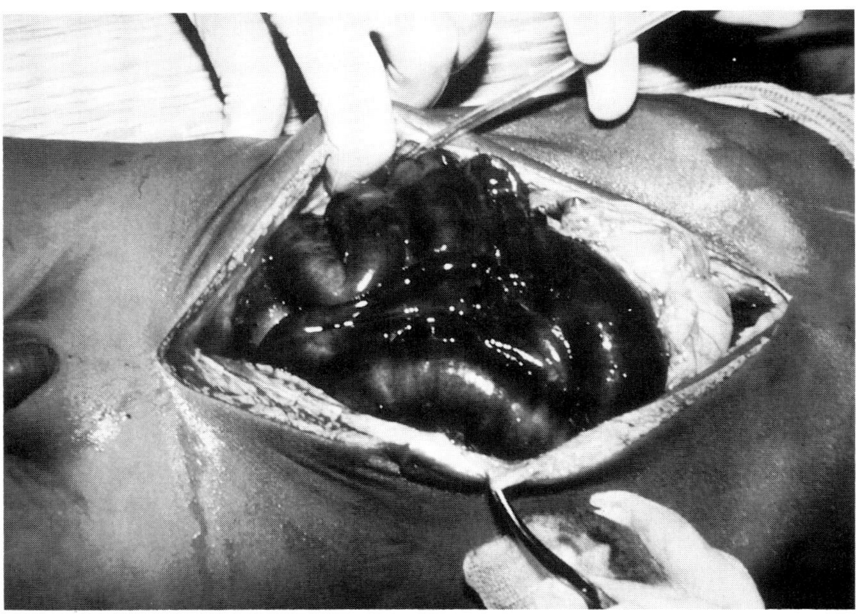

Figure 8.5 Extensive small bowel necrosis at autopsy in a young Khmer boy with enteritis necroticans.[83] Most cases have less extensive disease and segmental involvement of the small intestine with distended, thickened, red segments interspaced with normal appearing bowel are typically seen at laparotomy.[112] (See colour plate.)

addition, a subacute or chronic form of the disease has been described in Papua New Guinea manifested by intestinal strictures or malabsorption and probably represents the sequelae of a previous acute case of pigbel.[82,111]

C. Diagnosis and treatment

The diagnosis of enteritis necroticans is based entirely on the recognition of the syndrome. Cultural diagnosis usually requires sampling of the intestinal contents at the site of infection and *in vitro* toxin testing.[83] No laboratory assay has been developed to aid in the timely diagnosis of this disease. Davis summarized the management of pigbel which involves nasogastric suction, intravenous fluids, antimicrobial therapy, and in many cases, surgical resection of the involved intestinal segment.[111] Penicillin, chloramphenicol, or metronidazole have all been employed on an empirical basis. The critical aspect of management is the timing of the surgical procedure. Most patients are managed conservatively for a period of 24–48 h and if no improvement occurs, a laparotomy is performed. Surgery may be performed earlier in severe cases, but in the absence of toxicity, signs of perforation or other complications, surgery is usually delayed. The extent of the process is felt to be defined by the time the patient presents to the hospital and the operation does not prevent progression of the lesions. Delay beyond the initial assessment period may, however, be deleterious and once bowel obstruction has occurred, early surgery reduces mortality.[113] Adjunctive therapy with antisera (passive anti-β-toxin administration) is not effective.[114]

V. *CLOSTRIDIUM SEPTICUM* ENTEROCOLITIS

In addition to the well-described syndrome of non-traumatic myonecrosis and gas gangrene (see Chapter 9), *C. septicum* has been associated with typhlitis or neutropenic enterocolitis.[115] Distinct from *C. perfringens*-associated enteritis necroticans, the lesions of neutropenic enterocolitis predominate in the distal ileum and cecum. This distal enterocolitis is typically seen in patients with hematologic malignancies and is usually, but not always, associated with low levels of circulating neutrophils. Affected patients present with fever and abdominal pain which may mimic appendicitis. Vomiting and diarrhea are also frequently associated symptoms. High rates of bacteremia are noted in most clinical series,[116] and a variety of Gram-negative and Gram-positive organisms have been recovered in addition to *C. septicum*. The pathogenesis involves mucosal damage the cause of which has been postulated to result from the effects of cytotoxic drugs (especially cytosine arabinoside) on rapidly dividing enterocytes,[116] or possibly, malignant infiltrates and other mechanisms.[115] Bacterial invasion of the bowel wall and dissemination to the blood stream may be a complication of mucosal disruption and lack of phagocytic defense mechanisms due to neutropenia. The variety of organisms associated with bacteremia in neutropenic enterocolitis (often polymicrobial bacteremia) support a secondary or opportunistic role for these pathogens.

Clostridium septicum has also been associated with enterocolitis in patients with cyclic neutropenia who have not received chemotherapy.[117] In contrast to the variety of bacterial species associated with enterocolitis in leukemia and cytotoxic chemotherapy, clostridial species were associated with five out of seven reports of enterocolitis in patients with cyclic neutropenia.[117] *Clostridium septicum* was reported in two out of the three cases in which the clostridia were speciated,[117] suggesting a primary role for *C. septicum* in this disease and neutropenia as a specific predisposition for *C. septicum* infection.[118] Histological findings in *C. septicum*-associated enterocolitis include submucosal edema with necrosis and hemorrhage, but without gas formation, similar to the pathology of *C. septicum*-associated 'braxy' in sheep and of malignant edema in broiler hens.[115] Early surgical intervention was originally advocated for the treatment of neutropenic enterocolitis because of the fulminant presentation and high mortality seen.[117,118] However, many patients can be treated with antibiotics and supportive care alone, particularly if the syndrome is recognized early and antimicrobial therapy is started empirically before the results of blood culture are known, a scenario more likely when the disease occurs during hospitalization.[119]

VI. SUMMARY

In conclusion, clostridia cause distinct gastrointestinal syndromes in humans as a result of the action of specific toxins on the intestinal mucosa. Host factors also predispose to different clinical syndromes and include: antimicrobial exposure (*C. difficile* diarrhea), trypsin deficiency (*C. perfringens* type C enteritis), and neutropenia (*C. septicum* enterocolitis). Clostridia are widely dispersed in the environment and the risk of developing a particular syndrome may also relate to the relative predominance of a particular species in that environment, such as in hospitals (*C. difficile* diarrhea) or the Highlands of Papua New Guinea (*C. perfringens* type C enteritis). Ingestion of contaminated foods in the case of *C. perfringens* type A diarrhea and *C. perfringens* type C enteritis may also determine the type of syndrome displayed. Other factors such as the site of involvement within the intestine have not been adequately explained, but may relate to other virulence factors of the clostridia or the presence of other copathogens or cofactors. A better understanding of the molecular biology and pathogenesis of the clostridia is needed to prevent these ubiquitous human infections.

VII. REFERENCES

1. Mentzer R.M. Jr, Golden G.T., Chandler J.G. and Horsley J.S. 3d (1975) A comparative appraisal of emphysematous cholecystitis. *Am J Surg* **129**: 10–15.
2. Yale C.E. and Balish E. (1992) The natural course of *Clostridium perfringens*-induced pneumatosis cystoides intestinalis. *J Med* **23**: 279–288.
3. Lyerly D.M., Krivan H.C. and Wilkins T.D. (1988) *Clostridium difficile* its disease and toxins. *Clin Microbiol Rev* **1**: 1–18.
4. Wilson K.H. (1993) The microecology of *Clostridium difficile. Clin Infect Dis* **16**(Suppl. 4): S214–S218.
5. Pavia A.T., Shipman L.D., Wells J.G., Puhr N.D., Smith J.D., McKinley T.W. and Tauxe R.V. (1990) Epidemiologic evidence that prior antimicrobial exposure decreases resistance to infection by antimicrobial-sensitive *Salmonella. J. Inf Dis* **161**: 255–260.
6. Gerding D.N., Olson M.M., Peterson L.R., Teasley D.G., Gebhard R.L., Schwartz M.L. and Lee J.T. Jr (1986) *Clostridium difficile*-associated diarrhea and colitis in adults: A prospective case-controlled epidemiologic study. *Arch Intern Med* **146**: 95–100.
7. Tedesco F.J., Barton R.W. and Alpers D.H. (1974) Clindamycin-associated colitis. *Ann Intern Med* **81**: 429–433.
8. Bartlett J.G. (1981) Antimicrobial agents implicated in *Clostridium difficile* toxin associated diarrhea or colitis. *Johns Hopkins Med J* **149**: 6–9.
9. Gerding D.N., Olson M.M., Johnson S., Peterson L.R. and Lee J.T. Jr (1990) *Clostridium difficile* diarrhea and colonization after treatment with abdominal infection regimens containing clindamycin or metronidazole. *Am J Surg* **159**: 212–217.
10. Pear S., Williamson T., Bettin K., Gerding D.N. and Galgiani J.N. (1994) Decrease in nosocomial *Clostridium difficile*-associated diarrhea by restricting clindamycin use. *Ann Int Med* **120**: 272–277.
11. Clabots C.R., Peterson L.R. and Gerding D.N. (1988) Characterization of a nosocomial *Clostridium difficile* outbreak by using plasmid profile typing and clindamycin testing. *J Inf Dis* **158**: 731–736.
12. Viscidi R., Wiley S. and Bartlett J.G. (1981) Isolation rates and toxigenic potential of *Clostridium difficile* isolates from various patient populations. *Gastroenterol* **81**: 5–9.
13. Johnson S., Clabots C.R., Linn F.V., Olson M.M., Peterson L.R. and Gerding D.N. (1990) Nosocomial *Clostridium difficile* colonization and disease. *Lancet* **336**: 97–100.
14. Clabots C.R., Johnson S., Olson M.M., Peterson L.R. and Gerding D.N. (1992) Acquisition of *Clostridium difficile* by hospitalized patients: Evidence for colonized new admissions as a source of infection. *J Inf Dis* **166**: 561–567.
15. Larson H.E., Barclay F.E., Honour P. and Hill I.D. (1982) Epidemiology of *Clostridium difficile* in infants. *J Inf Dis* **146**: 727–733.
16. Struelens M.J., Maas A., Nonhoff C., Deplano A., Rost F., Serruys E. and Delmee M. (1991) Control of nosocomial transmission of *Clostridium difficile* based on sporadic case surveillance. *Amer J Med* **91**(Suppl. 3B): 1385–1445.
17. McFarland L.V., Mulligan M.E., Kwok R.Y.Y. and Stamm W.E. (1989) Nosocomial acquisition of *Clostridium difficile* infection. *N Engl J Med* **320**: 204–210.
18. Olson M.M., Shanholtzer C.J., Lee J.T., Jr and Gerding D.N. (1994) Ten years of prospective *Clostridium difficile*-associated disease surveillance and treatment at the Minneapolis VA Medical Center, 1982–1991. *Infect Control Hosp Epidemiol* **15**: 371–381.

19. Hirschhorn L.R., Trnka Y., Onderdonk A., Lee M-L.T. and Platt R. (1994) Epidemiology of community-acquired *Clostridium difficile*-associated diarrhea. *J Inf Dis* **169**: 127–133.

20. Johnson S., Gerding D.N., Olson M.M., Weiler M.D., Hughes R.A., Clabots C.R. and Peterson L.R. (1990) Prospective, controlled study of vinyl glove use to interrupt *Clostridium difficile* nosocomial transmission. *Am J Med* **88**: 137–140.

21. Savage A.M. and Alford R.H. (1983) Nosocomial spread of *Clostridium difficile. Infect Control* **4**: 31–33.

22. Brooks S.E., Veal R.O., Kramer M., Dore L., Schupf N. and Adachi M. (1992) Reduction in the incidence of *Clostridium difficile*-associated diarrhea in an acute care hospital and a skilled nursing facility following replacement of electronic thermometers with single-use disposables. *Infect Cont Hosp Epidemiol* **13**: 98–103.

23. Clabots C.R., Johnson S., Bettin K.M., Mathie P.A., Mulligan M.E., Schaberg D.R., Peterson L.R. and Gerding D.N. (1993) Development of a rapid and efficient restriction endonuclease analysis typing system for *Clostridium difficile* and correlation with other typing systems. *J Clin Microbiol* **31**: 1870–1876.

24. Samore M.H., Bettin K.M., DeGirolami P.C., Clabots C.R., Gerding D.N. and Karchmer A.W. (1994) Wide diversity of *Clostridium difficile* types at a tertiary referral hospital. *J Inf Dis* **170**: 615–621.

25. Johnson S., Homann S.R., Bettin K.M., Quick J.N., Clabots C.R., Peterson L.R. and Gerding D.N. (1992) Treatment of asymptomatic *Clostridium difficile* carriers (fecal excretors) with vancomycin or metronidazole: A randomized, placebo-controlled trial. *Ann Intern Med* **117**: 297–302.

26. Simor A.E, Yake S.L., and Tsimidis K. (1993) Infection due to *Clostridium difficile* among elderly residents of a long-term-care facility. *Clin Infect Dis* **17**: 672–678.

27. Gerding D.N. and Johnson S. (1995) Use of the ratio of *Clostridium difficile*-associated diarrhea (CDAD) to asymptomatic colonization (Asx CD) to determine virulence of *C. difficile* (CD) strains in humans. ASM First International Conference on the Molecular Genetics and Pathogenesis of the Clostridia. Rio Rico, AZ, American Society for Microbiology. January 11–14, 1995. Abst. #B7.

28. Johnson S., Samore M.H., Killgore G.E., DeGirolami P.C., Tenover F.C. and Gerding D.N. (1995) First report of wide geographic distribution of an epidemic *Clostridium difficile* (CD) strain characterized by genetic typing. Programs and Abstracts of the Infectious Diseases Society of America 33rd Annual Meeting, San Francisco, CA, Infectious Diseases Society of America. 1995: 40. Abst. #93.

29. Lyerly D.M., Saum K.E., MacDonald D.K. and Wilkins T.D. (1985) Effects of *Clostridium difficile* toxins given intragastrically to animals. *Infect Immun* **47**: 349–352.

30. Lima A.A.M., Lyerly D.M., Wilkins T.D., Innes D.J. and Guerrant R.L. (1988) Effects of *Clostridium difficile* toxins A and B in rabbit small and large intestine in vivo and on cultured cells in vitro. *Infect Immun* **56**: 582–588.

31. Hecht G., Pothoulakis C., LaMont J.T. and Madara J.L. (1988) *Clostridium difficile* toxin A perturbs cytoskeletal structure and junction permeability in cultured human epithelial monolayers. *J Clin Invest* **82**: 1516–1524.

32. Teasley D.G., Gerding D.N., Olson M.M., Peterson L.R., Gebhard R.L., Schwartz M.L. and Lee J.T. Jr (1983) Prospective randomized trial of metronidazole versus vancomycin for *Clostridium difficile*-associated diarrhea and colitis. *Lancet* **ii**: 1043–1046.

33. Bartlett J.G. (1990) *Clostridium difficile*: Clinical considerations. *Rev Inf Dis* **12** (Suppl. 2): S243–251.

34. Johnson S., Adelmann A., Clabots C.R., Peterson L.R. and Gerding D.N. (1989) Recurrences of *Clostridium difficile* diarrhea not caused by the original infecting organism. *J Inf Dis* **159**: 340–343.

35. O'Neil G.L., Beaman M.H. and Riley T.V. (1991) Relapse versus reinfection with *Clostridium difficile. Epidemiol Infect* **107**: 627–635.

36. Dansinger M.L., Johnson S., Jansen P.C., Opstad N.L., Bettin K.M. and Gerding D.N. (1996) Protein-

losing enteropathy is associated with *Clostridium difficile* diarrhea, but not with asymptomatic colonization: Prospective, case-controlled study. *Clin Inf Dis* **22**: 932–937.

37. Gebhard R.L., Gerding D.N., Olson M.M., Peterson L.R., McClain C.J., Ansel H.J., Shaw M.J. and Schwartz M.L. (1985) Clinical and endoscopic findings in patients early in the course of *Clostridium difficile*-associated pseudomembranous colitis. *Am J Med* **78**: 45–48.
38. Tedesco F.J., Corless J.K. and Brownstein R.E. (1982) Rectal sparing in antibiotic-associated pseudomembranous colitis: A prospective study. *Gastroenterol* **83**: 1259–1260.
39. Morris J.B., Zollinger R.M. Jr and Stellato T.A. (1990) Role of surgery in antibiotic-associated pseudomembranous enterocolitis. *Am J Surg* **160**: 535–539.
40. Burke G.W., Wilson M.E. and Mehrez I.O. (1988) Absence of diarrhea in toxic megacolon complicating *Clostridium difficile* pseudomembranous colitis. *Am J Gastro* **83**: 304–307.
41. Cone J.B. and Wetzel W. (1982) Toxic megacolon secondary to pseudomembranous colitis. *Dis Colon Rectum* **25**: 478–482.
42. Levett P.N. (1986) *Clostridium difficile* in habitats other than the human gastrointestinal tract. *J Infect* **12**: 253–263.
43. Stieglbauer K.T., Gruber S.A. and Johnson S. (1995) Elevated serum antibody response to toxin A following splenic abscess due to *Clostridium difficile. Clin Infect Dis* **20**: 160–162.
44. Lofgen R.P., Tadlock L.M. and Soltis R.D. (1984) Acute oligoarthritis associated with *Clostridium difficile* pseudomembranous colitis. *Arch Intern Med* **144**: 617–619.
45. Peterson L.R. and Kelly P.J. (1993) The role of the clinical microbiology laboratory in the management of *Clostridium difficile*-associated diarrhea. *Infect Dis Clin North Am* **7**: 277–293.
46. Tucker K.D., Carrig P.E. and Wilkins T.D. (1990) Toxin A of *Clostridium difficile* is a potent cytotoxin. *J Clin Microbiol* **28**: 869–871.
47. Lyerly D.M., Sullivan N.M. and Wilkins T.D. (1983) Enzyme-linked immunosorbent assay for *Clostridium difficile* toxin A. *J Clin Microbiol* **17**: 72–78.
48. Merz C.S., Kramer C., Forman M., Gluck L., Mills K., Senft K., Steiman I., Wallace N. and Charache P. (1994) Comparison of four commercially available rapid enzyme immunoassays with cytotoxin assay for detection of *Clostridium difficile* toxin(s) from stool specimens. *J Clin Microbiol* **32**: 1142–1147.
49. Gerding D.N., Johnson S., Peterson L.R., Mulligan M.E. and Silva J., Jr (1995) Society for Healthcare Epidemiology of America position paper on *Clostridium difficile*-associated diarrhea and colitis. *Infect Control Hosp Epidemiol* **16**: 459–477.
50. Lashner B.A., Todorczuk J., Sahm D.F. and Hanauer S.B. (1986) *Clostridium difficile* culture-positive toxin-negative diarrhea. *Am J Gastroenterol* **81**: 940–943.
51. Lyerly D.M., Barroso L.A. and Wilkins T.D. (1991) Identification of the latex test-reactive protein of *Clostridium difficile* as glutamate dehydrogenase. *J Clin Microbiol* **29**: 2639–2642.
52. Gumerlock P.H., Tang Y.J., Weiss J.B. and Silva J. Jr. (1993) Specific detection of toxigenic strains of *Clostridium difficile* in stool specimens. *J Clin Microbiol* **31**: 507–511.
53. Manabe Y.C., Vinetz J.M., Moore R.D., Merz C., Charace P. and Bartlett J.G. (1995) *Clostridium difficile* colitis: An efficient clinical approach to diagnosis. *Ann Inter Med* **123**: 835–840.
54. Shanholtzer C.J., Peterson L.R., Olson M.M. and Gerding D.N. (1983) Prospective study of Gram-stained stool smears in diagnosis of *Clostridium difficile* colitis. *J Clin Microbiol* **17**: 906–908.
55. Marx C.E., Morris A., Wilson M.L. and Reller L.B. (1993) Fecal leukocytes in stool specimens submitted for *Clostridium difficile* toxin assay. *Diagn Microbiol Inf Dis* **16**: 313–315.
56. Silva J. Jr, Batts D.H., Fekety R., Plouffe J.F., Rifkin G.D. and Baird I. (1981) Treatment of *Clostridium difficile* colitis and diarrhea with vancomycin. *Am J Med* **71**: 815–822.
57. Bolton R.P. and Culshasw M.A. (1986) Fecal metronidazole concentrations during oral and intravenous therapy for antibiotic-associated colitis due to *Clostridium difficile. Gut* **27**: 1169–1172.
58. Hospital Infection Control Practices Advisory Committee (HICPAC). (1995) Recommendations for preventing the spread of vancomycin resistance. *Infect Control Hosp Epidemiol* **16**: 105–113.

59. deLalla F., Nicolin R., Rinaldi E., Scarpellini P., Rigoli R., Manfrin V. and Tramarin A. (1992) Prospective study of oral teicoplanin versus oral vancomycin for therapy of pseudomembranous colitis and *Clostridium difficile*-associated diarrhea. *Antimicrob Agents Chemother* **36**: 2192–2196.
60. Dudley M.N., McLaughlin J.C., Carrington G., Frick J., Nightingale C.H. and Quintiliani R. (1986) Oral bacitracin versus vancomycin therapy for *Clostridium difficile*-induced diarrhea: A randomized, double-blind trial. *Arch Intern Med* **146**: 1101–1104.
61. Young G.P., Ward P.B., Bayley N., Gordon D., Higgins G., Trapani J.A., McDonald M.I., Labrocy J. and Hecker R. (1985) Antibiotic-associated colitis due to *Clostridium difficile*: double-blind comparison of vancomycin with bacitracin. *Gastroenterol* **89**: 1038–1045.
62. Mogg G.A.G., Arabi Y., Youngs D., Johnson M., Bentley S., Burdon D.W. and Keighley M.R.B. (1980) Therapeutic trials of antibiotic associated colitis. *Scand J Infect Dis* (suppl. 22): 41–45.
63. Buggy B.P., Fekety R. and Silva J. Jr. (1987) Therapy of relapsing *Clostridium difficile*-associated diarrhea and colitis with the combination of vancomycin and rifampin. *J Clin Gastroenterol* **9**: 155–159.
64. Moncino M.D. and Falletta J.M. (1992) Multiple relapses of *Clostridium difficile*-associated diarrhea in a cancer patient: Successful control with long-term cholestyramine therapy. *Am J Pediatr Hematol Oncol* **14**: 361–364.
65. Pruksananonda P. and Powell K.R. (1989) Multiple relapses of *Clostridium difficile*-associated diarrhea responding to an extended course of cholestyramine. *Pediatr Infect Dis J* **8**: 175–178.
66. Tedesco F.J., Gordon D. and Fortson W.C. (1985) Approach to patients with multiple relapses of antibiotic-associated pseudomembranous colitis. *Am J Gastroenterol* **80**: 867–868.
67. Tvede M. and Rask-Madsen J. (1989) Bacteriotherapy for chronic relapsing *Clostridium difficile* diarrhea in six patients. *Lancet* **i**: 1156–1160.
68. Seal D., Borriello S.P., Barclay F., Welch A., Piper M. and Bonnycastle M. (1987) Treatment of relapsing *Clostridium difficile* diarrhea by administration of a non-toxigenic strain. *Eur J Clin Microbiol* **6**: 51–53.
69. Leung D.Y.M., Kelly C.P., Boguniewicz M., Pothoulakis C., LaMont J.T. and Flores A. (1991) Treatment with intravenously administered gamma globulin of chronic relapsing colitis induced by *Clostridium difficile* toxin. *J Pediatr* **118**: 633–637.
70. McFarland L.V., Surawicz C.M., Greenberg R.N., Fekety R., Elmer G.W., Moyer K.A., Melcher S.A., Bowen K.E., Cox J.L., Noorani Z., Harrington G., Rubin M. and Greenwald D. (1994) Randomized placebo-controlled trial of *Saccharomyces boulardii* in combination with standard antibiotics for *Clostridium difficile* disease. *JAMA* **271**: 1913–1918.
71. Genigeorgis D. (1975) Public Health Importance of *Clostridium perfringens*. *J Amer Vet Med Assoc* **167**: 821–827.
72. Shandera W.X., Tacket C.O. and Blake P.A. (1983) Food poisoning due to *Clostridium perfringens* in the United States. *J Inf Dis* **147**: 167–170.
73. Larson H.E. and Borriello S.P. (1988) Infectious diarrhea due to *Clostridium perfringens*. *J Inf Dis* **157**: 390–391.
74. Brett M.M., Rodhouse J.C., Donovan T.J., Tebbutt G.M. and Hutchinson D.N. (1992) Detection of *Clostridium perfringens* and its entertoxin cases of sporadic diarrhoea. *J Clin Pathol* **45**: 609–611.
75. Mpamugo O., Donovan T. and Brett M.M. (1995) Enterotoxigenic *Clostridium perfringens* as a cause of sporadic cases of diarrhoea. *J Med Microbiol* **43**: 442–445.
76. Samuel S.C., Hancock P. and Leigh D.A. (1991) An investigation into *Clostridium perfringens* enterotoxin-associated diarrhoea. *J Hosp Infect* **18**: 219–230.
77. McClane B.A. (1992) *Clostridium perfringens* enterotoxin: Structure, action and detection. *J Food Safety* **12**: 237–252.
78. Birkhead G., Vogt R.L., Heun E.M., Snyder J.T. and McClane B.A. (1988) Characterization of an outbreak of *Clostridium perfringens* food poisoning by quantitative fecal culture and fecal enterotoxin measurement. *J Clin Microbiol* **26**: 471–474.

79. McClane B.A. and Snyder J.T. (1987) Development and preliminary evaluation of a slide latex agglutination assay for determination of *Clostridium perfringens* type A enterotoxin. *J Immunol Methods* **100**: 131–136.
80. McClane B.A. and Strouse R.J. (1984) Rapid detection of *Clostridium perfringens* type A enterotoxin by enzyme-linked immunosorbent assay. *J Clin Microbiol* **19**: 112–115.
81. Fick K.A. and Wolken A.P. (1949) Necrotic jejunitis. *Lancet* **i**: 519–521.
82. Murrell T.G.C., Roth L., Egerton J., Samels J. and Walker P.D. (1966) Pig-bel: enteritis necroticans: A study in the diagnosis and management. *Lancet* **i**: 217–222.
83. Johnson S., Echeverria P., Taylor D.N., Paul S.R., Coninx R., Sakurai J., Eampokalap B., Jimakorn P., Cooke R.A., Lawrence G.W. and Walker P.D. (1987) Enteritis necroticans among Khmer children at an evacuation site in Thailand. *Lancet* **ii**: 496–500.
84. Lawrence G. and Cooke R. (1980) Experimental pigbel: the production and pathology of necrotizing enteritis due to *Clostridium welchii* type C in the guinea-pig. *Br J Exp Pathol* **61**: 261–271.
85. Lawrence G., Shann F., Freestone D.S. and Walker P.D. (1979) Prevention of necrotizing enteritis in Papua New Guinea by active immunization. *Lancet* **i**: 227–230.
86. Lawrence G.W., Lehmann D., Anian G., Coakley C.A., Saleu G., Barker M.J. and Davis M.W. (1990) Impact of active immunization against enteritis necroticans in Papua New Guinea. *Lancet* **336**: 1165–1167.
88. Murrell T.G.C. and Walker P.D. (1991) The pigbel story of Papua New Guinea. *Trans Royal Soc Trop Med Hyg* **85**: 119–122.
89. Millar J.S., Smellie S. and Coldman A.J. (1985) Meat consumption as a risk factor in enteritis necroticans. *Int J Epidemiol* **14**: 318–321.
90. Ashley D., Watson R., Andrew J.H., Banting S., Mackay J.R., Stillwell G. and Merrett M. (1991) Pig-bel but no pig: enteritis necroticans acquired in Australia. *Med J Aust* **155**: 47–50.
91. Egerton J.R. (1966) Bacteriology of enteritis necroticans in the New Guinea Highlanders. *Papua New Guinea Med J* **9**: 55–59.
92. Sterne M. and Warrack G.H. (1964) The types of *Clostridium perfringens. J. Pathol Bacteriol* **88**: 279–283.
93. Lawrence G., Walker P.D., Garap J. and Avusi M. (1979) The occurrence of *Clostridium welchii* type C in Papua New Guinea. *Papua New Guinea Med J* **22**: 69–73.
94. Wilson K.H., Silva J. and Fekety F.R. (1982) Fluorescent antibody test for detection of *Clostridium difficile* in stool specimens. *J Clin Microbiol* **16**: 464–468.
95. Hunter S.E., Brown J.E., Oyston P.C.F., Sakurai J. and Titball R.W. (1993) Molecular genetic analysis of beta-toxin of *Clostridium perfringens* reveals sequence homology with alpha-toxin, gamma-toxin, and leukocidin of *Staphylococcus aureus. Infect Immun* **61**: 3958–3965.
96. Lawrence G. and Walker P.D. (1976) Pathogenesis of enteritis necroticans in Papua New Guinea. *Lancet* **i**: 125–126.
97. Murrell T.G.C. (1967) Pigbel-epidemic and sporadic necrotizing enteritis in the highlands of New Guinea. *Aust Ann Med* **1**: 4–10.
98. Welch T.P. and Sumitswan S. (1975) Acute segmental ischaemic enteritis in Thailand. *Br J Surg* **62**: 716–719.
99. Vejparnich P. and Chongcharoenkamol B. (1979) Acute necrotizing enteritis in children. *J Pediatr Soc Thailand* **18**: 49–58.
100. Suksuthipnuth S. and Chupisalyoroj S. (1983) Acute segmental gangrenous enteritis. *J Pediatr Soc Thailand* **22**: 41–46.
101. Gan K.H., Biran D., Sukono, Satari S., Sujudi, Warsa R. and Njoo S.N. (1962) First outbreak of necrotizing enteritis caused by *Clostridium welchii* in Indonesia. *Madjalaii Kedokteran Indonesia* **12**: 9–15.
102. Pujari B.D. and Deodhare S.G. (1980) Necrotizing enteritis. *Br J Surg* **67**: 254–256.
103. Butler T., Dahms B., Lindpaintner K., Islam M., Azad M.A.K. and Anton P. (1987) Segmental

necrotizing enterocolitis: pathological and clinical features of 22 cases in Bangladesh. *Gut* **28**: 1433–1438.

104. Wright D.H. and Sanfield J.P. (1976) Enteritis necroticans in Uganda. *J Pediatr* **71**: 264–268.
105. van Enk A. and van der Esch E.P. (1973) Enteritis necroticans in Agogo, southern Ghana. *Trop Geogr Med* **25**: 313–324.
106. Severin W.P.J., de la Fuente A.A. and Stringer M.F. (1984) *Clostridium perfringens* type C causing necrotizing enteritis. *J Clin Pathol* **37**: 942–944.
107. Egerton J.R. and Walker P.D. (1964) The isolation of *Clostridium perfringens* type C from necrotic enteritis of man in Papua New Guinea. *J Pathol Bacteriol* **88**: 275–283.
108. van Loon F.L.P., van Schaik S.V., Banik A.K., Ahmed T., Zaman A. and Kay B.A. (1990) *Clostridium perfringens* type C in bloody and watery diarrhea in Bangladesh. *Trop Geogr Med* **42**: 123–127.
109. Clarke L.E., Diekmann-Guiroy B., McNamee W., Java D.J., Jr and Weiss S.M. (1994) Enteritis necroticans with midgut necrosis caused by *Clostridium perfringens. Arch Surg* **129**: 557–560.
110. Nuland S.B. (1995) The beast in the belly. *Discover* **16**: 58–73.
111. Davis M.W. (1985) A review of pigbel (necrotizing enteritis) in Papua New Guinea, 1961–1984. *Papua New Guinea Med J* **28**: 75–82.
112. Cooke R.A. (1979) Pigbel. In Rosenberg H.S. and Bolande R.P. (eds) *Perspectives in Pediatric Pathology*, vol 5, pp. 137–152. Masson Publications, New York.
113. Millar J.S. (1981) The surgical treatment of enteritis necroticans. *Br J Surg* **68**: 481–482.
114. Rooney J., Shepherd A. and Suebu A. (1979) *Clostridium welchii* type C antitoxin in the treatment of 'pigbel' (enteritis necroticans): a controlled trial in Papua New Guinea. *Papua New Guinea Med J* **22**: 62–68.
115. King A., Rampling A., Wight D.G.D. and Warren R.E. (1984) Neutropenic enterocolitis due to *Clostridium septicum* infection. *J Clin Pathol* **37**: 335–343.
116. Wade D.S., Nava H.R. and Douglass H.O. Jr. (1992) Neutropenic enterocolitis: Clinical diagnosis and treatment. *Cancer* **69**: 17–23.
117. Hopkins D.G. and Kushner J.P. (1983) Clostridial species in the pathogenesis of necrotizing enterocolitis in patients with neutropenia. *Am J Hematol* **14**: 289–295.
118. Anonymous (1987) *Clostridium septicum* and neutropenic enterocolitis. *Lancet* **ii**: 608.
119. Johnson S., Driks M.R., Tweten R., Ballard J., Stevens D.L., Anderson D.J. and Janoff E.N. (1994) Clinical courses of seven survivors of *Clostridium septicum* infection and their immunologic responses to a toxin. *Clin Infect Dis* **19**: 761–764.

Chapter 9

Necrotizing Clostridial Soft Tissue Infections

DENNIS L. STEVENS
Veterans Affairs Medical Center, Infectious Diseases Section, 500 West Fort Street, Boise, ID 83702, and University of Washington School of Medicine, Seattle, WA, USA

I. INTRODUCTION

The genus *Clostridium* comprises over 60 species of Gram-positive, anaerobic, spore-forming rods that cause a variety of infections in humans and other animals, usually by virtue of a myriad of protein exotoxins. *Clostridium tetani* and *C. botulinum* cause specific clinical diseases by elaborating single, but highly potent, toxins. Botulism results from ingestion of preformed toxin or multiplication of *C. botulinum* in the intestinal tract, while tetanus requires proliferation of *C. tetani* at the site of a penetrating injury. Signs of infection are frequently not apparent even with lethal exotoxemia. In contrast, other strains of clostridia, such as *C. perfringens* and *C. septicum*, cause aggressive necrotizing soft tissue infections, attributable in part to the elaboration of bacterial proteases, phospholipases, and other cytotoxins.

II. CLOSTRIDIAL GAS GANGRENE

A. Types

Smith described three types of clostridial soft tissue infections.[1] The first is simple wound contamination, in which infection *per se* does not develop owing to insufficient tissue devitalization. This is very common; MacLennan reported

The Clostridia: Molecular Biology and Pathogenesis
ISBN 0-12-595020-9

that 30–80% of open traumatic wounds are contaminated with clostridial species.[2] The second type, anaerobic cellulitis, occurs when there is a modest amount of devitalized tissue in a wound, yielding a sufficiently ischemic environment for the growth of *C. perferingens* or other clostridia. Gas is produced locally and extends along fascial planes, but invasion of healthy tissue and bacteremia do not occur. Appropriate medical and surgical management, including prompt removal of the devitalized tissue is the only treatment necessary, and mortality is generally nil.[1]

Clostridial gas gangrene or myonecrosis, defined as an acute invasion of healthy living muscle, undamaged by previous trauma or ischemia,[2] occurs in three different settings. Most commonly, traumatic gas gangrene develops after deep, penetrating injuries that compromise the blood supply (e.g., knife or gunshot wound, crush injury, or automobile accident) creating an anaerobic environment ideal for clostridial proliferation. This type of trauma accounts for about 70% of cases of gas gangrene and *C. perfringens* is found in about 80% of such infections.[1] The remaining cases are caused by *C. septicum*, *C. novyi*, *C. histolyticum*, *C. bifermentans*, *C. tertium* and *C. fallax*. Other conditions associated with traumatic gas gangrene are bowel and biliary tract surgery, intramuscular injection of epinephrine, improperly-performed (usually criminal) abortion, and various puerperal conditions, including retained placenta, prolonged delay in delivery following rupture of the amniotic membranes, and intrauterine fetal death. Spontaneous or nontraumatic gas gangrene is most commonly caused by the more aerotolerant *C. septicum*. As described later in this chapter, most of these cases occur in patients with gastrointestinal portals of entry, such as result from adenocarcinoma. Finally, recurrent gas gangrene caused by *C. perfringens* has been described in individuals with nonpenetrating injuries at sites of previous gas gangrene where spores of *C. perfringens* may remain quiescent in tissue for periods of 10–20 years then germinate when minor trauma provides conditions suitable for growth.[3]

B. Traumatic gas gangrene

1. *Clinical manifestations*

The first symptom is usually sudden onset of severe pain at the site of surgery or trauma.[2,4] The mean incubation period is less than 24 h, but ranges from 6 to 8 h to several days, probably depending on the degree of contamination and the extent of vascular compromise. The skin may initially appear pale, but quickly changes to bronze and then purplish red, and becomes tense and exquisitely tender. Bullae develop, which may be clear, red, blue, or purple. Gas in tissue may be detected by soft tissue radiographs, computerized tomography (CT) scan, or magnetic resonance imaging; however, none of these radiographic procedures are more specific or sensitive than physical examination revealing crepitus in soft tissue.[5] However, radiographic procedures are particularly helpful to demonstrate gas in deeper tissue such as the uterus. Rapidly-developing signs of systemic toxicity include tachycardia, low-grade fever, and diaphoresis, followed by shock and multiorgan failure. In one study shock was present in 50% of patients presenting and of those who developed shock at some point in their hospitalization, 40% died, compared with 20% in the group as a whole.[6] Bacteremia occurs in 15% of patients and may be associated with brisk hemolysis. One patient has been described wth a decrease in the hematocrit from 37% to 0% over a 24-h period.[7] Subsequently, despite transfusion with 10 units of packed red blood cells over a 4-h period, the hematocrit never exceeded 7.2%.[7] Results of studies with recombinant α- and θ-toxins show clearly that both contribute to this marked intravascular hemolysis. Not all cases of *C. perfringens* bacteremia have been associated with gas gangrene; thus in this setting, the presence of *C. perfringens* in the blood may be merely a transient phenomenon.[8] In contrast, results of another study suggested that 90% of *C. perfringens* isolates and 100% of *C. septicum* isolates from blood were associated with clinically significant infection.[9] Additional complications of clostridial myonecrosis include jaundice, renal failure, hypotension, and liver necrosis. Renal failure is largely due to hemoglobinuria and myoglobinuria but is complicated by acute tubular necrosis following hypotension. Toxins probably have direct effects on renal tubular cells, but this has not been proven.

2. Diagnosis

Increasing pain at the site of prior injury or surgery, together with signs of systemic toxicity and gas in the tissue support the diagnosis. Definitive diagnosis rests on demonstrating large, Gram-variable rods at the injury site. In fresh material, *C. perfringens* may appear encapsulated,[10] although this was not corroborated in gas gangrene associated with war-time trauma.[11] Surgical exploration is essential, and in bona fide cases will usually reveal muscle that does not bleed or contract when stimulated. In gross examination the muscle tissue is edematous and may have a reddish-blue to black coloration. Necrotizing fasciitis and cutaneous necrosis are also usually present. Microscopic evaluation of biopsy material invariably demonstrates organisms among degenerating muscle bundles, and a characteristic absence of acute inflammatory cells.[2,12]

3. Pathogenesis

The initiating trauma introduces organisms (either vegetative forms or spores) into the deep tissues, and produces an anaerobic niche with a sufficiently low redox potential and acid pH for optimal clostridial growth.[1,2] Necrosis of previously normal tissue progresses within hours, and no polymorphonuclear leukocytes (PMNL) are present at the site of active infection. In contrast, a modest PMNL influx occurs at the junction of necrotic and normal tissues. Margination of PMNL within capillaries and in small arterioles and post-capillary venules[13,14] is followed later in the course by leukostasis within larger vessels. Thus, the histopathology of clostridial gas gangrene is completely opposite from that seen in many soft tissue infections caused by organisms such as *Staphylococcus aureus*, in which an early luxuriant influx of PMNL localizes the infection without destruction of adjacent tissue or vessels.

Recent studies suggest that θ-toxin, when elaborated in high concentrations at the site of infection, destroys host tissues and inflammatory cells.[15] As the toxin diffuses into surrounding tissues or enters systemic circulation, θ-toxin promotes dysregulated PMNL/endothelial cell adhesive interactions and primes leukocytes for increased respiratory burst activity.[15] These actions are likely to lead to vascular leukostasis, endothelial cell injury, and regional tissue hypoxia. Such perfusion deficits expand the anaerobic environment and contribute to the rapidly advancing margins of tissue destruction that are characteristic of clostridial gangrene.[1]

Shock associated with gas gangrene may be attributable, in part, to direct and indirect effects of toxins. α-Toxin directly suppresses myocardial contractility *in vivo*[16] and may contribute to profound hypotension via a sudden reduction in cardiac output.[17] In experimental models, θ-toxin causes 'warm shock' which is defined as a markedly reduced systemic vascular resistance combined with a markedly increased cardiac output.[16,17] It is clear that θ-toxin accomplishes this indirectly by inducing endogenous mediators, such as the lipid autocoids, prostacyclin, or platelet-activating factor,[18] that cause relaxation of blood vessel wall musculature. Reduced vascular tone develops rapidly and, in order to maintain adequate tissue perfusion, a compensatory host response is required that either increases cardiac output or rapidly expands the intravascular blood volume. Patients with Gram-negative sepsis compensate for hypotension by markedly increasing cardiac output, but this adaptive mechanism may not be possible in *C. perfringens*-induced shock owing to direct suppression of myocardial contractility by α-toxin.[16] The role of other endogenous mediators such as cytokines (e.g., tumor necrosis factor, interleukin-1, interleukin-6) as well as the potent endogenous vasodilator bradykinin, have not been elucidated.

4. Treatment

Penicillin, clindamycin, tetracycline, chloramphenicol, metronidazole, and a number of cephalosporins have excellent *in vitro* activity against *C. perfringens* and other clostridia. No controlled clinical trials have been conducted to compare the efficacy of these agents in humans. Most textbooks state that penicillin is the drug of choice,[19,20] but these recommendations are largely based upon *in vitro* efficacy. Experimental studies in mice suggest that clindamycin has the greatest efficacy, and penicillin the least.[21,22] Other agents with greater efficacy than penicillin include erythromycin, rifampicin, tetracycline, chloramphenicol, and metronidazole.[21,22] Slightly greater survival was observed in animals receiving both clindamycin and penicillin; antagonism was observed when treatment included penicillin plus metronidazole.[22] Because some strains (2–5%) are resistant to clindamycin, a combination of penicillin and clindamycin is warranted. Based on his experimental studies and his vast clinical experience with gas gangrene, the late William Altemeier recommended tetracycline and penicillin.[23] Thus, given an absence of efficacy data from clinical trials in humans, the best treatment would appear to be clindamycin or tetracycline combined with penicillin. The failure of penicillin in experimental clostridial myonecrosis may be related to continued bacterial toxin production and filament formation.[24] In contrast, the efficacy of clindamycin and tetracycline may be related to their ability to rapidly inhibit toxin synthesis.[24]

Aggressive and thorough surgical debridement is mandatory to improve survival, preserve limbs and prevent complications.[19,20] The use of hyperbaric oxygen (HBO) is controversial, although some nonrandomized studies have reported excellent results with HBO therapy when combined with antibiotics and surgical debridement.[6,25,26] Some experimental studies demonstrate that HBO alone was an effective treatment if the inoculum was small and treatment was begun immediately.[27] Results of other studies suggest that HBO was only of slight benefit when combined with penicillin.[28] In these studies, however, survivals were greater with clindamycin alone than with either HBO alone, penicillin alone, or HBO plus penicillin.[28] The benefit of HBO, at least theoretically, is to inhibit bacterial growth,[29] preserve marginally perfused tissue, and inhibit toxin production.[30] Interestingly, Altemeier did not use HBO and was able to realize a mortality rate of less than 15% using surgical debridement and antibiotics (tetracycline plus penicillin).[23]

Therapeutic strategies directed against toxin expression *in vivo*, such as neutralization with specific antitoxin antibodies or inhibition of toxin synthesis, may be valuable adjuncts to traditional antimicrobial regimens; at present antitoxin is not available. Future strategies may target endogenous proadhesive molecules

(molecules which facilitate binding of neutrophils to endothelium) such that toxin-induced vascular leukostasis and resultant tissue injury are attenuated.

5. *Prognosis*

Patients presenting with gas gangrene of an extremity have a better prognosis than those with truncal or intra-abdominal gas gangrene, largely because it is difficult to debride adequately such lesions.[19,20,31] HBO therapy could be useful in such patients, yet there is little data on this subject. In addition to truncal gangrene, patients with associated bacteremia and intravascular hemolysis have the greatest likelihood of progressing to shock and death. As stated previously, patients who are in shock at the time diagnosis is made have the highest mortality.[31]

6. *Prevention*

Aggressive debridement of devitalized tissue and rapid repair of compromised vascular supply greatly reduce the frequency of gas gangrene in contaminated deep wounds. Intramuscular epinephrine, prolonged application of tourniquets, and surgical closure of traumatic wounds should be avoided. Patients with contaminated wounds should receive prophylactic antibiotics.

C. Spontaneous, nontraumatic gas gangrene caused by *Clostridium septicum*

1. *Clinical manifestations*

The onset of disease is abrupt, and the first symptom may be confusion or malaise. The patient may sense only heaviness or numbness, but excruciating pain is common, and extremely rapid progression of gangrene follows.[1,2,32–34] Swelling advances and bullae appear, filled with clear, cloudy, hemorrhagic, or purplish fluid. The skin around such bullae also has a purple hue, perhaps reflecting vascular compromise resulting from bacterial toxins diffusing into surrounding tissues.[32] Histopathology of muscle and connective tissues includes cell lysis and gas formation; remarkably, inflammatory cells are absent.[32]

Predisposing factors include colonic carcinoma, diverticulitis, gastrointestinal surgery, leukemia, lymphoproliferative disorders, cancer chemotherapy, radiation therapy, and more recently, AIDS.[32–34] Cyclic or other neutropenia is also associated with spontaneous gas gangrene caused by *C. septicum*, and in such cases necrotizing enterocolitis, cecitis, or distal ileitis are commonly found. These gastrointestinal pathologies permit bacterial access to the bloodstream;

consequently, the aerotolerant *C. septicum* can become established in normal tissues.[1] Patients surviving bacteremia or spontaneous gangrene caused by *C. septicum* should have appropriate diagnostic studies of the gastrointestinal tract to identify any underlying pathology.

2. Diagnosis

Unlike traumatic gas gangrene, bacteremia precedes cutaneous manifestations by several hours. In the absence of the usual cutaneous manifestations of gas gangrene, other causes of fever and extremity pain such as deep vein thrombophlebitis should be considered. Such evaluations tend to prolong establishment of the appropriate diagnosis, delaying treatment, thereby increasing mortality.

3. Pathogenesis

Clostridium septicum produces four toxins: α-toxin (lethal, hemolytic, necrotizing activity), β-toxin (DNase), γ-toxin (hyaluronidase) and δ-toxin (septicolysin, an oxygen-labile hemolysin), as well as a protease and a neuraminidase.[1] α-Toxin does not possess phospholipase activity and is thus distinct from the α-toxin of *C. perfringens*. Active immunization against α-toxin significantly protects against challenge with viable *C. septicum*.[35] The mechanism by which α-toxin contributes to *C. septicum* pathogenesis is unknown, but the recent cloning and sequencing of the gene for this toxin should facilitate studies in this area (see Chapter 16).

4. Treatment

Although no comparative human trials have evaluated the efficacy of antibiotics or HBO for treating clinical cases of spontaneous gas gangrene, *in vitro* data suggest that *C. septicum* is uniformly susceptible to penicillin, tetracycline, erythromycin, clindamycin, chloramphenicol, and metronidazole. The aerotolerance of *C. septicum* may reduce the efficacy of HBO therapy.[29]

5. Prognosis

The mortality of spontaneous gas gangrene ranges from 67% to 100% with the majority of deaths occurring within 24 h of onset. Risk factors include underlying malignancy and compromised immune status.

D. Other clostridial diseases

1. Necrotizing enteritis

Neutropenic enterocolitis is a fulminant form of necrotizing enteritis that occurs in neutropenic patients.[36] The degree of neutropenia is often profound and may be related to cyclic or other neutropenia, leukemia, aplastic anemia, or chemotherapy.[19,20,36] Symptoms include abdominal pain, chills and malaise. Copious watery diarrhea, abdominal distention, and pain localizing to the right lower quadrant develop, followed rapidly by signs of toxicity such as tachycardia, fever, and delirium. Radiographic examinations may reveal thickening of the wall of the colon or cecum, and in advanced cases, gas in the wall of the colon. Anecdotal reports suggest that CT scanning may be a superior means of diagnosing this condition. Complications include rupture of the bowel with peritonitis, bacteremia, and significant mortality. Aggressive supportive measures, surgical intervention, and appropriate antibiotics have reduced the mortality to 25%.[36]

Post-mortem examinations reveal that among children dying of leukemia, localized infection of the ileocecal region (typhlitis), is extremely common and may contribute to death in nearly 40% of patients.[36] Most of these children have profound neutropenia. *Clostridium septicum* is the most common organism isolated from the blood of such patients, and Gram stain and immunofluorescence studies demonstrate that in most cases these bacteria invade the bowel wall.

Other forms of necrotizing enteritis have occurred endemically in New Guinea (pigbel[37]), epidemically in Germany following World War II (Darmbrand[38]), and sporadically in Africa, Southeast Asia, and the United States (see Chapter 12).[19,20] All cases are associated with the ingestion of meats contaminated with *C. perfringens* type C. Clinical courses vary between abdominal pain, fever, and diarrhea which resolve spontaneously, to bloody diarrhea, ruptured bowel, and death; β-toxin has been implicated in the pathogenesis of these infections. The toxin paralyzes the intestinal villi, and causes friability and necrosis of the bowel wall. Predisposing factors include malnutrition, specifically in patients with diets low in protein and rich in trypsin inhibitors such as sweet potato or soy bean.[37,38] *Ascaris lumbricoides* is commonly found in such patients and it, too, secretes a trypsin inhibitor. These protease inhibitors protect β-toxin from intraluminal proteolysis.

Medical management should include aggressive fluid and electrolyte replacement, bowel decompression, and administration of penicillin or chloramphenicol. Surgical resection of necrotic bowel is necessary in 50% of patients and mortality rates as high as 40% have been reported. If peritonitis develops, broader antibiotic coverage may be necessary. In New Guinea immunization of children with a β-toxoid has dramatically reduced the incidence of this disease.[39]

2. *Clostridium sordellii* infections

Patients with *C. sordellii* infection present with unique clinical features including edema, absence of fever, leukemoid reaction, hemoconcentration, and, later, shock and multiorgan failure.[20] Often, *C. sordellii* infections develop after childbirth or after gynecologic procedures,[40] although some cases involve sites of minor trauma such as lacerations. Unlike *C. perfringens* and *C. septicum* infections, pain may not be a prominent feature. The absence of fever and a paucity of signs and symptoms of local infection make early diagnosis difficult.[20] The mechanisms of diffuse capillary leakage, massive edema, and hemoconcentration are not well established but clearly are related to elaboration of a potent toxin. Hematocrits of 75–80 have been described, and leukocytosis of 50–100 000 cells/mm^3 with a left shift is common.[12,40]

3. *Clostridium tertium* infections

Clostridium tertium causes bacteremia in compromised hosts who have received long courses of antibiotics, thus explaining the organism's relative resistance to penicillin, cephalosporins and clindamycin; however, *C. tertium* is usually quite sensitive to chloramphenicol, vancomycin, and metronidazole. Because this organism can grow aerobically, it may be mistakenly disregarded as a contaminant such as a diphtheroid or *Bacillus* species.[19,20]

III. REFERENCES

1. Smith L.D.S. (1975) Clostridial wound infections. In Smith L.D.S. (ed.) *The Pathogenic Anaerobic Bacteria*, pp. 321–324. C.C. Thomas, Springfield, IL, USA.
2. MacLennan J.D. (1962) The histotoxic clostridial infections of man. *Bacteriol Rev* **26**: 177–276.
3. Stevens D.L., Laposky L.L., Montgomery P. and Harris I. (1988) Recurrent gas gangrene at a site of remote injury: Localization due to circulating antitoxin. *West J Med* **148**: 204–205.
4. Weinstein L. and Barza M. (1972) Gas gangrene. *N Engl J Med* **289**: 1129.
5. Gozal D., Ziser A., Shupak A., Ariel A. and Melamed Y. (1986) Necrotizing fasciitis. *Arch Surg* **121**: 233–235.
6. Hart GB, Lamb RC and Strauss M.B. (1983) Gas gangrene: I. A collective review. *J Trauma* **23**: 991–1000.
7. Terebelo H.R., McCue R.L. and Lenneville M.S. (1982) Implication of plasma free hemoglobin in massive clostridial hemolysis. *JAMA* **248**: 2028–2029.
8. Gorbach S.L. and Thadepalli H. (1975) Isolation of clostridium in human infections: evaluation of 114 cases. *J Inf Dis* **131**: S81–S85.

9. Brook I. (1989) Anaerobic bacterial bacteremia: 12-year experience in two military hospitals. *J Inf Dis* **160**: 1071–1075.
10. Butler H.M. (1943) Pathogenicity of washed *Cl. welchii* and mode of development of *Cl. welchii* infections in man. *Med J Austr.* **2**: 224–226.
11. Keppie J. and Robertson M. (1944) The *in vitro* toxigenicity and other characters of strains of *Cl. welchii* type A from various sources. *J Pathol Bacteriol* **56**: 123–126.
12. Stevens D.L. (1995) Clostridial infections. In Stevens D.L. and Mandell G.L. (eds) *Atlas of Infectious Diseases*, pp. 13.1–13.9. Churchill Livingstone, New York.
13. McNee J.W. and Dunn B.S. (1917) The method of spread of gas gangrene into living muscle. *Br Med J* **1**: 727–729.
14. Robb-Smith A.H.T. (1945) Tissues changes induced by *Cl. welchii* type A filtrates. *Lancet* **ii** 362–368.
15. Bryant A.E., Bergstrom R., Zimmerman G.A., Salyer J.L., Hill H.R., Tweten R.K., Sato H. and Stevens D.L. (1993) *Clostridium perfringens* invasiveness is enhanced by effects of theta toxin upon PMNL structure and function: The roles of leukocytotoxicity and expression of CD1 1/CD 18 adherence glycoprotein. *FEMS Immunol Med Microbiol* **7**: 321–336.
16. Stevens D.L., Troyer B.E., Merrick D.T., Mitten J.E. and Olson R.D. (1988) Lethal effects and cardiovascular effects of purified alpha and theta toxins from *Clostridium perfringens*. *J Inf Dis* **157**: 272–279.
17. Asmuth D.A., Olson R.D., Hackett S.P., Bryant A.E., Tweten R.K., Tso J.Y., Zollman T. and Stevens D.L. (1995) Effects of *Clostridium perfringens* recombinant and crude phospholipase C and theta toxins on rabbit hemodynamic parameters. *J Inf Dis* **172**: 1317–1323.
18. Whatley R.E., Zimmerman G.A., Stevens D.L., Parker C.J., McIntyre T.M. and Prescott S.M. (1989) The regulation of platelet activating factor synthesis in endothelial cells – The role of calcium and protein kinase C. *J Biol Chem* **11**: 6325–6333.
19. Gorbach S.L. (1992) *Clostridium perfringens* and other clostridia. In Gorbach S.L., Bartlett J.G. and Blacklow N.R. (eds) *Infectious Diseases*, pp. 1587–1596. WB Saunders, Philadelphia.
20. Bartlett J.G. (1990) Gas gangrene (other clostridium-associated diseases). In Mandell G.L., Douglas R.G. and Bennett J.E. (eds) *Principles and Practice of Infectious Diseases*. pp. 1851–1860, Churchill Livingstone, New York.
21. Stevens D.L., Maier K.A., Lame B.M. and Mitten J.E. (1987) Comparison of clindamycin, rifampin, tetracycline, metronidazole, and penicillin for efficacy in prevention of experimental gas gangrene due to *Clostridium perfringens*. *J Inf Dis* **155**: 220–228.
22. Stevens D.L., Laine B.M. and Mitten J.E. (1987) Comparison of single and combination antimicrobial agents for prevention of experimental gas gangrene caused by *Clostridium perfringens*. *Antimicrob Agents Chemother* **31**: 312–316.
23. Altemeier W.A. and Fullen W.D. (1971) Prevention and treatment of gas gangrene. *J Amer Med Assoc* **217**: 806–813.
24. Stevens D.L., Maier K.A. and Mitten J.E. (1987) Effect of antibiotics on toxin production and viability of *Clostridium perfringens*. *Antimicrob Agents Chemother* **31**: 213–218.
25. Heimbach R.D., Boerema I., Brummelkamp W.H. and Wolfe W.G. (1977) Current therapy of gas gangrene. In Davis J.C. and Hunt T.K. (eds) *Hyperbaric Oxygen Therapy*, pp. 153–176. Undersea Medical Society, Bethesda, MD.
26. Bakker D.J. (1988) Clostridial myonecrosis. In Davis J.C. and Hunt T.K. (eds) *Problem Wounds: The Role of Oxygen*, pp. 153–172. Elsevier, New York.
27. Hill G.B. and Osterhout S. (1972) Experimental effects of hyperbaric oxygen on selected clostridial species: II. *In vivo* studies on mice. *J Inf Dis* **125**: 26–35.
28. Stevens D.L., Bryant A.E., Adams K. and Mader J.T. (1993) Evaluation of hyperbaric oxygen therapy for treatment of experimental *Clostridium perfringens* infection. *Clin Infect Dis* **17**: 231–237.

29. Hill G.B. and Osterhout S. (1972) Experimental effects of hyperbaric oxygen on selected clostridial species: I. *In vitro* studies. *J Inf Dis* **125**: 17–25.
30. van Unnik A.J.M. (1965) Inhibition of toxin production in *Clostridium perfringens in vitro* by hyperbaric oxygen. *Ant Leeuwenhoek* **31**: 181–186.
31. Hart G.B., Lamb R.C. and Strauss M.B. (1983) Gas gangrene: I. A collective review. *J Trauma* **23**: 991–1000.
32. Stevens D.L., Musher D.M., Watson D.A., Eddy H., Hamill R.J., Gyorkey F., Rosen H. and Mader J. (1990) Spontaneous, nontraumatic gangrene due to *Clostridium septicum. Rev Inf Dis* **12**: 286–296.
33. Johnson S., Driks M.R., Tweten R.K., Ballard J., Stevens D.L., Anderson D.J. and Janoff E.N. (1994) Clinical courses of seven survivors of *Clostridium septicum* infection and their immunologic responses to α toxin. *Clin Infect Dis* **19**: 761–764.
34. Alpern R.J. and Dowell V.R., Jr (1969) *Clostridium septicum* infections and malignancy. *J Amer Med Assoc* **209**: 385–388.
35. Ballard J., Bryant A., Stevens D. and Tweten R.K. (1992) Purification and characterization of the lethal toxin (alpha toxin) of *Clostridium septicum. Infect Immun* **60**: 784–790.
36. Farnell M.B. (1987) Neutropenic enterocolitis: A surgical disease? *Infect Surg* **6**: 120–132.
37. Lawrence G. and Walker P.D. (1976) Pathogenesis of enteritis necroticans in Papua, New Guinea. *Lancet* **i**: 125–128.
38. Guerrant R.L. (1990) Inflammatory enteritides. In Mandell G.L., Douglas R.G. and Bennett J.E. (eds) *Principles and Practice of Infectious Diseases*, pp. 870–880. Churchill Livingstone, New York.
39. Lawrence G., Shann F., Frestone D.S., *et al.* (1979) Prevention of necrotizing enteritis in Papua, New Guinea by active immunization. *Lancet* **i**: 227–230.
40. McGregor J.A., Soper D.E., Lovell G., and Todd J.K. (1989) Maternal deaths associated with *Clostridium sordellii* infection. *Am J Obstet Gynecol* **161**: 987–995.

Chapter 10

Clostridial Diseases of Animals

J. GLENN SONGER
Department of Veterinary Science, The University of Arizona, Tucson, AZ 85721, USA

I. INTRODUCTION

Clostridia are widely recognized as pathogens of domestic and wild animals (Tables 10.1 and 10.2). Clostridial infections take many forms, and remain a common problem for animal producers, veterinary practitioners, and diagnosticians.

II. *CLOSTRIDIUM PERFRINGENS*

Clostridium perfringens is thought by some to be the most widely-occurring bacterial pathogen,[1] and is without question the most important cause of clostridial enteric disease in domestic animals (Table 10.1).

A. Type A

Strains of type A (Table 10.1) are the most widespread in the intestines of warm-blooded animals and in the environment;[2–4] they are found in most soils that are specifically cultured for *C. perfringens*.[1] Lamb enterotoxemia (yellow lamb

The Clostridia: Molecular Biology and Pathogenesis
ISBN 0-12-595020-9

Table 10.1 Veterinary diseases produced by *Clostridium perfringens*

Type	Diseases in domestic animals	Major toxins
A	Myonecrosis, fowl necrotic enteritis, bovine and ovine enterotoxemia, porcine necrotic enterocolitis, equine colitis, canine hemorrhagic gastroenteritis	α
B	Dysentery, chronic enteritis in lambs, ovine hemorrhagic enterotoxemia, equine and bovine hemorrhagic enteritis	α,β,ε
C	Fowl necrotic enteritis, neonatal hemorrhagic or necrotic enterotoxemia (ovine, porcine, bovine, caprine, equine), ovine enterotoxemia	α,β
D	Enterotoxemia (lambs, calves), caprine enterocolitis, bovine enterotoxemia	α,ε
E	Bovine, ovine enterotoxemia, rabbit enteritis	α,ι

Table 10.2 Veterinary diseases produced by clostridia other than *Clostridium perfringens*

Organism	Major toxin(s)	Disease(s)
C. difficile	A (enterotoxin) B (cytotoxin)	Pseudomembranous colitis in laboratory animals, hemorrhagic necrotizing enterocolitis in foals, chronic diarrhea in dogs
C. spiroforme	ι (ADP-ribosylating)	Rabbit enteritis
C. colinum	None confirmed	Ulcerative enteritis in fowl
C. septicum	α	Ovine, bovine abomasitis
C. chauvoei	α,β	Blackleg
C. sordellii	Poorly characterized	Sudden death
C. novyi	α,β	Wound infections ('bighead'), infectious necrotic hepatitis
C. haemolyticum	β	Bacillary hemoglobinuria
C. botulinum	Botulinum toxins A, B, C_1, D, E, F, G C_2 (ADP-ribosylating)	Botulism in many species
C. tetani	Tetanus toxin Tetanolysin	Tetanus in many species

disease) is seen primarily in spring, when the population of nursing lambs is high, and has been most often reported from California and Oregon, USA.[5] A similar condition has been reported in goats[6] and calves.[7,8] The brief clinical course (6–12 h) terminates in death, with signs including depression, anemia, icterus, and hemoglobinuria; large numbers of *C. perfringens* (10^4–10^7 colony-forming units (cfu) per gram) can be found in intestinal contents. Type A strains have often been isolated in the western US from calves with tympany and hemorrhagic, necrotic abomasitis, sometimes accompanied by severe diarrhea.

Organisms resembling clostridia can be seen on the mucosa and in the submucosa.[9] Isolates of *C. perfringens* from some cases contain the *cpe* gene, which encodes an enterotoxin.[10]

Necrotic enteritis is an internationally important disease of domestic chickens.[11–14] It is usually caused by *C. perfringens* type A[12,13,15–18] or type C. Disease also occurs in at least some species of captive wild birds[19] and in turkeys.[20] Intestinal mucosal damage caused by high-fiber litter or concurrent infection with coccidia can predispose birds to development of necrotic enteritis.[14,21–23] Mild forms of disease result mainly in decreased rates of gain,[24] and depression, inappetence, anorexia, and diarrhea may be observed in some affected birds.[11,14,22,25] However, the course is usually short, with no signs observed until birds are found dead. Jejunal and ileal necrosis can extend the entire width of the mucosa,[26] although acute catarrh without necrosis is more common in the lower intestine. Gram-positive bacilli are often detected on the lamina propria and attached to cellular debris.

Both α- and β-toxins have been detected in feces and intestinal contents of chickens and other birds with necrotic enteritis,[12,13] and more α-toxin may be produced by isolates from birds with necrotic enteritis than by isolates from normal birds.[27] Crude preparations of α-toxin,[15,17,28] administered to conventional[29] or germ-free chicks,[30] yield typical necrotic lesions. α-Antitoxin may play a role in protection of fowl against enteritis.[30,31]

Adult horses occasionally present with 'intestinal clostridiosis', a condition characterized by profuse watery diarrhea and high mortality, with large numbers of type A organisms demonstrable in the gut.[32,33] *Clostridium perfringens* infection has also been described in horses with hemorrhagic cecitis, in horses dying peracutely without diarrhea,[34] and in a neonatal foal with hemorrhagic diarrhea.[35] In one experimental study, intravenous (i.v.) inoculation of ponies with a type A culture caused acute colic and hemorrhagic enterocolitis.[36]

Mild necrotizing enterocolitis and villous atrophy in suckling and feeder pigs was thought to have been caused by type A strains.[37–43] A similar syndrome was produced by oral inoculation of gnotobiotic colostrum-deprived pigs and also conventional weaner pigs.[44] However, α-toxin alone, inoculated into isolated loops of gut in pigs, did not produce significant lesions or fluid loss.[43,45]

Hemorrhagic canine gastroenteritis is sometimes associated with *C. perfringens*, often of unidentified type.[46] Peracute death is common, with hemorrhagic necrosis of the mucosa, although less severe forms present as watery-to-mucoid diarrhea with occasional blood. Organisms are common in and on necrotic tissue, but apparently do not invade normal tissue.[46–49]

Little is known about uptake of α-toxin from the intestine, and its role as a virulence factor in cases of type A enterotoxemia is unknown.[50,51] α-Toxin administered to neonatal piglets caused villous edema and mild enteritis, with little damage to the epithelium and blood vessels and no ultrastructural changes in villi, lymphatics, or other tissues.[38–40] Lambs[52] or calves[53] injected IV with sublethal doses of α-toxin developed transitory diarrhea and mucosal hyperemia. Large amounts of α-toxin can be found in gut contents in natural cases of disease in cattle. The syndrome in lambs and calves is often associated with massive intravascular hemolysis, capillary damage, inflammation, platelet aggregation, shock, and cardiac effects culminating in death.[4,54] α-Toxigenic *C. perfringens* caused enteropathy in neonatal pigs, and adherence or invasion accompanied proliferation.[38–40] Minor differences may exist in amino acid sequences of α-toxin from gas gangrene strains (possibly possessing a predilection for muscle tissue) and enterotoxemia strains (possibly capable of multiplication and disease production in the gut), conferring increased resistance to chymotrypsin upon the latter[55] and perhaps allowing accumulation in the gut with subsequent entry to the circulation.

B. Type B

Type B is the etiologic agent of dysentery in newborn lambs (Table 10.1). Disease occurs primarily in the UK, South Africa, and the Middle East.[4] Type B infection is apparently relatively rare in North America.[56] Older lambs may be involved in outbreaks, but lamb dysentery typically develops during the first few days of life.[57] The infection is acquired from the dam or the environment, and organisms in the gut increase in numbers, especially where there is heavy lactation by the dam. Enterotoxemia results, with enteritis and extensive hemorrhage and ulceration of the small intestine,[58,59] and the primary sign is sudden death. In acute cases, severe abdominal pain, bloody diarrhea, recumbency, and coma are followed by death in less than 24 h. Chronic disease (called 'pine') in older lambs manifests as chronic abdominal pain without diarrhea. Type B may also be associated with hemorrhagic enteritis in goats, calves, and foals.[58–60] Little is known about the pathogenesis of infection by type B strains.[61]

C. Type C

Clostridium perfringens type C infects pigs, cattle, sheep, horses, chickens, humans, and dogs, worldwide (Table 10.1).[62] Newborn animals are extraordinarily susceptible, perhaps owing to a lack of competitive gut flora; sudden dietary changes may also induce type C infections.[4]

Piglets may be more commonly affected by type C than are other domestic animals. Peracute disease, often with a clinical course less than 24 h, may affect piglets 1–2 days old.[62,63] Pigs display diarrhea and dysentery, with blood and necrotic debris in feces. The intestinal surface is often dark red, with gas accumulation in tissue and hemorrhagic exudate in the lumen; hemorrhagic necrosis of the mucosa, submucosa, and even muscularis mucosa can be extensive.[62,64–68] Up to 10^9 cfu of *C. perfringens* (mixed cultures of types A and C[10,66] or nearly pure cultures of type C[69]) can be isolated from 1 g of intestinal contents. Morbidity rates of 30–50%[63,66,67] and case fatality rates of 50–100% are not uncommon. Older piglets (1–2 weeks old) may have a longer course of infection, with nonbloody, yellowish diarrhea and necrosis of the jejunal mucosa.[62,63] Sows are thought to be a common source of the infection, but numbers of the organism in sow feces may be too low to be detected by culture.[66]

A similar disease may occur in neonatal calves,[70,71] lambs,[72] and goats. Hemorrhagic, necrotic enteritis and enterotoxemia develop in robust calves, usually less than 10 days old, accompanied by abdominal pain, and sometimes by nervous signs, including tetany and opisthotonos; type C infection in lambs resembles lamb dysentery.[73,74] Young adult sheep in certain geographic locales of the United Kingdom can be affected by struck, a name which derives from the rapid death which occurs in this condition, often leaving the impression that the animal has been struck by lightning. In endemic areas, most animals are probably colonized from the soil,[4] and the organism multiplies in the abomasum and small intestine, often following damage by poor-quality feed. Mucosal necrosis occurs in the absence of dysentery or diarrhea, with prominent evidence of toxemia.[75]

Type C enteritis in foals is characterized by depression, severe hemorrhagic diarrhea, dehydration, and, occasionally, colic.[76] Jejunal and ileal lesions usually consist of acute hemorrhagic enteritis with necrosis of villi, with large numbers of Gram-positive rods demonstrable in stained smears and sections.[35,76–78]

Acute hemorrhagic enteritis in dogs may be caused by type C infection.[79,80] Untyped isolates were obtained from a milder necrotic lesion in high proportion of dogs with parvovirus infection.[81] In the latter case, Gram-positive bacilli were observed in necrotic areas of the small intestine.

β-Toxin probably plays a major role in pathogenesis of type C infections. Its initial action in the jejunum is under conditions of curtailed proteolytic activity, possibly owing to a pancreatic secretion deficiency early in life, or to the ingestion of protease inhibitors which compromise the ability of the gut to detoxify β-toxin.[63,82,83] Damage to microvilli, mitochondria, and terminal capillaries prior to bacterial adhesion to the jejunal mucosa[64,84–86] progresses to extensive mucosal necrosis, and Gram-positive bacilli, and also spores,[87] can be demonstrated in association with the damaged intestine. Intestinal lesions can be impressive in extent and severity, but death results from β-toxemia.[63,87–89] It should be noted that experimental reproduction of type C disease in pigs has not been achieved by inoculation with toxin (crude culture supernatant fluids) alone.[63,90]

D. Type D

Type D (Table 10.1) causes enterotoxemia in sheep,[4] and is probably most prevalent in young lambs suckling heavily-lactating ewes. It is also the predominant cause of death in weaned animals up to nearly 1 year old, often in those fed rich rations in feedlots. The frequent association of enterotoxemia with upsets in the gut flora caused by an unmanaged change to a rich diet gives rise to the common name of the disease – overeating disease.[91] Multiplication of the organism, production of ε-toxin,[92] and absorption from the gut[83] lead to a toxemia with little enteritis. Peritoneal and pericardial effusions are typical.[93] ε-Toxin's effects on the central nervous system and other tissues result in sudden death, although some animals display dullness, opisthotonos, and convulsions before death.[83,91] Hyperglycemia and glycosuria are pathognomonic,[83,94] and pulpy kidney, another common name for the disease, derives from the commonly-found post mortem autolysis which rapidly occurs in toxin-damaged kidney tissue. Focal encephalomalacia may be a chronic manifestation of enterotoxemia.[95,96]

Type D infections also occur in calves,[97,98] in goats,[99,100] and occasionally in adult cattle and other animals.[60] In goats, fibrinous or mucohemorrhagic enterocolitis is common,[100-102] in the absence of other classical signs.

E. Type E

Type E is a rarely-reported cause of enterotoxemia of lambs, calves, and rabbits (Table 10.1).[103] First reported nearly 50 years ago in Britain, it is uncertain whether publications in the intervening years[75,104] represent genuine cases of ι-enterotoxemia. Recent evidence has shown that ι-toxin has ADP-ribosylating activity, and this finding could be the basis for studies of pathogenesis of type E infections. If detection of toxin in the gut is the sole diagnostic tool, infection by *Clostridium spiroforme* could be mistaken for *C. perfringens* type E infection, in that the former produces a toxin with significant structural and functional similarities to ι-toxin (see Chapter 20).

F. Enterotoxigenic *Clostridium perfringens*

Diagnosticians, in general, are only now becoming aware of the potential importance of enterotoxin (CPE)-producing strains of *C. perfringens* in enteric disease of domestic animals. Investigators have not usually examined clinical materials or isolates for CPE, making it difficult to interpret results in some reports of putative type A clostridial enteric disease. Nonetheless, evidence increasingly suggests that enterotoxigenic strains can play a role in the etiology of diarrheal disease in several animal species.[3,83,105–108] *cpe* was found in 14% of *C. perfringens* isolated from horses, 22% from cattle, and 10% from poultry,[109] and CPE was produced by 12% of isolates from cattle, sheep, and chickens with enteritis.[17] Polymerase chain reaction (PCR) assays revealed that nearly 5% of isolates from domestic animals were *cpe*-positive.[10,110]

CPE is frequently found in stools of diarrheic pigs,[37,41,43,44,105] and necropsy reveals superficial mucosal necrosis and villous atrophy.[37,41] Spore counts as high as 5×10^6/g of feces or intestinal contents have been reported in diarrheic pigs, compared with 2×10^4/g in nondiarrheic pigs.[105] Experimental disease following inoculation of hysterectomy-derived, colostrum-deprived (HDCD) piglets ranged from profuse, bloody diarrhea, enteritis, and death[42,45] to non-bloody diarrhea and intestinal gas accumulation with low mortality, the latter resembling infections in conventionally weaned pigs.[42] Enterotoxin caused fluid accumulation in ileal loops, and intragastric administration of CPE to HDCD piglets led to transient diarrhea.[43,45] Diarrhea and death were prevented in experimentally infected HDCD pigs given serum, milk, or colostrum from immunized sows;[45,111] pigs born to the same sows were protected against oral challenge with an enterotoxigenic strain.

CPE-associated enteropathy in dogs[48,49,112] often occurs as a nosocomial infection,[49] with affected animals displaying mild depression and soft-to-watery diarrhea with or without hemorrhage and catarrh. Fecal CPE was detected in more than 40% of diarrheic dogs and in only 7% of normal dogs.[46–49] In one study, 65% of isolates from dogs with fecal CPE also produced CPE *in vitro*.[49] However, some studies have revealed that many isolates of *C. perfringens* from dogs with fecal CPE (as determined by reverse passive latex agglutination test) do not contain *cpe* (our unpublished data).

Detection of large numbers of CPE-producing *C. perfringens* in the gut of affected horses[34,113] and lesions produced in experimental infections[114] suggest a role for CPE in pathogenesis of colitis X or other intestinal conditions of horses.[32,115,116] Challenge of horses with *C. perfringens* produced diarrhea and death[33] and ponies inoculated with CPE developed colic and hemorrhagic gastroenteritis.[115]

Clinical disease gives rise to serum antibodies in humans, pigs, sheep, cattle, and horses,[45,105,117] although protection by parenteral inoculation is apparently not protective. However, protection of suckling pigs by colostrum from immunized dams suggests that future commercial products for immunization of pigs could benefit from inclusion of CPE toxoid.[111]

G. Prophylaxis and therapy

Clostridium perfringens is susceptible to many antimicrobials, and prophylaxis in swine has been achieved by use of bacitracin,[118] lincomycin, and others and in chickens with virginiamycin, tylosin, and bacitracin.[46,119–121] Neither antimicrobial agents nor hyperimmune serum has much therapeutic value, because death quickly follows the onset of clinical signs.[122]

Commercial vaccines against *C. perfringens* infection are often multivalent, typically consisting of inactivated cells, toxoids, or both,[123,124] and requiring semi-annual boosters.[122,125,126]

H. Diagnosis

Diagnosis of *C. perfringens* infections in domestic animals requires evaluation of clinical signs, gross and microscopic lesions, bacteriological culture, and detection of toxins in pathologic specimens and in supernatant fluids of cultures.[2,127] There is little or no definitive evidence to support the quantitation of *C. perfringens* as an aid in diagnosis of enteric disease in animals.[4,75] Demonstration of toxins by *in vivo* assays[4,127] is no longer a common practice. Detection of a major toxin in clinical specimens can be useful, but does not necessarily confirm the existence of disease,[75] and failure to demonstrate toxins may be insignificant because of the lability of these proteins (especially β-toxin). Alternative methods for toxin detection include cytotoxicity assays for enterotoxin[105,128] and immunoassays for enterotoxin and enterotoxemia-associated toxins.[89,128–141] Gene probes and PCR assays for diagnosis of *C. perfringens* infections has also been reported.[10,50,51,108–110,142–145] However, with the exception of cytotoxicity and reverse-passive latex agglutination assays for CPE, none of these methods are yet in general use.

III. *CLOSTRIDIUM DIFFICILE*

Pseudomembranous colitis in humans results from overgrowth of the large bowel by *Clostridium difficile*, usually after a perturbation in bowel flora, caused by antibiotic therapy or other circumstances[146,147] (Table 10.2). An identical disease occurs in hamsters[148] and guinea pigs[149] after antibiotic treatment.

Disease has been diagnosed in domestic and wild animals, including a Kodiak bear,[150] a penguin,[151] captive ostriches,[152] rabbits,[153] hares,[154] and adult horses with colitis.[155] *Clostridium difficile*-associated disease is characterized by profuse watery diarrhea and dehydration in foals less than 1 week old.[155–158]

Affected foals may also display severe hemorrhagic necrotizing enterocolitis, with death following onset of clinical signs within 24 h.[157–159] Experimental reproduction of the disease has been achieved by oral inoculation.[157,158]

Clostridium difficile and toxin B have been demonstrated in gut contents of dogs (pups more commonly than adults) with chronic diarrhea. Infection of the same litter with different toxigenic phenotypes suggests transient infection with different strains.[160] *Clostridium difficile* has also been isolated from camels, horses, donkeys,[161] swine,[162] dogs and cats (up to 39% prevalence rate),[163–165] and other domestic species.[166] The importance of domestic animals as sources of infection for humans has not been established.[163,164]

IV. *CLOSTRIDIUM SPIROFORME*

Clostridium spiroforme, which has a distinct, loosely-coiled, spiral form when grown on blood agar,[167] causes ι-enterotoxemia of rabbits and other laboratory rodents (Table 10.2).[168–171] The organism is apparently not a normal inhabitant of the rabbit bowel, but is acquired from the environment, and disease follows disruption of bowel flora at weaning or following antibiotic treatment.[170,172,173] Poor hygiene,[174] stress,[175] and dietary factors[175–177] can have pronounced influence. The disease can be rapidly fatal, and gross lesions include cecal dilation with watery contents. The epithelial surface is necrotic, and an inflammatory infiltrate can be seen in the lamina propria. Isolates of *C. spiroforme* from diseased rabbits produce a lethal, dermonecrotic toxin that is neutralized by antibodies against *C. perfringens* ι-toxin[168,178] and enterotoxemia can be reproduced with gut filtrates from affected rabbits.[174] The toxin has recently been shown to have a binary structure and to possess ADP ribosyltransferase activity,[179] targeting monomeric actin and destroying the microfilament network.[180]

Definitive diagnosis is by demonstration of ι-like toxin by a lethality assay in mice or a cytotoxicity assay in Vero cells.[181–183] A change in diet can be an effective therapy, as can use of antibiotics such as metronidazole and penicillin G.[178,184] Protection of weanling rabbits against intraperitoneal challenge with ι-toxin can be achieved by vaccination with a toxoid.[185]

V. *CLOSTRIDIUM COLINUM*

Clostridium colinum causes ulcerative enteritis, a disease of sudden onset and rapidly-increasing mortality, originally described in bobwhite quail (*Colinus virginianus*),[186] but now often seen in chickens and other domestic fowl, including young turkeys; grouse, partridge, and other game birds are also affected (Table 10.2).[187,188] The mol% G+C is 43%, significantly higher than that of many pathogenic clostridia.[178,187]

Certain aspects of the natural history of *C. colinum* and ulcerative enteritis are reasonably clear. Natural infection is by the oral route.[189] Persistence of *C. colinum* in the environment is likely, and chronically-infected birds can be carriers.[189] In chickens, ulcerative enteritis often follows coccidiosis or some other stress.[187,190] *Clostridium colinum* establishes in the terminal one-third of the intestine, usually in birds less than 3 months old. Duodenal hemorrhage and enteropathy is accompanied by cecal necrosis and ulceration. A pseudomembrane sometimes develops, and perforation is not uncommon. Microscopic findings include mucosal desquamation, edema, hyperemia, and lymphocytic infiltration. Ulcers involve the villi and submucosa, and bacteria are numerous in necrotic tissue. The infection extends to the liver, producing diffuse centrilobular or pinpoint necrosis.[191] Splenic congestion, hemorrhage, and necrosis may also occur.[187] Birds often die after 1–2 days, and emaciation and 'razor breast' develop in those that survive longer. Mortality is frequently 10–30% in chickens, but up to 100% in young quail.[187] Culture filtrates of *C. colinum* are not toxic for mice[192] and little is known about virulence attributes of the organism or about the pathogenesis of ulcerative enteritis.

Diagnosis of ulcerative enteritis is by detection of typical lesions in the intestinal tract, liver, and spleen, supported by recovery of *C. colinum* from lesions. Subterminal spores can be seen *in vivo* but rarely *in vitro*. Bacitracin and streptomycin are commonly used therapeutically, although bacitracin-resistant strains have been reported.[193]

VI. *CLOSTRIDIUM SEPTICUM*

Clostridium septicum, the malignant edema bacillus, is commonly found in soil, and has been isolated from the feces of domestic animals and humans.[194–197] It is a frequent post mortem invader[4] and can apparently enter animal via liver flukes.[198] Iatrogenic infections appear to be more common in horses than in other species.[199–201]

Wound infections by *C. septicum* in animals are generally referred to as malignant edema and usually follow direct contamination of a traumatic wound,[4,90,202,203] incurred through castration, docking, or other procedures. Genital infections can be associated with difficult obstetrical presentations and mismanaged attempts at delivery, and umbilical infections are not uncommon in sheep. Spreading usually along fascial planes from the point of infection, hemorrhage, edema, and necrosis develop rapidly. Initially painful, warm, and pitting on pressure, the developing lesion becomes crepitant and cold, with loss of feeling. Fever, anorexia, and depression are followed by death, often in less than 24 h.

Clostridium septicum also causes enteric infections (Table 10.2).[204] Braxy or bradsot is a disease of sheep in which the organism penetrates the lining of the abomasum, resulting in a fatal bacteremia.[205] High mortality rates occur in yearling sheep in the United Kingdom, Norway, and Iceland, and cases have been reported in Europe, Australia,[206] the United States, and elsewhere. There are also published[204,205] and anecdotal reports of similar disease syndromes in calves.[93]

The pathogenesis of *C. septicum*-induced abomasitis is not well understood. Ingestion of frozen feed[204,205] may impair mucosal function, allowing *C. septicum* to multiply locally and disseminate, producing both local lesions and toxemia. Edema, hemorrhage, and necrosis can be found in the abomasum and the proximal small intestine.[206] Toxic products of *C. septicum*[178,207–210] cause increased capillary permeability and myonecrosis, perhaps encouraging spread of the infection.[211] Recently published information on the major toxin of *C. septicum*, α-toxin,[207,208] may provide a basis for clarification of its role in pathogenesis (see Chapter 16). β-Toxin (DNase), δ-toxin (hyaluronidase), γ-toxin (oxygen-labile hemolysin), and neuraminidase[209] are of uncertain importance.

The course of disease is rapid, so prevention of *C. septicum*-induced infections is preferable to treatment. Vaccines elicit antibody responses to both somatic and toxin antigens[123,202] and yield lifelong immunity.[212]

Diagnosis of *C. septicum* infection is based upon clinical signs, gross and microscopic findings at post mortem, Gram-straining direct smears, and bacteriological culture.[127] *Clostridium septicum* is likely to be the etiological agent if it predominates in specimens obtained immediately post mortem. *Clostridium chauvoei* infection should be ruled out first;[69] this can be accomplished by use of a rapid method such as a fluorescent antibody test.[213]

VII. *CLOSTRIDIUM CHAUVOEI*

Blackleg is an emphysematous, necrotizing myositis caused by *C. chauvoei* (Table 10.2), an inhabitant of the intestine, liver, and other tissues of cattle, sheep, and other species. Survival of *C. chauvoei* in soil may be a significant factor, since the disease occurs year-after-year on the same premises, usually in well-fed cattle less than 3 years old. The mechanism of infection is not clearly understood, but ingestion is probably the most common route of exposure in cattle; various tissues,[214,215] especially skeletal muscle, are seeded with spores from the intestine. The organism remains dormant until damage to the muscle provides conditions which favor germination, multiplication, and toxin production.[182,197] Exertion and acute indigestion may also be capable of initiating these events. In sheep, *C. chauvoei* infection more often presents as a wound infection resembling malignant edema or gas gangrene. Clinically, animals may display high fever, anorexia, and depression, often with lameness. Crepitant lesions may be detected in superficial areas, although many lesions are internal. However, sudden death, without display of clinical signs, is common. The common rancid-butter odor of the lesions results from bacterial production of butyric acid as an end-product. Peripheral areas of the resulting lesions are edematous and hemorrhagic, with evidence of myonecrosis, while the central areas of the lesions are typically dry and emphysematous. Microscopically, leukocytic infiltration is negligible, but degenerative changes occur in muscle fibers, and edema, emphysema, and hemorrhage are common.[4,172,200]

Clostridium chauvoei produces numerous toxic or putatively toxic factors,[216] including the necrotizing α-toxin, which is also hemolytic and apparently lethal. β-Toxin is a DNase which may be responsible for degeneration of muscle cell nuclei.[217,218] Hyaluronidase (γ-toxin), oxygen-labile hemolysin (δ-toxin), and neuraminidase are also produced, but are of uncertain importance in pathogenesis.

Vaccines for blackleg are typical of those used for many histotoxic clostridial infections, and consist of formalinized whole cultures or toxoids. Protection apparently arises from the immune response to a heat-labile soluble protective antigen.[219,220] Equine hyperimmune serum is useful for therapy and prophylaxis, and penicillin can be an effective prophylactic agent during outbreaks.

VIII. *CLOSTRIDIUM SORDELLII*

Clostridium sordellii is commonly found among the intestinal flora of domestic animals and in the soil (Table 10.2).[4] It has been isolated from cases of fatal myositis, liver disease, and sudden death in cattle, sheep, and horses.[4,221] Lesions consist of edema of the subcutaneous tissues and fascial planes, with subendocardial hemorrhage. Terminal septicemia has been observed in experimentally infected cattle.

The organism produces numerous putatively toxic substances, although most have not been adequately characterized. An edema-producing, lethal factor,[222] at least two phospholipases C, an oxygen-labile hemolysin, neuraminidase, and a DNase[223] have been described. It is assumed that these toxins account for most of the signs of the disease.

IX. *CLOSTRIDIUM NOVYI*

Clostridium novyi toxigenic types A, B, and C are well-accepted, while type D is often referred to as *C. haemolyticum* (Table 10.2).[224] Types A and B are common in soil, but type B can also be found normally in the intestine and liver of herbivorous animals.[225] Type C isolates are nontoxigenic and do not produce disease.

Strains of type A have been implicated as causes of gas gangrene in humans and wound infections in animals. Perhaps the hallmark of *C. novyi* type A-induced lesions is edema, probably mediated by vascular damage caused by α-toxin. 'Bighead' in young rams follows invasion by *C. novyi* type A of subcutaneous tissues damaged by fighting.[182] Edema rapidly spreads, involving head, neck, and cranial thorax; death is rapid.

Clostridium novyi type B causes infectious necrotic hepatitis of sheep and cattle, and occasionally infects horses and swine. Spores are often found dormant in Kupffer cells in normal liver. Liver injury, such as that accompanying migration of immature flukes (*Fasciola hepatica*), produces conditions that encourage germination. Production and dissemination of α-toxin lead rapidly to systemic effects,[226] which manifest as depression and anorexia, with acute or peracute death. Venous congestion resulting from pericardial edema produces the characteristic darkening of the underside of the skin, giving rise to the name 'black disease'. α-Toxin can be demonstrated in lesions,[227] which typically are characterized by edema, serosal effusion, and focal areas of hepatic necrosis.

α-Toxin is the major toxin mediating pathogenesis of *C. novyi* infections.[226,228] It has demonstrated cardio-, neuro- histo-, and hepatotoxicity,[229–231] although many reports have been based on toxin preparations that were almost certainly not of high enough purity to allow firm conclusions to be drawn. Genetically, production of α-toxin in *C. novyi* is phage mediated.[232] Roles of other toxins, including β (lecithinase), γ (necrotizing phospholipase D), δ (oxygen-labile hemolysin), and ε (lipase), are not known.

Typically, there is no effective treatment for *C. novyi* infections. Prophylaxis may be achieved with bacterin:toxoids or toxoids. Elimination of flukes is an important control measure.[4]

X. *CLOSTRIDIUM HAEMOLYTICUM*

Clostridium haemolyticum is often referred to as *C. novyi* type D (Table 10.2).[233,234] Phenotypically, it resembles type B,[235] with the exception that it produces no α-toxin and much more β-toxin than type B strains. The organism can be found in soil,[236,237] and ingestion leads to deposition of spores or vegetative cells in the digestive tract and liver. Under appropriate conditions, *C. haemolyticum*-induced disease, called bacillary hemoglobinuria, can develop. Bacillary hemoglobinuria occurs in North America, most often in the Rocky Mountains and along the Pacific and Gulf coasts. It can also be found in South America, Europe, and New Zealand.[238]

Bacillary hemoglobinuria typically occurs among well-nourished animals at least 1 year old. Migration of immature flukes through the liver causes hepatic necrosis, and the resulting hypoxia induces spore germination in Kupffer cells.[239] Production of β-toxin by vegetative cells leads to multiplication of the bacteria and further toxin production.[240] Local toxin action results in liver necrosis, while dissemination via the bloodstream causes intravascular hemolysis and hemorrhage.[241] Clinical signs include fever, pale mucus membranes, anorexia, abdominal pain, and hemoglobinuria (from which the disease derives its common name, 'redwater'). When hemoglobinuria appears, 40–50% of the red cells have been lysed. Serosal effusions and widespread hemorrhage are common, and the pathognomonic lesion is a large circumscribed liver infarct.[240] Large numbers of Gram-positive rods can be found in the sinusoids.[4] The case fatality rate is 90–95% in untreated animals, and death ultimately results from anoxia. Thus, the pathogenesis is similar to black disease of sheep, except that the primary toxin is β rather than α.

β-Toxin is a lethal, necrotizing, and hemolytic phospholipase C.[242,243] The phage which mediates α-toxin production in *C. novyi* type B[232] can transduce *C. haemolyticum* to α-toxin production.[244]

Formalinized whole cultures may be used for immunization, which must be repeated twice annually.[4] It has been suggested, but not confirmed, that immunity is primarily antibacterial.[245] Hyperimmune serum can be used for prophylaxis and therapy.[246]

XI. *CLOSTRIDIUM BOTULINUM*

Botulism is an intoxication which results in neuroparalysis, and can be caused by any of seven neurotoxins (Table 10.2).[247] While heterogenous in some respects, these toxins share the ability to block acetylcholine release from cholinergic nerve endings.[248] In the main, the toxins are serologically distinct.[249,250] The paralytic action of botulinum toxin on skeletal muscle is well-known (see Chapters 17 and 18).[251] Another toxin, designated C_2, has no neurotoxicity, but is a binary toxin, consisting of an H chain which mediates binding to cell membranes[252] and an L chain with ADP-ribosylating activity similar to ι-toxins of *C. perfringens* and *C. spiroforme*.[253]

Clostridium botulinum types A and B are found in soils, while types C, D, E, and F are more common in wet environments.[254] Germination of *C. botulinum* spores resident in the carcasses of dead animals or in rotting vegetation can yield sufficient toxin to cause outbreaks. Disease occurs mainly in ruminants, horses, mink, and fowl, although other carnivores, swine, and fish are occasionally affected. The most common types are C and D, with A, B, and E contributing sporadically. The predominant type varies between animal species and also geographically.[247] A multiplicity of distinctive names has been applied to the various conditions affecting animals, including loin disease and lamziekte (cattle), limberneck and western duck sickness (waterfowl), and spinal typhus and 'shaker foal syndrome' (horses).

Botulism in cattle can arise from many sources, including animal carcasses and feedstuffs. Phosphorus deficiency may encourage the development of pica in domestic ruminants, leading to consumption of carcasses and subsequent death due to botulism, called lamziekte (lame sickness) because of the lameness which often accompanies phosphorus deficiency.[251,255,256] Poultry litter is another common source of the toxin.[257,258] Because of degradation of botulinum toxin by bacteria in the rumen,[259] the oral lethal dose is much larger than that by other routes. Botulism in horses, caused by type C or D toxin, often involves adulteration of feed, as in the disease in cattle.[247] Another form of equine botulism, in which toxin is produced in the gut of the affected animal, is called 'shaker foal syndrome'.[260–262]

Vast outbreaks of type C botulism occur annually in wild waterfowl, perhaps more than in any other animal.[263] It has been suggested that the organism grows and produces toxin in tissues of dead invertebrates,[264] although outbreaks have been documented under other, widely-varied circumstances. It is accepted that vertebrate carcasses play a sustaining role in outbreaks,[265] with invertebrate larvae ingesting the toxin and being consumed by other fowl.[266] Outbreaks of type E botulism in wild birds are sporadic, perhaps associated with the consumption of toxin-carrying fish.[267,268] Type C botulism is also the most common form in poultry,[266,269] and intoxication from *C. botulinum* multiplying in the gut may be responsible for some outbreaks.[266,270–272] Botulism has also occurred frequently in farmed mink,[247] most often caused by type C.

Clinical signs of botulism include anorexia, incoordination, ataxia, flaccid paralysis, and labored breathing. Swallowing becomes difficult owing to paralysis of the tongue and pharynx, and animals often salivate excessively. Respiratory paralysis eventually causes death of the animal. Paralysis of facial muscles is common in cattle, and horses are subject to tremors. Waterfowl lose the ability to fly, cannot retract the nictitating membrane, and the head droops ('limberneck'). Chickens often have diarrhea.[4,251]

Toxoids of botulinum toxin are immunogenic, and have been employed for immunoprophylaxis in animals.[247,260,273] Antibodies have been found in the serum of carrion eaters,[274] but may not have been protective; the dogma is that sufficient toxin to immunize a host is more than enough to kill.[4,251] Vaccination is usually practiced only in populations at immediate risk.[273,275–279] Polyvalent antitoxins can be effective therapeutic agents, depending on the amount of toxin bound before treatment begins.

XII. *CLOSTRIDIUM TETANI*

Spores of *Clostridium tetani* are introduced to animal hosts, usually traumatically, where they germinate (Table 10.2)[280,281] and produce a neurotoxin which is responsible for all or most of the symptoms of tetanus. Soil is the usual source of *C. tetani* spores, especially soil rich in animal manure. Vegetative cells and spores are frequently detected in the digestive tract of animals, including cattle, horses, sheep, dogs, chickens, rats, guinea pigs, and humans.[282] Susceptibility to tetanus toxin and the circumstances under which animals are raised affect the rate of occurrence of tetanus. Thus, disease occurs frequently in horses, less often in other herbivores, and occasionally in pigs and carnivores.[255] Cattle may develop tetanus as a result of growth of *C. tetani* in the rumen.[283,284]

Tetanus usually results from contamination of a wound or the umbilicus by soil. Slight penetrating wounds or abrasions also commonly result in tetanus, as does surgery under primitive conditions, docking and castrating by banding or other means, ear tagging, injections, and post-partum reproductive tract infections. Often the wound is trivial, to the extent of going unnoticed, but necrosis, with reduced oxygen, is usually required. The incubation period may be as short as 24 hours or as long as 2 weeks, depending on the toxinogenicity of the strain, rate of transfer of toxin to the target tissues, and the relative sensitivity of the host.[285,286] In ascending tetanus, toxin is transported intra-axonally to the neurons in the CNS, traveling retrograde via the peripheral motor nerve endings. Passing across the synapse, toxin then binds to presynaptic axonal terminals, and the resulting hyperactivity of motor neurons causes sustained spasms in the innervated muscles. Traveling within the spinal cord, toxin can affect other muscle groups. Descending tetanus is the result of vascular dissemination of toxin, with effects often beginning in sites distant from the nidus of infection.

Early clinical signs include muscular tremor and increased stimulus response. This is followed by impaired muscle function in the head and neck, with difficulty in chewing and swallowing sometimes caused by lockjaw. In horses, nostrils may appear flared, the third eyelid is retracted, and the ears are often stiffly erect. Tetanic spasms give way to permanent rigidity. Muscles of the back and tail become rigid, with orthotonus. Respiration becomes increasingly difficult, leading eventually to respiratory failure and death within a few days in young animals and in less than 2 weeks in adults. Incubation is often longer in carnivores and progression of the disease is slower than in ungulates. Mortality is at least 50% and is highest in the young.[4,251]

Clostridium tetani produces a neurotoxin, tetanus toxin, and an oxygen-labile hemolysin, tetanolysin. Very little tetanolysin is produced by virulent strains of *C. tetani*, but it may cause necrosis, creating better conditions for multiplication of *C. tetani*. Strains not producing tetanus toxin are avirulent, and antibodies to the toxin are protective.[4]

Some species are naturally resistant to tetanus, although antitoxic antibodies are usually not demonstrated in their tissues.[287] Acquired resistance to tetanus is based upon circulating antitoxin, and among susceptible species, widespread vaccination has dramatically lessened the impact of tetanus on animal production. Passive immunity acquired by neonates from the dam (especially in the case of foals and mares) protects for 2–3 months, after which active immunization with toxoid can begin. Boosters, given at 1–5-year intervals, are commonly recommended.

Therapy is directed toward neutralization of preformed toxin, elimination of toxin production, and symptomatic treatment of the patient. Tetanus antitoxin provides immediate passive protection, but is much more effective when used prophylactically than therapeutically.[288] Attention to the wound, if any is apparent, and administration of large doses of penicillin can be effective in halting production of toxin. Muscle relaxants and sedatives are usually useful, and artificial feeding may be necessary.

XIII. REFERENCES

1. Smith L. and Williams B.L. (1984) *The Pathogenic Anaerobic Bacteria*, 3rd edn. CC Thomas, Springfield, IL.
2. Carter G.R. and Chengappa M.M. (1991) *Essentials of Veterinary Bacteriology and Mycology*, 4th edn. Lea and Febiger, Philadelphia, PA.
3. Niilo L. (1980) *Clostridium perfringens* in animal disease: a review of current knowledge. *Can Vet J* **21**: 141–148.
4. Timoney J.F., Gillespie J.H., Scott F.W. and Barlough J.E. (1988) *Hagan and Bruner's Microbiology and Infectious Diseases of Domestic Animals*, 8th edn. Comstock Publishing Associates, Ithaca, NY.
5. McGowan G., Moulton J.E. and Rood S.E. (1958) Lamb losses associated with *Clostridium perfringens* type A. *J Am Vet Med Assoc* **133**: 219–221.
6. Russell W.C. (1970) Type A enterotoxemia in captive wild goats. *J Am Vet Med Assoc* **157**: 643–646.

7. Fleming S. (1985) Enterotoxemia in neonatal calves. *Vet Clin North Am Food Anim Pract* **1**: 509–514.
8. Rose A.L. and Edgar O. (1936) Enterotoxaemic jaundice of cattle and sheep. A preliminary report on the aetiology of the disease. *Aust Vet J* **12**: 212–220.
9. Roeder B.L., Chengappa M.M., Nagaraja T.G., Avery T.B. and Kennedy G.A. (1988) Experimental induction of abomasal tympany, abomasitis, and abomasal ulceration by intraruminal inoculation of *Clostridium perfringens* type A in neonatal calves. *Am J Vet Res* **49**: 201–207.
10. Songer J. Glenn, Hunter S.E.C. and Chen F. (1993) PCR method for detection of alpha, beta, and epsilon toxin genes of *Clostridium perfringens. 74th Annual Conference of Research Workers in Animal Diseases*, in Chicago, Il. (Abstract 71), p. 15.
11. Bains B.S. (1968) Necrotic enteritis of chickens. *Aust Vet J* **44**: 40.
12. Köhler B., Marx G., Kolbach S. and Böttcher E. (1974) Untersuchungen zur nekrotischen enteritis der Hühner. 1. Diagnostik und Bekampfung. *Monatsh Veterinaermed* **29**: 380–384.
13. Köhler B., Kolbach S. and Meine J. (1974) Untersuchungen zur nekrotischen enteritis der Hühner. 2. Microbiologische Aspekte. *Monatsh Veterinaermed* **29**: 385–391.
14. Nairn M.E. and Bamford V.W. (1967) Necrotic enteritis of broiler chickens in western Australia. *Aust Vet J* **43**: 49–54.
15. Al-Sheikhly F. and Truscott R.B. (1977) The interaction of *Clostridium perfringens* and its toxins in the production of necrotic enteritis of chickens. *Avian Dis* **21**: 256–263.
16. Long J.R., Pettit J.R. and Barnum D.A. (1974) Nectoric enteritis in chickens. II Pathology and proposed pathogenesis. *Can J Comp Med* **38**: 467–474.
17. Niilo L. (1978). Enterotoxigenic *Clostridium perfringens* type A isolated from intestinal contents of cattle, sheep, and chickens. *Can J Comp Med* **42**: 357–363.
18. Truscott R.B. and Al-Sheikhly F. (1977) Reproduction and treatment of necrotic enteritis in broilers. *Am J Vet Res* **38**: 857–861.
19. Stuve G., Hofshagen M. and Holt G. (1992) Necrotizing lesions in the intestine, gizzard, and liver in captive capercaillies (*Tetrao urogallus*) associated with *Clostridium perfringens. J Wildlife Dis* **28**: 598–602.
20. Gazdzinski P. and Julian R.J. (1992) Necrotic enteritis in turkeys. *Avian Dis* **36**: 792–798.
21. Al-Sheikhly F. and Al-Saieg A. (1980) Role of coccidia in the occurrence of necrotic enteritis of chickens. *Avian Dis* **24**: 324–333.
22. Helmboldt C.F. and Bryant E.S. (1971) The pathology of necrotic enteritis in domestic fowl. *Avian Dis* **15**: 775–780.
23. Shane S.M., Gyimah J.E., Harrington K.S. and Snider T.G. (1985) Etiology and pathogenesis of necrotic enteritis. *Vet Res Comm* **9**: 269–287.
24. Kaldhusdal M. and Hofshagen M. (1992) Barley inclusion and avoparcin supplementation in broiler diets. 2. Clinical, pathological, and bacteriological findings in a mild form of necrotic enteritis. *Poult Sci* **71**: 1145–1153.
25. Long J.R. (1973) Necrotic enteritis in broiler chickens. I. A review of the literature and the prevalence for the disease in Ontario. *Can J Comp Med* **37**: 302–308.
26. Murakami S., Okazaki Y., Kazama T., Suzuki T., Iwabuchi I. and Kirioka K. (1989) A dual infection of *Clostridium perfringens* and *Escherichia coli* in broiler chicks. *J Japan Vet Med Assoc* **42**: 405–409.
27. Hofshagen M. and Stenwig H. (1992) Toxin production by *Clostridium perfringens* isolated from broiler chickens and capercaillies (*Tetrao urogallus*) with and without necrotizing enteritis. *Avian Dis* **36**: 837–843.
28. Al-Sheikhly F. and Truscott R.B. (1977) The pathology of necrotic enteritis of chickens following infusion of crude toxins of *Clostridium perfringens* into the duodenum. *Avian Dis* **21**: 241–255.
29. Al-Sheikhly F. and Truscott R.B. (1977) The pathology of necrotic enteritis of chickens following infusion of broth cultures of *Clostridium perfringens* into the duodenum. *Avian Dis* **21**: 230–240.

30. Fukata T., Hadate Y., Baba E. and Arakawa A. (1991) Influence of bacteria on *Clostridium perfringens* infections in young chickens. *Avians Dis* **35**: 224–247.
31. Nadas U.G., Berkoglu A. and Alp R. (1991) Detection of *Clostridium perfringens* alpha antitoxin in the serum of healthy fowls and determination of its relationship with enterotoxaemia. *Pendik Hayvan Hastaliklari Merkez Arastirma Enstitusu Dergisi* **22**: 94–106.
32. Wierup M. and DiPietro J.A. (1981) Bacteriologic examination of equine fecal flora as a diagnostic tool for equine intestinal clostridiosis *Clostridium perfringens. Am J Vet Res* **42**: 2167–2169.
33. Wierup M. (1977) Equine intestinal clostridiosis. An acute disease in horses associated with high intestinal counts of *Clostridium perfringens* type A. *Acta Vet Scand Suppl* **62**: 1–182.
34. Radostits O.M., Blood D.C. and Gay C.C. (1994) *Veterinary Medicine*, 8th edn. Baillière Tindall, London.
35. Drolet R., Higgins R. and Cécyre A. (1990) Necrohemorrhagic enterocolitis caused by *Clostridium perfringens* type C in a foal. *Can Vet J* **31**: 449–450.
36. Ochoa R. and Kern S.R. (1980) The effects of *Clostridium perfringens* type A enterotoxin in Shetland ponies – clinical, morphologic and clinicopathologic changes. *Vet Pathol* **17**: 738–747.
37. Collins J.E., Bergeland M.E., Bouley D., Ducommun A.L., Francis D.H. and Yeske P. (1989) Diarrhea associated with *Clostridium perfringens* type A enterotoxin in neonatal pigs. *J Vet Diagn Invest* **1**: 351–353.
38. Johannsen U., Arnold P., Köhler B. and Selbitz H.J. (1993) Studies into experimental *Clostridium perfringens* type A enterotoxaemia of suckled piglets: Experimental provocation of the disease by *Clostridium perfringens* type A intoxication and infection. *Monatsh für Veterinaermed* **48**: 129–136.
39. Johannsen U., Menger S., Arnold P., Köhler B. and Selbitz H.J. (1993) Experimental *Clostridium perfringens* type A enterotoxaemia in unweaned piglets. II. Light- and electron-microscopic investigations on the pathology and pathogenesis of experimental *C. perfringens* type A infection. *Monatsh für Veterinarmed* **48**: 299–306.
40. Johannsen U., Menger S., Arnold P., Köhler B. and Selbitz H.J. (1993) Experimental *Clostridium perfringens* type A enterotoxaemia in unweaned piglets. *Monatsh für Veterinarmed* **48**: 267–273.
41. Nabuurs M.J.A., Haagsma J., van der Molen E.J. and van der Heijden P.J. (1983) Diarrhea in one to three week-old piglets associated with *Clostridium perfringens* type A. *Ann Recherch Veterinair* **14**: 408–411.
42. Olubunmi P.A. and Taylor D.J. (1985) *Clostridium perfringens* type A in enteric diseases of pigs. *Trop Vet* **3**: 28–33.
43. Popoff M.R. and Jestin A. (1985) Enteropathogenicity of purified *Clostridium perfringens* enterotoxin in the pig. *Am J Vet Res* **46**: 2147–2148.
44. Jestin A., Popoff M.R. and Mahe S. (1985) Epizootiologic investigations of a diarrheic syndrome in fattening pigs. *Am J Vet Res* **46**: 2149–2151.
45. Estrada-Correa A.E. (1986) *Studies of* Clostridium perfringens *Type A Enteritis in the Pig.* PhD Thesis. Glasgow University, United Kingdom.
46. Prescott J.F., Johnson J.A., Patterson J.M. and Bulmer W.S. (1978) Haemorrhagic gastroenteritis in the dog associated with *Clostridium welchii. Vet Rec* **103**: 116–117.
47. Burrows C.F. (1977) Canine hemorrhagic gastroenteritis. *J Am An Hosp Assoc* **13**: 451–458.
48. Carman R.J. and Lewis J.C.M. (1983) Recurrent diarrhoea in a dog associated with *Clostridium perfringens* type A. *Vet Rec* **112**: 342–343.
49. Kruth S.A., Prescott J.F., Welch M.K. and Brodsky M.H. (1989) Nosocomial diarrhea associated with enterotoxigenic *Clostridium perfringens* infection in dogs. *J Am Vet Med Assoc* **195**: 331–334.
50. Daube G., China B., Simon P., Hvala K. and Mainil J. (1994) Typing of *Clostridium perfringens* by *in vitro* amplification of toxin genes. *J Appl Bacteriol* **77**: 650–655.
51. Daube G., Simon P., Limbourg B., Renier K. and Kaeckenbeeck A. (1994) Molecular typing of *Clostridium perfringens. Ann Med Veterinair* **138**: 183–191.

52. Niilo L. (1971) Mechanism of action of the enteropathogenic factor of *Clostridium perfringens* type A. *Infect Immun* **3**: 100–106.

53. Niilo L. (1973) Effect on calves of the intravenous injection of the enterotoxin of *Clostridium welchii* type A. *J Comp Pathol* **83**: 265–269.

54. Stevens D.L., Troyer B.E., Merrick D.T., Mitten J.E. and Olson R.D. (1988) Lethal effects and cardiovascular effects of purified α and θ toxins. *J Inf Dis* **157**: 272–279.

55. Ginter A., Williamson E.D., Dessy F., Coppe P., Fearn A. and Titball R.W. (1995) Comparison of the α-toxin produced by bovine enteric and gas gangrene strains of *C. perfringens International Conference on the Molecular Genetics and Pathogenesis of the Clostridia, 1995.* (Abstract A17), p. 19. American Society for Microbiology, Washington, D.C..

56. Tunnicliff E.A. (1933) A strain of *Clostridium welchii* producing fatal dysentery in lambs. *J Inf Dis* **52**: 407–412.

57. Dalling T. (1926) Lamb dysentery. *J Comp Pathol* **39**: 148–163.

58. Frank F.W. (1956) *Clostridium perfringens* type B from enterotoxemia in young ruminants. *Am J Vet Res* **17**: 492–494.

59. Roberts R.S. (1958) Clostridial diseases. In Stableforth A.W. and Galloway I.A. (eds) *Infectious Diseases of Animals, Diseases Due to Bacteria,* Vol. 1, pp. 160–228. Academic Press, New York.

60. Stubbings D.P. (1990) *Clostridium perfringens* enterotoxaemia in two young horses. *Vet Rec* **127**: 431.

61. Jansen B.C. (1961) The beta toxin of *Clostridium welchii* type B, Wilsdon, in relation to the production of a vaccine against lamb dysentery. *Onderstepoort J Vet Res* **28**: 495–549.

62. Mackinnon J.D. (1989) Enterotoxaemia caused by *Clostridium perfringens* type C. *Pig Vet J* **22**: 119–125.

63. Niilo L. (1988) *Clostridium perfringens* type C enterotoxemia. *Can Vet J* **29**: 658–664.

64. Arbuckle J.B.R. (1972) The attachment of *Clostridium welchii* (*Cl. perfringens*) type C to intestinal villi of pigs. *J Pathol* **106**: 65–72.

65. Field H.I. and Gibson E.A. (1955) Studies on piglet mortality. 2. *Clostridium welchii* infection. *Vet Rec* **67**: 31–34.

66. Ohnuna Y., Kondo H., Saino H., Taguchi M., Ohno A. and Matsuda T. (1992) Necrotic enteritis due to *Clostridium perfringens* type C in newborn piglets. *J Japan Vet Med Assoc* **45**: 738–741.

67. Okazaki Y., Inage M., Iwabuchi I., Suzuki T., Okada N., Watanabe K. and Yamamoto T. (1993) Mass outbreak of necrotic enteritis in newborn piglets caused by *Clostridium perfringens* type C on a pig farm. *J Japan Vet Med Assoc* **46**: 214–217.

68. Vanderkop M.A. and Sutmoller H. (1993) Porcine clostridial enteritis. *Can Vet J* **34**: 246–247.

69. Møller K. and Ahrens P. (1996) Comparison of toxicity neutralization–, ELISA– and PCR tests for typing of *Clostridium perfringens* and detection of the enterotoxin gene by PCR. *Anaerobe* **2**: 103–110.

70. Griner L.A. and Bracken F.K. (1953) *Clostridium perfringens* (type C) in acute hemorrhagic enteritis of calves. *J Am Vet Med Assoc* **122**: 99–102.

71. Niilo L., Harries W.N. and Jones G.A. (1974) *Clostridium perfringens* type C in hemorrhagic enterotoxemia of neonatal calves in Alberta. *Can Vet J* **15**: 224–226.

72. Griner L.A. and Johnson H.W. (1954) *Clostridium perfringens* type C in hemorrhagic enterotoxemia of lambs. *J Am Vet Med Assoc* **125**: 125–127.

73. Nakabayashi D., Ogino H., Watanabe T., Nabeya M., Murayama J. and Ishikawa M. (1990) Serological survey of *Clostridium perfringens* type A infection in cattle using an indirect haemagglutination test. *J Vet Med Japan* **43**: 593–597.

74. Tominaga K., Takeya G. and Okada K. (1984) A first case report in Yamaguchi prefecture of the outbreak of haemorrhagic enteritis necroticans of dairy cattle caused by *Clostridium perfringens* type A. *Yamaguchi J Vet Med* **11**: 71–76.

75. Sterne M. and Thomson A. (1963) The isolation and identification of clostridia from pathological conditions of animals. *Bull Off Int Epizoot* **59**: 1487–1489.

76. Howard-Martin M., Morton R.J., Qualls C.W. Jr and MacAllister C.J. (1986) *Clostridium perfringens* type C enterotoxemia in a newborn foal. *J Am Vet Med Assoc* **189**: 564–565.

77. Pearson E.G., Hedstrom O.R., Sonn R. and Wedam J. (1986) Hemorrhagic enteritis caused by *Clostridium perfringens* type C in a foal. *J Am Vet Med Assoc* **188**: 1309–1310.

78. Sims L.D., Tzipori S., Hazard G.H. and Carroll C.L. (1985) Haemorrhagic necrotizing enteritis in foals associated with *Clostridium perfringens. Aust Vet J* **62**: 194–196.

79. Argenti L., Coiro R. and Ciorba A. (1987) Haemorrhagic enteritis in dogs associated with the presence of *Clostridium perfringens* type C. *Summa* **4**: 279–281.

80. Lage M., Jose-Galo A., de Barros-Amado A.A. and Almeida I. (1992) Hemorrhagic gastroenteritis in a dog with isolation of *Clostridium perfringens. Repos Trabalhos Laboratorio Nacional Investigacao Veterinaria* **15**: 45–49.

81. Turk J., Fales W.H., Miller M., Pace L., Fischer J., Johnson G., Kreeger J., Turnquist S., Pittman L., Rottinghaus A. and Gosser H. (1992) Enteric *Clostridium perfringens* infection associated with parvoviral enteritis in dogs: 74 cases (1987–1990). *J Am Vet Med Assoc* **200**: 991–994.

82. Eason R.J. and van Rij R. (1984) Pigless pigbel: enteritis necroticans in the Solomon Islands. *Papua New Guinea Med J* **27**: 42–44.

83. Niilo L. (1993) *Clostridium perfringens.* In Gyles C.L. and Thoen C.O. (eds) *Pathogenesis of Bacterial Infections in Animals*, pp. 114–123, 2nd edn. Iowa State University Press, Ames, IA.

84. Johannsen U., Menger S., Erwerth W. and Köhler B. (1986) *Clostridium perfringens* type C enterotoxemia (necrotizing enteritis) of suckling piglets. 3. Light and electron microscopic investigations on the pathology and pathogenesis of experimental *Clostridium perfringens* type C infection. *Arch Exp Veterinarmed* **40**: 895–909.

85. Johannsen U., Menger S., Erwerth W. and Köhler B. (1986) *Clostridium perfringens* type C enterotoxemia (necrotizing enteritis) of suckling piglets. 2. Light and electron microscopic studies on the pathology and pathogenesis of experimental *Clostridium perfringens* type C intoxication. *Arch Exp Veterinarmed* **40**: 881–894.

86. Johannsen U., Erwerth W., Kunz G. and Köhler B. (1986) *Clostridium perfringens* type C enterotoxemia (necrotizing enteritis) of suckling piglets. 1. Attempts at experimental induction of disease by *Clostridium perfringens* type C intoxication and infection. *Arch Exp Veterinarmed* **40**: 811–825.

87. Kubo M. and Watase H. (1985) Electron microscopy of *Clostridium perfringens* in the intestine of neonatal pigs with necrotic enteritis. *Japan J Vet Sci* **47**: 497–501.

88. Brooks M.E., Sterne M. and Warrack G.H. (1957) A reassessment of the criteria used for type differentiation of *Clostridium perfringens. J. Pathol Bacteriol* **74**: 185–195.

89. El-Idrissi A.H., Ward G.E., Johnson D.W., Benkirane A. and Fassi-Fehri M.M. (1992) Bacteriological investigation of sudden sheep mortality in Morocco. *Prev Vet Med* **12**: 35–46.

90. Bergeland M.E. (1981) Clostridial infections. In Leman A.D., Glock R.D., Mengeling W.L., Penny R.H.C., Scholl E. and Straw B. (eds) *Diseases of Swine*, 5th edn, pp. 418–431. Iowa State University Press, Ames, IA.

91. Popoff M.R. (1984) Bacteriological examination in enterotoxaemia of sheep and lamb. *Vet Rec* **114**: 324.

92. Bullen J.J. (1970) Role of toxins in host-parasite relationships. In Ajl S.J., Kadis S. and Montie T.C. (eds) *Microbial Toxins*, Vol. 1, pp. 233–276. Academic Press, New York.

93. Jubb K.V.F., Kennedy P. and Palmer N. (1993) *Pathology of Domestic Animals*, vol. 2, 4th edn. Academic Press, London.

94. Gardner D.E. (1973) Pathology of *Clostridium welchii* type D enterotoxaemia. 3. Basis of the hyperglycaemic response. *J Comp Pathol* **83**: 525–529.

95. Buxton D., MacLeod N.S.M. and Nicolson T.B. (1981) Focal symmetrical encephalomalacia in young cattle, caused by *Clostridium welchii. Vet Rec* **108**: 459.

96. Hartley W.J. (1956) A focal encephalomalacia of lambs. *NZ Vet J* **4**: 129–135.
97. Griner L.A., Aichelman W.W. and Brown G.D. (1956) *Clostridium perfringens* type D (Epsilon) enterotoxemia in Brown Swiss dairy calves. *J Am Vet Med Assoc* **129**: 375–376.
98. Munday B., Mason R.W. and Cumming R. (1973) Observations on diseases of the central nervous system of cattle in Tasmania. *Aust Vet J* **49**: 451–455.
99. Oxer D.T. (1956) Enterotoxaemia in goats. *Aust Vet J* **32**: 62–66.
100. von Rotz A., Corboz L. and Waldvogel A. (1984) *Clostridium perfringens* type D enterotoxaemia of goats in Switzerland. Pathological and bacteriological investigations. *Schweiz Arch Tierheilk* **126**: 359–364.
101. Blackwell T.E. and Butler D.G. (1992) Clinical signs, treatment, and postmortem lesions in dairy goats with enterotoxemia: 13 cases (1979–1982). *J Am Vet Med Assoc* **200**: 214–217.
102. Blackwell T.E., Butler D.G., Prescott J.F. and Wilcock B.P. (1991) Differences in signs and lesions in sheep and goats with enterotoxemia induced by intraduodenal infusion of *Clostridium perfringens* type D. *Am J Vet Res* **52**: 1147–1152.
103. Baskerville M., Wood M. and Seamer J.H. (1980) *Clostridium perfringens* type E enterotoxaemia in rabbits. *Vet Rec* **107**: 18–19.
104. Hart B. and Hooper P.T. (1967) Enterotoxemia of calves due to *Clostridium welchii* type E. *Aust Vet J* **43**: 360–363.
105. Estrada-Correa A.E. and Taylor D.J. (1989) Porcine *Clostridium perfringens* type A spores, enterotoxin, and antibody to enterotoxin. *Vet Rec* **124**: 606–610.
106. van Baelen D. and Devriese L.A. (1987) Presence of *Clostridium perfringens* enterotoxin in intestinal samples from farm animals with diarrhoea of unknown origin. *Zentralbl Veterinarmed Ser B* **34**: 713–716.
107. van Damme-Jongsten M., Haagsma J. and Notermans S. (1990) Testing strains of *Clostridium perfringens* type A isolated from diarrhoeic piglets for the presence of the enterotoxin gene. *Vet Rec* **126**: 191–192.
108. van Damme-Jongsten M., Rodhouse J., Gilbert R. and Notermans S. (1990) Synthetic DNA probes for detection of enterotoxigenic *Clostridium perfringens* strains isolated from outbreaks of food poisoning. *J Clin Microbiol* **28**: 131–133.
109. Tschirdewahn B., Notermans S., Wernars K., and Untermann F. (1992) The presence of enterotoxigenic *Clostridium perfringens* strains in faeces of various animals. *Int J Food Microbiol* **14**: 175–178.
110. Songer J.G. and Meer R.R. (1966) Genotyping of *Clostridium perfringens* by polymerase chain reaction in a useful adjunct to diagnosis of clostridial enteric disease in animals. *Anaerobe* **2**: 197–203.
111. Estrada-Correa A.E. and Taylor D.J. (1984) The studies of immunity to enteric *Clostridium perfringens* type A infections in pigs. *Proc. 8th Int. Pig. Vet. Soc. Congr., 1989.* (Abst. 138.) IPVS.
112. Werdeling F., Amtsberg G. and Tewes S. (1991) Enterotoxin producing strains of *Clostridium perfringens* in faeces of dogs and cats. *Berl Münch Tierarztl Wochenschr* **104**: 228–233.
113. Schiefer H.B. (1981) Equine colitis X, still an enigma. *Can Vet J* **22**: 162–165.
114. Prescott J.F., Staempfli H.R., Barker I.K., Bettoni R. and Delaney K. (1988) A method for reproducing fatal idiopathic colitis (colitis X) in ponies and isolation of a clostridium as a possible agent. *Equine Vet J* **20**: 417–420.
115. Kanoe M., Inoue S., Abe T., Anzai T., Kamada M., Imagawa H. and Kanemaru T. (1990) Isolation of *Clostridium perfringens* from foals. *Microbios* **64**: 153–158.
116. Staempfli H.R., Townsend H. and Prescott J.F. (1991) Prognostic features and clinical presentation of acute idiopathic enterocolitis in horses. *Can Vet J* **32**: 11–17.
117. Niilo L. and Cho H.J. (1985) Clinical and antibody responses to *Clostridium perfringens* type A enterotoxin in experimental sheep and calves. *Can J Comp Med* **49**: 145–148.
118. Schultz R.A., Rock F. and Cue T. (1986) Use of bacitracin methylene disalicylate (BMD) in sow

rations for control of clostridial enteritis in suckling pigs. *Proc. Annu. Mtg. Am. Assoc. Swine. Pract.* pp. 383–384, Minneapolis, MN.

119. Smith H.W. (1972) The antibacterial activity of nitrovin in vitro: the effect of this and other agents against *Clostridium welchii* in the alimentary tract of chickens. *Vet Rec* **90**: 310–312.
120. Stutz M.W., Johnson S.L. and Judith F.R. (1983) Effects of diet and bacitracin on growth, feed efficiency, and populations of *Clostridium perfringens* in the intestine of broiler chicks. *Poult Sci* **62**: 1619–1625.
121. Stutz M.W., Johnson S.L., Judith F.R. and Miller B.M. (1983) *In vitro* and *in vivo* evaluations of the antibiotic efrotomycin. *Poult Sci* **62**: 1612–1618.
122. Ripley P.H. and Gush A.F. (1983) Immunization schedule for the prevention of infectious necrotic enteritis caused by *Clostridium perfringens* type C in piglets. *Vet Rec* **112**: 201–202.
123. Hjerpe C.A. (1990) Clostridial disease vaccines. *Vet Clin North Amer Food Animal Pract* **6**: 222–234.
124. Walker P.D. (1992) Bacterial vaccines: old and new, veterinary and medical. *Vaccine* **10**: 977–990.
125. Kennedy K.K., Norris S.J., Beckenhauer W.H. and White R.G. (1977) Vaccination of cattle and sheep with combined *Clostridium perfringens* type C and type D toxoid. *Am J Vet Res* **38**: 1515–1518.
126. Odendaal M.W., Visser J.J., Bergh N. and Botha W.J.S. (1989) The effect of passive immunization on active immunity against *Clostridium perfringens* type D in lambs. *Onderstepoort J Vet Res* **56**: 251–255.
127. Carter G.R. (1984) *Diagnostic Procedures in Veterinary Microbiology*. C.C. Thomas, Springfield, IL.
128. Berry P.R., Rodhouse J.C., Hughes S., Bartholomew B.A. and Gilbert R.J. (1988) Evaluation of ELISA, RPLA, and Vero cell assays for detecting *Clostridium perfringens* enterotoxin in faecal specimens. *J Clin Pathol* **41**: 458–461.
129. Bartholomew B.A., Stringer M.F., Watson G.N. and Gilbert R.J. (1985) Development and application of an enzyme linked immunosorbent assay for *Clostridium perfringens* type A enterotoxin. *J Clin Pathol* **38**: 222–228.
130. Beh K.J. and Buttery S.H. (1978) Reverse phase passive haemoagglutination and single radial immunodiffusion to detect epsilon antigen of *Clostridium perfringens* type D. *Aust Vet J* **54**: 541–544.
131. El-Idrissi A.H. and Ward G.E. (1992) Development of double sandwich ELISA for *Clostridium perfringens* beta and epsilon toxins. *Vet Microbiol* **31**: 89–99.
132. El-Idrissi A.H. and Ward G.E. (1992) Evaluation of enzyme-linked immunosorbent assay for diagnosis of *Clostridium perfringens* enterotoxemias. *Vet Microbiol* **31**: 389–396.
133. Harmon S.M. and Kautter D.A. (1986) Evaluation of a reversed passive latex agglutination test kit for *Clostridium perfringens* enterotoxin. *J Food Protect* **49**: 523–525.
134. Holdsworth R.J. and Parratt D. (1994) Development of an ELISA assay for *Clostridium perfringens* phospholipase C (alpha toxin). *J Immunoassay* **15**: 293–304.
135 Jackson S.G., Yip-Chuck D.A. and Brodsky M.H. (1986) Evaluation of the diagnostic application of an enzyme immunoassay for *Clostridium perfringens* type A enterotoxin. *Appl Environ Microbiol* **52**: 969–970.
136. Martin P.K. and Naylor R.D. (1994) A latex agglutination test for the qualitative detection of *Clostridium perfringens* epsilon toxin. *Res Vet Sci* **56**: 259–261.
137. Martin, P.K., Naylor R.D. and Sharpe R.T. (1988) Detection of *Clostridium perfringens* beta toxin by enzyme-linked immunosorbent assay. *Res Vet Sci* **44**: 270–271.
138. McClane B.A. and Strouse R.J. (1984) Rapid detection of *Clostridium perfringens* type A enterotoxin by enzyme-linked immunosorbent assay. *J Clin Microbiol* **19**: 112–115.
139. Nagahama M., Kobayashi K., Ochi S. and Sakurai J. (1991) Enzyme-linked immunosorbent assay for rapid detection of toxins from *Clostridium perfringens*. *FEMS Microbiol Lett* **68**: 41–44.

140. Naylor R.D., Martin P.K. and Sharpe R.T. (1987) Detection of *Clostridium perfringens* epsilon toxin by ELISA. *Res Vet Sci* **42**: 255–256.
141. Notermans S., Heuvelman C., Beckers H., and Uemura T. (1984) Evaluation of the ELISA as tool in diagnosing *Clostridium perfringens* enterotoxins. *Zentralbl Bakteriol Mikrobiol Hyg B* **179**: 225–234.
142. Fach P. and Guillou J.P. (1993) Detection by in vitro amplification of the alpha-toxin (phospholipase C) gene from *Clostridium perfringens. J Appl Bacteriol* **74**: 61–66.
143. Kokai-Kun, J.F., Songer J. G., Czeczulin J.R., Chen F. and McClane B.A. (1994) Comparison of western immunoblots and gene detection assays for identification of potentially enterotoxigenic isolates of *Clostridium perfringens. J Clin Microbiol* **32**: 2533–2539.
144. Møller K., Ahrens P. and Meyling A. (1995) Comparison of toxicity neutralization, ELISA, and PCR tests for typing of *Clostridium perfringens* and for detection of the enterotoxin gene. *Int. Conf on the Molecular Genetics and Pathogenesis of the Clostridia, 1995.* (Abst. B16), p. 23. American Society for Microbiology, Washington, D.C..
145. Songer J.G. (1995) Diagnosis of clostridial infections of domestic animals. Int. Conf. on the Molecular Genetics and Pathogenesis of the Clostridia, 1995. p. 9. American Society for Microbiology, Washington, D.C.
146. Arson H.E. and Welch A. (1993) *In vitro* and *in vivo* characterization of resistance to colonization with *Clostridium difficile. J Med Microbiol* **38**: 103–108.
147. Rifkin G.D. and Fekety J.D., Jr. (1985) Antibiotic-induced enterocolitis in animals. In Pfeiffer C.J. (ed.) *Animal Models for Intestinal Disease*, pp. 123–133. CRC Press, Boca Raton, FL.
148. Libby J.M., Jortner B.S. and Wilkins T.D. (1982) Effects of the two toxins of *Clostridium difficile* in antibiotic-associated cecitis in hamsters. *Infect Immun* **36**: 822–829.
149. Lowe B.R., Fos J.G. and Bartlett J.G. (1980) *Clostridium difficile*-associated cecitis in guinea pigs exposed to penicillin. *Am J Vet Res* **41**: 1277–1279.
150. Orchard J.C., Fekety R. and Smith J.R. (1983) Antibiotic-associated colitis due to *Clostridium difficile* in a Kodiak bear. *Am J Vet Res* **44**: 1547–1548.
151. Hines R.S. and Dickerson S. (1993) Pseudomembranous enteritis associated with ciprofloxacin and *Clostridium difficile* in a penguin (*Eudyptes chrysolophus*). *J Zoo Wildlife Med* **24**: 553–556.
152. Frazier K.S., Herron, A.J., Hines M.E. II, Gaskin J.M. and Altman J.M. (1993) Diagnosis and enterotoxemia due to *Clostridium difficile* in captive ostriches (*Struthio camelus*). *J Vet Diagn Invest* **5**: 623–625.
153. Rehg J.E. and Shoung L.Y. (1981) *Clostridium difficile* colitis in a rabbit following antibiotic therapy for pasteurellosis. *J Am Vet Med Assoc* **179**: 1296.
154. Debard J., Dubos F., Martinet L. and Ducluzeau R. (1979) Experimental reproduction of neonatal diarrhea in young gnotobiotic hares simultaneously associated with *Clostridium difficile* and other *Clostridium* strains. *Infect Immun* **24**: 7–11.
155. Traub-Dargatz J.L. and Jones R.L. (1993) Clostridia-associated enterocolitis in adult horses and foals. *Vet Clin North Am Equine Pract* **9**: 411–421.
156. Beier R., Amtsberg G. and Peters M. (1994) Bacteriological investigation of the occurrence and significance of *Clostridium difficile* in horses. *Pferdeheilkunde* **10**: 3–7.
157. Jones R.L., Shideler R.K. and Cockerell G.L. (1988) Association of *Clostridium difficile* with foal diarrhea. *Proc. 5th Int. Conf. Equine Infect. Dis*, Lexington, KY. pp. 71–74. The University Press of Kentucky, Lexington, KY.
158. Jones R.L., Adney W.S., Alexander A.F., Shideler R.K. and Traub-Dargatz J.L. (1988) Hemorrhagic necrotizing enterocolitis associated with *Clostridium difficile* infection in four foals. *J Am Vet Med Assoc* **193**: 76–79.
159. Jones R.L., Adney W.S. and Shideler R.K. (1987) Isolation of *Clostridium difficile* and detection of cytotoxin in the feces of diarrheic foals in the absence of antimicrobial treatment. *J Clin Microbiol* **25**: 1225–1227.

160. Perrin J., Buogo C., Gallusser A., Burnens A.P. and Nicolet J. (1993) Intestinal carriage of *Clostridium difficile* in neonate dogs. *J Vet Med Ser B* **40**: 222–226.

161. Hafiz S. and Oakley C.L. (1976) *Clostridium difficile*; isolation and characteristics. *J Med Microbiol* **9**: 129–136.

162. Jones M.A. and Hunter D. (1983) Isolation of *Clostridium difficile* from pigs. *Vet Rec* **112**: 253.

163. O'Neill G., Adams J.E., Bowman R.A. and Riley T.V. (1993) A molecular characterization of *Clostridium difficile* isolates from humans, animals and their environments. *Epidemiol Infect* **111**: 257–264.

164. Riley T.V., Adams J.E., O'Neill G.L. and Bowman R.A. (1991) Gastrointestinal carriage of *Clostridium difficile* in cats and dogs attending veterinary clinics. *Epidemiol Infect* **107**: 659–665.

165 Weber A., Kroth P. and Heil G. (1989) Occurrence of *Clostridium difficile* in faeces of dogs and cats. *J Vet Med* **36**: 568–576.

166. Levett P.N. (1986) *Clostridium difficile* in habitats other than the human gastro-intestinal tract. *J Infect* **12**: 253–263.

167. Borriello S.P., Davies H.A. and Carman R.J. (1986) Cellular morphology of *Clostridium spiroforme*. *Vet Microbiol* **11**: 191–195.

168. Borriello S.P. and Carman R.J. (1983) Association of iota-like toxin and *Clostridium spiroforme* with both spontaneous and antibiotic-associated diarrhea and colitis in rabbits. *J Clin Microbiol* **17**: 414–418.

169. Boriello S.P. and Carman R.J. (1988) Other clostridial causes of diarrhea and colitis in man and animals. In Rolfe R.G. and Finegold S.M. (eds), Clostridium difficile: *Its Role in Intestinal Disease*, pp. 65–98. Academic Press, London.

170. Carman R.J. and Borriello S.P. (1984) Infectious nature of *Clostridium spiroforme*-mediated rabbit enterotoxaemia. *Vet Microbiol* **9**: 497–502.

171. Harris I.E. and Portas B.H. (1985) Enterotoxaemia in rabbits caused by *Clostridium spiroforme*. *Aust Vet J* **62**: 342–343.

172. Carman R.J., Boriello S.P. and Waynforth H.B. (1985) Microbiological changes during experimental *Clostridium spiroforme* mediated diarrhoea (iota enterotoxaemia) of rabbits. In Archibald J., Pitchfield J. and Roswell H.C. (eds), *The Contribution of Laboratory Animal Science to the Welfare of Man and Animals – Past, Present and Future, 8th Symposium of ICLAS/CALAS (Vancouver, 1983)*, pp. 471–476. Gustav Fischer Verlag, Stuttgart.

173. Katz L., Lamont J.T., Trier J.S., Sonnenblick E.B., Rothman S.W., Broitman S.A. and Reith S. (1978) Experimental clindamycin-associated colitis in rabbits. *Gastroenterology* **74**: 246.

174. Rai R.B. and Ahlawat S.P.S. (1992) Enteric diseases in rabbits: Aetiopathological studies. *Indian Vet J* **70**: 211–214.

175. Peeters J.E. and Charlier G.J. (1985) The enteritis complex of rabbits kept for meat production. *Vlaams Diergeneeskundig Tijdschrift* **54**: 482–497.

176. Haffar A., Laval A. and Guillou J.P. (1988) *Clostridium spiroforme* enterotoxaemia in adult rabbits. *Point Veterinair* **20**: 99–102.

177. Peeters J.E., Orsenigo R., Maertens L., Gallazzi D. and Colin M. (1993) Influence of two isoenergetic diets (starch vs fat) on experimental colibacillosis (EPEC) and iota-enterotoxaemia in early weaned rabbits. *World Rabbit Sci* **1**: 53–66.

178. Cato E.P., George W.L. and Finegold S.M. (1986) Genus *Clostridium*. In Sneath P.H.A., Mair N.S., Sharpe M.E. and Holt J.G. (eds), *Bergey's Manual of Systematic Bacteriology*, Vol. 2, pp. 1141–1200. Williams and Wilkins, Baltimore, MD.

179. Popoff M.R. and Boquet P. (1988) *Clostridium spiroforme* toxin is a binary toxin which ADP-ribosylates cellular actin. *Biochem Biophys Res Commun* **152**: 1361–1368.

180. Aktories K. and Wegner A. (1992) Mechanisms of the cytopathic action of actin-ADP-ribosylating toxins. *Mol Microbiol* **6**: 2905–8.

181. Carman R.J. and Borriello S.P. (1983) Laboratory diagnosis of *Clostridium spiroforme*-mediated diarrhoea (iota enterotoxaemia) of rabbits. *Vet Rec* **113**: 184–185.
182. Sterne M. and Batty I. (1975) *Pathogenic Clostridia*. Butterworths, London.
183. Yonushonis W.R., Carman R.J. and Sims R.E. (1987) Diagnosis of spontaneous *Clostridium spiroforme* iota enterotoxemia in a barrier rabbit breeding colony. *Lab Anim Sci* **37**: 69–71.
184. Carman R.J. and Wilkins T.D. (1991) In vitro susceptibility of rabbit strains of *Clostridium spiroforme* to antimicrobial agents. *Vet Microbiol* **28**: 391–397.
185. Ellis T.M., Gregory A.R. and Logue G.D. (1991) Evaluation of a toxoid for protection of rabbits against enterotoxaemia experimentally induced by trypsin-activated supernatant of *Clostridium spiroforme*. *Vet Microbiol* **28**: 93–102.
186. Berkhoff G.A. (1985) *Clostridium colinum* sp. nov., nom. rev., the causative agent of ulcerative enteritis (quail disease) in quail, chickens, and pheasants. *Int J Syst Bacteriol* **35**: 155–159.
187. Berkhoff H.A. (1991) Ulcerative enteritis (quail disease). In Calnek B.W., Barnes H.J., Beard C.W., Reid W.M. and Yoder H.W. Jr. (eds), *Diseases of Poultry*, 9th edn, pp. 258–264. Iowa State University Press, Ames, IA.
188. Peckham M.C. (1984) Ulcerative enteritis (quail disease) [*Clostridium colinum* infection]. In Hofstad M.S. (ed.) *Diseases of Poultry*, 8th edn, pp. 242–250. Iowa State University Press, Ames, IA.
189. Bass C.C. (1939) Observation on the specific cause and nature of 'quail disease' or ulcerative enteritis in quail. *Proc Soc Exp Biol Med* **42**: 377–380.
190. Davis R.B. (1973) Ulcerative enteritis in chickens: Coccidiosis and stress as predisposing factors. *Poult Sci* **52**: 1283–1287.
191. Berkhoff G.A., Campbell S.G. and Naylor H.B. (1974) Etiology and pathogenesis of ulcerative enteritis ('quail disease'): Isolation of the causative anaerobe. *Avian Dis* **18**: 186–194.
192. Berkhoff G.A., Campbell S.G., Naylor H.B. and Smith L. (1974) Etiology and pathogenesis of ulcerative enteritis ('quail disease'): characterization of the causative anaerobe. *Avian Dis* **18**: 195–204.
193. Scanlan C.M. and Hargis B.M. (1991) Bacitracin resistant *Clostridium colinum* associated with ulcerative enteritis in commercially raised Bobwhite quail (*Colinus virginianus*) in Texas. *Poult Sci* **70**(Suppl. 1): 181.
194. Finegold, S.M., Sutter V.L. and Mathisen, G.E. (1983) Normal indigenous intestinal flora, p. 3–31: In D. Hentges (ed.), *Human Intestinal Microflora in Health and Disease*. Academic Press, New York.
195. Kahn C.M. (1924) Anaerobic spore-bearing bacteria of the human intestine in health and in certain diseases. *J Inf Dis* **35**: 423–478.
196. Princewill T.J.T., Agba M.I. and Jemitola S.O. (1984) Isolation of *Clostridium* species from paddock soil. *Bull An Hlth Produc Africa* **32**: 263–268.
197. Princewill, T.J.T. (1965) Sources of clostridial infections in Nigerian livestock and poultry. *Bull An Hlth Produc Africa* **33**: 323–326.
198. Gariev B.G. (1986) Role of freshwater snails in the epidemiology of certain clostridial infections of sheep. *Veterinariya Moscow USSR* **10**: 39.
199. Harwood D.G. (1984) Apparent iatrogenic clostridial myositis in cattle. *Vet Rec* **115**: 412.
200. Mullaney T.P., Brown C.M. and Flint-Taylor R. (1984) Clostridial myositis in horses following intramuscular administration of ivermectin. *Proc Annu Meet Am Assoc Vet Lab Diagn* **27**: 171–177.
201. van Nie G.J. (1984) Liability, prostaglandins and Clostridium. Malignant oedema in a cow after injection of fenprostalene. *Tijdschrift Diergeneeskunde* **109**: 669–671.
202. Gyles C.L. (1993) Histotoxic clostridia. In Gyles C.L. and Thoen C.O. (eds), *Pathogenesis of Bacterial Infections in Animals*, 2nd edn pp. 106–113. Iowa State University Press, Ames, IA.
203. MacLennan J.D. (1962) The histotoxic clostridial infections of man. *Bacteriol Rev* **26**: 177–276.

204. Schamber G.J., Berg I.E. and Molesworth J.R. (1986) Braxy or bradsot-like abomasitis caused by *Clostridium septicum* in a calf. *Can Vet J* **27**: 194.

205. Saunders G. (1986) Diagnosing braxy in calves and lambs. *Vet Med* **81**: 1050–1052.

206. Ellis T.M., Rowe J.B. and Lloyd J.M. (1983) Acute abomasitis due to *Clostridium septicum* infection in experimental sheep. *Aust Vet J* **60**: 308–309.

207. Ballard J., Bryant A., Stevens D. and Tweten R.K. (1992) Purification and characterization of the lethal toxin (alpha-toxin) of *Clostridium septicum. Infect Immun* **60**: 784–790.

208. Ballard J., Sokolov Y., Yuan W.L., Kagan B.L. and Tweten R.K. (1993) Activation and mechanism of *Clostridium septicum* alpha toxin. *Mol Microbiol* **10**: 627–634.

209. Gadalla M.S.A. and Collee J.G. (1968) The relationship of the neuraminidase of *Clostridium septicum* to the haemagglutinin and other soluble products of the organism. *J Pathol Bacteriol* **96**: 169–185.

210. Zenz K.I., Roggentin P. and Schauer R. (1993) Isolation and properties of the natural and the recombinant sialidase from *Clostridium septicum. Glycoconj J* **10**: 50–56.

211. Riddell S.W., Miles B.L., Skitt B.L. and Allen S.D. (1993) Histopathologic effects of culture filtrates of *Clostridium septicum* in rabbit intestinal loops. *Clin Infect Dis 16 Suppl* **4**: S211–213.

212. Morgan K.L., Green D.S., Green N.J. and Hillyer M.H. (1988) The injection site reaction and antibody response in sheep and goats following the use of multivalent clostridial vaccines. In Smith O.B. and Osman H.G. (eds) *Proc Workshop Goat Produc Humid Tropics*, pp. 117–124. University of Ife.

213. Batty I. and Walker P.D. (1963) Fluorescent-labelled clostridial antisera as specific reagents. *Bull Off Int Epizoot* **59**: 1499–1513.

214. Kerry J.B. (1964) A note on the occurrence of *Clostridium chauvoei* in the spleen and livers of normal cattle. *Vet Rec* **76**: 396–397.

215. Shaw I.G. (1958) Black-leg. *Agriculture* **65**: 138–141.

216. Ramachandran S. (1969) Haemolytic activities of *Cl. chauvoei. Indian Vet J* **46**: 754–768.

217. Dixit S.N. and Khera S.S. (1968) Pathogenesis of *Clostridium chauvoei* infection in guinea pigs. *Indian J Exp Biol* **6**: 80–83.

218. Jayaraman M.S., Lal R. and Dhanda M.R. (1962) Toxin production by *Clostridium chauvoei. Indian Vet J* **39**: 481–484.

219. Verpoort J.A., Joubert F.J. and Jansen B.C. (1966) Studies on the soluble antigen and haemolysin of *Clostridium chauvoei* strain 64. *S Afr J Agric Sci* **9**: 153–172.

220. Chankler H.M. and Gulasekhuram J. (1970) An evaluation of characteristics of *Clostridium chauvoei* which possibly indicate a highly protective strain. *Aust J Exp Biol Med Sci* **48**: 187–197.

221. Coleman J.D., Hill J.S., Bray H.T., Armstrong D.A. and Morgan C.O. (1975) Prevention of sudden death caused by *Clostridium sordellii* in feedlot cattle. *Vet Med Small Anim Clin* **70**: 191–195.

222. Arseculäratne S.N., Panabokke R.G. and Wijesundra S. (1969) The toxins responsible for the lesions of *Clostridium sordellii* gas gangrene. *J Med Microbiol* **2**: 37–53.

223 Smith L. (1975) *Clostridium sordellii. The Pathogenic Anaerobic Bacteria*, 2nd edn, pp. 291–298. CC Thomas, Springfield, IL.

224. Smith L.D. (1975) *Clostridium novyi. The Pathogenic Anaerobic Bacteria*, 2nd edn, pp. 257–270. CC Thomas, Springfield, IL.

225. Niilo L., Dorward W.J. and Avery R.J. (1969) The role of *Clostridium novyi* in bovine disease in Alberta. *Can Vet J* **10**: 159–169.

226. Elder J.M. and Miles A.A. (1957) The action of the lethal toxins of gas gangrene clostridia on capillary permeability. *J Pathol Bacteriol* **74**: 133–145.

227. Williams B.M. (1962) Black disease of sheep: Observations on the disease in mid-Wales. *Vet Rec* **74**: 1536–1542.

228. Cotran R.S. (1967) Studies on inflammation. Ultrastructure of the prolonged vascular response induced by *Clostridium oedematiens* toxin. *Lab Invest* **17**: 39–60.

229. Aikat B.K. and Dible, J.H. (1960) The local and general effects of cultures and culture-filtrates of *Clostridium oedematiens, Cl. septicum, Cl. sporogenes*, and *Cl. histolyticum. J Pathol Bacteriol* **79**: 227–241.
230. Rutter J.M. and Collee J.G. (1967) Studies on the soluble antigens of *Clostridium oedematiens* (*Cl. novyi*). *J Med Microbiol* **2**: 395–417.
231. Pemberton, J.R., Matson, R.L., Claus, K.D. and Macheak, M.F. (1971) Changes in plasma enzyme levels of sheep infected with *Clostridium novyi* type B. *Clin Chim Acta* **34**: 431–436.
232. Eklund M.W., Poysky F.T., Peterson M.E. and Meyers J.A. (1976) Relationships of bacteriophages to alpha toxin production in *Clostridium novyi* types A and B. *Infect Immun* **14**: 793–803.
233. Oakley C.L. and Warrack G.H. (1959) The soluble antigens of *Clostridium oedematiens* type D (*Cl. haemolyticum*). *J Pathol Bacteriol* **78**: 543–551.
234. Nakamura S., Kimura I., Yamakawa K. and Nishida S. (1983) Taxonomic relationships among *Clostridium novyi* types A and B, *Clostridium haemolyticum* and *Clostridium botulinum* type C. *J Gen Microbiol* **139**: 1473–1479.
235. Cato E.P., Holdeman L.V. and Moore W.E.C. (1982) Electrophoretic study of *Clostridium* species. *J Clin Microbiol* **15**: 688–702.
236. Safford J.W. and Smith L.D. (1954) A study of the epizootiology of bacillary hemoglobulinuria. *Proc 91st Ann Mtg Am Vet Med Assoc Seattle, WA*. pp. 159–162.
237. Smith L. and Jasmin A.M. (1956) The recovery of *Clostridium hemolyticum* from the livers and kidneys of apparently normal cattle. *J Am Vet Med Assoc* **129**: 68–71.
238. Smith L. (1954) *Introduction to Pathogenic Anaerobes.* University of Chicago Press, Chicago, IL.
239. Olander H.J., Hughes J.P. and Biberstein E.L. (1966) Bacillary hemoglobinuria: Induction by liver biopsy in naturally and experimentally infected animals. *Pathol Vet* **3**: 421–450.
240. van Kampen K.R. and Kennedy P.C. (1969) Experimental bacillary hemoglobinuria. II. Pathogenesis of the hepatic lesion in rabbits. *Pathol Vet* **6**: 59–75.
241. Erwin B.G. (1977) Experimental induction of bacillary hemoglobinuria in cattle. *Am J Vet Res* **38**: 1625–1627.
242. Smith L. (1975) *Clostridium haemolyticum. The Pathogenic Anaerobic Bacteria*, 2nd edn, pp. 271–280. CC Thomas, Springfield, IL.
243. Durakhashan H. and Lauerman C.H. (1981) Some properties of beta toxin produced by *Clostridium hemolyticum* strain IRP135. *Comp Immunol Microbiol Infect Dis* **4**: 307–316.
244. Shallehm G. and Eklund M.W. (1980) Conversion of *Clostridium novyi* type D (*Cl. hemolyticum*) to alpha toxin production by phages of *Cl. novyi* type A. *FEMS Microbiol Lett* **7**: 83–86.
245. Claus K.D. and Macheak M.E. (1965) Nonantigenic nature of *Clostridium haemolyticum* toxoid. *Am J Vet Res* **26**: 353–356.
246. Vawter L.R. and Records E. (1929) Immunization of cattle against bacillary hemoglobinuria. *J Am Vet Med Assoc* **75**: 210–214.
247. Smith L.D. and Sugiyama H. (1988) *Botulism, the Organism, its Toxins, the Disease.* CC Thomas, Springfield, IL.
248. Simpson L.L. (1981) The origin, structure, and pharmacological activity of botulinum toxin. *Pharmacol Rev* **33**: 155–188.
249. Oguma K., Murayama S., Syuto B., Iida H. and Kubo S. (1984) Analysis of antigenicity of *Clostridium botulinum* type C_1 and D toxins by polyclonal and monoclonal antibodies. *Infect Immun* **43**: 584–588.
250. Kozaki S., Kamata Y., Nagai Y., Ogasawara J. and Sakaguchi G. (1986) The use of monoclonal antibodies to analyze the structure of *Clostridium botulinum* type B derivative toxin. *Infect Immun* **52**: 786–791.
251. Rocke T.E. (1993) *Clostridium botulinum.* In Gyles C.L. and Thoen C.O. (eds) *Pathogenesis of Bacterial Infections in Animals*, 2nd edn, pp. 86–96. Iowa State University Press, Ames IA.

252. Ohishi I. (1983) Lethal and vascular permeability activities of botulinum C_2 toxin induced by separate injections of the two toxin components. *Infect Immun* **40**: 336–339.
253. Simpson L.L. (1989) Peripheral actions of the botulinum toxins. In Simpson L.L. (ed.) *Botulinum Neurotoxin and Tetanus Toxin*, pp. 153–178. Academic Press, New York.
254. Wobeser G., Marsden S. and MacFarlane R.J. (1987) Occurrence of toxigenic *Clostridium botulinum* type C in the soil of wetlands in Saskatchewan. *J Wildl Dis* **23**: 67–76.
255. Mason J.H. (1968) *Clostridium botulinum* type D in mud of lakes of the Zululand Game Parks. *J S Afr Med Assoc* **39**: 37–38.
256. Theiler A., Viljoen P.R., Green H.H., du Toit P.J., Meier K. and Robinson E.M. (1927) Lamziekte. In *11th and 12th Annual Reports Direct Vet Educ Res, Union of S Afr*, vol. 2, pp. 819–821.
257. Egyed M.N. (1987) Outbreaks of botulism in ruminants associated with ingestion of feed containing poultry waste. In Eklund M.W. and Dowell V.R. Jr (eds) *Avian Botulism: An International Perspective*, pp. 317–380. CC Thomas, Springfield, IL.
258. Smart J.L., Jones T.O., Clegg F.A. and McMurtry M.J. (1987) Poultry waste associated type C botulism in cattle. *Epidemiol Infec* **98**: 73–79.
259. Allison M.J., Maloy S.E. and Matson R.R. (1976) Inactivation of *Clostridium botulinum* toxin by ruminal microbes from cattle and sheep. *Appl Environ Microbiol* **32**: 685–688.
260. Johnston J. and Whitlock R.H. (1987) Botulism. In Robinson N.B. (ed.), *Current Therapy in Equine Medicine – Two*, pp. 367–370. WB Saunders, Philadelphia, PA.
261. Swerczek T.W. (1980) Toxicoinfectious botulism in foals and adult horses. *J Am Vet Med Assoc* **176**: 217–220.
262. Swerczek T.W. (1980) Experimentally induced toxicoinfectious botulism in horses and foals. *Am J Vet Res* **41**: 348–350.
263. Jensen W.I. and Price J.I. (1987) The global importance of type C botulism in wild birds. In Eklund M.W. and Dowell V.R. Jr. (eds), *Avian Botulism: An International Perspective*, pp. 33–54. CC Thomas, Springfield, IL.
264. Bell F.G., Sciple G.W. and Hubert A.A. (1955) A microenvironment concept of the epizoology of avian botulism. *J Wildl Mgmt* **19**: 352–357.
265. Reed T.M. and Rocke T.B. (1992) The role of avian carcasses in the epizootiology of avian botulism. *Wildl Soc Bull* **20**: 175–182.
266. Dohms I.B. (1987) Laboratory investigation of botulism in poultry. In Eklund M.W. and Dowell V.R. Jr. (eds), *Avian Botulism: An International Perspective*, pp. 296–314. CC Thomas, Springfield, IL.
267. Brand C.J., Schmitt S.M., Duncan, R.M. and Cooley, T.M. (1988) An outbreak of type B botulism among common loons (*Gavia immer*) in Michigan's upper peninsula. *J Wildl Dis* **24**: 471–476.
268. Eklund M.W., Poysky F.T., Peterson M.B., Peck L.W. and Brunson W.D. (1984) Type B botulism in salmonids and conditions contributing to outbreaks. *Aquaculture* **41**: 293–309.
269. Haagsma J. and ter Laak E.A. (1977) Aetiology and epidemiology of botulism in broiler fowls. *Tijdschrift Diergeneeskunde* **102**: 429–439.
270. Miyazaki S. and Sakaguchi G. (1978) Experimental botulism in chickens: the cecum as the site of production and absorption of botulinum toxin. *Jpn J Med Sci Biol* **31**: 1–15.
271. Sakaguchi G. and Kaji R. (1980) Avian botulism. *J Jpn Soc Poult Dis* **16**: 37–49.
272. Huyn S. and Sakaguchi G. (1989) Implication of coprophagy in pathogenesis of chicken botulism. *Jpn J Vet Sci* **51**: 582–586.
273. Jansen B.C., Knoetze P.C. and Visser F. (1976) The antibody response of cattle to *Clostridium botulinum* types C and D toxoids. *Onderstepoort J Vet Res* **43**: 165–173.
274. Ohishi I., Sakaguchi G., Riemann H., Behymer D. and Hurvell B. (1979) Antibodies of *Clostridium botulinum* toxins in free-living birds and mammals. *J Wildl Dis* **15**: 3–9.
275. Appleton G.S. and White P.O. (1959) Field evaluation of *Clostridium botulinum* type C toxoids in mink. *Am J Vet Res* **20**: 166–169.

276. White P.G. and Appleton G.S. (1969) Botulinum toxoid. *J Am Vet Med Assoc* **137**: 652–653.
277. Boroff D.A. and Reilly J.R. (1959) Studies of the toxin of *Clostridium botulinum*. V. Prophylactic immunization of pheasants and ducks against avian botulism. *J Bacteriol* **77**: 142–146.
278. Bennetts H.W. and Hall H.T.B. (1937) The control of toxic paralysis (botulism) in sheep and cattle. *J Dept Agric W Aust* **14**: 381–386.
279. Larsen B.C., Nicholes P.S. and Gebhardt L.P. (1955) Successful immunization of mink with a toxoid against *Clostridium botulinum* type C. *Am J Vet Res* **16**: 573–575.
280. Shoesmith, J.G. and Holland K.T. (1972) The germination of spores of *Clostridium tetani*. *J Gen Microbiol* **70**: 253–261.
281. Wells C.L. and Balish B. (1983) *Clostridium tetani* growth and toxin production in the intestines of germfree rats. *Infect Immun* **41**: 826–828.
282. TenBroeck C. and Bauer J.H. (1922) The tetanus bacillus as an intestinal saprophyte in man. *J Exp Med* **36**: 261–266.
283. Herd R.P. and Riches W.R. (1964) An outbreak of tetanus in cattle. *Aust Vet J* **40**: 356–357.
284. Wallis A.S. (1963) Some observations on the epidemiology of tetanus in cattle. *Vet Rec* **75**: 188–191.
285. Kryzhanovsky G.N. (1981) Pathophysiology. In *Tetanus – Important New Concepts*, Vol. 1, pp. 109–182. Excerpta Medica, Amsterdam.
286. Wellhöner H.H. (1982) Tetanus neurotoxin. In Adrian R.H. (ed.) *Reviews of Physiology, Biochemistry, and Pharmacology*, Vol. 93, p. 18. Springer Verlag, Berlin.
287. Metchnikov E.L. (1898) Recherches sur l'influence de l'organisme sur les toxines. *Ann Inst Pasteur (Paris)* **12**: 81–90.
288. Chang T.W. and Weinstein L. (1957) Effect of cortisone on treatment of tetanus with antitoxin. *Proc Soc Exp Biol Med* **94**: 431–433.

D: Pathogenesis of Infections

Chapter 11

The Pathogenesis of Gas Gangrene

AMY E. BRYANT AND DENNIS L. STEVENS
Veterans Affairs Medical Center, Infectious Diseases Section, 500 West Fort Street, Boise, ID 83702 and University of Washington School of Medicine, Seattle, WA, USA

I. INTRODUCTION

Clostridium perfringens type A is the most common organism isolated from patients with trauma-induced gas gangrene.[1] Trauma provides a low oxidation–reduction potential and a readily accessible supply of amino acids and peptides necessary for bacterial growth.[2] Initially growth occurs within the devitalized, anaerobic milieu. However, acute invasion and destruction of healthy, living tissue rapidly ensues, with margins often advancing several inches per hour despite appropriate antibiotic therapy.[1,3] While clinicians and surgeons have realized for centuries that prompt and thorough surgical removal of affected tissues is essential to survival,[1,3] the pathogenic mechanisms underlying the profound shock and rapidly progressing destruction of viable tissue are only now being elucidated. Although several theories have been proposed to explain these pathologies,[4,5] it is the general consensus that the local and systemic manifestations of gas gangrene are in fact related to the elaboration of potent extracellular protein toxins.[1,6,7]

The Clostridia: Molecular Biology and Pathogenesis
ISBN 0-12-595020-9

II. THE ORGANISM AND ITS ARMAMENT

Clostridium perfringens is a Gram-positive, spore-forming, non-motile, rod-shaped organism commonly found in soil and in the intestines of humans and other animal species. Although classified as an anaerobe, *C. perfringens* is aerotolerant. Under optimal conditions, the generation time for *C. perfringens* can be as little as 8–10 min, and growth is accompanied by abundant gas production.[6] The species produces at least 12 extracellular toxins, of which α, β, ε and θ are considered major lethal toxins. Based on production of these major toxins, the species has been divided into five types, A–E.[6] Of these subgroups, type A causes the majority of human infections, and is the subject of this text.

Only *C. perfringens*, type A produces the α-toxin, a phospholipase C (PLC). This exotoxin has the distinction of being the first bacterial toxin to which an enzymatic activity, i.e., lecithinase, was ascribed. α-Toxin is hemolytic, destroys platelets and leukocytes, and increases capillary permeability, effects which are likely related to its ability to cleave sphingomyelin and the phosphoglycerides of choline, ethanolamine and serine present in eukaryotic cell membranes; it requires divalent cations such as calcium and magnesium for optimal activity.[5] Zinc enhances α-toxin production in culture[8] and is essential for its activity *in vivo*. Histidine residues have been shown to be essential for the coordination of zinc ions.[9] Basak and coworkers recently crystallized α-toxin, providing preliminary X-ray diffraction analysis of the protein.[10] Titball *et al.* determined that the protein is composed of two domains; the *N*-terminal domain possesses the PLC activity while the *C*-terminal domain confers the cytolytic properties (see Chapter 14).[11]

α-Toxin's role as the major lethal factor in *C. perfringens* infections is supported by numerous studies utilizing a variety of approaches. Both active and passive immunization of animals against α-toxin[12,13] or its *C*-terminal fragment[12] are protective in experimental wild-type infections. Similarly, experimental infections established with genetic mutants of *C. perfringens* lacking α-toxin[14] or with strains which produce less α-toxin[15] were markedly less fulminant and mortality was significantly reduced.

The importance of theta toxin in the pathogenesis of gas gangrene has been controversial despite the early knowledge of its hemolytic nature,[16] and its serological and antigenic relationship to the cholesterol-binding, thiol-activated cytolysins (TACs) from *Streptococcus pyogenes*, *Streptococcus pneumoniae*, and *Listeria monocytogenes*.[17–20] Recently, the amino acid and nucleotide sequences of pneumolysin, streptolysin O and θ-toxin have been determined.[21–24] Impressive homology exists among the amino acid sequences of these TAC toxins, particularly in the region containing the cysteine residue near the carboxy terminus, where a highly conserved segment of 12 amino acids is identical for all three toxins.[22,23] Some of these TAC toxins, including θ-toxin, have been shown to facilitate the growth of TAC-toxin-producing organisms within mammalian phagocytic cells.[25] Furthermore, experimental animal studies have demonstrated protective efficacy of several antibody preparations against TAC toxins.[26–28] These studies support a principal role for TACs in the pathogenesis of their respective diseases.

Other exotoxins produced by *C. perfringens* type A include collagenase, hyaluronidase, and two possible leukocidins, the ν-antigen and the 'non-α-δ-θ' hemolysin.[6] Investigations into the roles of collagenase and hyaluronidase in the pathogenesis of gas gangrene are only now being undertaken. Similarly, little is known regarding the roles of the υ antigen or the non-α-δ-θ hemolysin because the former is without lethal or necrotizing activity and the latter is rare in *C. perfringens* type A.[6]

III. PATHOLOGY OF CLOSTRIDIAL MYONECROSIS

Descriptions of gangrenous infections with gas in the tissues date to the Middle Ages. World War I surgeons stationed at the allied casualty clearing stations recognized that wounds which severed main blood vessels were associated with massive or 'group' gangrene, which meant that bacterial proliferation and muscle destruction easily and rapidly ensued throughout all muscle groups which had been cut off from the blood supply.[3] However, it was the careful observations of McNee and Dunn that provided insight into the mechanisms whereby gas gangrene advanced into healthy tissue: Impressed by the natural course of this infection they wrote, '. . . the rapidity of spread of gas gangrene into living muscle, once the disease has begun, is so remarkable as to demand some definite explanation'.[3] Their 1917 report in the *British Medical Journal* attempted to provide that explanation based on the histopathology of serial sections of single muscle bundles from freshly amputated limbs of soldiers with gas gangrene.[3] They noted that few organisms were present at the advancing edge of the infection, that the fibers appeared ischemic, and that a toxic fluid, formed in the gangrenous tissue behind, spread between the fibers, breaking down the tissue and providing an optimal environment for the organisms to proliferate. They also noted that '. . . leucocytes are generally conspicuous by their absence . . .' from involved tissues, but were seen in the interfascial planes. These gross and microscopic observations have stood the test of time.[29] Subsequent experimental histopathology studies demonstrated that the tissue pathologies characteristic of clostridial gas gangrene could be reproduced by injection of either a crude clostridial toxin preparation[30] or recombinant clostridial α-toxin[31] or θ-toxin[28] into healthy muscle.

Thus, the histopathology of gas gangrene is unique, and distinctly different from infections caused by bacteria such as *Staphylococcus aureus*, *Haemophilus influenzae*, or *Streptococcus pneumoniae*. In these cases, a luxuriant pyogenic response occurs at the site of infection. With clostridial myonecrosis, leukocytes, when present, are localized between fascial planes[3] and are often amassed within small vessels near the demarcation between healthy and necrotic tissues.[28] Leukocytes in these areas exhibit altered morphology and karyolysis,[30] suggesting that they are directly affected *in vivo* by the presence of clostridial exotoxins.

IV. THE ANTI-INFLAMMATORY EFFECTS OF α- AND θ-TOXINS

There are several plausible explanations for the lack of a tissue inflammatory response in clostridial gas gangrene. An absence of bacterial- or host-derived chemoattractants seems unlikely to account for the paucity of leukocytes in the tissues, since both an extracellular clostridial product and serum incubated with killed bacilli were potent chemoattractants.[28] The α- and θ-toxins are cytolytic for leukocytes in high concentrations,[32,33] and destruction of any infiltrating phagocytes at the site of active bacterial proliferation and toxin elaboration may contribute to the marked reduction in numbers of inflammatory cells in these areas. However, if this were the only mechanism responsible for the absence of a tissue inflammatory response, one would expect to observe abundant phagocytes approaching the nidus of infection but abruptly halted at the point in the diffusion gradient where toxin concentrations reached cytolytic proportions. Furthermore, leukocyte extravasation from vessels distal to the focus of infection would proceed unimpaired. This scenario, then, does not fully account for the paucity of inflammatory cells in the tissues and the marked leukostasis in adjacent vasculature which is classically observed in both human and experimental cases of gas gangrene.

Another possible mechanism involves the effects of sublytic amounts of α- and θ-toxins on the function and interaction of leukocytes and endothelial cells. Toxin-induced dysregulation of the normal, physiological mechanisms of leukocyte accumulation, adherence and extravasation, which orchestrate the pyogenic responses in other infections, could explain, in part, the leukostasis and anti-inflammatory response characteristic of gas gangrene. Further, these dysregulated events could lead to local and regional ischemia, thereby extending the optimal region for clostridial proliferation.

Successful transmigration of leukocytes through the vessel and to the site of infection is the culmination of a complex cascade of both leukocyte- and endothelial cell (EC)-dependent adherence and activational events (Figure 11.1).[34] Initially, the circulating, unactivated leukocyte is tethered to the

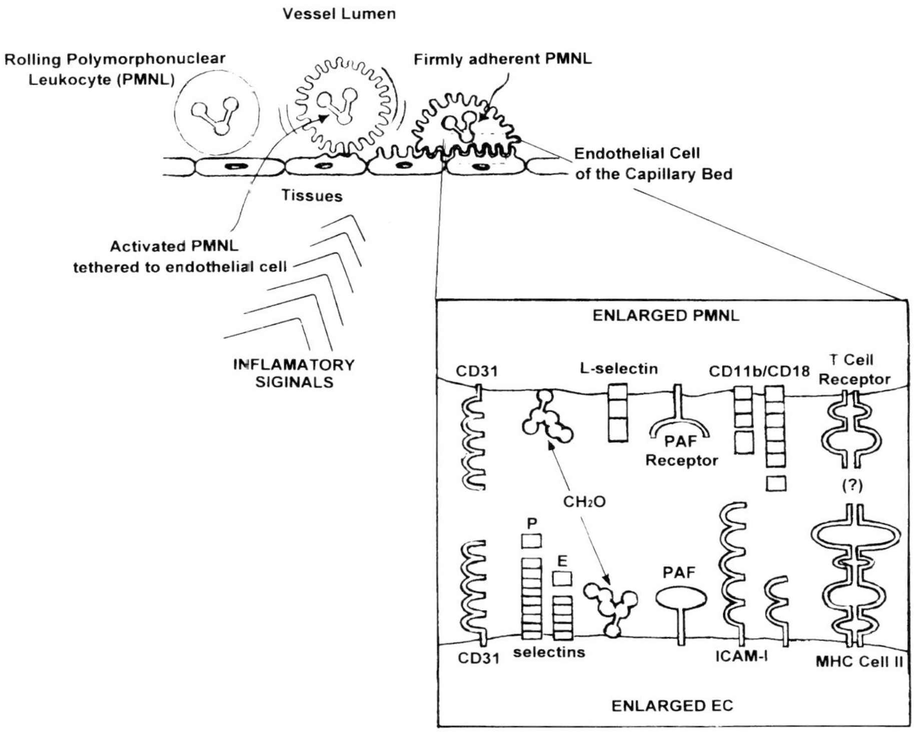

Figure 11.1 Schematic of proinflammatory leukocyte/endothelial cell receptor–ligand interactions. Successful transmigration of leukocytes through the vessel and to the site of infection is the culmination of a complex cascade of both leukocyte- and endothelial cell (EC)-dependent adherence and activational events. Initially, the circulating, unactivated leukocyte is tethered to the activated EC and rolls along the vessel's luminal surface (processes mediated by selectins). Tethering results in activation of the leukocyte by PAF, which in turn causes shedding of L-selectin, and functional upregulation of leukocyte CD11b/CD18 promoting firm adhesion to ICAM-1 constitutively expressed on the EC. Local production of inflammatory mediators (e.g., tumor necrosis factor, TNF; interleukin-1; interleukin-8) augments the inflammatory response by stimulating EC to increase ICAM-1 expression and to transiently express E-selectin. Strongly adherent, activated leukocytes move to EC junctions and emigrate between endothelial cells, aided by platelet–endothelial cell adhesion molecule-1 (PECAM-1/CD31). (Illustration by A.E. Bryant and Michael Wyett.)

activated EC and rolls along the vessel's luminal surface – processes mediated by selectins. Tethering results in juxtacrine activation of the leukocyte by platelet activating factor (PAF), functional upregulation of leukocyte CD11b/CD18, and firm adhesion to intercellular adhesion molecule-1 (ICAM-1), which is constitutively expressed on the EC.[35] Local production of cytokines (e.g., tumor necrosis factor, TNF; interleukin-1, IL-1) augments the inflammatory response by stimulating EC to produce the neutrophil chemoattractant/activator interleukin-8 (IL-8), to increase ICAM-1 expression, and transiently to express endothelial leukocyte adhesion molecule-1 (ELAM-1). Strongly adherent, activated leukocytes move to EC junctions and migrate between endothelial cells, aided by platelet–endothelial cell adhesion molecule-1 (PECAM-1/CD31).

Work done in our laboratory has shown that α-toxin strongly induces the expression of ELAM-1 and ICAM-1 on cultured human umbilical vein endothelial cells.[36] The magnitude of these responses was similar to that reported in other studies using either TNF or IL-1,[37] or bacterial lipopolysaccharide (LPS).[38]

α-Toxin was also a potent inducer of endothelial cell-derived IL-8.[36] As with ELAM-1 expression, the dynamics of PLC-induced IL-8 synethsis were comparable to that induced by TNF or LPS.[39] The local production of IL-8 in gas gangrene could amplify the recruitment of leukocytes and prime them for enhanced respiratory burst activity. Interestingly, Smith *et al.*[40] have shown that neutrophils exposed to 100 nM IL-8 for 30 min lost the ability to transmigrate through an endothelial cell monolayer in response to an IL-8, but not an *N*-formyl-methionyl-leucyl-phenylalanine chemoattractant gradient, a process termed 'homologous desensitization'.[40] Exposure of neutrophils to higher IL-8 concentrations (>1000 nM) resulted in heterologous desensitization.[40] Whether homologous and/or heterologous IL-8-induced desensitization of neutrophils contributes to the lack of phagocyte emigration into tissues infected with *C. perfringens* remains to be determined.

In contrast to α-toxin, θ-toxin caused a modest, but significant, increase in ICAM-1, had no effect on ELAM-1 expression, and did not induce detectable IL-8 synthesis.[36] These findings support the concept that endothelial cell responses are dictated by the nature of the inflammatory signal.[34] However, intramuscular injection of θ-toxin in mice produces marked vascular leukostasis adjacent to the site of toxin injection,[28] suggesting that other mechanisms may be operative *in vivo*. For instance, we have previously shown that θ-toxin caused functional upregulation of leukocyte CD11b/CD18.[28] In addition, θ-toxin stimulated cultured endothelial cells to synthesize PAF,[41] a proadhesive lipid autocoid that contributes to early neutrophil adherence to the endothelium.[42] Further, θ-toxin has potent degranulating activity and could potentially stimulate EC to express P-selectin, a preformed adherence molecule stored within Weibel–Palade bodies.[43]

V. THE ROLES OF α- AND θ-TOXINS IN VASCULAR INJURY AND THE PROGRESSION OF TISSUE DESTRUCTION

Toxin-induced hyperadhesion of leukocytes, with enhanced respiratory burst activity (due to direct toxin action[28,32] or to toxin-induced IL-8[36] or PAF[41] synthesis by host cells), and toxin-induced chemotaxis deficits[28,32] could result in neutrophil-mediated vascular injury. Direct toxin-induced cytopathic effects on EC may also contribute to vascular abnormalities associated with gas gangrene. Over prolonged incubation periods, α-toxin at sublytic concentrations causes EC to undergo profound morphological changes[36] similar to those described following prolonged TNF or IFN-γ exposure.[44] *In vivo*, conversion of EC to this fibroblastoid morphology could contribute to the localized vascular leakage and massive swelling observed clinically with this infection. Similarly, the direct cytotoxicity of θ-toxin could disrupt endothelial integrity and contribute to progressive edema. These cytopathic effects could expose the subendothelium and initiate coagulation and platelet aggregation, thus compounding microvascular occlusion.

Thus, via the mechanisms outlined above, both α- and θ-toxins may cause local, regional and systemic vascular dysfunction. For instance, local absorption of exotoxins within the capillary beds could affect the function of the endothelium lining the post-capillary venules, resulting in impairment of phagocyte delivery to the site of infection. Toxin-induced endothelial dysfunction and microvascular injury could also cause loss of albumin, electrolytes, and water into the interstitial space, resulting in edema. These events, combined with leukostasis within the venule, would increase venous pressures and favor further loss of fluid and protein in the distal capillary bed. Ultimately, a reduced arteriolar flow would impair oxygen delivery, thereby attenuating phagocyte oxidative killing and facilitating anaerobic glycolysis of muscle tissue. The resultant drop in tissue pH, together with reduced oxygen tension, might further decrease the redox potential of viable tissues to a point suitable for growth of this anaerobic bacillus.[6] As infection progresses and additional toxin is absorbed, larger venous channels could be affected, causing regional vascular compromise, increased compartment pressures, and rapid anoxia and necrosis of large muscle groups. Systemic shock, multiorgan failure, and death follow multiplication of *C. perfringens* in tissues, possibly as a result of toxin dispersal via arterial circulation.[45]

VI. THE ROLES OF α- AND θ-TOXINS IN SHOCK

The mechanisms of hemodynamic collapse in patients with Gram-positive bacterial sepsis are poorly understood. *Clostridium perfringens* is a worthy organism for study because shock is a common terminal event in gas gangrene, the organism does not possess endotoxin, and because two well-characterized exotoxins, α-toxin and θ-toxin, may contribute to the dramatic clinical course of patients with this infection.[46]

The pattern of hemodynamic changes provides insights into the mechanisms of septic shock associated with *C. perfringens* infection. A prompt reduction in cardiac index (CI) occurred in rabbits receiving either α-toxin or a crude toxin preparation.[45] Although many physiological mechanisms could contribute to such a reduction, we subsequently demonstrated a direct reduction in myocardial contractility (d*F*/d*t*) in isolated atrial strips bathed with recombinant α (r-α)-toxin.[45] As reflected by the increased mortality in the rabbits receiving r-α-toxin and crude toxin, a greater change in CI was measured in these groups than in those receiving recombinant θ (r-θ)-toxin or normal saline.[45] Similarly, a marked decline in mean arterial pressure (MAP) was observed in rabbits treated with r-α-toxin and crude toxin, although these effects were delayed.[45] Thus, rabbits receiving α toxin-containing preparations maintained MAP in the face of a falling CI by as yet uncharacterized compensatory mechanisms for a brief period before hypotension ultimately occurred. The physiologic mechanisms which maintained MAP did not include significant changes in heart rate or central venous pressure until the terminal stages.

In contrast, r-θ-toxin-treated rabbits demonstrated changes most characteristic of 'warm shock', with profound decreases in peripheral vascular resistance after approximately 1 h of toxin infusion.[7,45] Interestingly, rabbits receiving crude toxin (containing both α- and θ-toxin activity) demonstrated peripheral vascular resistance (PVR) pressures intermediate to the other toxin groups. This observation provides insights into possible antagonistic interactions between r-θ-toxin and r-α-toxin in terms of PVR. Support for such 'antagonism' is derived from experiments performed by Sakurai *et al.*[47,48] who demonstrated that when sublethal doses of purified α-toxin were administered to rats, increases in blood pressure were observed over the course of the experiment.[47] However, in an *ex vivo* model using rat skeletal muscle, purified α-toxin inhibited electrically stimulated muscle contraction after prolonged incubation in a dose-dependent manner.[48] The results of these experiments suggest that α-toxin initially augments cardiac function, counteracting the vasodilatory effect of r-θ-toxin; decreased cardiac output and death occur later.

Hemolysis and/or hypoxemia have been described clinically, yet in experimental studies these factors alone were not responsible for mortality in either the r-α-toxin- or crude toxin-treated rabbits since rabbits which survived r-θ-toxin treatment had degrees of hemolysis and hypoxemia similar to those that did not survive.

These data suggest that the cardiovascular dysfunction caused by r-α-toxin may be mediated by direct myocardial toxicity. This is supported by *in vitro* data,[49] which demonstrate inhibition of the inotropic cardiac response in isolated embryonic chick heart preparations through a calcium-dependent enzyme.

In contrast, r-θ-toxin may contribute to septic shock through more indirect routes including augmented release of endogenous mediators such as the cytokines TNF, IL-1, and IL-6[50,51] and the lipid autocoids, PAF and prostaglandin I_2.[41] Perhaps θ-toxin-induced synthesis of nitric oxide by host cells such as macrophages or endothelial cells could also play a role in early hypotension.

VII. SUMMARY

The classic features of gas gangrene caused by *C. perfringens* type A can be attributed to the action of potent extracellular bacterial toxins on host cell function. Both α- and θ-toxins likely contribute to the progression of vascular injury and tissue destruction through their ability to prime leukocyte functional activity and to stimulate soluble and cell-associated proadhesive molecule expression in both leukocytes and endothelial cells.

Both α- and θ-toxins have unique and profound effects on cardiovascular function. Our studies[7,14,45] and those of others[52–54] support the conclusion that α-toxin is the major lethal factor associated with experimental *C. perfringens* infections. Further, we have described one physiological mechanism, i.e., direct inhibition of myocardial contractility, whereby PLC mediates cardiovascular collapse.[7,45] Hypotension caused by θ-toxin is not associated with direct cardiotoxicity, but could be related to induction of endogenous mediators such as cytokines.

Thus, therapeutic strategies directed against bacterial toxin activity *in vivo*, such as neutralization of toxins with specific antibodies[55,56] or inhibition of toxin synthesis with antibiotics,[57] may be valuable adjuncts to traditional antimicrobial regimens. While the effects of these and other clostridial toxins on proadhesive and transmigrational cell–cell interactions have yet to be investigated, a more complete understanding of the pathophysiology of these vascular events may provide a rationale for anti-adhesion therapy. Such novel treatment strategies may greatly reduce the necessity for amputation of limbs and may improve patient survival.

VIII. REFERENCES

1. MacLennan J.D. (1962) The histotoxic clostridial infections of man. *Bacteriologic Rev* **26**: 177–276.
2. Smith L.D.S. (1975) Clostridial wound infections. In Smith L.D.S. (ed.) *The Pathogenic Anaerobic Bacteria*, pp. 321–324. CC Thomas, Springfield, IL.
3. McNee J.W. and Dunn J.S. (1917) The method of spread of gas gangrene into living muscle. *Br Med J* **1**: 727–729.
4. Bullen J.J. (1970) Role of toxins in host-parasite relationships. In Montie T.C., Kadis S. and Ajl S.J. (eds), *Microbial Toxins*, pp. 233–275. Academic Press, London.

5. Ispolatovskaya M.V. (1972) Type A *Clostridium perfringens* toxin. In Kadis S., Montie T.C. and Ajl S.J. (eds), *Microbial Toxins*, pp. 109–158. Academic Press, New York.
6. Smith L.D.S. (1975) *Clostridium perfringens*. In Smith L.D.S. (ed.) *The Pathogenic Anaerobic Bacteria*, pp. 115–176. CC Thomas, Springfield, IL.
7. Stevens D.L., Troyer B.E., Merrick D.T., Mitten J.E. and Olson R.D. (1988) Lethal effects and cardiovascular effects of purified alpha- and theta-toxins from *Clostridium perfringens*. *J Inf Dis* **157**: 272–279.
8. Murata R., Soda, S., Yamamoto, A., Sato H. and Ito A. (1969) The effect of zinc on the production of various toxins of *Clostridium perfringens*. *Japan J Med Sci Biol* **22**: 133–148.
9. Titball R.W. and Rubidge T. (1990) The role of histidine residues in the alpha toxin of *Clostridium perfringens*. *FEMS Microbiol Lett* **56**: 261–265.
10. Basak A.K., Stuart D.I., Nikura T. *et al.* (1994) Purification, crystallization and preliminary x-ray diffraction studies of alpha-toxin of *Clostridium perfringens*. *J Mol Biol* **244**: 648–650.
11. Titball R.W., Fearn A.M. and Williamson E.D. (1993) Biochemical and immunological properties of the *C*-terminal domain of the alpha-toxin of *C. perfringens*. *FEMS Microbiol Lett* **110**: 45–50.
12. Williamson E.D. and Titball R.W. (1993) A genetically engineered vaccine against alpha-toxin of *Clostridium perfringens* protects against experimental gas gangrene. *Vaccine* **11**: 1253–1258.
13. Stevens D.L., Gibbons A.E., Bergstrom R.A. and Sato H. (1988) Immunoprophylaxis of experimental gas gangrene caused by *Clostridium perfringens*. *International Symposium on Anaerobic Bacteriol Infections*. (Abstr. 46) Monte Carlo, Monaco.
14. Awad M.M., Bryant A.E., Stevens D.L. and Rood J.I. (1995) Virulence studies on chromosomal α-toxin and θ-toxin mutants constructed by allelic exchange provide genetic evidence for the essential role of α-toxin in *Clostridium perfringens*-mediated gas gangrene. *Mol Microbiol* **15**: 191–202.
15. Ninomiya M., Matsushita O., Minami J., Sakamoto H., Nakano M. and Okabe A. (1994) Role of alpha-toxin in *Clostridium perfringens* infection determined by using recombinants of *C. perfringens* and *Bacillus subtilis*. *Infect Immun* **62**: 5032–5039.
16. Bernheimer A.W. (1947) Comparative kinetics of hemolysis induced by bacterial and other hemolysins. *J Gen Physiol* **30**: 337–353.
17. Bernheimer A.W. (1976) Sulfhydryl activated toxins. In Bernheimer A.W. (ed.) *Mechanisms in Bacterial Toxinology*, pp. 85–97. John Wiley and Sons, New York.
18. Geoffroy C. and Alouf J.E. (1984) Antigenic relationships between sulfhydryl-activated toxins. In Alouf J.E. (ed.) *Bacterial Protein Toxins*, pp. 241–243. Academic Press, London.
19. Todd E.W. (1941) The oxygen-labile haemolysin for θ toxin of *Clostridium welchii*. *Br J Exp Pathol* **22**: 172–178.
20. Cowell J.L. and Bernheimer A.W. (1977) Antigenic relationships among thiol-activated cytolysins. *Infect Immun* **16**: 397–399.
21. Tweten R.K. (1988) Cloning and expression in *Escherichia coli* of the perfringolysin O (theta-toxin) gene from *Clostridium perfringens* and characterization of the gene product. *Infect Immun* **56**: 3228–3234.
22. Tweten R.K. (1988) Nucleotide sequence of the gene for perfringolysin O (theta-toxin) from *Clostridium perfringens*: Significant homology with the genes for streptolysin O and pneumolysin. *Infect Immun* **56**: 3235–3240.
23. Kehoe M.A., Miller L., Walker J.A. and Boulnois G.J. (1987) Nucleotide sequence of the streptolysin O (SLO) gene: structural homologies between SLO and other membrane-damaging, thiol-activated toxins. *Infect Immun* **55**: 3228–3232.
24. Paton J.C., Merry A.M., Lock R.A., Hansman D. and Manning P.A. (1986) Cloning and expression in *Escherichia coli* of the *Streptococcus pneumoniae* gene encoding pneumolysin. *Infect Immun* **54**: 50–55.
25. Portnoy D.A., Tweten R.K., Kehoe M. and Bielecki J. (1992) Capacity of lysteriolysin O, streptolysin

O, and perfringolysin O to mediate growth of *Bacillus subtilis* within mammalian cells. *Infect Immun* **60**: 2710–2717.

26. Paton J.C., Lock R.A. and Hansman D.J. (1983) Effect of immunization with pneumolysin on survival time of mice challenged with *Streptococcus pneumoniae*. Infect Immun **40**: 548–552.
27. Bailey J.T., Halsey N.A. and Johnson M.K. (1987) Prevention of *S. pneumoniae* bacteremia and mortality in rats by immunization with pneumolysin toxoid. *Interscience Conference on Antimicrobial Agents and Chemotherapy*, New York, NY. (Abstr.895) American Society for Microbiology Press, Washington, D.C.
28. Bryant A.E., Bergstrom R., Zimmerman G.A., Salyer J.L., Hill H.R., Tweten R.K., Sato H. and Stevens D.L. (1993) *Clostridium perfringens* invasiveness is enhanced by effects of theta toxin upon PMNL structure and function: The roles of leukocytotoxicity and expression of CD11/CD18 adherence glycoprotein. *FEMS Immunol Medical Microbiol* **7**: 321–336.
29. Stevens D.L. (1995) Clostridial infections. In Stevens D.L. and Mandell G.L. (eds) *Atlas of Infectious Diseases*, pp. 13.1–13.9. Churchill Livingstone, Philadelphia, P.A.
30. Robb-Smith A.H.T. (1945) Tissues changes induced by *C. welchii* type A filtrates. *Lancet* **ii**: 362–368.
31. Bryant A.E., Awad M.M., Lyristis M., Rood, J.I. and Stevens D.L. (1995) Perfringolysin O and phospholipase C from *Clostridium perfringens* impair host inflammatory cell defense mechanisms. *ASM First International Conf on Molecular Genetics and Pathogenesis of the Clostridia.* (Abstr. B6) p. 21. Rio Rico, AZ.
32. Stevens D.L., Mitten J. and Henry C. (1987) Effects of alpha and theta toxins from *Clostridium perfringens* on human polymorphonuclear leukocytes. *J Inf Dis* **156**: 324–333.
33. Stevens D.L., Gibbons A.E. and Bergstrom R.A. (1989) Ultrastructural changes in human granulocytes induced by purified exotoxins from *Clostridium perfringens. American Society for Microbiology Annual Meeting*, New Orleans. (Abstr. J17). American Society for Microbiology Press, Washington, D.C.
34. Bevilacqua M.P. (1993) Endothelial-leukocyte adhesion molecules. *Annu Rev Immunol* **11**: 767–804.
35. Patel K.D., Lorant E., Jones D.A., Prescott M., McIntyre T.M. and Zimmerman G.A. (1993) Juxtacrine interactions of endothelial cells with leukocytes: tethering and signaling molecules. *Behring Inst Mitt* **92**: 144–164.
36. Bryant A.E. and Stevens D.L. (1996) Phospholipase C and perfringolysin O from *Clostridium perfringens* upregulate endothelial cell-leukocyte adherence molecule 1 and intercellular leukocyte adherence molecule 1 expression and induce interleukin-8 synthesis in cultured human umbilical vein endothelial cells. *Infect Immun* **64**: 358–362.
37. Pober J.S., Gimbrone Jr. M.A., Lapierre L.A., Mendrick D.L., Fiers W., Rothlein R. and Springer T.A. (1986) Overlapping patterns of activation of human endothelial cells by interleukin 1, tumor necrosis factor, and immune interferon. *J Immunol* **137**: 1893–1896.
38. Bevilacqua M.P., Pober J.S., Mendrick D.L., Cotran R.S. and Gimbrone M.A. (1987) Identification of an inducible endothelial-leukocyte adhesion molecule. *Proc Natl Acad Sci USA* **84**: 9238–9242.
39. Sticherling M., Hetzel F., Schroder J.M. and Christophers E. (1993) Time- and stimulus-dependent secretion of NAP-1/IL-8 by human fibroblasts and endothelial cells. *J Invest Derm* **101**: 573–576.
40. Smith W.B., Gamble J.R., Clark-Lewis I. and Vadas M.A. (1993) Chemotactic desensitization of neutrophils demonstrates interleukin-8 (IL-8)-dependent and IL-8-independent mechanisms of transmigration through cytokine-activated endothelium. *Immunology* **78**: 491–497.
41. Whatley R.E., Zimmerman G.A., Stevens D.L., Parker C.J., McIntyre T.M. and Prescott S.M. (1989) The regulation of platelet activating factor synthesis in endothelial cells – The role of calcium and protein kinase C. *J Biol Chem* **11**: 6325–6333.
42. Zimmerman G.A., McIntyre T.M., Prescott S.M. and Otsuka K. (1990) Brief review: Molecular

mechanisms of neutrophil binding to endothelium involving platelet activating factor and cytokines. *J Lipid Mediators* **2**: S31–S43.
43. McEver R.P., Beckstead J.H., Moore K.L., Marshall-Carlson L. and Bainton D.F. (1989) GMP-140, a platelet α-granule membrane protein, is also synthesized by vascular endothelial cells and is localized in Weibel–Palade bodies. *J Clin Invest* **84**: 92–99.
44. Stolpen A.H., Guinan E.C., Fiers W. and Pober J.S. (1986) Recombinant tumor necrosis factor and immune interferon act singly and in combination to reorganize human vascular endothelial cell monolayers. *Am J Pathol* **123**: 16–24.
45. Asmuth D.A., Olson R.D., Hackett S.P., Bryant A.E., Tweten R.K., Tso J.Y., Zollman T. and Stevens D.L. (1995) Effects of *Clostridium perfringens* recombinant and crude phospholipase C and theta toxins on rabbit hemodynamic parameters. *J Inf Dis* **172**: 1317–1323.
46. Smith L.D.S. (1979) Virulence factors of *Clostridium perfringens. Rev Infect Dis* **1**: 254–260.
47. Sakurai J., Oshita Y. and Fujii Y. (1985) Effect of *Clostridium perfringens* α-toxin on the cardiovascular system of rats. *Toxicon* **23**: 905–912.
48. Sakurai J., Tsuchiya Y., Ochi S. and Fujii Y. (1991) Effect of *Clostridium perfringens* α-toxin on contraction of isolated guinea-pig diaphram. *Microbiol Immunol* **35**: 481–486.
49. Regal J.F. and Shigeman F.E. (1980) The effect of phospholipase C on the responsiveness of cardiac receptors. I. Inhibition of the adrenergic inotropic response. *J Pharmacol Exp Ther* **214**: 282–290.
50. Hackett S.P. and Stevens D.L. (1992) Streptococcal toxic shock syndrome: Synthesis of tumor necrosis factor and interleukin-1 by monocytes stimulated with pyrogenic exotoxin A and streptolysin O. *J Inf Dis* **165**: 879–885.
51. Houldsworth S., Andrew P.W. and Mitchell T.J. (1994) Pneumolysin stimulates production of tumor necrosis factor alpha and interleukin-1 beta by human mononuclear phagocytes. *Infect Immun* **62**: 1501–1503.
52. Anderson B.R. and van Epps D.E. (1972) Suppression of chemotactic activity of human neutrophils by streptolysin O. *J Inf Dis* **125**: 353–359.
53. Doebber T.W., Wu M.S., Robbins J.C., Choy B.M., Chang M.N. and Shen T.Y. (1985) Platelet activating factor (PAF) involvement in endotoxin induced hypotension in rats. Studies with PAF receptor antagonist kadsurenone. *Biochem Biophys Res Commun* **127**: 799–808.
54. Duncan J.L. and Schlegel R. (1975) Effect of streptolysin O on erythrocyte membranes, liposomes, and lipid dispersions. A lipid-cholesterol interaction. *J Cell Biol* **67**: 160–173.
55. Stevens D.L., Gibbons, A.E., Bergstrom, R.A. and Sato, H. (1988) Immunoprophylaxis of experimental gas gangrene caused by *Clostridium perfringens. Int Symp on Anaerobic Bacterial Infections, Monte Carlo, Monaco* (Abstract) *September 25–28.*
56. Tempest P.R., White P., Williamson E.D., Titball R.W., Kelly D.C., Kemp G.J., Gray P.M., Forster S.J., Carr F.J. and Harris W.J. (1994) Efficient generation of a reshaped human mAb specific for the alpha toxin of *Clostridium perfringens. Protein Eng* **7**: 1501–1507.
57. Stevens D.L., Maier K.A. and Mitten J.E. (1987) Effect of antibiotics on toxin production and viability of *Clostridium perfringens. Antimicrob Agents Chemother* **31**: 213–218.

Chapter 12

The Pathogenesis of Enteritis Necroticans

GREGOR W. LAWRENCE
Queensland Institute of Medical Research, Post Office Royal Brisbane Hospital, Queensland 4029, Australia

I INTRODUCTION

A. Background

Enteritis necroticans (EN) is a clostridial enterotoxaemia of man caused by *Clostridium perfringens* type C. It was first recognized as a major health problem in Europe at the end of World War II. In north-eastern Germany EN was called darmbrand. There was an epidemic in the area around Lübeck lasting from 1944 to 1949; this was carefully studied[1] and the causative role of C. *perfringens* type C recognized.[2] Subsequently, EN has been recognized in Africa, the Indian subcontinent and South-east Asia. However, it was the recognition of endemic EN in the highland areas of Papua New Guinea (PNG) in 1961,[3] that rekindled interest in the disease. Subsequent work in PNG led to some understanding of this very interesting disease, the pathogenesis of which is a result of the interplay of characteristics of the organism, of its ß-toxin and of the way of life of the Highland subsistence farming community it affects.[4] 'Pigbel', as EN is known in PNG, was a very common disease, second only to pneumonia as a cause of death in hospital in children after weaning,[5] and, it caused more than half the deaths in

The Clostridia: Molecular Biology and Pathogenesis
ISBN 0-12-595020-9

older children aged from 5 to 10 years.[6] Trying to understand the reasons for the appearance and disappearance of darmbrand in Germany and of the persistence of pigbel in PNG has led to some general understanding of the pathogenesis of the disease, but there are still many deficiencies in the story, some of which may be clarified now that the nature and action of ß-toxin, and of the other histotoxic clostridial toxins, are being investigated.

EN is still a problem in parts of South East Asia, however, it tends to occur in restricted areas and is frequently not recognized owing to a lack of experienced rural surgical services. In Vietnam, EN has been recognized and carefully studied for more than 10 years. In some regions in the country it is just as common as in the high incidence areas of PNG. In Vietnam, the pathogenesis also seems to follow the same pattern as in PNG, where the disease is most common in children eating a low-protein diet in which sweet potato is a staple. This review deals with what is known and inferred about the pathogenesis of EN in PNG: it may not apply completely to the situation in other locations.

B. Clinical

Abdominal pain is the cardinal symptom in pigbel. Most patients develop severe abdominal pain 1–5 days after eating a high-protein meal; in PNG the meat is usually pig. Frequently, there is vomiting and often initial diarrhea, the diarrhea is short lived and the stool usually contains blood and necrotic intestinal mucosal material. Those with severe disease develop intestinal obstruction because of patchy full-thickness necrosis of the small intestine, with many requiring surgery, while milder cases usually recover without treatment, or after treatment with antibiotics and nasogastric suction. The most severe cases die quickly from toxemia or are found at post mortem after sudden unexplained death. Some children present late with malnutrition and chronic obstruction from post-pigbel stricture or adhesions.[7]

In general, the more severe the disease, the closer the illness to the preceding protein containing meal. Clinical observation developed the understanding that the disease process that causes the intestinal damage in EN happens, and is largely completed, before the patient presents for medical care. Evidence for this conclusion is that good surgical management of EN is expectant waiting: patients are resuscitated and then observed to see whether they will recover or require surgery later. There is no concern that the pathological process is continuing with increasing damage to the gut. Early operation does not permit good judgment of how much of the damaged bowel is adequately viable, and what must be removed, as well as what can safely be left. *C. perfringens* type C multiplies rapidly in the gut in the hours after the meal is eaten. There is some experimental evidence based on passive immunization to support the contention that most of the damage to the gut is caused by the production of ß-toxin in that early period of rapid organism growth.[8] It follows that the clinical course of the disease reflects the attempts at recovery of the damaged bowel from the damage done by the toxin. The pathological findings largely support this concept.

C. Pathology

EN is patchy, it involves the small bowel, mostly the jejunum, but it frequently also affects the ileum. The colon is never directly involved. There is mucosal necrosis and variable necrosis of the bowel wall, up to full thickness. The worst patches are often antimesenteric. The mesentery itself becomes thickened, shortened and odematous and often hemorrhagic. The mesenteric nodes are enlarged and hemorrhagic. At surgery, which is usually 5–10 days after consumption of the meat containing meal, there is obvious lymphatic obstruction in the damaged gut wall and mesentery. Adherent mesentery binds the thick damaged loops of bowel together. The damage in the gut is visually suggestive of vascular involvement: transverse stripes of hemorrhagic injury and patches of necrosis alternate with relatively normal areas. In Thailand, the disease was called segmental infarcts of the small intestine.[9] Many pathologists have described the pathology as infarction, implying death of tissue due to interrupted blood supply. There is no physical evidence for this – the lesions appear to be a direct effect of ß-toxin.

Dead mucosa, stained with bile often persists, presumably reflecting the low level of proteolytic activity in the lumen. Sections through areas of damage demonstrate areas of necrosis and hemorrhage surrounded by a dense mainly polymorphonuclear leucocyte infiltrate. In established lesions at surgery there are inflammatory changes around the venules, and not the arterial vessels. These changes around the veins are considered to be pathognomonic of pigbel by Cooke,[10] but are a late feature, not being present early in experimental lesions even when the necrotic lesions are fully established.[8] Careful examination of the mucosal surface often shows small patches of involvement 2–3 mm diameter. When sections pass through these areas there is a core of necrosis a few millimetres in diameter passing from the mucosal to the serosal surface of the gut wall. These patches are surrounded by normal mucosa. An intriguing question is what pathological mechanism leads to this peculiar distribution of lesions? Cooke specifically sought lymphoid antigen processing organelles in the small spots of necrosis described above, however, none was seen.

II. OVERVIEW OF EPIDEMIOLOGY AND PATHOGENESIS

Pigbel is largely a disease of children. It is exceptionally rare in infants under 1 year of age. In Goroka, the age-specific incidence was highest in 4-year-olds. About 70% of the cases are children of 10 and under. This age distribution probably reflects the effects of exposure to the causes of pigbel and acquired immunity. It seems likely that the relative immunity to pigbel in the first year of life is a result of good nutrition while suckling, and also of transfer of maternal immunity. The age distribution of cases of darmbrand was quite different, it particularly affected the very young, including breast fed babies, and the very old.

In PNG, EN is largely confined to the Highland areas of the country. It is rare among Melanesian people on the coast and is virtually unknown in Europeans living in the Highlands, although many of the Europeans are of similar genetic background to the people in Germany and Norway where darmbrand was common. The reasons for the persistence of pigbel in the highlands and for its geographic distribution lie in the nature of ß-toxin and in the habits and the diet of Highlanders.

ß-Toxin is a protein that is very sensitive to proteolysis by trypsin.[11] Highland people have a very low protein diet as do many subsistence farmers in the third world.[12] Low protein diets lead to low production of proteolytic pancreatic enzymes. The dietary staple in the Highlands is the sweet potato *Ipomoea batatas*. Like many plant foods sweet potatoes contain inhibitors of trypsin, but the sweet potato inhibitor is unusual in being quite heat stable[13] and able to survive limited cooking. Meat is infrequently eaten in the Highlands, and usually in association with some type of celebration. When a child has a meat-containing meal contaminated with *C. perfringens* type C and containing sweet potato coupled with a background of protein subnutrition, the trypsin secreted into the gut may be unable to deal with the sudden production of ß-toxin. Thus, mucosal damage and perhaps clinical pigbel may result. Europeans in PNG are protected by having a high-protein diet and plenty of proteases. Coastal people are usually protected because they do not consume large amounts of a staple containing heat-stable trypsin inhibitor. In a similar fashion, darmbrand disappeared at the time when the nutrition and the protease levels of the population were improving after wartime and post-war deprivation. The susceptibility of breastfed German babies may have been a result of good pre-war nutrition and hygiene of the mothers. High gut protease activity would have protected the women from exposure to ß-toxin. When war-time conditions removed their proteolytic protection the mothers had no antibody to transfer to their offspring. This brought about a situation similar to that which occurs with piglets of a non-immune sow in a contaminated farrowing pen: where proteolytic activity is suppressed by colostral trypsin inhibitor, present to protect immunoglobulin being transferred across the neonatal gut.[14] In PNG, maternal antibody transferred and good nutrition, protects infants in the first year or so.

This theory explaining the pathogenesis of EN in PNG, fits the observed facts fairly well. It is consistent with the pathogenesis of the veterinary type C enterotoxaemias and shares elements of the pathogenesis of the type C diseases of neonatal animals, that depend on the presence of a dietary trypsin inhibitor in colostrum (see Chapter 10). The fact that virtually identical conditions occur in the parts of Vietnam, where EN is also common in children, and where the medical personnel are well trained to recognize the disease, provides strong support.

III. PATHOGENIC FACTORS

A. The organism

1. Occurrence

Clostridium perfringens type C was readily recovered from darmbrand cases because of its heat-resistant spores. In PNG *C. perfringens* type C was initially found only in pigbel cases. Later, using a fluorescent antibody technique, type C organisms were demonstrated in, but not isolated from, the feces of a very high proportion of the Highland people, and in the soil and pig feces.[15] In PNG *C. perfringens* type C is very much less common than type A organisms, as is the case in the UK,[16] and so, although very easily seen using the fluorescent stain, it is very difficult to isolate, because of the preponderance of type A. Some type C organisms from nonpigbel sources in PNG have been isolated using an affinity technique and have been shown to produce ß-toxin.[17]

Poor hygiene and the close association of people with their pigs increases the opportunities for food contamination with *C. perfringens* type C in PNG. In the case of darmbrand there was also environmental contamination. The German organism was extremely heat resistant, the spores withstanding boiling for over 4 h. This made isolation of the organism easy, but also contributed significantly to the contamination of the environment, and particularly of cooked food with *C. perfringens* type C. Kloos reported the spread of cases of EN along railways and roads, perhaps reflecting the spread of the heat-resistant strain.[18] Enterotoxaemia in piglets is often associated with heavy contamination of farrowing pens with *C. perfringens* type C. Control strategies reduce or eliminate the contact of sows with areas of high *C. perfringens* type C concentration. Vaccination is also effective.

2. Attachment and mucosal damage in EN

Sections of material obtained at surgery sometimes show Gram-positive rods attached to necrotic villi. Large numbers of bacteria are not seen attached to the surrounding more normal mucosa, however, it must be remembered that surgical specimens are obtained well after the initiating process in EN. Walker and Murrell,[19] using SEM on surgical specimens, demonstrated rod-shaped organisms on the surface of damaged bowel. However, the bacteria were largely in areas of denuded mucosa and this should not be interpreted as evidence of specific attachment as is seen with some types of pathogenic *Escherichia coli*. Attachment might be important in pathogenesis and it may follow initial damage by ß-toxin. This would then be comparable to the initial toxin mediated damage seen in the type A enteritis of chickens, that is thought to be mediated by α-toxin.[20] At present, the role of attachment in the pathogenesis of EN remains unclear.

Mucosal damage from other infections could play an initial part in the pathogenesis of EN. There is an association between *Ascaris lumbricoides* and pigbel; in Goa in India, EN is sometimes known as ascaris enteritis. It has been suggested that the trypsin inhibitor[21] that the worm secretes to protect itself in the intesintal lumen may contribute to the pathogenesis of EN.

B. Toxins

1. β-Toxin

β-Toxin is central to the pathogenesis of EN. Direct injection of toxic filtrates from *C. perfringens* type C cultures into the small intestine of guinea pigs was shown to produce the typical pathology of darmbrand.[22] This was also the case with experimental pigbel using the organisms isolated in PNG. Luminal toxin, either injected alone or as part of a growing culture, produced mucosal and gut wall necrosis, and the typical banded pattern of hemorrhage and necrosis seen in the human disease. Absorption of β-toxin causing immediate local cell death seems to be central to the pathology of EN. The lability of the toxin is also important as this may restrict and localize the damage. The narrow cores of full-thickness wall necrosis mentioned in the previous Section, are presumably the result of toxin being absorbed along the lymphatic path. The hemorrhagic enlarged nodes and the thickened and hemorrhagic mesentery of surgical cases possibly represent the limits of toxin progression in these cases. Death results from any substantial passage of toxin into the circulation.

β-Toxin is very toxic having an LD_{50} for mice of <100 μg/kg.[11] Its action is very short lived, either because of destruction by tissue proteases, or perhaps it is quickly and tightly bound. Injection of purified β-toxin into the tail vein of mice is rapidly fatal. If the drainage of the tail vessels is impeded by a tourniquet for 10 min, a similar intravenous dose results in necrosis of most of the tail tissues without evidence of illness or the death of the mouse.[23]

Although antibody protects against the systemic effects of β-toxin, and against the production of EN in the experimental models in guinea pigs,[8] protection is related to the local concentrations of toxin and of antibody. High local concentrations of toxin can overcome the protection given by amounts of antibody sufficient to protect the whole animal against substantial amounts of toxin. When purified, β-toxin was carefully injected, as a final boosting step, into the marginal vein of the ears of rabbits previously immunized with a toxoid of purified β-toxin, the animals were completely protected against any systemic effects. However, there was local, full-thickness, apparently painless necrosis of the ear edge along and around the injected vein. The necrosis had the typical appearance of the patches of intestinal necrosis in EN. There was central tissue death with a surrounding border of hyperemic damaged, but recovering, tissue. This demonstrates the toxicity of β-toxin and how local its effects can be. The intestinal lesions in EN may be a reflection of this combination of toxicity and lability.

In cultures of *C. perfringens* type C designed to maximize the production of β-toxin, toxin appears early and is at a maximum early during the period of logarithmic growth.[24] Levels begin to drop because of the production of proteases. The dynamics of β-toxin production help to explain the relationship of EN to a large meal and the associated burst of bacterial growth. The animal models developed to experimentally produce EN with orally administered agents all included the organism, nutrient for bacterial growth in the gut, and protease inhibitors.[8]

2. Other toxins

Other toxins may contribute to the pathogenesis and the clinical picture. *Clostridium perfringens* enterotoxin may be responsible for the diarrhea that is often a feature of the first stages of pigbel. P.D. Walker found that all the *C. perfringens* type C isolates tested from pigbel cases were enterotoxin producers (personal communication). The role of α-toxin is also not clear.

3. Antitoxic immunity to β-toxin

Active or passive immunity protects against the veterinary type C enterotoxemias. Active immunity produced by a type C toxoid has reduced the hospital admissions for pigbel in the PNG Highlands by 80% following the introduction of immunization as a public health measure in 1980.[25] In contrast, serotherapy with antisera raised in horses had no effect on established pigbel in a double-blind controlled trial.[26] These findings demonstrate the fundamental role of toxin and add weight to the contention that the pathological process is usually fully developed when the patient first presents for hospital care.

C. Trypsin inhibitors

The sweet potato inhibitor inhibits trypsin and not chymotrypsin. The inhibitor is relatively heat stable, readily surviving the low temperatures achieved in usual PNG Highland cooking practice. In well-nourished individuals the concentration of the two enzymes in the pancreatic juice is about equal. However, in studies in monkeys, when protein intake was low, there was a marked drop in trypsin production and almost complete cessation of chymotrypsin production.[27] This suggests that sweet potato inhibitor would only induce susceptibility to the damaging effects of β-toxin in protein-deficient individuals. Animal experiments in the guinea pig model of pigbel support this suggestion.[8] However, even in the protein-deficient animal exposed to dietary inhibitors and β-toxin, immunization against the β-toxin protects against the organism–toxin–trypsin-inhibitor combination.

D. Diet and nutrition

1. The meat-containing meal

Millar and Smellie clearly showed the association of meat-eating with subsequent pigbel. Similarly, a large and often protein-containing meal was a frequent feature preceding darmbrand. Dietary change, from a poor to a richer diet, is also recognized as a factor in struck, the *C. perfringens* type C enterotoxaemia of sheep (see Chapter 10). Dietary change apparently provides an environment for rapid clostridial growth and toxin production within the gut. The protein content of the diet affects the bowel bacterial flora. Mansson and Olsson,[29] demonstrated an increase in bowel *C. perfringens* levels of about three-log following a change of diet from a low protein to 20% protein content.[29] In some cases of pigbel the meat meal may also result in an element of substrate excess for the available trypsin, for example in a child who normally has a very low protein diet and has a generous protein-rich meal. Most parents admit to their child eating pork prior to pigbel, but many claim that the amount was small. The small amount of meat may be correct, or its reported size may partly be a reflection of the universally recognized connection between eating pig and pigbel in Highland PNG.

Contaminated pork is probably the usual source of the *C. perfringens* type C that causes pigbel, but this is not always the case. There have been cases of pigbel in which the meat was tinned and recently opened. Raw peanuts have also been implicated in a number of cases. In addition to their high protein content, raw peanuts contain heat-labile trypsin inhibitors. In clostridial enterotoxaemia of sheep *C. perfringens* is resident in the gut and proliferates in the changed environment resulting from dietary change. It is possible that in some cases of pigbel *C. perfringens* is already present in the gut of children and proliferates following the high protein meal and produces ß-toxin. This mechanism may explain the pigbel cases following meals of probably uncontaminated tinned meat. Suggestions have been made that *Ascaris lumbricoides*, itself a resident of the upper gut, could be a source of gut contamination with *C. perfringens* type C through having the organism in its own bowel flora, thus contributing to the pathogenesis of some cases of EN through another mechanism.

2. Diet in the PNG Highlands

Highland people traditionally consume sweet potato as their staple. About 75% of energy intake comes from this source. The protein intake of Highlanders eating a traditional diet is extremely low. Although frank malnutrition is unusual in Highland children, the children are short for their age and have been shown to grow with a dietary supplement containing protein. Stool trypsin levels are low.[30]

IV. SUMMARY

EN, like the veterinary enterotoxaemias, is a sporadic disease that depends on the interplay of a number of pathogenic factors. In EN the organism must be available and is, presumably, usually a contaminant of food during a dietary change that encourages bacterial growth. The proteolytic activity in the gut must be inadequate, owing to low dietary protein levels and/or through the consumption of dietary inhibitors. There must be little or no antibody against ß-toxin available to protect the mucosa or to limit damage to the bowel wall.

That so many factors are needed to cause a clinical episode of EN helps explain the very low attack rate seen after PNG Highland pig feasts. Changes in any of these factors may interfere with the disease process. In Germany, improving nutrition provided proteolytic protection against β-toxin and darmbrand disappeared. In PNG, since 1980, immunization with a *C. perfringens* type

C toxoid has altered another factor – immunity to β-toxin – and reduced pigbel dramatically.[25] The vaccine can provide protection against clinical EN until other changes impinging on the pathogenesis of the disease, such as improving nutrition, make immunization unnecessary.

V. REFERENCES

1. Hansen K., Jeckeln E., Jochims J., Lezius A., Meyer-Burgdorf H. and Schütz F. (eds) (1949) *Darmbrand: Enteritis Necroticans.* Georg Thieme, Stuttgart.
2. Zeissler J. and Rassfeld-Sternberg L. (1949) Enteritis necroticans due to *Clostridium welchii* type F. *Br Med J* **i**: 267–269.
3. Murrell T.G.C. and Roth L. (1963) Necrotizing jejunitis: a newly discovered disease in the highlands of New Guinea. *Med J Aust* **i**: 61–69.
4. Lawrence G.W. and Walker P.D. (1976) Pathogenesis of enteritis necroticans in Papua New Guinea. *Lancet* **i**: 125–127.
5. Linnemann V. and Shann F. (1984) Review of pigbel at Goroka Hospital. In Davis M.W. (ed.), *Pigbel-necrotising Enteritis in Papua New Guinea. PNG Institute of Medical Research Monograph Series* No. 6, pp. 11–15. PNG IMR, Goroka.
6. Smith D. (1979) The mortality of pigbel (enteritis necroticans) in children in Tari 1971–1976. *PNG Med J* **22**: 74–78.
7. Shepherd A. (1979) Clinical features and operative treatment of pigbel – enteritis necroticans. *PNG Med J* **22**: 18–23.
8. Lawrence G. and Cooke R. (1980) Experimental pigbel: The production and pathology of necrotizing enteritis due to *Clostridium welchii* type C in the guinea-pig. *Br J Exp Path* **61**: 261–271.
9. Headington J T., Sathornsumathi S., Simark S. and Sujatanond W. (1967) Segmental infarcts of the small intestine and mesenteric adenitis in Thai children. *Lancet* **i**: 802–806.
10. Cooke R.A. (1979) The pathology of Pig Bel. *Per Pediat Pathol* **5**: 137–152.
11. Sakurai J. and Duncan C.L. (1978) Some properties of beta toxin produced by *Clostridium perfringens* type C. *Infect Immun* **21**: 678–680.
12. Heywood P.F. and Jenkins C. (1992) Nutrition in Papua New Guinea. In Attenborough R.D. and Alpers M.P. (eds), *Human Biology in Papua New Guinea: The Small Cosmos, Research monographs on human population biology series, No 10,* pp. 249–267. Oxford Scientific Publications, Oxford.
13. Sugiura M., Ogiso T., Kazuyuki T., Sumihiro T. and Akira I. (1973) Studies on trypsin inhibitors in sweet potato, 1. Purification and some properties. *Biochim Biophys Acta* **328**: 407–417.
14. Griner L.A. (1963) Some factors influencing the incidence of enterotoxaemia in domestic animals. *Bull Off Int Epiz* **59**: 1443–1451.
15. Lawrence G., Walker P.D., Garap J. and Avusi M. (1979) The occurrence of *Clostridium perfringens* type C in Papua New Guinea. *PNG Med J* **22**: 69–73.
16. Taylor A.W. and Gordon W.S. (1940) A survey of the types of *Clostridium welchii* in soil and the intestinal contents of animals and man. *J Path Bact* **50**: 271–277.
17. Lawrence G., Brown R., Bates J., Saul A., Davis M., Spark R. and Anian G. (1984) An affinity technique for the isolation of *Clostridium perfringens* Type C from man and pigs in Papua New Guinea. *J Appl Bact* **57**: 333–338.
18. Kloos K. and Brummund W. (1951) Epidemiolische Studien über die Enteritis gravis in Nordholstein. *Zeitsch Hygiene* **132**: 64–96.

19. Walker P.D., Murrell T.G.C. and Nagy L.K. (1980) Scanning electronmicroscopy of the jejunum in enteritis necroticans. *J Med Microbiol* **13**: 445–450.
20. Al-Sheikly F. and Truscott R.B. (1977) The pathology of necrotic enteritis of chickens following the infusion of crude toxins of *Clostridium perfringens* into the duodenum. *Avian Dis* **21**: 241–247.
21. Kucich U. and Peanasky R.J. (1969) Trypsin inhibitors from *Ascaris lumbricoides* var. *suis. Biochim Biophys Acta* **200**: 47–57.
22. Schütz F. (1949) Aetologie and Pathogenese. In Hansen K., Jeckeln E., Jochims J., Lezius A., Meyer-Burgdorf H. and Schütz F. (eds), *Darmbrand: Enteritis Necroticans*, pp. 186–212. Georg Thieme, Stuttgart.
23. Lawrence G., Watt S. and Basten A. (1979) Purification and some effects of *Clostridium perfringens* type C beta toxin. *PNG Med J* **22**: 79–86.
24. Jansen B.C. (1961) The beta toxin of *Clostridium welchii* type B (Wilsdon), in relation to the production of a vaccine against lamb dysentery. *Onderspoort J* **28**: 495–549.
25. Lawrence G.W., Lehmann D., Anian G., Coakley C.A., Saleu G., Barker M.J. and Davis M.W. (1990) Impact of active immunisation against enteritis necroticans in Papua New Guinea. *Lancet* **336**: 1165–1167.
26. Rooney J., Shepherd A. and Suebu A. (1979) *Clostridium welchii* type C antitoxin in the treatment of 'pig bel' (enteritis necroticans): A controlled trial in Papua New Guinea. *PNG Med J* **22**: 57–59.
27. Gyr K., Wolf R.H., Imondi A.R. and Felsenfeld O. (1975) Exocrine pancreatic function in protein deficient Patas monkeys studied by means of a test meal and an indirect pancreatic function test. *Gastroenterology* **68**: 488–494.
28. Millar J.S., Smellie S. and Coldman A.J. (1985) Meat consumption as a risk factor in enteritis necroticans. *Int J Epidem* **14**: 318–321.
29. Mansson I and Olsson B. (1961) The intestinal flora of pigs. 1. *Acta Agric Scand* **11**: 197–210.
30. Davis M.W. (1984) Intestinal trypsin levels and susceptibility to pigbel. In Davis M.W. (ed.), *Pigbel-necrotising Enteritis in Papua New Guinea, PNG Institute of Medical Research Monograph series No. 6*, pp. 35–38. PNG IMR, Goroka.

E: Membrane Active Toxins and Enzymes

Chapter 13

The Thiol-Activated Clostridial Toxins

RODNEY K. TWETEN
Department of Microbiology and Immunology, University of Oklahoma Health Sciences Center, Oklahoma City, OK 73190, USA

I. INTRODUCTION

The 'thiol-activated' cytolysins are a fascinating group of toxins with several unique characteristics that are not found in other cytolytic toxins. It is likely that these toxins have diverged from a single primordial gene based on the relatively high degree of homology at the level of their primary structure (reviewed in ref. 1). Although they have retained a remarkable degree of structural similarity many have changed in ways that facilitate the pathogenic process of the bacterial species that produce them. Although these toxins have historically been referred to as 'thiol-activated' or 'oxygen-labile' toxins it is evident from recent investigations that these toxins are not necessarily thiol-activated or oxygen-labile. However, because the use of these terms have become entrenched in the literature it is unlikely that a new, more accurate description would be accepted.

The thiol-activated cytolysins have been found to be present in most members of the pathogenic clostridia, although only few have been studied to any extent. Those which have been purified include perfringolysin O or θ-toxin (*Clostridium perfringens*),[2–4] tetanolysin (*C. tetani*)[5] and botulinolysin (*Clostridium botulinum*).[6] Of these toxins, only perfringolysin O (PFO or θ-toxin) has been studied

The Clostridia: Molecular Biology and Pathogenesis
ISBN 0-12-595020-9

extensively, although a few salient studies have been carried out with tetanolysin. Therefore, the primary focus of this article will be the more recent studies that have been carried out with PFO. PFO is a member of a family of cytolytic proteins, collectively termed the 'thiol-activated toxins', that are produced by various Gram-positive pathogenic bacteria. Most of these toxins are actively secreted by the host cell with the exception of pneumolysin, which is released from the cell by lysis. PFO is actively secreted by the cell during exponential growth and is cytolytic to most types of erythrocytes and probably most nucleated mammalian cells. This wide range of activity may be explained by the fact that the receptor for these toxins is thought to be cholesterol. Although most pathogenic clostridia are thought to produce a related toxin, most of the work on clostridia-derived thiol-activated toxins has been carried out on PFO derived from gangrene-causing type A *C. perfringens*. Similar toxins are found in pathogenic species of *Streptococcus*, *Listeria*, *Bacillus* and other clostridia. Because of the close relationship between these toxins it is difficult to discuss one without referring to data from the studies of others. Therefore, although the primary focus of this article will be perfringolysin O (PFO), experimental results from the study of related toxins will also be included.

II. TOXIN STRUCTURES AND GENES

To date, no crystal structure of a thiol-activated cytolysin has been reported. Therefore, much of what is known is inferred from structure–function studies of the various toxins of this family. Most of the work on thiol-activated toxins produced by the clostridia has been performed on PFO with a few studies carried out with tetanolysin. Botulinolysin has been purified but to date there has been no further analysis of this toxin. Of these toxins only the gene for PFO has been cloned and sequenced.[3,7] Therefore, much of what is known about these toxins produced by the clostridia has been obtained primarily from the study of PFO.

The PFO structural gene, *pfoA*, was originally cloned and sequenced by Tweten.[3,7] The DNA-derived primary structure of the clostridial toxin was determined to exhibit significant homology with the two other thiol-activated toxins: streptolysin O from *Streptococcus pyogenes*[8] and pneumolysin from *S. pneumoniae*.[9] Since then several other genes for members of this family of toxins have been cloned and shown to exhibit 40–70% homology with *pfoA* (reviewed in ref. 1). The primary structure of PFO exhibits the hallmark 11-mer sequence of the thiol-activated cytolysins of ECTGLAWEWWR which is completely conserved in all of the thiol-activated toxins sequenced to date with the exception of LSO (see ligeriolysin O from *Listeria seeligeri*)[10] which has a phenylalanine substituted for the alanine. This sequence contains the only cysteine in the secreted form of PFO and three of the seven tryptophan residues in the molecule. Modification of the cysteine results in the inactivation of the toxin which can be restored by the addition of a reducing agent. Presumably, the slow loss of activity in crude preparations of these toxins results from the formation of disulfides with the sulfhydryl of this cysteine since the decrease in activity can be restored by the addition of a reducing reagent. Hence, the term 'thiol-activated' was applied to these toxins and has become the defining term for this family of toxins. Unfortunately, the sulfhydryl function does not appear to be essential to the *in vitro* cytolytic activity of PFO, or other related toxins, and so this term is no longer descriptive of this family of toxins (although it is unlikely to be abandoned soon). Substitution of this cysteine with alanine in SLO,[11] pneumolysin[12] or PFO (R.K. Tweten, unpublished results) does not appreciably alter the hemolytic activity of these toxins. The remainder of the primary structure of PFO is essentially unremarkable in that it does not exhibit any structures which are indicative of specific functions. There are no extended regions of hydrophobic amino acids that might be responsible for interaction of the toxin with the membrane.

The regulation of the *pfoA* gene expression has been the subject of intense investigation in recent years. The regulation of PFO and other extracellular products will be dealt with in more detail elsewhere in this book (see Chapter 26) and therefore will only be briefly summarized here. A few years after the *pfoA* gene was cloned and sequenced[3,7] it was cloned by Shimizu *et al.*[13] together with much of the flanking DNA. In the region of DNA upstream of the *pfoA* gene Shimizu *et al.* found a regulatory gene, *pfoR*, that could increase the expression of *pfoA* in *Escherichia coli*. More recently Lyristis *et al.*[14] and Shimizu *et al.*[15] reported the presence of a virulence gene regulator that controlled the expression of a variety of extracellular toxins and enzymes. Both groups identified a gene, termed *virR*, the product of which resembled the response regulator of a family of two-component regulatory systems. These systems are composed of a histidine kinase and response regulator which respond to various environmental stimuli and transmit these signals via phosphorylation reactions.[16] In addition to the response regulator analog, Lyristis *et al.*[14] also identified the histidine kinase analog of the two-component regulatory system in *C. perfringens*, termed *virS* by these workers, which was located immediately downstream of the *virR* gene. When *virS* was knocked out by transposon mutagenesis[14] a complete loss of PFO production was observed.

III. THE CYTOLYTIC MECHANISM

PFO is probably the most well-studied toxin of this family of thiol-activated toxins with respect to its cytolytic mechanism, although a few salient studies have been carried out with tetanolysin. A significant body of work has also been carried out with other members of this toxin family which are produced by other bacterial genera. The results of studies from other related toxins must be included here for a coherent picture of the mechanism and structure of these toxins. Although there has been considerable work done on these toxins it should not mislead the reader into thinking that the cytolytic mechanism of PFO is well understood (nor is the cytolytic mechanism of other related toxins). There is still a considerable degree of ignorance that surrounds the mechanism of PFO and related toxins.

A. Membrane binding

There are three hallmark traits of the thiol-activated toxins: the loss of activity when the sensitive thiol is oxidized or modified, the requirement for cholesterol in the membrane for cytolytic activity and the oligomerization of the toxin monomers into large complexes on the membrane. The fact that PFO requires membrane cholesterol for activity suggests that it and others of this family of cytolysins evolved to target eukaryotic cells because few prokaryotic cells contain sterols in their membranes. Since PFO only attacks membranes containing cholesterol it has been assumed that it is the receptor for PFO and related toxins. Prigent and Alouf[17] initially described the structure of the sterol molecule that was required for binding of the streptolysin molecule. They found that the 3-β-hydroxyl group of cholesterol was absolutely required for toxin binding and that any changes in this group significantly decreased the activity of the toxin. Direct interaction of PFO with cholesterol has been demonstrated by Ohno-Iwashita *et al.*[4] using a method analogous to Western blotting. These workers separated erythrocyte lipids by thin-layer chromatography and then incubated the plate with radiolabeled PFO. After removing unbound PFO, by repeatedly washing the thin layer plate, the only membrane component that was bound by the radiolabeled PFO was cholesterol.

More recently, Nakamura *et al.*[18] determined that the interaction of cholesterol with PFO induced a significant increase in the intensity of the tryptophan emission of PFO, suggesting that the cholesterol was either interacting with one or more of the tryptophan residues of PFO or indirectly altering the environment of some of the tryptophan residues. It was suggested that the cholesterol bound at or near the highly conserved ECTGLAWEWWR peptide since this region is rich in tryptophan residues (three out of seven in PFO). Also, earlier work by Iwamoto *et al.*[19] suggested that modification of the unique cysteine in this peptide decreased membrane binding of PFO. However, studies on pneumolysin (PLY) by Boulnois *et al.*[20] indicated that the interaction of these toxins with erythrocyte membranes may be more complex than initially thought. These workers found that when PLY was modified with diethypyrocarbonate (which primarily targets histidines) it lost cytolytic activity and also the ability to bind to erythrocyte membranes. Histidine-367 of pneumolysin was the only histidine that was conserved in all of the related toxins. When His-367 of PLY was replaced with arginine[20] the same characteristics (loss of hemolytic activity and membrane binding) were observed in the mutated PLY as were observed with the chemically modified PLY. Based on these studies Boulnois *et al.*[20] suggested that cholesterol binding may be a post-binding event. Inherent in this suggestion is the possibility that the receptor for PLY and related toxins is not cholesterol, although currently there is insufficient data to confirm this possibility.

Although it is clear that the presence of cholesterol in the membrane is absolutely required for activity of PFO, it also appears that the packing of cholesterol in the membrane may be important. Early studies with various thiol-activated cytolysins[17,21–26] have shown that the presence of cholesterol is necessary for the activity of these toxins. In addition, it appears that the absolute concentration of cholesterol in the membrane is also important.[27] Rottem *et al.*[27] determined that when cholesterol concentration in liposome membranes was greater than 30 mol%, maximum binding of tetanolysin to liposomes was achieved. This observation was independent of the total cholesterol content in the assay. Sharp increases in binding of tetanolysin to the liposomes were observed when the liposomes contained 20 mol% and 30 mol% cholesterol. It was clear from these studies that the total amount of cholesterol in the assay was unimportant, but that the relative concentration of cholesterol in the membrane was a critical factor in tetanolysin binding. More recently, Ohno-Iwashita *et al.*[28] determined that PFO preferentially binds to cholesterol surrounded by C_{18} phospholipids in a cholesterol-rich region of the membrane. Interestingly, earlier studies on the packing structure of cholesterol in membranes determined that at a membrane concentration of approximately 20–40 mol% cholesterol is thought to undergo the formation of cholesterol-rich domains.[29–30] The intriguing possibility that tetanolysin and related toxins recognize a specific packing of cholesterol in the membrane is suggested by these data. However, the precise nature of the toxin–cholesterol interaction has not been elucidated.

The presence of a single cysteine in the conserved ECTGLAWEWWR motif of PFO and related toxins has been the subject of considerable interest for several reasons. One of the most curious aspects of this residue is the fact that modification of the cysteine eliminates activity, but the substitution of alanine for the cysteine does not appreciably change the cytolytic activity of the toxin. Why none of the sequenced toxins from this family of toxins have been found to contain an alanine substitution is curious because alanine-substituted toxins would be more stable than the cysteine-containing wild type. However, the modification of the sole cysteine of PFO (Cys-458) with sulfhydryl-specific reagents clearly has a dramatic effect on the toxin, considering that more than 99.9% of the activity is lost upon modification of this residue. Iwamoto *et al.*[19] found that modification of the thiol of Cys-458 with a reversible sulfhydryl reagent caused the loss of hemolytic activity and membrane binding. These investigators also found that PFO underwent a reversible conformational change after modification of Cys-458 with a reversible sulfhydryl reagent. These data suggest that the cysteine may be located at a critical region of the toxin that triggers a conformational change in the toxin, perhaps to facilitate an additional step of the cytolytic process which initiates toxin oligomerization.

Whether cholesterol interacts directly with the ECTGLAWEWWR peptide remains unclear but the work of Iwamoto *et al.*[19] and more recently that of Nakamura *et al.*[18] (see above) suggest that this possibility may be true. In contrast to these findings, however, Saunders *et al.*[12] and Pinkney *et al.*[11] found that substitution of the 'essential cysteine' of pneumolysin and streptolysin O with serine greatly decreased the hemolytic activity of these two toxins, but did not apparently affect membrane binding and oligomerization. The role of this highly conserved site therefore remains controversial. Why the cysteine residue has been retained in all thiol-activated toxins sequenced to date also remains a mystery. The fact that alanine could substitute for the cysteine and eliminate the oxidative inactivation of these toxins suggests that the sulfhydryl group of the cysteine may have an undiscovered function. It is possible that the sulfhydryl group functions in some other way that is not connected with the cytolytic activity of the toxin.

B. Oligomerization

The oligomerization of PFO on the membrane is a major step in the cytolytic mechanism of PFO. Once PFO has bound to the membrane of a target cell it is clear that the monomers diffuse laterally in the membrane until they encounter additional monomers with which they then oligomerize into large complexes. The process by which the monomers prepare for oligomerization and the forces involved in oligomerization remain unclear, although certain steps in the mechanism can be inferred from the available data. After the monomeric toxin binds to the membrane certain changes must occur in the structure which facilitate the next step in the cytolytic process, the oligomerization of the membrane-bound monomers into a supramolecular complex. PFO does not readily oligomerize in solution and can be concentrated to high levels (>25 mg/ml; R.K. Tweten, personal observation) without any appreciable oligomerization occurring. However, it appears that preparation of PFO or tetanolysin for electron microscopy can induce the formation of oligomers in the absence of membranes.[27,31] It is possible that the sites that facilitate monomer oligomerization are obscured in solution, but become exposed after membrane binding. However, it is also possible that these sites are always exposed and membrane binding simply concentrates the toxin to an extent that increases the probability that monomers collide and oligomerize.

Oligomerization is essential to the cytolytic process and probably is responsible for the formation of a cytolytic complex. Harris *et al.*[32] examined the oligomerization process using fluorescence energy transfer with donor (fluorescein) and acceptor labeled (rhodamine) PFO. The results of this work determined that oligomerization preceded cytolysis, that oligomerization was temperature-dependent and that the activation energy of oligomerization was approximately 78.5 kJ/mol. Subsequent work by Tweten *et al.*[33] and Iwamoto *et al.*[34] determined that a trypsin-derived carboxy-terminal fragment of PFO (T2), that began at Asn-304,[4] contained the receptor binding domain and one oligomerization site. This fragment inhibited native PFO by preventing oligomerization, apparently by forming inactive complexes of PFO and T2. From this work the general location of at least one oligomerization site and the receptor-binding region of PFO can be surmised to be present in the carboxy terminal region of PFO.

IV. THE ROLE OF PFO IN PATHOGENESIS

The role(s) of PFO in pathogenesis, specifically gas gangrene, has been recently explored by Awad *et al.*[35] using isogenic strains of *C. perfringens* type A, one of which carries an inactive *pfoA* gene. Two derivative strains of a type A *C. perfringens* were generated by homologous recombination of the chromosomal *pfoA* gene or α-toxin gene, *plc*, with a gene carrying a erythromycin-resistance determinant inserted into the respective toxin gene. The primary result of this work suggested that α-toxin was the major pathogenic factor for *C. perfringens*-induced gas gangrene and that PFO had a lesser role that was not clearly distinguished. Remarkably, although PFO is quite cytolytic *in vitro* it did not appear to contribute significantly to the development of tissue necrosis. Also, one of the hallmark characteristics of gas gangrene, the lack of infiltrating PMNLs into the necrotic tissue, was apparently due to the activity of the α-toxin and not PFO. Therefore, although this study did confirm the earlier notions that α-toxin was the main pathogenic factor during *C. perfringens* gas gangrene it did not significantly increase our understanding of the role of PFO. Although much has been written about the *in vivo* and *in vitro* effects of PFO[36–40] it is difficult to determine whether these effects contribute to the establishment or maintenance of the disease itself. The role of PFO and other clostridial thiol-activated cytolysins in clostridial disease may be difficult to elucidate if the contributions of these toxins are subtle. In a multifactorial disease process such as gas gangrene these effects will be difficult to isolate from the myriad effects of other factors produced by the organism and host.

V. SUMMARY

Certain features of the cytolytic and pathogenic mechanisms of PFO and other related clostridial thiol-activated toxins have been elucidated, but many other aspects of these processes in the thiol-activated toxins remain unclear. One of the most interesting problems of the PFO cytolytic mechanism is still not understood: the transition of PFO from a water-soluble protein to a membrane-bound complex. It is possible that PFO and related toxins form a porin-like structure in the membrane[41] with each of the monomers contributing to the overall formation of a amphipathic β-barrel that interacts with the lipid bilayer. This distribution of the hydrophobic face of the complex among the monomers may explain the relatively high solubility of the monomers compared with the oligomerized complex. Because each contributes only a small percentage of the overall lipid interface of the oligomer, each monomer could retain the properties of a soluble protein. Alternatively, these amphipathic domains may be obscured in the monomer and exposed only after membrane binding.

There still remains many undefined and intriguing features of the clostridial thiol-activated toxins. For example, cholesterol clearly plays an important role in the initial phase of the cytolytic process of these toxins, but its role is still incompletely understood. Is it the actual receptor or does it function in some post-binding event as suggested by Boulnois *et al.*?[42] Based on the studies with tetanolysin it also appears that these toxins discriminate between cholesterol packing motifs in membranes containing different concentrations of cholesterol; why a toxin would utilize a receptor such as cholesterol, but recognize a specific packing of the sterol in the membrane is an enigma. One possibility may be that the specific packing of cholesterol is required for the cytolytic process, perhaps the thiol-activated cytolysins oligomerize along the borders of cholesterol-rich domains and the bulk lipid.

The actual mechanism that these toxins use to disrupt membranes is not completely understood. A definable pore is not apparently produced by these toxins, as has been found for toxins such as aerolysin, *Clostridium septicum* α-toxin or *Staphylococcus aureus* α-toxin. From the planar membrane studies on tetanolysin it is clear that the membrane perturbation does not resemble the formation of a uniform pore or even a heterogeneous mixture of pores. The degree to which PFO must oligomerize to form a lytic complex is also unclear. Many of the electron micrographs that have been taken of the oligomerized structures of PFO and related toxins over the years by numerous researchers often show many more arcs than complete circles of oligomerized toxin. To date, it has not been determined if the arcs or rings are the functionally lytic complex. Therefore, the minimum size of oligomer that defines a lytic complex is unclear.

The role of PFO in pathogenesis remains somewhat obscure in view of the most recent gene knock-out experiments of Awad *et al.*.[35] These experiments clearly demonstrated that the major clinical syndromes of gas gangrene caused by *C. perfringens* are due mainly to the α-toxin. Any role for PFO in the disease was not obvious in mutants that lacked the *pfoA* gene, suggesting that PFO plays a more subtle role in the disease. This role may not be identified using current approaches because subtle effects in a multifactorial process would be difficult to dissect. Nevertheless, the importance of the clostridial thiol-activated cytolysins, and related cytolysins in general, cannot be underestimated since this toxin gene has been more widely disseminated among the Gram-positive bacterial pathogens than any other known bacterial toxin gene.

VI. REFERENCES

1. Tweten R.K. (1995) Pore-forming toxins of the gram positive bacteria. In Roth J.A., Bolin C.A., Brogden K.A., Minion C. and Wannemuehler M.J. (eds), *Virulence Mechanisms of Bacterial Pathogens*, pp. 207–230. American Society for Microbiology, Washington D.C.
2. Smyth C.J. (1975) The identification and purification of theta-haemolysin (theta-toxin) of *Clostridium perfringens* type A. *J Gen Microbiol* **87**: 219–238.
3. Tweten R.K. (1988) Cloning and expression in *Escherichia coli* of the perfringolysin O (theta-toxin) gene from *Clostridium perfringens* and characterization of the gene product. *Infect Immun* **56**: 3228–3234.
4. Ohno-Iwashita Y., Iwamoto M., Mitsui K., Kawasaki H. and Ando S. (1986) Cold-labile hemolysin produced by limited proteolysis of theta-toxin from *Clostridium perfringens. Biochemistry* **25**: 6048–6053.
5. Mitsui N., Mitsui K. and Hase J. (1980) Purification and some properties of tetanolysin. *Microbiol Immunol* **24**: 575–584.
6. Haque A., Sugimoto N., Horiguchi Y., Okabe T., Miyata T., Iwanaga S. and Matsuda M. (1992) Production, purification, and characterization of botulinolysin, a thiol-activated hemolysin of *Clostridium botulinum. Infect Immun* **60**: 71–78.
7. Tweten R.K. (1988) Nucleotide sequence of the gene for perfringolysin O (theta-toxin) from *Clostridium perfringens*: significant homology with the genes for streptolysin O and pneumolysin. *Infect Immun* **56**: 3235–3240.
8. Kehoe M.A., Miller L., Walker J.A. and Boulnois G.J. (1987) Nucleotide sequence of the streptolysin O (SLO) gene: structural homologies between SLO and other membrane-damaging, thiol-activated toxins. *Infect Immun* **55**: 3228–3232.
9. Walker J.A., Allen R.L., Falmagne P., Johnson M.K. and Boulnois G.J. (1987) Molecular cloning, characterization, and complete nucleotide sequence of the gene for pneumolysin, the sulfhydryl-activated toxin of *Streptococcus pneumoniae. Infect Immun* **55**: 1184–1189.

10. Haas A., Dumbsky M. and Kreft J. (1992) Listeriolysin genes: complete sequence of ILO from *Listeria ivanovii* and of ISO from *Listeria seeligeri*. *Biochim Biophys Acta* **1130**: 81–84.
11. Pinkney M., Beachey E. and Kehoe M. (1989) The thiol-activated toxin streptolysin O does not require a thiol group for activity. *Infect Immun* **57**: 2553–2558.
12. Saunders K.F., Mitchell T.J., Walker J.A., Andrew P.W. and Boulnois G.J. (1989) Pneumolysin, the thiol-activated toxin of *Streptococcus pneumoniae*, does not require a thiol group for *in vitro* activity. *Infect Immun* **57**: 2547–2552.
13. Shimizu T., Okabe A., Minami J. and Hayashi H. (1991) An upstream regulatory sequence stimulates expression of the perfringolysin O gene of *Clostridium perfringens*. *Infect Immun* **59**: 137–142.
14. Lyristis M., Bryant A.E., Sloan J., Awad M.M., Nisbet I.T., Stevens D.L. and Rood J.I. (1994) Identification and molecular analysis of a locus that regulates extracellular toxin production in *Clostridium perfringens*. *Mol Microbiol* **12**: 761–777.
15. Shimizu T., Bathein W., Tamaki M. and Hayashi H. (1994) The virR gene, a member of a class of two-Component response regulators, regulates the production of perfringolysin o, collagenase, and hemagglutinin in *Clostridium perfringens*. *J Bacteriol* **176**: 1616–1623.
16. Swanson R.V., Alex L.A. and Simon M.I. (1994) Histidine and aspartate phosphorylation: two-component systems and the limits of homology. *Trends Biochem Sci* **19**: 485–490.
17. Prigent D. and Alouf J.E. (1976) Interaction of streptolysin O with sterols. *Biochem Biophys Acta* **433**: 422–428.
18. Nakamura M., Sekino N., Iwamoto M. and Ohno I.Y. (1995) Interaction of theta-toxin (perfringolysin O), a cholesterol-binding cytolysin, with liposomal membranes: change in the aromatic side chains upon binding and insertion. *Biochemistry* **34**: 6513–6520.
19. Iwamoto M., Ohno-Iwashita Y. and Ando S. (1987) Role of the essential thiol group in the thiol-activated cytolysin from *Clostridium perfringens*. *Eur J Biochem* **167**: 425–430.
20. Boulnois G.J., Mitchell T.J., Saunders F.K., Mendez F.J. and Andrew P.W. (1990) Structure and function of pneumolysin, the thiol-activated toxin of *Streptococcus pneumoniae*. In Rappuoli R., Alouf J.E., Falmagne P., Fehrenbach F.J., Freer J., Gross R., Jeljaszewicz J., Montecucco C., Tomasi M., Wadström T. and Witholt B. (eds), *Bacterial Protein Toxins*, pp. 43–51. Gustav Fischer Verlag, Stuttgart.
21. Watson K.C., Rose T.F. and Kerr E.J. (1972) Some factors influencing the effect of cholesterol on streptolysin O activity. *J Clin Pathol* **25**: 885–891.
22. Shany S., Bernheimer A.W., Grushoff P.S. and Kim K.S. (1974) Evidence for membrane cholesterol as the common binding site for cereolysin, streptolysin O and saponin. *Mol Cell Biochem* **3**: 179–186.
23. Shany S., Bernheimer A.W., Grushoff P.S. and Kim K.S. (1975) Proceedings: Interaction of cereolysin and streptolysin O with red blood cell membrane cholesterol. *Isr J Med Sci* **11**: 1188–1189.
24. Duncan J.L. and Schlegel R. (1975) Effect of streptolysin O on erythrocyte membranes, liposomes, and lipid dispersions A protein–cholesterol interaction. *J Cell Biol* **67**: 160–174.
25. Hase J., Mitsui K. and Shonaka E. (1976) *Clostridium perfringens* exotoxins. IV Inhibition of the theta-toxin induced hemolysis by steroids and related compounds. *Japan J Exp Med* **46**:45–50.
26. Cowell J.L. and Bernheimer A.W. (1978) Role of cholesterol in the action of cereolysin on membranes. *Arch Biochem Biophys* **190**: 603–610.
27. Rottem S., Cole R.M., Habig W.H., Barile M.F. and Hardegree M.C. (1982) Structural characteristics of tetanolysin and its binding to lipid vesicles. *J Bacteriol* **152**: 888–892.
28. Ohno-Iwashita Y., Iwamoto M., Mitsui K., Ando S. and Iwashita S. (1991) A cytolysin, theta-toxin, preferentially binds to membrane cholesterol surrounded by phospholipids with 18-carbon hydrocarbon chains in cholesterol-rich region. *J Biochem (Tokyo)* **110**: 369–375.
29. Haberkorn R.A., Griffin R.G., Meadows M.D. and Oldfield E. (1977) Deuterium nuclear magnetic

resonance investigation of the dipalmatoyl lecithin-cholesterol-water system. *J Am Chem Soc* **99**: 7353–7355.

30. Gershfeld N.L. (1978) Equilibrium studies of lecithin-cholesterol interactions I Stoichiometry of lecithin-cholesterol complexes in bulk systems. *Biophys J* **22**: 469–488.
31. Mitsui K., Sekiya T., Okamura S., Nozawa Y. and Hase J. (1979) Ring formation of perfringolysin O as revealed by negative stain electron microscopy. *Biochem Biophys Acta* **558**: 307–313.
32. Harris R.W., Sims P.J. and Tweten R.K. (1991) Kinetic aspects of the aggregation of *Clostridium perfringens* theta toxin on erythrocyte membranes: A fluorescence energy transfer study. *J Biol Chem* **266**: 6936–6941.
33. Tweten R.K., Harris R.W. and Sims P.J. (1991) Isolation of a tryptic fragment from *Clostridium perfringens* θ-toxin that contains sites for membrane binding and self-aggregation. *J Biol Chem* **266**: 12449–12454
34. Iwamoto M., Ohnoiwashita Y. and Ando S. (1990) Effect of isolated C-terminal fragment of theta-toxin (perfringolysin-O) on toxin assembly and membrane lysis. *Eur J Biochem* **194**: 25–31.
35. Awad M.M., Bryant A.E., Stevens D.L. and Rood J.I. (1995) Virulence studies on chromosomal α-toxin and θ-toxin mutants constructed by allelic exchange provide genetic evidence for the essential role of α-toxin in *Clostridium perfringens*-mediated gas gangrene. *Mol Microbiol* **15**: 191–202.
36. Saito M., Ando S., Mitsui K., Homma Y. and Takenawa T. (1986) Activation of phosphatidic acid metabolism of human erythrocyte membranes by perfringolysin O. *Biochem Biophys Res Commun* **137**: 23–28.
37. Bremm K.D., König W., Pfeiffer P., Rauschen I., Theobald K., Thelestam M. and Alouf J.E. (1985) Effect of thiol-activated toxins (streptolysin O, alveolysin, and theta toxin) on the generation of leukotrienes and leukotriene-inducing and –metabolizing enzymes from human polymorphonuclear granulocytes. *Infect Immun* **50**: 844–851.
38. Saito M. (1983) Activation of the calcium permeability of erythrocyte membranes by perfringolysin O. *J Biochem (Tokyo)* **94**: 323–326.
39. Stevens D.L., Mitten J. and Henry C. (1987) Effects of alpha and theta toxins from *Clostridium perfringens* on human polymorphonuclear leukocytes. *J Infect Dis* **156**: 324–333.
40. Bryant A.E., Bergstrom R., Zimmerman G.A., Salyer J.L., Hill H.R., Tweten R.K., Sato H. and Stevens D.L. (1993) *Clostridium perfringens* invasiveness is enhanced by effects of theta toxin upon PMNL structure and function: the roles of leukocytotoxicity and expression of CD11/CD18 adherence glycoprotein. *FEMS Immunol Med Microbiol* **7**: 321–336.
41. Parker M.W., Buckley J.T., Postma J.P.M., Tucker A.D., Leonard K., Pattus F. and Tsernoglou D. (1994) Structure of the Aeromonas toxin proaerolysin in its water-soluble and membrane-channel states. *Nature* **367**: 292–295.
42. Boulnois G.J., Mitchell T., Saunders K., Mendez X. and Andrew P. (1991) Structure, function and role in disease of pneumolysin, the thiol-activated toxin of *Streptococcus-Pneumoniae*. In Ron E.Z. and Rottem S. (eds), *Microbial Surface Components and Toxins in Relation to Pathogenesis*, pp. 161–169. Plenum, New York.

Chapter 14

Clostridial Phospholipases

RICHARD W. TITBALL

Defence Evaluation and Research Agency, CBD Porton Down, Salisbury, Wiltshire, SP4 0JQ, UK

1. INTRODUCTION

In 1941 MacFarlane and Knight[1] reported a breakthrough in the understanding of the biochemical basis of bacterial disease by demonstrating that the main extracellular product of *Clostridium perfringens*, the α-toxin, was a phospholipase C enzyme. The finding that bacterial enzymes could possess toxic properties stimulated activity to identify toxic enzymes produced by other bacteria. The phospholipases C from *Clostridium novyi* were shown to be toxic.[2,3] However, the possibility that all phospholipases C were toxic was soon disproved. Miles and Miles[4] showed that the *Clostridium bifermentans* phospholipase C was, by their criteria, nontoxic and later Otnæss *et al.*[5] showed that the phospholipase C of *Bacillus cereus* was nontoxic. The basic tenet, that bacterial enzymes could act as potent toxins remained intact but clearly, enzymatic activity alone, was not sufficient for toxicity.

With a clarity of vision MacFarlane[6] reviewed the phospholipases C produced by the clostridia and developed a key hypothesis; that the toxicity of the *C. perfringens* and *C. bifermentans* enzymes correlated with hemolytic activity rather than with phospholipase C activity. This chapter aims to review the

The Clostridia: Molecular Biology and Pathogenesis
ISBN 0-12-595020-9

properties of the clostridial phospholipases and to revisit the suggestion of MacFarlane, with particular reference to the α-toxin of *C. perfringens.*

II. THE CLOSTRIDIAL PHOSPHOLIPASES

All of the clostridial phospholipases which have been characterized have phospholipase C activity (Table 14.1). The activity of these enzymes can be detected as a zone of turbidity surrounding colonies of bacteria cultured on a solid medium containing egg-yolk, indeed this forms the basis of the 'Nagler reaction', a diagnostic test for *C. perfringens.* Much of the pioneering work with these enzymes was carried out in the 1940s. The subsequent decline in interest in these enzymes has been reversed over the past five years and significant progress has been made in the understanding of the relationships of clostridial phospholipases C.

A. *C. perfingens* α–toxin

The *C. perfringens* α-toxin is the best studied of the clostridial phospholipases. The toxin has been shown to play a key role in the pathogenesis of experimental gas gangrene.[19–21] Like the other clostridial phospholipases C characterized, the enzyme is exported into the growth medium, and the deduced amino acid sequence of the protein reveals a typical signal sequence.[7] The α-toxin produced by strain NCTC8237 (the type strain) has been purified by several groups of workers.[7–10,22] Estimations of molecular weight, determined by gel filtration chromatography,[8,23,24] are generally in agreement with the predicted molecular weight from the deduced amino acid sequence, indicating that the protein exists as a monomer in solution. Not surprisingly, several groups used strain NCTC8237 as the source of the gene encoding the α-toxin (alternatively termed *cpa* or *plc*) for gene cloning and sequencing analysis.[7,12,25,26] More recently, the *cpa* gene has been isolated from a variety of other type A strains of *C. perfringens* and biotype B and C strains. In all biotypes, the gene has been shown to be present as a single copy on the chromosome, close to the origin of replication.[27,28] The deduced amino acid sequences of these toxins are not identical: three amino acid substitutions (Ala-174 to Asp-174, Thr-177 to Ala-177 and Ser-335 to Pro-335) are found in the deduced amino acid sequences of the α-toxins from strains 13, CER89L43 and CER89L1216,[29] strain 8-6[30] and strain NCIB10662.[31] Zinc-saturated forms of the α-toxins from these strains are more resistant to chymotrypsin than the α-toxin from strain NCTC8237.[29] Recently, α-toxin produced within the gut has been associated with a variety

Table 14.1. Phospholipases C produced by the clostridia

Clostridium species	Biotype	Name	Mol wt	pI	Substrate specificity	Ion requirements	Hemolysis	Lethality
C. perfringens	A–E	α-Toxin	42,500[a]	5.4–5.6[7–9]	PC,SPM,PS,LPC[8,10]	Zn^{2+}, Ca^{2+}[8]	+[b]	+[7]
C. bifermentans		Phospholipase C	42,776[c]	NR	NR	Ca^{2+}[13]	±[12]	–[12]
C. novyi	A	γ-Toxin	30,000[d]	7.0[15]	PC,SPM,LPC,PE,PS, PG[14,15]	Zn^{2+}, Ca^{2+},	+[e]	+[f]
	A	PI-PLC	30,000[15]	5.6, 6.0[15]	PI[15]	Mg^{2+}[14] None[g]	–[15]	NR
	B, D	β-Toxin	32,000[s]	NR	PC[17]	NR	+[i]	+[17]
C. sordellii		Phospholipase C	NR	NR	NR	NR	NR	NR
C. paraperfringens		Phospholipase C	NR	NR	NR	NR	+/–[j]	–[18]
C. absonum		Phospholipase C	NR	NR	NR	NR	+/–[k]	NR[18]

Hemolytic activity was considered to be complete (+), weak (±) or non-hemolytic (-). Enzymes were considered to be toxic if the lethal dose was less than 10 μg/mouse. NR = not reported.

[a] Calculated from deduced amino acid sequence of strain NCTC8237 mature protein without metal ions.[7]

[b] Sensitivity of erythrocytes from different species; cattle and mice > rabbit, man, sheep. Sheep erythrocytes show hot-cold lysis.[11]

[c] Calculated from deduced amino acid sequence of strain ATCC638 mature protein, without metal ions.[12]

[d] Determined by gel filtration chromatography.[14]

[e] Horse erythrocytes were more sensitive to hemolysis than sheep erythrocytes.[6,14]

[f] Toxicity of *C. haemolyticum* culture filtrate attributed to γ-toxin.[16]

[g] Zn^{2+}, Ca^{2+} and Mg^{2+} tested.[15]

[h] Determined by gel filtration chromatography (h).

[i] Hemolytic for sheep erythrocytes.[17]

[j] Strains of *C. paraperfringens* reported to vary from nonhemolytic to hemolytic on human blood agar.[18]

[k] Strains of *C. absonum* reported to be hemolytic on human blood agar.[18]

of diseases of animals and humans including enterotoxaemia and sudden death in certain breeds of calves[32] and piglets[33] and in necrotic enteritis in chickens.[34,35] The intragastric administration of a crude preparation of the α-toxin caused a disease in piglets with similar symptoms to *C. perfringens* type A infection.[32] The α-toxin has also been associated with some cases of sudden infant death syndrome in humans.[36] It seems likely that the enhanced chymotrypsin resistance of the α-toxin produced by some strains of *C. perfringens* may be important for the development of these diseases.

B. *Clostridium bifermentans* phospholipase C

Comparisons of the properties of the *C. bifermentans* phospholipase C and the *C. perfringens* α-toxin were the subject of many investigations during the early 1940s, although some of these studies require cautious interpretation because the enzymes used were almost certainly, by modern criteria, impure. Although the enzyme was immunologically related to the α-toxin, cross-neutralization studies showed that neutralization of the homologous enzyme by antiserum was at least 10 times more efficient.[4] The specific activities of the *C. bifermentans* strain ATCC638 phospholipase C, measured using purified phosphatidylcholine or sheep erythrocytes, has been reported to be 50 and 45 times, respectively, lower than the comparable activities of the α-toxin.[12] Miles and Miles[37] found that the toxicity of the *C. bifermentans* enzyme varied from strain to strain; the most potent enzyme was at least nine times less toxic than *C. perfringens* α-toxin. This observation raises the possibility that the phospholipases C produced by different strains of *C. bifermentans* are not identical. In the author's laboratory it has been found that DNA from two different strains (LPHL2825 and LPHL2534) reacted differently with a gene probe derived from the α-toxin open reading frame (Table 14.2) or oligonucleotide probes from within this region (unpublished data) also suggesting that different forms of the *C. bifermentans* enzyme exist.

C. *C. sordellii* phospholipase C

Clostridium sordellii can be distinguished from *C. bifermentans* by the absence of mannose in the cell walls of the latter[18] and the ability of *C. sordellii* to cause gangrenous disease in man and enterotoxaemia in sheep and cattle. There are no reports of work to characterize the phospholipase C produced by *C. sordellii*. However, the results of Southern blotting experiments suggest that the *C.*

Table 14.2. Southern hybridization of the *Clostridium perfringens* α-toxin probe with *Hin*dIII digested DNA from clostridia

Clostridium species	Biotype	Strain	Fragment size detected after hybridization of α-toxin probe with *Hin*dIII digest of DNA (kb)
C. perfringens	A	NCTC8237	3.1
	B	NCTC6121	3.1
	C	NCTC3180	3.1
	D	NCTC8346	3.1
	E	CN 5063	3.1
C. bifermentans		LPHL2825	3.6
		LPHL2534	–
C. absonum		ATCC27555	–
		ATCC27636	–
C. novyi	A	NCTC6737	3.1
	D	NCTC8350	3.1
C. sordellii		DBL1339	–

sordellii is not identical to the *C. perfringens* α-toxin, or the phospholipase C from all strains of *C. bifermentans* (Table 14.2).

D. *Clostridium novyi* phospholipases C

1. Clostridium novyi *γ-toxin*

The *C. novyi* γ-toxin, produced by type A strains of the bacterium, is immunologically distinct from the *C. novyi* ß-toxin and the *C. perfringens* α-toxin.[38] The relationship between this enzyme and the *C. perfringens* α-toxin awaits clarification. However, Southern hybridization studies indicate that a sequence with a high degree of homology to the *C. perfringens* α-toxin is present in DNA from *C. novyi* type A (Table 14.2). This observation, together with the demonstration that this enzyme, like the α-toxin, requires zinc and calcium ions for activity,[14] suggests that the γ-toxin is related to the *C. perfringens* α-toxin. The substrate specificity and hemolytic activity of the γ-toxin are not identical to those of the α-toxin and the further characterization of this enzyme could provide new insights into the molecular basis of substrate specificity.

2. Clostridium novyi *β-toxin*

The ß-toxin which is produced by type B and type D (formerly *C. haemolyticum*) strains of *C. novyi* is toxic[17] for mice, but the precise LD_{50} dose is not known.

The toxin is stable at 37°C but not at 60°C.[17] The relationship of this toxin to other clostridial (or bacterial) phospholipases C awaits investigation.

3. Clostridium novyi *PI-PLC*

Taguchi and Ikezawa[15] reported the purification of a phosphatidylinositol hydrolyzing enzyme (PI-PLC) by a type A strain of *C. novyi* which could be separated from the γ-toxin by isoelectric focusing. The enzyme did not require Zn^{2+}, Ca^{2+} or Mg^{2+} ions for activity and was not inhibited by EDTA or *o*-phenanthroline. The *C. novyi* PI-PLC may be related to the type I enzymes produced by *Staphylococcus aureus*, *Bacillus cereus*, and *Listeria monocytogenes* (PLC-A) which share significant amino acid sequence homology.[39] Like these enzymes, the *C. novyi* PI-PLC may be able to cleave the phosphatidylinositol-glycan-ethanolaine anchor for membrane proteins, which leads to the release of enzymes such as alkaline phosphatase and acetylcholinesterase.[40] The role of the *C. novyi* enzyme in disease has not been reported and, although the *L. monocytogenes* enzyme has been shown to play a role in mediating escape from the phagosome[41,42] it should be borne in mind that the latter is considered to be an intracellular pathogen.

E. *Clostridium absonum* phospholipase C

The production of a phospholipase C by *C. absonum* has been reported by Nakamura *et al.*[18] The enzyme has not been characterized, but antiserum against the *C. perfringens* α-toxin is partially neutralizing. Our finding, that an α-toxin gene probe did not react with *C. absonum* DNA under conditions of high stringency (Table 14.2) would suggest that the *C. absonum* phospholipase C is only distantly related to the *C. perfringens* α-toxin.

F. *Clostridium paraperfringens* phospholipase C

The *C. paraperfringens* phospholipase C is neutralized by a tenfold excess (compared with neutralization of the homologous enzyme) of antiserum against the *C. perfringens* α-toxin.[18] The finding that *C. paraperfringens* DNA reacted with an α-toxin gene probe (Table 14.2) also suggests a high degree of homology between the *C. paraperfringens* phospholipase C and the *C. perfringens* α-toxin.

III. GENE CLONING AND SEQUENCING REVEALS A FAMILY OF PHOSPHOLIPASES C

The comparison of immunological crossreactivities, and the reactivities of gene probes with DNA from phospholipase C producing clostridia suggested that the enzymes produced by these bacteria were related, but not identical. The first confirmation of this possibility was the finding that the deduced amino acid sequence of the *C. bifermentans* phospholipase C was 51% identical with the *C. perfringens* α-toxin.[12] The identity was greatest in the *N*-terminal regions of the proteins (Figure 14.1). The genes encoding other clostridial phospholipases C have not been isolated and characterized. However, the genes encoding the

```
CPA        W---DGKIDGTGTHAMIVTQGVSILENDLSKNEPESVRKN-LEILKENMHELQLGSTYPD
CBPLC      W---DGKKDGTGTHSLIAEHGLSMLNNDLSGNEYQQVKDN-IKILNEYLGDLKLGSTYPD
PC-PLC     WSAEDKHKEGVNSHLWIVNRAIDIMSRNT-----TLVKQDRVAILNEWRTELENGIYAAD
PLC-B      WSADNPTNTDVNTHYWLFKQAEKILAKDV-----NHMRANLMNELKKFDKQIAQGIYDAD
           *   .   .....*  . ... ......        .. . .  *..   ..  *   .*

CPA        YDKNAYDL--YQDHFWDPDTDNNFSKDNSWYLAYSIPDTGESQIRKFSALARYEWQRGNY
CBPLC      YDPNAYDL--YQDHFYDPDTGNNFTIDNSWYASYPIYDTSRNSVRKFATLAKNEWEKGNF
PC-PLC     YENPYYDNSTFASHFYDPDNGKTYI---------PFAKQAKETGAKYFKLAGESYKNKDM
PLC-B      HKNPYYDTSTFLSHFYNPDRDNTYL---------PGFANAKITGAKYFNQSVTDYREGKF
           ..   **   . .**..** ....          .    .     *.   .  ..   .

CPA        KQATFYLGEAMHYFGDIDTPYHPANVTAVDSAG--HVKFETFAEERKEQYKINTAGCKTN
CBPLC      KEATFLLGQGLHYLGDLNTPYHASNVTAVDSPG--HVKYETFVEERKDNYALNTSGNDTT
PC-PLC     KQAFFYLGLSLHYLGDVNQPMHAANFTNLSYPQGFHSKYENFVDTIKDNYKV------TD
PLC-B      DTAFYKLGLAIHYYTDISQPMHANNFTAISYPPGYHCAYENYVDTIKHNYQA------TE
           . * . ** ..** .*.. * *..* *... .   *  .*..... *..*        *.

CPA        EAFYTDILKNKDFNAWSKEYAR----GFAKTGKSI---YYSHASMSHSW-DDWDYAAKVT
CBPLC      SGVYKEAMENPSFNKWMTQNSI----KYAKIAKDL---YYSHSTMSHSW-DDWDYSGREA
PC-PLC     GNGYWN-WKGTNPEEWIHGAAVVAKQDYSGIVNDNTKDWFVKAAVSQEYADKWRAEVTPM
PLC-B      DMVAKR-FCSDDVKDWLYENAKRAKADYPKIVNAKTKKSYLVGN------SEWKKDTVEP
           .        . . . *    .      .. . ..     .  ..      ..*  .

CPA        ----LANSQKGTAGYIYRFLHDVSEGNDPSVGKNVKELVAYISTSGEKDAGTDDYMYFGI
CBPLC      ----IKNSQVCTAGFLYRFMNEVSNGNTGDNISLTNEFNIVLKTADNKYAGTDDNVYFGF
PC-PLC     TGKRLMDAQRVTAGYIQLWFD--TYGDR
PLC-B      TGARLRDSQQTLAGFLEFWSK--K-TNE
               . ..*   **..  . .  . ..  .  . ..*.   ..*...* ***** .***.

CPA        KTKDGKTQEWEMDNPGNDFMTGSKDTYTFKLKD-ENLKIDDIQNMWIRKRKYTAFSDAYK
CBPLC      ETNEGKKFEWKLDNAGNDFERNQVDNYILKTKDGEEVDINNISNYWIRKERLTSISDDWE
           .*..**. **..**.****  .  *.*..* ** *...*..* * **** . *..**...

CPA        PENIKIIANGKVVVDKDINEWISGNSTYNI-K
CBPLC      LSNFKLIANGKVIQQQDVNKVFTGNETYYINK
            .*.*.******. ..*.*. ..**.** * *
```

Figure 14.1 Alignment of the deduced amino acid sequences of the *Clostridium perfringens* α-toxin,[7] *Clostridium bifermentans* phospholipase C,[12] *Bacillus cereus* PC-PLC[43] and the *Listeria monocytogenes* PLC-B.[44] Residues known to be involved in the coordination of zinc ions in the *Bacillus cereus* PC-PLC[49] are shaded. From *Microb Rev* **57**: 347–366, with permission from the American Society for Microbiology.

phosphatidylcholine preferring phospholipases C (PC-PLC) from *B. cereus*[43] and the *L. monocytogenes* PLC-B[4] have been isolated, and the encoded proteins shown to have significant homology with the *C. perfringens* and *C. bifermentans* enzymes.[7,25,39,44]

IV. THE α-TOXIN COMPRISES TWO DOMAINS

A comparison of the amino acid sequences of the zinc-metallophospholipases C revealed that the PC-PLC and PLC-B proteins lacked a 120 amino acid region which was located at the *C*-terminus of the clostridial enzymes. This finding raised the possibility that the *C. perfringens* α-toxin and the *C. bifermentans* phospholipase C were composed of two domains. The crystallization of the α-toxin has been reported by Basak *et al.*[45] and a preliminary low resolution structure has revealed that the α-toxin is composed of two domains (unpublished data). The precise determination of the structure will allow a detailed comparison of the structure of the *N*-terminal domain with the *B. cereus* PC-PLC and should provide insight into the possible role of the *C*-terminal domain.

A. The *N*-terminal domain of the α-toxin has phospholipase C activity

The observation that the *N*-terminal region of the α-toxin showed homology with the phospholipases C from *B. cereus* and *L. monocytogenes* suggested that the *N*-terminal 250 amino acids of the *C. perfringens* and *C. bifermentans* enzymes possessed phospholipase C activity. This was confirmed by Titball *et al.*[46] who showed that a fragment of the α-toxin (*N*-terminal domain; Cpa_{1-249}) retained phospholipase C activity but did not possess hemolytic or lethal activities and had a much reduced sphingomyelinase activity. This result is in accordance with the observations that the *B. cereus* and *L. monocytogenes* enzymes were non-hemolytic and nontoxic phospholipases C.[5,47,48] The relationship between the *N*-terminal domain of the α-toxin and the *B. cereus* PC-PLC is further confirmed by comparing the locations of amino acids known, from the crystal structure, to be involved in the coordination of catalytically essential zinc ions in the *B. cereus* PC-PLC.[49] Additional evidence for the role of zinc ions in the α-toxin is derived from studies in which zinc-chelating agents, such as EDTA and *o*-phenanthroline, have been shown to inactivate the phospholipase C activity[8] and the chemical modification of histidine residues by diethylpyrocarbonate inactivates the enzyme only after pretreatment with EDTA,[50] which would expose otherwise protected imidazole groups. The precise function of individual histidine residues within the *N*-terminal domain has been determined by generating site-directed mutants of the toxin.[51] Replacement of those residues predicted to play an active site role ($His_{68,126,136,148}$) with glycine abolished phospholipase C, sphingomyelinase, hemolytic and lethal activities. Of these residues, His_{148} appeared to bind, with high affinity, a zinc ion(s) which played a key role in catalytic activity. Modification of $His_{46,207,121,241}$ did not affect the activity of the alpha-toxin. A His_{11} mutant would also be expected to have a reduced activity but this mutant was unstable and could not be characterized.

Therefore, it appears likely that the molecular architecture of the *N*-terminal domain of the α-toxin is similar to the *B. cereus* PC-PLC,[49] a seven-helix protein which forms a twisted barrel structure and contains three zinc ions within the putative active site region (Figure 14.2). The *C. perfringens* α-toxin has been shown to contain two zinc atoms[8,51] and this apparent discrepancy may be explained by considering that the third zinc atom in the *B. cereus* PC-PLC is weakly bound (this may be the zinc ion removed by EDTA treatment of the α-toxin)[51] and is present only under zinc-saturated conditions.[8] Histidine residues at positions 68, 126 and 136 appear to play a role in the binding of this zinc ion

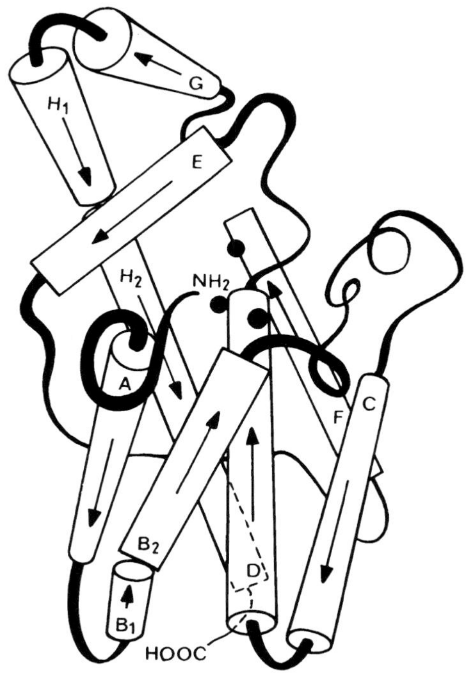

Figure 14.2 Schematic representation of the crystal structure of the *Bacillus cereus* PC-PLC. The alpha helical regions of the protein are shown as cylinders (identified with the letters A to G) and zinc ions are shown as filled circles. Redrawn from Hough *et al.* (1989), with permission from *Nature*.[49]

and replacement of this ion with other divalent cations such as manganese, cobalt or cadmium has been demonstrated.[51] The zinc ions within the α-toxin may serve three functions; by extrapolation from the *B. cereus* PC-PLC and from the data provided above it is likely that they contribute directly to the catalytic activity of the enzyme;[8] they also appear to play a role in stabilizing the protein, particularly against the effects of proteases;[52] and finally zinc ions may play a role in the binding of the α-toxin to cell membranes.[51]

B. The *C*-terminal domain of the α-toxin is responsible for hemolytic and lethal activity

The finding that the $His_{68,126,136,148}$ mutants of the α-toxin were devoid of enzyme, hemolytic and lethal activity[51] along with the finding that only the phospholipase C activity could be attributed directly to the *N*-terminal domain[46] suggested that the phospholipase C activity of the α-toxin was essential for hemolytic and lethal activity of the toxin. However, it is clear that phospholipase C activity alone is not sufficient for hemolysis and lethality. Of particular interest was the finding that the *N*-terminal domain of the α-toxin (Cpa_{1-249}) had reduced sphingomyelinase activity[46] since it had previously been shown that the sphingomyelinase activity of *B. cereus* was due to a separate enzyme from the PC-PLC and that these enzymes could act synergistically to yield a hemolytic complex (cereolysin A–B).[43] The observation that the PC-PLC and sphingomyelinase genes formed an operon, with the sphingomyelinase gene located 3′ to the PC-PLC gene might have indicated that the α-toxin gene arose from a fusion of the *B. cereus* genes.[30] Therefore, it was suggested that the *C*-terminal domain of the α-toxin possessed a sphingomyelinase activity which acted synergistically with the phospholipase C activity in the *N*-terminal domain to yield a hemolytic and lethal toxin. To investigate this possibility further, Titball *et al.*[53] isolated a fragment of the α-toxin ($Cpa_{247-370}$) which was devoid of any detectable enzyme or toxic activity.[53] While this result might have indicated that the $Cpa_{247-370}$ was incorrectly folded, this seems unlikely since the fragment was immunologically similar to the holotoxin. In addition, the mixing of Cpa_{1-249} and $Cpa_{247-370}$ partially restored hemolytic activity, presumably as a result of the binding of the individual domains in solution. This experiment clearly demonstrated that the *C*-terminal domain of the α-toxin is responsible for conferring hemolytic (and possibly lethal) properties on the α-toxin.

The function of the *C*-terminal domain of the α-toxin awaits clarification. The deduced amino acid sequence of $Cpa_{247-370}$ does not show significant homology with the deduced amino acid sequence of sphingomyelinases. However, significant homology has been reported with the *N*-terminal region of human arachidonate-5-lipoxygenase (HA5L),[46] an enzyme which is involved in the conversion of arachidonic acid into leukotrienes.[54,55] This sequence similarity, which includes five aligned tyrosine residues, may indicate the similar functions of these polypeptides in the binding to fatty acyl chains. In this respect it is of interest to note that the chemical modification of tyrosine residues in the α-toxin has been reported to abolish hemolytic activity.[56]

We may now consider why the *C. bifermentans* phospholipase C, which has a *C*-terminal domain, is nontoxic and only weakly hemolytic.[12] This apparent anomaly can be explained by considering that the *C. bifermentans* enzyme has a low turnover rate[12] and, taking this difference into consideration, the toxicity and hemolytic activity of the *C. bifermentans* enzyme are in line with expectations.

V. THE MOLECULAR BASIS OF TOXICITY

A. Activation of intracellular pathways

1. The arachidonic acid cascade

The observation that the *C. perfringens* α-toxin caused hemolysis led to the suggestion by early workers that generalized membrane damage would lead to edema, shock, hemorrhage and local tissue necrosis.[6] There has been a report of massive hemolysis, attributed to α-toxin, in a case of *C. perfringens* septicemia[57] but such gross effects of the toxin are considered to be exceptional. More recently, various workers have shown that the α-toxin is able to elicit a variety of subtle effects on cells, which leads to tissue dysfunction, shock and death (Figure 14.3). The hydrolysis of membrane phospholipids leads to the accumulation of diacylglycerol[58,59] within cells which can activate the arachidonic acid cascade.[58,59] The subsequent generation of leukotrienes and thromboxanes would result in inflammation, edema, platelet aggregation, muscle contraction and increases in vascular permeability.[55,60] All of these effects have been reported to be induced by the α-toxin[61–64] and the production of thromboxane A_2 in rat aorta cells treated with α-toxin has been demonstrated by Fujii and Sakurai.[8]

2. Protein kinase C

The activation of protein kinase C by diacylglycerol results in the activation of intracellular phospholipases,[65] establishing a positive feedback loop and potentiating the effects described above. Indeed, the production of arachidonic acid metabolites in cultured intestinal cells treated with α-toxin has been attributed

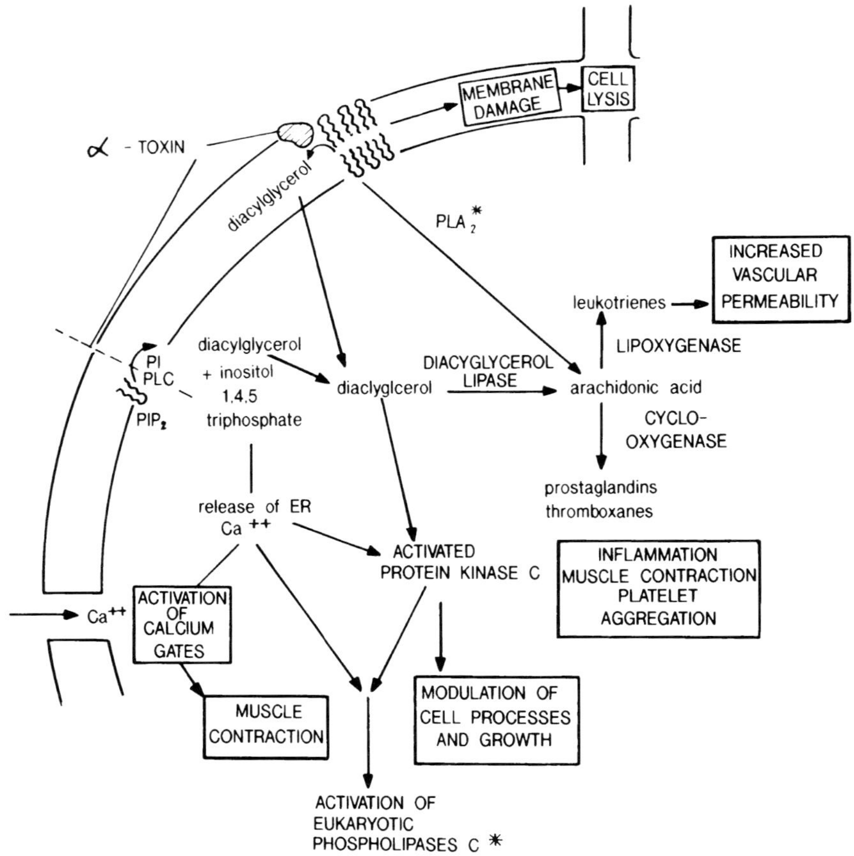

Figure 14.3 Biochemical pathways which are activated after treatment of cells with *Clostridium perfringens* α-toxin. Eukaryotic phospholipases which might be activated are shown starred. PLA2 = phospholipase A2; PI PLC = phosphatidylinositol phospholipase C; ER = endoplasmic reticulum; PIP2 = phosphatidylinositol biophosphate.

mainly to activated cell phospholipases A_2.[66] Sakurai *et al.*[67,68] have shown that treatment of rabbit erythrocytes with α-toxin leads to generation of diacylglycerol, inositol triphosphate and phosphatidic acid. While the diacylglycerol may be generated by the hydrolysis of membrane phospholipids, the formation of inositol triphosphate and phosphatidic acid would indicate activation of the phosphatidylinositol cycle.[69] These changes in rabbit erythrocyte membranes were accompanied by a decrease in phosphatidylinositol bisphosphate levels which would suggest that membrane phospholipases C were activated (since α-toxin is considered to be unable to hydrolyze phosphatidylinositol). In addition, treatment of erythrocyte membranes with pertussis toxin, which is known to ADP-ribosylate GTP-binding proteins involved in membrane phospholipase C activation, resulted in the abolition of phosphatidic acid formation.[68] These results suggest that the activation of membrane phospholipases may contribute to degradation of membrane phosphatidylinositol. The activation of membrane phospholipases is usually mediated by agonist binding to a cell surface receptor. The binding of the α-toxin to this receptor might lead to membrane phospholipase C activation.[68] Alternatively, it should be borne in mind that the products of phospholipid hydrolysis (e.g. phosphorylcholine and diacylglycerol) can lead to the activation of membrane phospholipases.[69]

3. Calcium mobilization

The hydrolysis of membrane phosphatidylinositol-4,5 bisphosphate, by the α-toxin or by the activated endogenous cellular phospholipases, would result in the generation of inositol 1,4,5 triphosphate,[65] which is known to stimulate the release of intracellular stores of calcium. Inositol 1,3,4,5 tetraphosphate, derived from inositol 1,4,5 triphosphate, is thought to lead to the activation of calcium gates, with the influx of calcium into the cell.[65] These changes in intracellular calcium levels lead to a variety of effects on cells, including smooth muscle contraction. It is likely that the contraction of the rat aorta and the rat ileum, after treatment with α-toxin, were induced by the activation of calcium gates since calcium entry blockers such as verapamil and nifedipine blocked the contraction.[63,64] The activation of calcium gates might also be responsible for the marked changes in frog muscle membrane potentials after treatment with α-toxin, which were reported by Boethius *et al.*[70] It seems likely that the prolonged exposure of smooth muscle to the α-toxin would lead to the loss of contractile ability and this may explain why the α-toxin has such a potent cardiotoxic activity leading to reduced heart rate, blood pressure and cardiac output.[71] The specific effects described above, which would lead to muscle contraction, inflammation and reduced cardiac activity may have an overall effect of reducing blood supply and promoting anoxia in the tissues adjacent to the focus of gas gangrene infection. This would promote the further growth of *C. perfringens* and may, in part, explain why the infection can spread rapidly through tissues.

B. The α-toxin is able to hydrolyze membrane phospholipids

Central to any of these mechanisms of toxicity is the ability of the α-toxin to cleave, or to induce cleavage of, membrane phospholipids. In this respect, the measurement of hemolysis can serve as an indicator of membrane phospholipid hydrolysis. The phospholipases C produced by the clostridia, and the *B. cereus* PC-PLC, show different hemolytic activities and can serve as models for

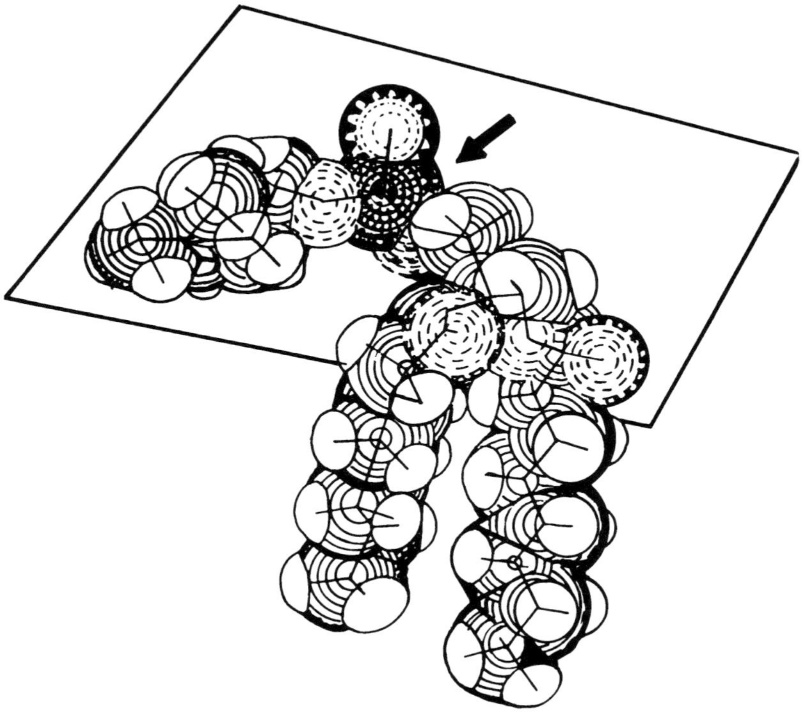

Figure 14.4 Model of membrane-integrated phosphatidylcholine, showing the interfacial area (boxed) which would be accessible to phospholipases C and the site of cleavage (arrowed). Charged regions on the phospholipid head group are also depicted (redrawn from El Sayed *et al.* (1985) with permission from Elsevier Science).[73]

examining the molecular basis of membrane interaction. It has been proposed that hemolytic phospholipases are able to enter membrane bilayers maintained at high lateral pressures,[9,72] suggesting that the hemolytic enzymes possessed mechanisms for promoting membrane insertion. The architecture of the membrane bilayer indicates that the head group of phospholipids would be exposed and accessible for cleavage by phospholipases C (Figure 14.4). However, phospholipid recognition by the *B. cereus* PC-PLC involves not only the head group but also the fatty acyl chains and the associated ester bonds,[73] both of which would be buried in the tightly packed membrane bilayer.

Other factors, such as membrane phospholipid composition may also influence the hemolytic potential of phospholipases C. The distribution of phospholipids in the membrane bilayer shows a high degree of asymmetry with phosphatidylcholine and sphingomyelin the main components of the outer leaflet.[74] These phospholipids are the preferential substrates for the α-toxin[8] whereas the non-hemolytic *B. cereus* PC-PLC is unable to hydrolyze sphingomyelin.[48,75] It should also be noted that a mixture of the *B. cereus* PC-PLC and sphingomyelinase causes hemolysis.[43]

The role of calcium in promoting the interaction of the α-toxin with membrane phospholipids has been reported by a number of workers. Calcium ions are known to be essential for phospholipase C activity[76] and for haemolytic activity of the α-toxin.[10] The role of calcium ions appears to be different from the role of zinc ions (which play a structural and catalytic role in the protein) and has been investigated by Moreau *et al.*[9] using phospholipid monolayers. In this system, calcium ions were shown to play an essential role in promoting the binding of the toxin to lipids. It has been proposed that calcium ions reduce the overall negative charge on the phospholipid and thus permit binding of the α-toxin; the finding that other cations are also able to promote lipid binding provides credence to this hypothesis.[77] The calcium ions may bind to the phosphate group of the phospholipid[11] thereby neutralizing the charge in this region and permitting binding of the α-toxin. However, other workers have suggested that calcium ions bind to the α-toxin first and may induce a conformational change in the protein prior to lipid binding.[78,79] This hypothesis would suggest the presence of a calcium-binding site within the α-toxin but the location of this site has not yet been determined.

VI. CONCLUSIONS

A number of the clostridial species produce phospholipases C and, in some cases, these enzymes have been characterized. However, a number of fundamental questions remain unanswered; what are the relationships of the *C. novyi* enzymes to the *C. perfringens* and *C. bifermentans* enzymes? Are the enzymes produced by different strains of *C. bifermentans* dissimilar? Many of these enzymes appear to be related and the cloning and sequencing of the genes encoding these enzymes will permit more detailed comparisons to be made. The properties of these enzymes are different with respect to substrate specificity and toxicity. Increasing opportunity exists for the rational comparison of the structure and function of these enzymes. Indeed, the range of clostridial enzymes which are available provides the researcher with a unique opportunity to explore structure–function relationships. These comparisons will provide new insights into the molecular basis of substrate specificity and the mechanisms by which proteins interact with membranes. Many of these insights will be applicable to other prokaryotic and eukaryotic phospholipases C. For example, there is an increasing awareness of the role of eukaryotic phospholipases C in the etiology of a wide range of inflammatory diseases.

With respect to the α-toxin of *C. perfringens*, many questions remain unanswered. The suggestion by MacFarlane[6] that toxicity of the clostridial phospholipases C correlates with hemolytic rather than phospholipase C activity appears well founded but the molecular basis of toxicity is still unclear. Clearly, the protein must interact with the cell membrane but the molecular basis of this interaction is not known. The determination of the crystal structure of the α-toxin will aid studies in this area. By whatever mechanism the α-toxin interacts with the membrane it is possible that activation of cell membrane phospholipases C, in addition to the inherent phospholipase C activity of the α-toxin, lead to membrane perturbation and activation of PKC and the arachidonic acid cascade.

VII. REFERENCES

1. MacFarlane M.G. (1948) The biochemistry of bacterial toxins. 3. The identification and immunological relations of lecithinases present in *Clostridium oedematiens* and *Clostridium sordellii* toxins. *Biochem J* **42**: 587–590.
2. Jasmin A.M. (1947) Enzyme activity in *Clostridium hemolyticum* toxin. *Am J Vet Res* **8**: 289–293.
3. Oakley C.L., Warrack G.H. and Clarke P.H. (1947) The toxins of *Clostridium oedematiens* (*Cl. novyi*). *J Gen Microbiol* **1**: 91–107.
4. Miles E.M. and Miles A.A. (1947) The lecithinase of *Clostridium bifermentans* and its relation to the α-toxin of *Clostridum welchii. J Gen Microbiol* **1**: 385–399.
5. Otnæss A-B., Giercksky K.E. and Prydz H. (1976) Parenteral administration of phospholipase C in the rat. Distribution, elimination, and lethal doses. *Scand J Clin Lab Invest* **36**: 553–559.
6. MacFarlane M.G. (1955) On the biochemical mechanism of action of gas-gangrene toxin. In *Mechanisms of Microbial Pathogenicity (5th Symposium of the Society for General Microbiology)*, pp. 57–77. Cambridge University Press, Cambridge.
7. Titball R.W., Hunter S.E.C., Martin K.L., Morris B.C., Shuttleworth A.D., Rubidge T., Anderson D.W. and Kelly D.C. (1989) Molecular cloning and nucleotide sequence of the alpha-toxin (phospholipase C) of *Clostridium perfringens. Infect Immun* **57**: 367–376.
8. Krug E.L. and Kent C. (1984) Phospholipase C from *Clostridium perfringens*: preparation and characterisation of homogenous enzyme. *Arch Biochem Biophys* **231**: 400–410.
9. Moreau H., Pieroni G., Jolivet-Raynaud C., Alouf J.E. and Verger R. (1988) A new kinetic approach for studying phospholipase C (*Clostridium perfringens* α-toxin) activity on phospholipid monolayers. *Biochem* **27**: 2319–2323.
10. Takahashi T., Sugahara T. and Ohsaka A. (1981) Phospholipase C from *Clostridium perfringens. Method Enzymol* **71**: 710–725.
11. McDonel J.L. (1986) Toxins of *Clostridium perfringens* types A, B, C, D and E. In Dorner F. and Drews J (eds) *Pharmacology of Bacterial Toxins*, pp. 477–517. Pergamon Press, Oxford.
12. Tso J.Y. and Siebel C. (1989) Cloning and expression of the phospholipase C gene from *Clostridium perfringens* and *Clostridium bifermentans. Infect Immun* **57**: 468–476.
13. Lewis G.M. and MacFarlane M.G. (1953) The lecithinase of *Clostridium bifermentans* toxin. *Biochem J* **54**: 138–142.
14. Taguchi R. and Ikezawa H. (1975) Phospholipase C from *Clostridium novyi* type A.I. *Biochim Biophys Acta* **409**: 75–85.
15. Taguchi R. and Ikezawa H (1978) Phosphatidyl inositol-specific phospholipase C from *Clostridium novyi* type A. *Arch Biochem Biophys* **186**: 196–201.

16. MacFarlane M.G. (1950) The biochemistry of bacterial toxins. 4. The lecithinase activity of *Clostridium haemolyticum* toxin. *Biochem J* **47**: 267–270.
17. Darakhshan H. and Lauerman L.H. (1981) Some properties of beta-toxin produced by *Clostridium haemolyticum* strain IRP-135. *Comp Immun Microbiol Infect Dis* **4**: 307–316.
18. Nakamura S., Shimamura T., Hayase M. and Nishida S. (1973) Numerical taxonomy of saccharolytic clostridia, particularly *Clostridium perfringens*-like strains: descriptions of *Clostridium absonum* sp.n. and *Clostridium paraperfringens*. *Int J Syst Bacteriol* **23**: 419–429.
19. Awad M.M., Bryant A.E., ,Stevens D.L. and Rood J.I. (1995) VIrulence studies on chromosomal α-toxin and θ-toxin mutants constructed by allelic exchange provide genetic evidence for the essential role of α-toxin in *Clostridium perfringens*-mediated gas gangrene. *Mol Microbiol* **15**: 191–202.
20. Ninomiya M., Matsustuta O., Minami J., Sakamoto H., Nakano M. and Okabe A. (1994) Role of alpha-toxin in *Clostridium perfringens* infection determined using recombinants of *C. perfringens* and *Bacillus sulphilis*. *Infect Immun* **62**: 5032–5039.
21. Williamson E.D. and Titball R.W. (1993) A genetically engineered vaccine against the alpha-toxin of *Clostridium perfringens* also protects mice against experimental gas gangrene. *Vaccine* **11**: 1253–1258.
22. Smyth C.J. and Arbuthnott J.P. (1974) Properties of *Clostridium perfringens* (*welchii*) type-A alpha-toxin (phospholipase C) purified by electrofocusing. *J Med Micribiol* **7**: 41–67.
23. Mitsui K., Mitsui N. and Hase J. (1973) *Clostridium perfringens* exotoxins. I. Purification and properties of the α-toxin. *Japan J Exp Med* **43**: 65–80.
24. Möllby R. and Wadström T. (1973) Purification of phospholipase C (alpha-toxin) from *Clostridium perfringens*. *Biochim Biophys Acta* **321**: 569–584.
25. Leslie D., Fairweather N., Pickard D., Dougan G. and Kehoe M. (1989) Phospholipase C and haemolytic activities of *Clostridium perfringens* alpha-toxin cloned in *Escherichia coli*: sequence and homology with a *Bacillus cereus* phospholipase C. *Mol Microbiol* **3**: 383–392.
26. Okabe A., Shimizu T. and Hayashi H. (1989) Cloning and sequencing of a phospholipase C gene of *Clostridium perfringens*. *Biochem Biophys Res Comm* **160**: 33–39.
27. Canard B. and Cole S.T. (1989) Genome organisation of the anaerobic pathogen *Clostridium perfringens*. *Proc Natl Acad Sci USA* **86**: 6676–6680.
28. Canard B., Saint-Joanis B. and Cole S.T. (1992) Genomic diversity and organisation of virulence genes in the pathogenic anaerobe *Clostridium perfringens*. *Mol Microbiol* **6**: 1421–1429.
29. Ginter A., Williamson E.D., Dessy F., Coppe Ph., Fearn A. and TItball R.W. (1996) Molecular variation between the α-toxins from the type strain (NCTC8237) and clinical isolates of *Clostridium perfringens* associated with disease in man and animals. *Microbiology* **142**: 191–198.
30. Saint-Joanis B., Garnier T. and Cole S.T. (1989) Gene cloning shows the alpha-toxin of *Clostridium perfringens* to contain both sphingomyelinase and lecithinase activities. *Mol Gen Genetics* **219**: 453–460.
31. Katayama S-., Matsushita O., Minami J., Mizobuchi S. and Okabe A. (1993) Comparison of the alpha-toxin genes of *Clostridium perfringens* type A and C strains: evidence for extragenic regulation of transcription. *Infect Immun* **61**: 457–463.
32. Cygan Z. and Buczek J. (1993) Nonenterotoxigenic anaerobe *C. perfringens* A as a cause of calf diarrhoea. *Medycyna Wet* **49**: 469–471.
33. Johannsen U., Arnold P., Kohler B. and Selbitz H.J. (1993) Untersuchungen zur experimentellen *Clostridium-perfringens*-Typ-A-Enterotoxamie der Saugferkel. *Mh Vet-Med* **48**: 129–136.
34. Long J.R. and Truscott R.B. (1976) Necrotic enteritis in broiler chickens. III Reproduction of the disease. *Can J Comp Med* **40**: 53–59.
35. Baba E., Fuller A.L., Gilbert J.M., Thayer S.G. and McDougald L.R. (1992) Effects of *Eimeria brunetti* infection and dietary zinc on experimental induction of necrotic enteritis in broiler chickens. *Avian Dis* **36**: 59–62.

36. Murrell W.G., Stewart B.J., O'Neill C., Siarakas S. and Kariks S. (1993) Enterotoxigenic bacteria in the sudden infant death syndrome. *J Med Microbiol* **39**: 114–127.
37. Miles E.M. and Miles A.A. (1950) The relation of toxicity and enzyme activity in the lecithinases of *Clostridium bifermentans* and *Clostridium welchii. J Gen Microbiol* **4**: 22–35.
38. MacFarlane M.G. and Knight B.C.J.G. (1941) The biochemistry of bacterial toxins. I. Lecithinase activity of *C. welchii* toxins. *Biochem J* **35**: 884–902.
39. Titball R.W. (1993) Bacterial phospholipases C. *Microbiol Rev* **57**: 347–366.
40. Ikezawa H. (1986) The physiological action of bacterial phosphatidylinositol-specific phospholipase C. The release of ectoenzymes and other effects. *J Toxicol Toxin Rev* **5**: 1–24.
41. Camilli A., Goldfine H. and Portnoy D.A. (1991) *Listeria monocytogenes* mutants lacking phosphatidylinositol-specific phospholipase C are avirulent. *J Exp Med* **173**: 751–754.
42. Mengaud J., Braun-Breton C. and Cossart P. (1991) Identification of a phosphatidylinositol-specific phospholipase C in *Listeria monocytogenes*: a novel type of virulence factor? *Mol Microbiol* **5**: 367–372.
43. Gilmore M.S., Cruz-Rodz A.L., Leimeister-Wachter M., Kreft J. and Goebel W. (1989) A *Bacillus cereus* cytolytic determinant, cereolysin AB, which comprises the phospholipase C and sphingomyelinase genes; nucleotide sequence and genetic linkage. *J Bacteriol* **171**: 744–753.
44. Vazquez-Boland J-A., Kocks C., Dramsi S., Ohayon H., Geoffroy C., Mengaud J. and Cossart P. (1992) Nucleotide sequence of the lecithinase operon of *Listeria monocytogenes* and possible role of lecithinase in cell–cell spread. *Infect Immun* **60**: 219–230.
45. Basak A., Stuart D.I., Nikura T., Bishop D.H.L., Kelly D.C., Fearn A. and Titball R.W. (1994) Purification, crystallisation and preliminary X-ray diffraction studies of alpha-toxin of *Clostridium perfringens. J Mol Biol* **244**: 648–650.
46. Titball R.W., Leslie D.L., Harvey S. and Kelly D.C. (1991) Haemolytic and sphingomyelinase activities of *Clostridium perfringens* alpha-toxin are dependent on a domain homologous to that of an enzyme from the human arachidonic acid pathway. *Infect Immun* **59**: 1872–1874.
47. Geoffroy C., Raveneau J., Beretti J-L., Lechroisey A., Vaquez-Boland J-A., Alouf J.E. and Berche P. (1991) Purification and characterisation of an extracellular 29-kilodalton phospholipase C from *Listeria monocytogenes. Infect Immun* **59**: 2382–2388.
48. Otnæss A-B., Little C., Sletten K., Wallin R., Johnsen S., Flensgrud R. and Prydz H. (1977) Some characteristics of phospholipase C from *Bacillus cereus. Eur J Biochem* **79**: 459–468.
49. Hough E., Hansen L.K., Birknes B., Jynge K., Hansen S., Hordvik A., Little C., Dodson E. and Derewenda Z. (1989) High resolution (1.5Å) crystal structure of phospholipsae C from *Bacillus cereus. Nature* **338**: 357–360.
50. Titball R.W and Rubidge T. (1990) The role of histidine residues in the alpha toxin of *Clostridium perfringens. FEMS Microbiol Lett* **68**: 261–266.
51. Nagahama M., Okagawa Y., Nakayama T., Nishioka E. and Sakurai J. (1995) Site-directed mutagenesis of histidine residues in *Clostridium perfringens* alpha-toxin. *J Bacteriol* **177**: 1179–1185.
52. Sato H., Yamakawa Y., Ito A. and Murata R. (1978) Effect of zinc and calcium ions on the production of alpha-toxin and proteases by *Clostridium perfringens. Infect Immun* **20**: 325–333.
53. Titball R.W., Fearn A.M. and Williamson E.D. (1993) Biochemical and Immunological properties of the C-terminal domain of the alpha-toxin of *Clostridium perfringens. FEMS Microbiol Lett* **110**: 45–50.
54. Pelech S.L. and Vance D.E. (1989) Signal transduction via phosphatidylcholine cycles. *TIBS* **14**: 28–30.
55. Samuelsson B. (1983) Leukotrienes: mediators of immediate hypersensitivity reactions and inflammation. *Science* **220**: 568–575.
56. Sakurai J., Fujii Y., Torii K. and Kobayashi K. (1989) Dissociation of various biological activities of *Clostridium perfringens* alpha toxin by chemical modification. *Toxicon* **27**: 317–323.
57. Hübl W., Mostbeck B., Hartleb H., Pointner H., Kofler K. and Bayer P.M. (1993) Investigation of the

pathogenesis of massive hemolysis in a case of *Clostridium perfringens* septicemia. *Ann Hematol* **67**: 145–147.

58. Fujii Y. and Sakurai J. (1989) Contraction of the rat isolated aorta caused by *Clostridium perfringens* alpha toxin (phospholipase C): evidence for the involvement of arachidonic acid metabolism. *Br J Pharmacol* **97**: 119–124.
59. Diener M., Eglème C. and Rummel W. (1991) Phospholipase C-induced anion secretion and its interaction with carbachol in the rat colonic mucosa. *Eur J Pharmacol* **200**: 267–276.
60. Wasserman M.A., Smith III E.F., Underwood D.C. and Barnette M.A. (1991) Pharmacology and pathophysiology of 5-lipoxygenase products. In Crooke S.T. and Wong A. (eds), *Lipoxygenases and Their Products*, pp. 1–50. Academic Press Inc, San Diego, CA.
61. Sugahara T., Takahashi T., Yamaya S. and Ohsaka A. (1976) *In vitro* aggregation of platelets induced by alpha-toxin (phospholipase C) of *Clostridium perfringens. Jap J Med Sci Biol* **29**: 255–263.
62. Sugahara T., Takahashi T., Yamaya S. and Ohsaka A. (1977) Vascular permeability increase by α-toxin (phospholipase C) of *Clostridium perfringens. Toxicon* **15**: 81–87.
63. Fujii Y., Nomura S., Oshita Y. and Sakurai J. (1986) Excitatory effect of *Clostridium perfringens* alpha toxin on the rat isolated aorta. *Br J Pharmacol* **88**: 531–539.
64. Sakurai J., Fujii Y. and Shirotani M. (1990) Contraction induced by *Clostridium perfringens* alpha toxin in the isolated rat ileum. *Toxicon* **28**: 411–418.
65. Berridge M.J. (1987) Inositol triphosphate and diacylglycerol: two interacting secondary messengers. *Ann Rev Biochem* **56**: 159–193.
66. Gustafson C. and Tagesson C. (1990) Phospholipase C from *Clostridium perfringens* stimulates phospholipase A_2-mediated arachidonic acid release in cultured intestinal epithelial cells (INT 407). *Scand J Gastroenterol* **25**: 363–371.
67. Sakurai J., Ochi S. and Tanaka H. (1990) Evidence for coupling of *Clostridium perfringens* alpha-toxin-induced hemolysis to stimulated phosphatidic acid formation in rabbit erythrocytes. *Infect Immun* **61**: 3711–3718.
68. Sakurai J., Ochi S. and Tanaka H. (1994) Regulation of *Clostridium perfringens* alpha-toxin-activated phospholipase C in rabbit erythrocyte membranes. *Infect Immun* **62**: 717–721.
69. Waite M. (1987) *Handbook of Lipid Research. 5. The Phospholipases.* Plenum Press, New York.
70. Boethius J., Rydqvist B., Mölby R. and Wadström T. (1973) Effect of a highly purified phospholipase C on some electrophysiological properties of the frog muscle fibre membrane. *Life Sciences* **13**: 171–176.
71. Siarakas S., Damas E. and Murrell W.G. (1995) Is cardiorespiratory failure induced by bacterial toxins the cause of sudden infant death syndrome? Studies with an animal model (the rabbit). *Toxicon* **33**: 635–649.
72. van Deenen L.L.M., Demel R.A., Guerts van Kessel W.S.M., Kamp H.H., Roelofsen B., Verkleij A.J., Wietz K.W.A. and Zwaal R.F.A. (1976) Phospholipases and monolayers as tools in studies on membrane structure. In Hatefi Y. and Djavadi-Ohaniance L. (eds), *The Structural Basis of Membrane Function*, pp. 21–28. Academic Press, London.
73. El-Sayed M.Y., DeBose C.D., Coury L.A. and Roberts M.F. (1985) Sensitivity of phospholipase C (*Bacillus cereus*) activity to phosphatidylcholine structural modifications. *Biochim Biophys Acta* **837**: 325–335.
74. Verkleij A.J., Zwaal R.F.A., Roelofsen B., Comfurius P., Kastelijn D. and van Deenen L.L.M. (1973) The asymmetric distribution of phospholipids in the human red cell membrane. A combined study using phospholipases and freeze-etch electron microscopy. *Biochim Biophys Acta* **323**: 178–193.
75. Little C., Aurebekk B. and Otnæss A-B. (1975) Purification by affinity chromatography of phospholipase C from *Bacillus cereus. FEBS Lett* **52**: 175–179.
76. Kurioka S. and Matsuda M. (1976) Phospholipase C assay using p-nitrophenylphosphorylcholine

and its application to studying the metal and detergent requirement of the enzyme. *Anal Biochem* **75**: 281–289.

77. Bangham A.D. and Dawson R.M.C. (1962) Electrokinetic requirements for the reaction between *Cl. perfringens* α-toxin (phospholipase C) and phospholipid substrates. *Biochim Biophys Acta* **59**: 103–115.
78. Young P.R., Snyder W.R. and McMahon F. (1992) Kinetic mechanism of *Clostridium perfringens* phospholipase C. *Biochem J* **280**: 407–410.
79. Klein R., Miller N., Kemp P. and Laser H. (1975) The activation of phospholipase C from *Clostridium perfringens* by quinine: an absolute requirement for calcium ions. *Chem Phys Lipids* **15**: 15–26.

Chapter 15

The *Clostridium perfringens* β-toxin

SOPHIE E.C. LEARY AND RICHARD W. TITBALL
Defence Evaluation and Research Agency, CBD Porton Down, Salisbury, Wiltshire SP4 0JQ, UK

I. INTRODUCTION

Clostridium perfringens produces several exotoxins, including four major lethal toxins, which play an important part in the pathogenesis of infection of this organism. The differential production of the major toxins is also used to classify the bacterium into five toxigenic types, A, B, C, D and E. β-Toxin is produced by types B and C, and its association with diseases of domestic livestock has been recognized since the early 1930s.[1] It is now recognized as the major virulence determinant in enterotoxaemia and necrotic enteritis of a variety of animal species, particularly sheep, lambs, calves, piglets and fowl (see Chapter 10).

It is also known to cause a similar disease in humans called enteritis necroticans or pig-bel, which is endemic in the Highlands of Papua New Guinea (see Chapter 12).[2] The conditions leading to the onset of the human disease are curious, and are associated with the consumption of under-cooked meat contaminated with *C. perfringens* type C spores by individuals who have a reduced level of intestinal proteases resulting from malnourishment, or who consume other foods containing protease inhibitors.[3] The importance of ß-toxin and the role of these contributing factors in causing pig-bel was demonstrated using a guinea-pig model.[4]

The Clostridia: Molecular Biology and Pathogenesis
ISBN 0-12-595020-9

The diseases in both humans and animals can be controlled effectively by the administration of β-toxoid, prepared from *C. perfringens* type C culture filtrates. For example, the incidence of pig-bel in children in Papua New Guinea was reduced eightfold by vaccination.[5] Despite the significance of the disease in veterinary medicine and some human populations, relatively little is known about the biophysical properties of the toxin, or its mode of action. However, the recent application of molecular techniques has yielded further information which should contribute significantly to our understanding of this toxin.

II. MOLECULAR GENETICS

The gene encoding β-toxin (*cpb*) has been isolated and cloned from two *C. perfringens* strains: the type B strain NCTC8533[6] and the type C strain CPC-Cph1.[7] The DNA sequences of the cloned fragments, which included the *cpb* open reading frame and 332 bp (type B) or 198 bp (type C) upstream, differed by only two nucleotides. Both of the changes occurred within the *cpb* gene, although only one affected the amino acid sequence (at position 144 in the mature protein), resulting in the substitution of lysine (type C) for glutamic acid (type B).[7]

Loss of β-toxin production by *C. perfringens* type C strains has been linked to the disappearance of a 75 MDa plasmid,[8] thus indicating that the gene may be plasmid-encoded. This was supported by the observation that a *cpb* probe hybridized only with DNA from the ß-toxin-producing types, B and C.[6] A PCR assay using *cpb* primers confirmed that the gene was found only in types B and C; no cross-reaction was detected with nine other species of *Clostridium* or a range of intestinal pathogenic bacteria.[9] A diagnostic polymerase chain reaction (PCR) assay also specifically detected the *cpb* gene in *C. perfringens* strains isolated from cases of necrotic enteritis in piglets.[9]

III. TOXICITY

Clostridium perfringens is probably unique not only in terms of the number of protein toxins that it produces, but also in terms of their toxicity and lethal activity, and β- and ε-toxins are the most potent. The LD_{50} of purified β-toxin in mice has been reported to be between 0.31 and 12.5 μg/kg by the intravenous route and between 4.5 and 93.5 μg/kg when administered intraperitoneally.[10–12]

Interestingly, recombinant β-toxin expressed in *Escherichia coli* as a fusion with glutathione S-transferase (GST) was nontoxic in mice when administered intravenously or intraperitoneally at doses of up to 25 μg or 1.25 μg, respectively, and lethal activity was not restored after cleavage of the toxin from the carrier protein.[7] However, polyclonal antiserum raised against the GST–β-toxin fusion neutralized the lethal activity of native β-toxin in type C culture filtrates, indicating that the antigenicity of the recombinant protein was retained.

IV. BIOLOGICAL PROPERTIES

Although the gross pathologic effects of ß-toxin in humans and livestock have been described,[13,14] the biological activities underlying these effects are not understood. It has been demonstrated that 40–100 μg of purified β-toxin causes hemorrhagic necrosis and severe destruction of intestinal villi and mucosae in guinea-pig intestinal loops,[15] which is similar to the effects caused by type C culture filtrates.[4,15] Parnas[16] also demonstrated that β-toxin may affect the mobility of villi in the intestines of rabbits. Clearly, the ability of β-toxin to induce these types of biological effects in the intestine can be linked to the lesions encountered in the natural disease. However, it has also been shown that β-toxin has wider effects both *in vitro* and *in vivo*. The toxin is known to be dermonecrotic in guinea-pigs[11] and it has a number of pharmacological effects. Following intravenous inoculation, purified β-toxin caused an increase in the blood pressure of rats, and a corresponding constriction of the arteries.[17,18] It was proposed that this hypertensive effect was caused by the action of toxin on the autonomic nervous system and the release of catecholamines. The effect was inhibited by the co-injection of antitoxin and the use of α-adrenergic and ganglionic blocking agents.

The *in vitro* effects of β-toxin on cultured cells have been documented to a limited extent. Allen[19] reported that the toxin was cytotoxic to primary monocytes from guinea-pigs. Jolivet-Reynaud *et al.*[15] studied the effects of β-toxin on a number of cell lines, including Vero cells, mouse and human fibroblasts, transformed intestine cells, Hep cells and Chinese hamster ovary (CHO) cells. Cytotoxic activity was recorded only in CHO cells which had been treated with a relatively low dose of the toxin (0.1–10 μg), although subsequent cell lysis was not observed even after incubation for 5 h. The significance or mechanism of this cytotoxic activity is unknown.

V. PHYSICAL PROPERTIES

Molecular cloning of the *cpb* gene indicated that the toxin consisted of a single polypeptide chain of 336 amino acids. A signal peptide of 27 amino acids was located at the *N*-terminus, and when this fragment was cleaved, the predicted molecular mass of the mature protein was approximately 34.9 kDa.[6] Previously, the molecular mass of native β-toxin has been reported to be 20–47.5 kDa,[10,15,20,21] although when highly purified, it was estimated to be 30–32 kDa by gel filtration[11,12,22] and 40 kDa by sodium dodecyl sulfate–polyacrylamide gel electrophoresis (SDS–PAGE).[11] The protein purified by Sakurai and Fujii[11] lacked detectable phospholipase C activity and was shown to be homogenous by native PAGE and SDS–PAGE. In addition, it gave a single precipitin line with diagnostic *C. perfringens* types B and C antiserum, and only one peak of lethal activity at pH 5.6 was observed after isoelectric focusing. Interestingly, ß-toxin purified by the method of Sakurai and Fujii[11] was subsequently shown by Western blotting with an anti β-toxin monoclonal antibody to consist of two forms with molecular masses of approximately 34 and 118 kDa.[6] The 118 kDa form was detected even after boiling in the presence of 2.5% 2-mercaptoethanol.

Recombinant β-toxin purified from the periplasm of *E. coli* also contained the 34 and 118 kDa forms, and it was suggested that the smaller component corresponded to the monomeric gene product and the larger component represented an oligomer consisting of three to four toxin subunits.[6] When recombinant β-toxin was fused to GST, only the monomeric form of the fusion protein with a molecular mass of 62 kDa was detected.[7]

Sakurai and Duncan[12] showed that the dermonecrotic activity of β-toxin was reduced by 75% after incubating at 50°C for 5 min, indicating that it is relatively heat-labile. In addition, they demonstrated that it was sensitive to proteolysis, and when treated with trypsin at 37°C, toxin activity was almost completely lost within 30 min. The susceptibility of β-toxin to trypsin digestion would normally prevent it from reaching high concentrations in the human intestine and may explain why β-toxin-induced disease is linked to unusual dietary conditions affecting the level of intestinal proteases.[3]

VI. STRUCTURE–FUNCTION RELATIONSHIPS

In their purification method, Sakurai and Fujii[11] used an affinity column of zinc-chelate Sepharose to bind β-toxin, indicating that the protein contained exposed histidine and/or cysteine residues. From the deduced amino acid sequence of the mature protein, it is known that the toxin contains four histidine residues at positions 85, 175, 293 and 304 and a single cysteine at position 265.[6]

Treatment of β-toxin with thiol-modifying or oxidizing agents, which target the single cysteine, abolished dermonecrotic and lethal activity, although this reaction was reversed by reduction.[23,24] Thus, it was concluded that cysteine may be an essential residue for toxin activity. A similar effect has been reported in another family of bacterial toxins, the sulfydryl-activated cytolysins, which also contain a single cysteine. However, in three members of this family, streptolysin O, listeriolysin O, and pneumolysin, site-directed mutagenesis of cysteine to alanine yielded mutant toxins that retained at least 75% of the wild-type haemolytic activity, suggesting that the cysteine was not an essential residue.[25] It is therefore interesting that when serine was substituted for cysteine in β-toxin, and the mutant toxin was expressed in *Bacillus subtilis*, the protein recovered was similar to the native toxin in terms of lethal activity (V. Steinþórsdóttir, personal communication).[26] This preliminary data also suggests that cysteine is not an essential residue for β-toxin activity, although this cannot be confirmed until the activity of wild-type recombinant toxin, produced in *B. subtilis*, has been tested. In this context, the reason for the nontoxicity of recombinant β-toxin purified from *E. coli* remains unclear.[7]

The amino acid sequence of β-toxin was homologous to a family of membrane-active *Staphylococcus aureus* toxins; the greatest degree of homology (28–29%) was observed with α-toxin, the Fast components of leukocidin and leukocidin R, and the H1gB component of gamma-haemolysin (Figure 15.1).[6] Alignment of these toxins also indicated that the amino acid identity was highest in the *C*-terminal two-thirds of the proteins. The *S. aureus* toxins are cytolytic, and the α-toxin has been shown to achieve its effects by forming membrane-spanning pores consisting of toxin hexamers.[6] Although there is some evidence that ß-toxin may form oligomers,[6] it is unlikely to have cytolytic activity. Jolivet-Reynaud *et al.*[15] were unable to detect lysis of CHO cells, although they did observe cytotoxic activity, and Sakurai and Fujii[11] demonstrated that 25 μg of toxin failed to disrupt mast cells or lysosomes (as shown by the absence of extracellular histamine or acid phosphatase, respectively), despite treatment for 30 minutes. In addition, recombinant ß-toxin was not haemolytic for rabbit erythrocytes (unpublished observations).[6]

The amino acid sequence homology with *S. aureus* α-toxin indicates that β-toxin is likely to be a 'membrane-active' toxin, but the absence of cytolytic activity suggests that it is unable to adopt the conformational changes which are predicted to occur in α-toxin during pore-formation.[26] It is possible that the toxins interact with the cell membrane in a similar way, perhaps by binding to related membrane components, although it is not known how β-toxin achieves its effects after these initial stages of membrane interaction.

VII. FUTURE PERSPECTIVES

The studies on the molecular biology of β-toxin, which have been carried out over the past few years, will now permit a more detailed investigation of structure–function relationships using site-directed mutants. In addition, a more detailed

```
Cpb         NDIGKTTTITRNKTSDGYTII-TQNDKQIISYQSVDSSSKNEDGFTASI     48
Hla         AD-SDI---NIKTGTTDIGSNTTVKTGDLVTY-------DKENGMHKKV
LukF        -E-GKITPVSVKKVDDKVTLYKTTATADSDKF-------K----ISQIL
LukF-R      AE-GKITPVSVKKVDDKVTLYKTTATADSDKF-------K----ISQIL
HlgB        AE-GKITPVSVKKVDDKVTLYKTTATADSDKF-------K----ISQIL
                                  *

Cpb         DARFIDDK-YSSEMTTLINLTGFMSSKKEDVIKKYNLHDVTNSTAINFP     96
Hla         FYSFIDDKNHNKKLLVIRTKGTIAGQYRVYSEEGANK------SGLAWP
LukF        TFNFIKDKSYDKDTLVLKATGNINSGFVKPNPNDYDF------SKLYWG
LukF-R      TFNFIKDKSYDKDTLVLKAAGNINSGYERPNPKDYDF------SKIYWG
HlgB        TFNFIKDKSYDKDTLVLKATGNINSGFVKPNPNDYDF------SKLYWG
               ** **

Cpb         VRYSISILNESINENVKIVDSIPKNTISQKTVSNTMGYKIGGSIEIEEN     145
Hla         SAFKVQLQLPD-NEVAQISDYYPRNSIDTKEYMSTLTYGFNGNVTGDDT
LukF        AKYNVSISSQS-NDSVNAVDYAPKNQNEEFQVQNTLGYTFGGDISISN-
LukF-R      AKYNVSISSQS-NDSVNVVDYAPKNQNEEFQVQNTLGYTFGGDISISN-
HlgB        AKYNVSISSQS-NDSVNVVDYAPKNQNEEFQVQNTLGYTFGGDISISN-
                        *      *  * *         *  *   *

Cpb         KPKASIESEYAE-SSTIEYVQPDFSTIQTDHSTSKA-SWDT---KFTET     189
Hla         GKIGGLIGANVSIGHTLKYVQPDFKTILESPTD-KKVGWKVIFNNMVNQ
LukF        GLSGGLNG-NTAFSETINYKQESYRT-LSRNTNYKNVGWGVEAHKIMN-
LukF-R      GLSGGLNG-NTAFSETINYKQESYRTTLSRNTNYKNVGWGVEAHKIMNN
HlgB        GLSGGLNG-NTAFSETINYKQESYRTTLSRNTNYKNVGWGVEAHKIMNN
                           *  * *    *        *   *

Cpb         TRGNYNLKSNNPVYGNEMFMYGRYTNVPATENIIPDYQMSKLITGGLNP     238
Hla         NWGPYDRDSWNPVYGNQLFMKTRNGSMKAADNFLDPNKASSLLSSGFSP
LukF        GWGPYGRDSFHPTYGNELFLAGRQSSAYAGQNFIAQHQMPLLSRSNFNP
LukF-R      GWGPYGRDSFHPTYGNELFLAGRQSSAYAGQNFIAQHQMPLLSRSNFNP
HlgB        GWGPYGRDSFHPTYGNELFLAGRQSSAYAGQNFIAQHQMPLLSRSNFNP
              * *   *  * ***  *   *     *  *         *      *

Cpb         NMSVVLTAP--NGTEESIIKVKMERERNCYYLNWNGANWVGQVYSRLAF     285
Hla         DFATVITMDRKASKQQTNIDVIYERVRDDYQLHWTSTNWKGTNTKD--K
LukF        EFLSVLS-HRQDRAKKSKITVTYQREMDLYQIRWNGFYWAGANYKN--F
LukF-R      EFLSVLS-HRQDGAKKSKITVTYQREMDLYQIRWNGFYWAGANYKN--F
HlgB        EFLSVLS-HRQDGAKKSKITVTYQREMDLYQIRWNGFYWAGANYKN--F
                *             * *   *    *   *    * *

Cpb         DTPNVDSHIFTFKINWLTHKVTAI                              309
Hla         WTDRSSE---RYKIDWEKEEMT----------N
LukF        KTRTFKS---TYEIDWENHKVKLLDTKETENNK
LukF-R      KTRTFKS---TYEIDWENHKVKLLDTKETENNK
HlgB        KTRTFKS---TYEIDWENHKVKLLDTKETENNK
             *           * *
```

Figure 15.1 Alignment of the secreted form of *Clostridium perfringens* β-toxin (Cpb) with the secreted forms of *Staphylococcus aureus* α-toxin (Hla), fast components of leukocidin (LukF) and leukocidin R (LukF-R), and gamma-toxin component B (HlgB). Identical residues are shown below the aligned sequences (*).

comparison of the properties of *S. aureus* α-toxin and *C. perfringens* ß-toxin could provide a fresh insight into the molecular basis of toxicity of both proteins. These studies may also result in the development of a second-generation ß-toxin vaccine for human and veterinary use, which is both safer and cheaper to produce than the current vaccine.

VIII. REFERENCES

1. McEwen A.D. and Roberts R.S. (1930) 'Struck': enteritis and peritonitis of sheep caused by a bacterial toxin derived from the alimentary canal. *J Comp Pathol* **44**: 26–49.
2. Murrell T.G.C. and Roth L. (1963) Necrotising jejunitis: a newly discovered disease in the HIghlands of New Guinea. *Med J Aust* **i** (3): 61–69.
3. Lawrence G. and Walker P.D. (1976) Pathogenesis of enteritis necroticans in Papua New Guinea. *Lancet* **i**: 125–126.
4. Lawrence G. and Cooke R. (1980) Experimental pig-bel: the production and pathology of necrotising enteritis due to *Clostridium welchii* type C in the guinea-pig. *Br J Exp Pathol* **61**: 261–271.
5. Lawrence G., Shann F., Freestone D.S. and Walker P.D. (1979) Prevention of necrotising enteritis in Papua New Guinea by active immunisation. *Lancet* **i**: 227–230.
6. Hunter S.E.C., Brown J.E., Oyston P.C.F., Sakurai J. and Titball R.W. (1993) Molecular genetic analysis of beta-toxin of *Clostridium perfringens* reveals sequence homology with alpha-toxin, gamma-toxin and leukocidin of *Staphylococcus aureus. Infect Immun* **61**: 3958–3965.
7. Steinþórsdóttir V., Friðriksdóttir V., Gunnarsson E. and Andrésson O.S. (1995) Expression and purification of *Clostridium perfringens* beta-toxin glutathione S-transferase fusion protein. *FEMS Microbiol Lett* **130**: 273–278.
8. Duncan C.L., Rokos E.A., Christenson C.M. and Rood J.I. (1978) Multiple plasmids in different toxigenic types of *Clostridium perfringens*: possible control of beta-toxin production. In Schlessinger D. (ed.), *Microbiology – 1978*, pp. 246–248. American Society for Microbiology, Washington D.C.
9. Buogo C., Capaul S., Häni H., Frey J. and Nicolet J. (1995) Diagnosis of *Clostridium perfringens* type C enteritis in pigs using a DNA amplification technique (PCR). *J Vet Med* **42**: 51–58.
10. Lawrence G., Watt S. and Basten A. (1979) Purification and some effects of *Clostridium welchii* type C beta-toxin. *Papua New Guinea Med J* **22**: 79–86.
11. Sakurai J. and Fujii Y. (1987) Purification and characterisation of *Clostridium perfringens* beta-toxin. *Toxicon* **25**: 1301–1310.
12. Sakurai J. and Duncan C.L. (1978) Some properties of beta-toxin produced by *Clostridium perfringens* type C. *Infect Immun* **21**: 678–680.
13. Cooke R. (1979) The pathology of pig bel. *Papua New Guinea Med J* **22**: 35–38.
14. Buxton A. and Fraser G. (1977) *Animal MIcrobiology*, Vol. 1. Blackwell Scientific Publications, Oxford.
15. Jolivet-Reynaud C., Popoff M.R., Vinit M-A., Ravisse P., Moreau H. and Alouf J.E. (1986) Enteropathogenicity of *Clostridium perfringens* beta-toxin and other clostridial toxins. In Falmagne P. *et al.* (eds), *Bacterial Protein Toxins: 2nd European Workshop*, pp. 145–151. Gustav Fischer Verlag, Stuttgart.
16. Parnas J. (1976) The effects of *Clostridium perfringens* beta-toxin (type C) on the mobility of intestinal segments *in vitro. Abstr Zentralblat Bacteriol Hyg* **234**: 243–246.
17. Sakurai J., Fujii Y., Matsuura M. and Endo K. (1981) Pharmacological effect of beta-toxin of *Clostridium perfringens* type C on rats. *Microbiol Immunol* **25**: 423–432.
18. Sakurai J., Fujii Y., Dezake K. and Endo K. (1984) Effect of *Clostridium perfringens* beta-toxin on blood pressure of rats. *Microbiol Immunol* **28**: 23–32.
19. Allan D. (1963) Cytotoxic effects produced by the beta-toxin of *Clostridium welchii* on guinea-pig monocytes. *Immunology* **6**: 3–14.
20. McDonel J.L. (1986) Toxins of *Clostridium perfringens* types A, B, C, D and E. In Dorner F. and Drews J. (eds), *Pharmacology of Bacterial Toxins*, pp. 477–517. Pergamon Press, Oxford.
21. Worthington R.W. and Mülders M.S.G. (1975) The partial purification of *Clostridium perfringens* beta-toxin. *Onderstepoort J Vet Res* **42**: 91–98.

22. Sakurai J. and Duncan C.L. (1977) Purification of beta-toxin from *Clostridium perfringens* type C. *Infect Immun* **18**: 741–745.
23. Sakurai J., Fujii Y. and Matsuura M. (1980) Effect of oxidising and sulfhydryl group reagents on beta-toxin from *Clostridium perfringens* type C. *Microbiol Immunol* **24**: 595–601.
24. Sakurai J., Fujii Y. and Nagahama M. (1992) Effect of p-chloromercuribenzoate on *Clostridium perfringens* beta-toxin. *Toxicon* **30**: 323–330.
25. Alouf J.E. and Geoffroy C. (1991) The family of antigenically related cholesterol-binding ('sulphydryl-activated') cytolytic toxins. In Alouf J.E. and Freer J.H. (eds), *Sourcebook of Bacterial Protein Toxins*, pp. 147–186. Academic Press, London.
26. Bhakdi S. and Tranum-Jensen J. (1991) Alpha-toxin of *Staphylococcus aureus. Microbiol Rev* **55**: 733–751.

Chapter 16

Clostridium septicum α-toxin

JIMMY D. BALLARD AND RODNEY K. TWETEN*

*Microbiology and Molecular Genetics, Harvard Medical School, Boston, MA 02115 and *Department of Microbiology and Immunology, University of Oklahoma Health Sciences Center, Oklahoma City, OK 73190, USA*

I. INTRODUCTION

Following its relatively early description by Pasteur and Joubert,[1] *Clostridium septicum* remained largely uninvestigated for almost 100 years. During the last decade an increase in the morbidity of *C. septicum*-associated diseases has focused interest in the study of this organism. A seminal report by Stevens *et al.*[2] brought greater awareness of *C. septicum* and the main disease it causes, namely atraumatic gas-gangrene. As reported by Stevens, atraumatic gas-gangrene occurs primarily in patients with predisposing conditions such as colonic cancer, leukemia and diabetes. Because of the fulminant nature of this disease and rapid time to death (often less than 24 h), it had been suggested that one or more lethal toxins may play a major role in the disease process.

Studies of *C. septicum* exoproteins suggest the organism produces at lease five extracellular proteins, which include hemolysin(s), deoxyribonuclease, hyaluronidase, and a fibrinolysin. Of the limited studies on *C. septicum* toxins, a 1944 report by Bernheimer[3] described a putative lethal factor, termed α-toxin, that was lethal in mice and also hemolytic. At that time the protein was never fully purified owing to the limitations of purification technology, so definitive analysis

The Clostridia: Molecular Biology and Pathogenesis
ISBN 0-12-595020-9

of this toxin was not possible. In retrospect, the early work of Bernheimer was accurate and insightful since, as will be discussed in this chapter, the α-toxin of *C. septicum* is both lethal and cytolytic.

II. α-TOXIN ISOLATION AND CHARACTERIZATION

A single lethal protein was isolated from *C. septicum* strain BX96 crude culture supernatants.[4] A combination of preparative anion exchange, cation exchange, gel filtration and high-resolution cation exchange chromatography was utilized to isolate a single protein with an overall yield of 50%. Although mouse lethality assays were first used to identify the toxin-containing fractions, it was soon found that hemolytic activity copurified with the lethal fraction. It has subsequently been shown that both activities could be attributed to one protein. This has simplified purification of the toxin by allowing α-toxin containing fractions to be determined by hemolysis assay. In keeping with Bernheimer's report,[3] this lethal factor was termed α-toxin.

The initial characterization of α-toxin[4] revealed that purified α-toxin was ≈ 49 kDa by sodium dodecyl sulphate-polyacrylamide gel electrophoresis (SDS-PAGE) analysis and has a pI of ≈8.4. α-Toxin has a LD_{50} of approximately 10 μg/kg body weight for mice and exhibits a hemolytic activity of 10^6–10^7 hemolytic units/mg. The first pure preparations of α-toxin exhibited a smaller protein of ≈44 kDa that was present in variable amounts with the 49 kDa form. Analysis of the two proteins by two-dimensional peptide map analysis and the use of specific antibody against the 49 kDa protein revealed that the 44 kDa protein was a degradation product of the former.

Rabbit polyclonal antibody against α-toxin reacted with a single cross-reactive 49 kDa protein in all *C. septicum* strains examined. However, there are no detectable cross-reactive extracellular proteins from other clostridial species. This finding initially suggested that *C. septicum* α-toxin was unique among the clostridia.

Mice immunized with sublethal doses of α-toxin are partly protected when challenged with the *C. septicum* whole organism.[4] It has also been found that some patients who survived the disease mounted a strong immunological response to α-toxin.[5] Most recently, using the rat model it has been shown that a rat, treated solely with α-toxin, demonstrates many of the same physiological conditions as patients with atraumatic gas-gangrene (J.D. Ballard, R.K. Tweten and D. Bracket, unpublished results). These data suggest a significant role for α-toxin in the overall disease process. Because *C. septicum* has not been successfully transformed, evaluation of isogenic derivative strains of *C. septicum* attenuated in α-toxin production has not been possible.

III. PROTEOLYTIC ACTIVATION

As indicated above, batches of purified α-toxin contained variable amounts of a nicked product, which was estimated to be about 4 kDa smaller than the full-length protein. It was found that stored toxin (−70°C) was gradually converted to the smaller form. When protease inhibitors were included in the first three steps of the purification the 49 kDa form of the protein was the sole product. However, this preparation of the protein did not exhibit detectable hemolytic activity.[4] This observation suggested that the 44 kDa protein might be the active form of α-toxin and that proteolytic activation was required for cytolytic activity. This hypothesis was tested by incubating the 49 kDa form of the protein with various proteases at different ratios of toxin to protease, then testing for hemolytic activity. Interestingly, it was found that α-toxin hemolytic activity was significantly enhanced when the protein was first cleaved with trypsin, chymotrypsin, subtilisin or proteinase-K.[6]

Trypsin generated a single cleavage product which most closely resembled naturally nicked toxin and, therefore, it has been used for all subsequent proteolytic activation studies. The 49 kDa form was termed AT^{pro} (α-toxin protoxin form) and the 44 kDa form was termed AT^{act} (α-toxin activated form). Both AT^{pro} and AT^{act} had the same amino-terminus sequence, suggesting that the proteolytic cleavage required for activation occurred proximal to the carboxy terminus of the protein.

Although it remains unclear when or where proteolytic activation takes place during the course of the disease, we do know that AT^{pro} can be activated even after binding to the cell surface of erythrocytes. AT^{pro} bound to erythrocytes is rapidly induced to lyse the cells only following the addition of trypsin. The trypsin-induced lysis generally occurred within 5 min, while in the absence of trypsin the toxin took more than 12 h to lyse the cells.

When *C. septicum* is grown to stationary phase in defined media (free of any media-associated proteases or inhibitory peptides), AT^{pro} is processed to AT^{act}, suggesting that the organism produces a factor that can activate the toxin. To date, attempts to isolate this protease have not been successful. Whether a specific protease which activates AT^{pro} is secreted into the medium by *C. septicum* is currently unknown. The possibility that the protease comes from an intracellular source due to cell lysis has also not been ruled out. In addition, we have found that AT^{act} is extremely unstable in solution where it tends to aggregate to a cytolytically inactive form. Although the evidence remains anecdotal, it would appear that activation at the target cell surface would be the more efficient mode of action for α-toxin. The structure of the proteolytic activation site suggests that this may be true since, as shown below, the primary structure of this region exhibits some definite characteristics of various eukaryotic protease substrates.

It was thought initially that processing may lead to structural changes that expose buried hydrophobic regions of the protein, thereby, making membrane insertion possible. Using both phase-partitioning and fluorescence analysis (ANS probe) assays, there appeared to be no readily detectable changes in the hydrophobic nature of the protein following processing (J.D. Ballard, R.K. Tweten and B.R. Sellman, unpublished results). The first significant hint to the role of the processing event was obtained by analyzing the nature of the two forms of α-toxin after binding to the erythrocyte membrane. Immunoblots of erythrocyte membranes treated with either AT^{pro} or AT^{act} revealed that AT^{act} formed a high-molecular-weight, SDS-resistant complex, whereas, AT^{pro} remained in the monomeric form.[6] This high-molecular-weight complex was similar to the oligomerized form of other pore-forming toxins such as aerolysin[7,8] and *Staphylococcus aureus* α-toxin.[9]

IV. α-TOXIN PORE FORMATION

Cytolytic proteins, whose mechanism of lysis are based on pore formation and osmotic lysis, typically exhibit a prelytic release of potassium ions. This has been observed for α-toxin.[6] Erythrocytes treated with AT^{act} exhibit a nearly complete release of potassium prior to any detectable hemolysis. As was expected there was no release of potassium in cells treated with AT^{pro}. The onset of potassium release generally occurs within 40 s of the addition of the toxin, and is largely complete by 150 s. Hemolysis, however, is not detected until almost 360 s following the addition of the toxin to the erythrocytes.

Planar membrane analysis has been used to evaluate the potential channel forming activity of both AT^{pro} and AT^{act}.[6] Addition of AT^{act} (2.1 nM) to the *cis*-side of a planar membrane resulted in discrete increases in membrane conductance. No change in conductance was observed when AT^{pro} was added to the membranes, which was consistent with its lack of activity in the cytolytic assays. The single conductance steps were the order of 140–180 pS (pico Siemens) in a solution of 100 mM KCl. The current–voltage curve for AT^{act} is linear in the range of −100 to +100 mV. The pores demonstrate a tendency to close in the range 120–150 mV at both positive and negative voltages. The channels show a reversal potential of about 18–20 mV in a ten-fold gradient of KCl concentration, indicating that the channel is slightly anion-selective.

The diameter of the pore formed by α-toxin has been estimated using various sizes of osmotic stabilizers such as polyethylene glycol. Erythrocytes are osmotically stabilized to lysis in the presence of PEG 600 or larger molecules. This result is consistent with a pore size of at least 1.3–1.6 nm in diameter. Taken together, the pore size and the prelytic release of potassium both suggest the erythrocytes are lysed by a colloid-osmotic mechanism.

V. NUCLEATED CELLS TARGETED BY α-TOXIN

While much of the characterization of α-toxin has been carried out on erythrocytes, it remains to be determined if these are the cells targeted *in vivo* by the toxin. To date, we have examined the susceptibility of several other cell types to lysis by α-toxin. All the nucleated cell lines examined, except one (L929), are sensitive to α-toxin. Additionally, α-toxin lysed the cells much more efficiently following proteoytic activation, confirming that the activation event was not unique to the toxin's hemolytic activity.

VI. CLONING AND EXPRESSION OF α-TOXIN ENCODING GENE

The α-toxin encoding gene has been cloned in *Escherichia coli* and its genetic sequence has been determined.[10,11] The recombinant protein is expressed in *E. coli* and appears to be functionally the same as the wild-type protein. Sequence analysis predicts a protein of 49 755 kDa, with a 32 amino acid signal sequence that is apparently cleaved during secretion after a AYA sequence.[11]

The trypsin cleavage site, which converts AT^{pro} to AT^{act}, apparently occurs on the amino-terminal side of the serine residue in the sequence of R_{367}SVD. The sequence within the cleavage site has two interesting features: it is rich in basic amino acids (P_{388}LPDKKRRGKRSVD) and includes a furin consensus site of $RGKR_{367}$. Several dibasic residues are present in this region just downstream of a putative β turn of PLP. It is possible that one or more of these sites may be targets of eukaryotic proteases. Several proteases which cleave prohomones have recognition sites which include one or more dibasic amino acid pairs downstream of a putative β-turn.[12]

No extended regions of hydrophobic residues exist in the primary structure of α-toxin, suggesting that the membrane interactive regions may be composed of amphipathic β-sheets. A similar finding has been reported for aerolysin based on the crystal structure of the monomeric protoxin.[13] Sequence comparisons reveal no similarity at the amino acid level with any other known clostridial protein. α-Toxin does appear to exhibit similarity with aerolysin from *Aeromonas hydrophila*; it exhibits greater than 27% identity and approximately 72% similarity over a contiguous stretch of 387 amino acids.[11] Of further interest is the functional similarities between the two proteins.[6] Aerolysin is a pore forming toxin that requires carboxy-terminal proteolytic cleavage for activation. The position of the proteolytic cleavage site is also positionally conserved between the two toxins when they are aligned for maximal similarity. Like α-toxin, aerolysin aggregates on membranes and forms anion selective pore in lipid bilayers.[7,8] Since the Gram-positive anaerobic *C. septicum* is phylogenetically distant from the Gram-negative waterborne pathogen *A. hydrophila* it is not clear how these two organisms came to acquire the same toxin gene. Although strong similarity exist between the two proteins, there is no similarity between the G-C content of the two genes; the α-toxin gene exhibits a low G-C content (29.3%), whereas aerolysin has almost twice the G-C content (58.6%). This appears to be the first example of extensive homology between a Gram-positive and a Gram-negative toxin. It is likely that the two genes diverged very early from a common source.

VII. SUMMARY

Although the exact cellular target of α-toxin remains to be determined, many of the events leading to cell lysis are clear. Following secretion from the bacterium, α-toxin is activated to its cytolytic form by the proteolytic removal of a 4 kDa carboxy-terminus peptide. Protelolytic activation could potentially occur prior to or following cell binding, however, the most likely scenario is that activation occurs at the cell surface. Following activation, α-toxin aggregates into a high-molecular-weight complex and forms anionic selective channels in the target membrane. Following channel formation, the cell becomes osmotically imbalanced and lysis. At present, work continues on the structure–function aspects of the toxin and also its cellular receptor and role in pathogenesis.

VIII. REFERENCES

1. Joubert P.A. (1877) Charbon et speticemie. *Bull Acad Med (Paris)* **6**: 781.
2. Stevens D.L., Musher D.M., Watson D.A., Eddy H., Hamill R.J., Gyorkey F., Rosen H. and Mader J. (1990) Spontaneous, nontraumatic gangrene due to *Clostridium septicum. Rev Infect Dis* **12**: 286–296.
3. Bernheimer A.W. (1944) Parallelism in the lethal and hemolytic activity of the toxin of *Clostridium septicum. J Exp Med* **80**: 309–320.
4. Ballard J., Bryant A., Stevens, D. and Tweten R.K. (1992) Purification and characterization of the lethal toxin (alpha toxin) of *Clostridium septicum. Infect Immun* **60**: 784–790.
5. Johnson S., Driks M.R., Tweten R.K., Ballard J., Stevens D.L., Anderson D.J. and Janoff E.N. (1994) Clinical courses of seven survivors of *Clostridium septicum* infection and their immunological responses to α-toxin. *Clin Infect Dis* **19**: 761–764.
6. Ballard J., Sokolov, Y., Yuan, W.-L., Kagan B.L. and Tweten R.K. (1993) Activation and mechanism of *Clostridium septicum* alpha toxin. *Mol Microbiol* **10**: 627–634.

7. Garland W.J. and Buckley J.T. (1988) The cytolytic toxin aerolysin must aggregate to disrupt erythrocytes, and aggregation is stimulated by human glycophorin. *Infect Immun* **56**: 1249–1253.
8. Howard S.P. and Buckley J.T. (1985) Activation of the hole-forming toxin aerolysin by extracellular processing. *J Bacteriol* **163**: 336–340.
9. Korchev Y.E., Alder G.M., Bakhramov A., Bashford C.L., Joomun B.S., Sviderskaya, E.V., Usherwood P.N.R. and Pasternak C.A. (1995) *Staphylococcus aureus* alpha-toxin-induced pores: channel-like behavior in lipid bilayers and patch clamped cells. *J. Membrane Biol* **143**: 143–151.
10. Imagawa T., Dohi Y. and Higashi Y. (1994) Cloning nucleotide sequence and expression of a hemolysin gene of *Clostridium septicum. FEMS Microbiol Lett* **117**: 287–292.
11. Ballard J., Crabtree J., Roe B.A. and Tweten R.K. (1995) The primary structure of *Clostridium septicum* alpha-toxin exhibits similarity with that of *Aeromonas hydrophila* aerolysin. *Infect Immun* **63**: 340–344.
12. Brackch N., Rholam M., Boussetta H. and Cohen P. (1993) Role in B-turn in proteolytic processing of peptide hormone precursors at dibasic sites. *Biochemistry* **32**: 4925–4930.
13. Parker M.W., Buckley J.T., Postma J.P.M, Tucker A.D., Leonard K., Pattus F. and Tsernoglou D. (1994) Structure of the aeromonas toxin proaerolysin in its water-soluble and membrane-channel states. *Nature* **367**: 292–295.

F: Neurotoxins

Chapter 17

The Genetic Basis of Toxin Production in *Clostridium botulinum* and *Clostridium tetani*

IAN HENDERSON, TOM DAVIS, MICHAEL ELMORE AND NIGEL P. MINTON

Department of Molecular Microbiology, Centre for Applied Microbiology and Research, Porton Down, Salisbury, Wiltshire SP4 0JG, UK

I. INTRODUCTION

Certain bacterial species belonging to the genus *Clostridium* are characterized by their ability to produce extremely potent neurotoxins. Classified as either tetanus neurotoxin (TeNT) or botulinum neurotoxin (BoNT), they are synthesized as single polypeptide chains, which are subsequently cleaved by bacterial/host proteases to produce di-chain toxins, each comprising a light chain (50 kDa) and a heavy chain (100 kDa) linked by a single disulphide bridge. They act by inhibiting the exocytosis of neurotransmitters at CNS synapses (TeNT) and neuromuscular junctions (BoNT), leading to spastic (tetanus) or flaccid (botulism) paralysis (see Chapter 7). The deadly nature of the diseases these toxins cause has assured them some considerable attention from the scientific community. As a result, much is now known of their mode of action (see Chapter 18). The molecular analysis of the toxin structural genes has played a significant part in this process. Genetic analysis has also revealed other information on toxin production, gene transfer, diagnosis and prevention.

The Clostridia: Molecular Biology and Pathogenesis
ISBN 0-12-595020-9

II. CHARACTERIZATION OF THE NEUROTOXINS

To date, eight different neurotoxin types have been recognized; these are the single form of TeNT, and seven (A–G) serologically distinct BoNTs.[1] One of the most important factors to emphasize is that while production of TeNT is confined to a single clostridial species, *Clostridium tetani*, BoNT is produced by a range of species. Historically any organisms producing BoNT were designated *Clostridium botulinum*. Despite this, it soon became apparent that four distinct groupings could be made based on physiological parameters,[2] and that each of these groupings contained atoxinogenic counterparts which had lost the ability to produce toxin.[3] The authenticity of these groups was reinforced through phylogenetic analysis of the 16S rRNA sequences of the various members of these groups.[4,5] The taxonomic standing of *C. botulinum* has been further undermined by the isolation of clostridial strains which could be readily categorized as strains of *Clostridium butyricum*[6,7] and *Clostridium barati* (see Chapter 1).[8,9] This demonstrates that although toxinogenicity is an important clinical attribute, it is not a sufficient basis for defining a bacterial species. Thus, distinct species names should be assigned for each physiological/genetic group, and BoNT production and type should be indicated separately.[4,5,10] However, such an approach has been adopted only for group IV *C. botulinum* strains, which are sometimes referred to as *Clostridium argentinense*.[11] Some of the attributes of currently recognized neurotoxinogenic clostridia are summarized in Table 17.1.

A. The neurotoxin genes

Representative genes of all of the clostridial neurotoxins, from the various serotypes and physiological groups, have been isolated and sequenced (Table 17.2). The % G+C ratio of these genes is characteristically low (ranging from 24.51% to 27.84%). They therefore exhibit an extreme codon bias common to clostridial genes.[31,32] After typical prokaryotic ribosome-binding sites, all of these genes utilize an AUG translational start codon and all terminate with an ochre (UAA) stop codon. The one exception is the BoNT/F gene of *C. barati* ATCC 43756, which makes use of an amber (UGA) stop codon. For those genes that have been sequenced for a sufficient distance downstream of their stop codon (the BoNT/A genes of *C. botulinum* strains NCTC 2916 and 62A; the TeNT gene of *C. tetani*), sequence motifs typical of transcriptional termination signals are apparent.

Table 17.1. Characteristics of the various neurotoxingenic clostridial groups[12,13]

Clostridium species	Group	Toxin type	Complex[a] form	Proteolysis[b]	Glucose fermentation	Lipase production	Optimum growth temperature (°C)	Related[c] species
C. botulinum	I	A	M, L, LL	+	+	+	30–37	*C. sporogenes*
		B	M, L	+	+	+	30–37	*C. sporogenes*
		F	M	+	+	+	30–37	*C. sporogenes*
C. botulinum	II	B	M, L	–	+	+	25–30	'E-like' organisms
		E	M	–	+	+	25–30	'E-like' organisms
		F	M	–	+	+	25–30	'E-like' organisms
C. botulinum	III	C	M, L	±	–	–	30–37	*C. novyi*
		D	M, L	±	–	–	30–37	*C. novyi*
C. argentinense	IV	G	M, L	+	–	–	30–37	*C. argentinense*
C. butyricum	V	E	M	[d]	–	–	30–37	*C. butyricum*
C. barati	VI	F	M	–	+	–	20–45	*C. barati*
C. tetani		TeNT		[e]	–	–	37	*C. tetanomorphum*

[a] The three recognized forms of progenitor toxin with molecular weights of 300, 500 and 900 kDa for the M, L and LL forms, respectively.
[b] Digestion of milk proteins.
[c] 'E-like' organisms refers to clostridial strains that share characteristics with *C. botulinum* group II.
[d] *C. butyricum* partially digests milk protein.
[e] *C. tetani* is obligately proteolytic but cannot degrade complex proteins such as those in milk.

Table 17.2. Currently characterized clostridial neurotoxin genes

Toxin	Clostridial strain	Group[a]	% G+C	Ribosome binding site	Amino acids	Reference
BoNT/A	*C. botulinum* NCTC 2916	I	26.2	TTATAAGAGGTGTTAAAT	1296	14
BoNT/A	*C. botulinum* 62A	I	26.3	TTATAAGAGGTGTTAAAT	1296	15
BoNT/A	*C. botulinum* Kyoto-F	I	26.8	TTATAAGAGGTGTTAAAT	1296	16
BoNT/B	*C. botulinum* Danish	I	25.5	ATAAGGAGGAGAATATTT	1291	17
BoNT/B	*C. botulinum* Eklund	II	25.5	ATAAGGAGGAGAATATTT	1291	18
BoNT/B[a]	*C. botulinum* NCTC 2916	I	25.3	ATAAGGAGGAGAATATTT	1287[b]	This laboratory
BoNT/C	*C. botulinum phage* C468	III	26.2	TTAGGAGATGTTAGTATT	1291	19
BoNT/C	*C. botulinum* phage C-st	III	26.2	TTAGGAGATGTTAGTATT	1291	20
BoNT/D	*C. botulinum* phage DVD/-3	III	25.6	TTAGGAGATGTTAGTATT	1276	21
BoNT/CD	*C. botulinum* phage C6813	III	26.1	TTAGGAGATGTTAGTATT	1280	22
BoNT/E	*C. botulinum* NCTC 11219	II	25.0	AATTAGGAGATGCTGTAT	1252	23
BoNT/E	*C. botulinum* Beluga	II	25.1	AATTAGGAGATGCTGTAT	1251	24
BoNT/E	*C. butyricum* ATCC 43181	V	24.8	AATTAGGAGATGCTGTAT	1251	24
BoNT/F	*C. botulinum* Langeland	I	26.5	AAGTATAGGGGGGATTTT	1278	25
BoNT/F	*C. botulinum* ATCC 23387	II	26.9	AGGTATAGGGGGATTTTT	1274	26
BoNT/F	*C. barati* ATCC 43756	VI	24.5	TATAGGAGGTTGATATAT	1268	27
BoNT/G	*C. argentinense* 89	IV	27.1	AATTAGGAGGTATATATT	1297	This laboratory
BoNT/G	*C. botulinum* NCFB 3012	IV	27.1	Not determined	1297	28
TeNT	*C. tetani* plasmid pE88	–	27.8	AATTAGGAGATGATACGT	1315	29
TeNT	*C. tetani* CN 3911	–	27.8	AATTAGGAGATGATACGT	1315	30

[a] This is a 'silent' BoNT/B gene found in a type A *C. botulinum* strain.

[b] This BoNT/B is four amino acids shorter than that of the type B strain Danish due to mutation of Glu-127 to a stop codon, the loss of a single base from codon Ile-964 and a 6 bp deletion which removes amino acids Tyr-328 and Lys-329.

B. DNA-based detection systems for neurotoxin genes

The availability of gene sequences has opened up the possibility of using DNA based methods for detecting the presence of clostridial strains with the potential to cause botulism. The application of such technology could reduce the need for the mouse bioassay in toxicity testing of suspected isolates. To date, several groups have been involved in the development of oligonucleotide primers for the detection of neurotoxin genes using polymerase chain reaction (PCR) based methods.[33–36] Most groups have concentrated on the design of primers that will detect those toxin serotypes of particular hazard to the human population, namely, types A, B, E and F. Design of the required primers has been complicated by the similarity between genes, and particularly the shared homology with the TeNT gene. Despite these difficulties, several different primer sets have been developed by sequence alignment of the toxin serotypes.

Work in this laboratory has concentrated on developing pairs of primers which amplify a DNA fragment of a size specific to the BoNT gene targeted, and which can be used in the form of a cocktail in a multiplex PCR. An example of the use of one such cocktail is illustrated in Figure 17.1. This particular mixture of primers is capable of detecting the presence of genes encoding either type A, B, E or F BoNT, and resulting in the amplification of DNA fragments of 600 bp, 550 bp, 1300 bp or 800 bp, respectively. The use of such a primer cocktail allows a single PCR reaction to be undertaken for diagnosis of one of the four toxins responsible for human botulism. The system is also, of course, able simultaneously to diagnose the presence of two toxin genes in the same bacterial host (see below).

The DNA detection systems developed have been assessed in artificially contaminated foodstuffs or by using purified DNA. However, such systems have yet to be adopted as routine diagnostic tools for the typing/detection of botulinum-producing clostridial strains. The laboratory investigation of suspected cases of botulism still, therefore, relies on immunological testing of samples for toxin, culture and selection of lipase positive bacteria, and testing of individual colonies for toxicity in the mouse bioassay.

C. Multiple toxin phenotypes/genotypes

Clostridial strains that express more than one toxin serotype are rare. The first published example was *C. botulinum* strain 84 isolated in Argentina,[37] which was shown, using appropriate antitoxin sera, to express toxin with features common to both BoNT/A and BoNT/F. Similarly, strains have been isolated that express BoNT/A and BoNT/B,[38,39] and BoNT/B and BoNT/F.[40] Subsequent

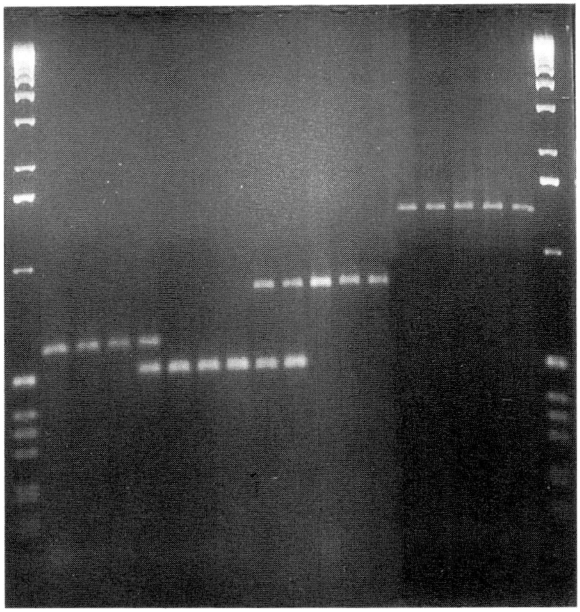

1 2 3 4 5 6 7 8 9 10 11 12 13 14 15 16 17 18 19

Figure 17.1 PCR detection of BoNT genes using 'Multiplex' ABEF primer mix. The primer mix employed contains four oligonucleotide primer pairs specific for four different BoNT genes. These generate fragments of 600, 500, 1300 and 800 bp if the gene encodes BoNT/A, BoNT/B, BoNT/E or BoNT/F, respectively. Lanes (from left to right), template DNA was derived from (1) kb ladder; (2) A, 17862; (3) A, NCTC 7272; (4) A, NCTC 19397; (5) A, NCTC 2916 (inc. 'silent' B gene); (6) B, NCTC 7273 (group I); (7) B, Danish (group I); (8) B, NCTC 17844 (group II); (9) B_f, ATCC 43757; (10) B_f, ATCC 43758; (11) F, Langeland (group I); (12) F, 10281 (group I); (13) F, 202-F (group II); (14) E, 11219; (15) E, NCTC 8550; (16) E, Vancouver Herring; (17) E, *Clostridium butyricum* ATCC 43755; (18) E. *C. butyricum* ATCC 43181; (19) kb ladder. In each case, the BoNT type is indicated prior to the strain name.

investigations of these strains has shown that two individual toxin serotypes are expressed. Typically, more of one toxin serotype is produced than the other, which is indicated through the use of upper and lower case in the strain designation, e.g., Af (strain 84 – a ratio of 10:1, i.e. there is 10 times as much BoNT/A as BoNT/F produced by this strain), Ba and Bf.[41,42]

The development of primers for the specific detection of neurotoxin genes by PCR has facilitated a more thorough investigation of strains which produce more than one toxin. Thus, in a recent study, PCR screening indicated that out of the 79 *C. botulinum* type A strains, 49 also carried sequences specific to a BoNT/B gene.[35] Only one of these strains was found to produce BoNT/B, although only at extremely low levels (25 mouse lethal doses, MLD) compared to BoNT/A (500 000 MLD). A more recent study has also detected these unexpressed or 'silent' genes in type A *C. botulinum* strains.[43] These particular strains were further categorized on the basis of whether the strain had been isolated from cases of either food-borne or infant botulism, and on the basis of which two restriction fragment polymorphism (RFLP) patterns was generated following *Rsa*I cleavage of a DNA fragment derived by PCR-amplification of a region of the gene encoding a part of the heavy chain. The silent genes were found only in those strains that expressed BoNT/A genes of one specific RFLP type, regardless of the type of disease caused. The meaning of this finding is unclear but it would suggest that some form of exclusion or sorting mechanism is taking place in those bacteria that express a specific 'type' of BoNT/A gene.

The study by Cordoba *et al.*[43] confirmed our previously reported finding[44] that the *C. botulinum* strain NCTC 2916 harbors a silent B gene. We have isolated and characterized this gene by nucleotide sequencing,[45] and demonstrated the presence of an entire structural gene coding for a variant of BoNT/B exhibiting 97% nucleotide sequence identity to the equivalent gene of the type B *C. botulinum* strain Danish.[17] The copy of the gene identified was, however, defective. Thus, translation of the nucleotide sequence revealed the presence of an amber translational stop codon (UAG) in place of a Glu codon at position 127. In addition, a further single base deletion in the 3′-end of the gene resulted in an independent translational frameshift. Preliminary RNA analysis has indicated that the gene is transcribed normally (T.O. Davis, I. Henderson and N.P. Minton, unpublished data).

D. Hybrid (mosaic) toxins

Neurotoxins with the properties of more than one toxin serotype have also been identified. To date, such a phenomenon has only been detected for proteins with properties common to both BoNT/C and BoNT/D.[46–48] One such example is the hybrid isolated from the *C. botulinum* strain C6813 which was demonstrated to possess antigenic characteristics common to the BoNT/C light chain and the BoNT/D heavy chain.[48] The gene encoding this hybrid has recently been characterized.[22] It codes for a protein of 1280 amino acids (147,813 D) consisting of sequences common to the light chain and the N-terminal region of the heavy chain of BoNT/C to up to amino acid position 901, at which point sequence similarity changes over to BoNT/D for the remainder of the molecule. The light chain has 97% and 48% identity with the BoNT/C and BoNT/D light chains, respectively. The heavy chain shares 66% and 85% identity with BoNT/C and BoNT/D heavy chains, respectively.

E. Neurotoxin homologies

As is evident from the alignment presented in Figure 17.2, clostridial neurotoxins share regions of extensive amino acid sequence similarity. Typically, toxins of the same serotype isolated from the same physiological group are practically identical. For example, the BoNT/A genes from *C. botulinum* strains NCTC 2916 and 62A differ by only two nucleotides. The one exception to this is the BoNT/A of the *C. botulinum* strain Kyoto-F, which possesses 22 and 107 dissimilar amino acids compared with the light and heavy chain, respectively, of the *C. botulinum* NCTC 2916 BoNT/A. The extent of diversity between neurotoxins of the same serotype from different physiological groups is greater. This is most easily observed by comparison of the group I, II and VI BoNT/F proteins, where the group II and VI molecules differ from the group VI molecule by 151 and 341 amino acids, respectively. The neurotoxins also have conserved functional domains, relating to the metalloprotease activity (HELxHxxH, BoNT/A position 223 to 230) in the light chain, the single disulphide linkage (conserved Cys-430, light chain; Cys-454, heavy chain), and membrane spanning region (PYxGxAL, BoNT/A positions 635–645) in the heavy chain. This contrasts with regions that show comparatively high sequence dissimilarity, the most conspicuous of which is the C-terminal portion of the heavy chain. As this region of the toxin is involved in receptor binding, its uniqueness may reflect the fact that different toxin serotypes target different neuronal receptors (see Chapter 18).

Despite the regions of similarity, the overall amino acid identity between the neurotoxin serotypes is low, exceeding 40% in only a few cases, i.e., the similarity seen between the light chains of BoNT/B, BoNT/G and TeNT, and between the heavy chains of BoNT/C and BoNT/D, and that shared by both chains of BoNT/E and BoNT/F. In this context, the mosaic BoNT/CD is a special case. The degree of relatedness may provide a measure of the historical timescale of toxin gene dissemination among clostridial species. A distance matrix analysis of the neurotoxins that calculates the relative rates of sequence divergence on the basis of sequence similarity (assuming identical rates of mutation)

Figure 17.2 (on pages 269–275) Alignment of botulinum and tetanus neurotoxin amino acid sequences using Clustal V.[49] Boxed sequences indicate identical or similar amino acids. Asterisks below the sequences indicate absolutely conserved amino acids. Numbering above the alignment corresponds to BoNT/A. By reference to Table 17.1, the sequenced toxins where derived from the following strains: BoNT/A, *C. botulinum* NCTC 2916; BoNT/A (Inf), *C. botulinum* Kyoto-F; BoNT/B (GpI), *C. botulinum* Danish; BoNT/B (GpII), *C. botulinum* Eklund; BoNT/C, *C. botulinum* C-St/468C; BoNT/E, *C. botulinum* NCTC 11219; BoNT/E (buty), *C. butyricum* ATCC 43181/43755; BoN/F (GpI), *C. botulinum* Langeland; BoNT/F (GpII), *C. botulinum* ATCC 23387; BoNT/F (bara), *C. barati* ATCC 43756; BoNT/G, *C. botulinum* 89G/NCFB 3012; TeNT, *C. tetani* CN3911.

```
                 1
BoNT/A           MQFVNKQFNYKDPVNGVDIAYIKIPNV-GQMQPVKAFKIHNKIWVIPERD   49
BoNT/A (inf)     MPFVNKQFNYKDPVNGVDIAYIKIPNA-GQMQPVKAFKIHNKIWVIPERD   49
BoNT/B (GpI)     MPVTINNFNYNDPIDNNNIIMMEPPFARGTGRYYKAFKITDRIWIIPERY   50
BoNT/B (GpII)    MPVTINNFNYNDPIDNDNIIMMEPPFARGTGRYYKAFKITDRIWIIPERY   50
BoNT/C           MPITINNFNYSDPVDNKNILYLDTHLNTLANEPEKAFRITGNIWVIPDRF   50
BoNT/D           MTWPVKDFNYSDPVNDNDILYLRIPQNKLITTPVKAFMITQNIWVIPERF   50
BoNT/CD          MPITINNFNYSDPVDNKNILYLDTHLNTLANEPEKAFRIIGNIWVIPDRF   50
BoNT/E           MP-KINSFNYNDPVNDRTILYIKPG---GCQEKYKSFNIMKNIWIIPERN   46
BoNT/E (buty)    MP-TINSFNYNDPVNNRTILYIKPG---GCQQFYKSFNIMKNIWIIPERN   46
BoNT/F (GpI)     MPVVINSFNYNDPVNDDTILYMQIPYEEKSKKYYKAFEIMRNVWIIPERN   50
BoNT/F (GpII)    MPVAINSFNYNDPVNDDTILYMQIPYEEKSKKYYKAFEIMRNVWIIPERN   50
BoNT/F (bara)    MPVNINNFNYNDPINNTTILYMKMPYYEDSNKYYKAFEIMDNVWIIPERN   50
BoNT/G           MPVNIKNFNYNDPINNDDIIMMEPFNDPGPGTYYKAFRIIDRIWIVPERF   50
TeNT             MPITINNFRYSDPVNNDTIIMMEPPYCKGLDIYYKAFKITDRIWIVPERY   50
                 *      * * **                    * * *    *  * *

                 50
BoNT/A           TF-TNPEEGDLNPPPEAKQVPVSYYDSTYLSTDNEKDNYLKGVTKLFERI   98
BoNT/A (inf)     TF-TNPEEGDLNPPPEAKQVPVSYYDSTYLSTDNEKDNYLKGVTKLFERI   98
BoNT/B (GpI)     TFGYKPEDFN-KSSGIFNRDVCEYYDPDYLNTNDKKNIFLQTMIKLFNRI   99
BoNT/B (GpII)    TFGYKPEDFN-KSSGIFNRDVCEYYDPDYLNTNDKKNIFFQTLIKLFNRI   99
BoNT/C           SRNSNPNLNK--PPRVTSPKSG-YYDPNYLSTDSDKDPFLKEIIKLFKRI   97
BoNT/D           SSDTNPSLSK--PPRPTSKYQS-YYDPSYLSTDEQKDTFLKGIIKLFKRI   97
BoNT/CD          SRDSNPNLNK--PPRVTSPKSG-YYDPNYLSTDSEKDTFLKEIIKLFKRI   97
BoNT/E           VIGTTPQDFH--PPTSLKNGDSSYYDPNYLQSDEEKDRFLKIVTKIFNRI   94
BoNT/E (buty)    VIGTIPQDFL--PPTSLKNGDSSYYDPNYLQSDQEKDKFLKIVTKIFNRI   94
BoNT/F (GpI)     TIGTDPSDFD--PPASLENGSSAYYDPNYLTTDAEKDRYLKTTIKLFKRI   98
BoNT/F (GpII)    TIGTNPSDFD--PPASLKNGSSAYYDPNYLTTDAEKDRYLKTTIKLFKRI   98
BoNT/F (bara)    IIGKKPSDFY--PPISLDSGSSAYYDPNYLTTDAEKDRFLKTVIKLFNRI   98
BoNT/G           TYGFQPDQFN-ASTGVFSKDVYEYYDPTYLKTDAEKDKFLKTMIKLFNRI   99
TeNT             EFGTKPEDFN--PPSSLIEGASEYYDPNYLRTDSDKDRFLQTMVKLFNRI
                      *                 ***  **     *        *  * **

                 100
BoNT/A           YSTDLGRMLLTSIVRGIPFWGGSTIDTELKVIDTNCINVI----QPDGSY   144
BoNT/A (inf)     YSTDLGRMLLTSIVRGIPFWGGSTIDTELKVIDTNCINVI----QPDGSY   144
BoNT/B (GpI)     KSKPLGEKLLEMIINGIPYLGDRRVPLEEFNTNIASVTVNKLISNPGEVE   149
BoNT/B (GpII)    KSKPLGEKLLEMIINGIPYLGDRRVPLEEFNTNIASVTVNKLISNPGEVE   149
BoNT/C           NSREIGEELIYRLSTDIPFPGNNNTPINTFDFDVDFNSVDVKTRQGNNWV   147
BoNT/D           NERDIGKKLINYLVVGSPFMGDSSTPEDTFDFTRHTTNIAVEKFENGSWK   147
BoNT/CD          NSREIGEELIYRLATDIPFPGNNNTPINTFDFDVDFNSVDVKTRQGNNWV   147
BoNT/E           NNNLSGGILLEELSKANPYLGNDNTPDNQFHIGDASA-VEIKFSNGSQDI   143
BoNT/E (buty)    NDNLSGRILLEELSKANPYLGNDNTPDGDFIINDASA-VPIQFSNGSQSI   143
BoNT/F (GpI)     NSNPAGEVLLQEISYAKPYLGNEHTPINEFHPVTRTTSVNIKSSTNVKSS   148
BoNT/F (GpII)    NSNPAGKVLLQEISYAKPYLGNDHTPIDEFSPVTRTTSVNIKLSTNVESS   148
BoNT/F (bara)    NSNPAGQVLLEEIKNGKPYLGNDHTAVNEFCANNRSTSVEIKESNGTTDS   148
BoNT/G           NSKPSGQRLLDMIVDAIPYLGNASTPPDKFAANVANVSINKKIIQPGAED   149
TeNT             KNNVAGEALLDKIINAIPYLGNSYSLLDKFDTNSNSVSFNLLEQDPSGAT   148
                      *   *       *   *

                        150
BoNT/A           RSEEL--NLVIIGPSADIIQFECKSFGH-----EVLNLTRNGYGSTQYIR   187
BoNT/A (inf)     RSEEL--NLVIIGPSADIIQFECKSFGH-----DVLNLTRNGYGSTQYIR   187
BoNT/B (GpI)     RKKGIFANLIIFGPGPVLNENETIDIGI-----QNHFASREGFGGIMQMK   194
BoNT/B (GpII)    RKKGIFANLIIFGPGPVLNENETIDIGI-----QNHFASREGFGGIMQMK   194
BoNT/C           KTGSINPSVIITGPRENIIDPETSTFKLT----NNTFAAQEGFGALSIIS   193
BoNT/D           VTNIITPSVLIFGPLPNILDYTASLTLQG----QQSNPSFEGFGTLSILK   193
BoNT/CD          KTGSINPSVIITGPRENIIDPETSTFKLT----NNTFAAQEGFGALSIIS   193
BoNT/E           ----LLPNVIIMGAEPDLFETNSSNISLR----NNYMPSNHGFGSIAIVT   185
BoNT/E (buty)    ----LLPNVIIMGAEPDLFETNSSNISLR----NNYMPSNHGFGSIAIVT   185
BoNT/F (GpI)     ----IILNLLVLGAGPDIFENSSYPVRKLMDSGGVYDPSNDGFGSINIVT   194
BoNT/F (GpII)    ----MLLNLLVLGAGPDIFESCCYPVRKLIDPDVVYDPSNYGFGSIQLMS   194
BoNT/F (bara)    ----MLLNLVILGPGPNILECSTFPVRIFPNNIA-YDPSEKGFGARMMSR   193
BoNT/G           QIKGLMTNLIIFGPGPVLSDNFTDSMIM-----NGHSPISEGFGARMMIR   194
TeNT             TKSAMLTNLIIFGPGPVLNKNEVRGIVLRVDNK-NYFPCRDGFGSIMQMA   197
                             *                            * *
```

```
                   200                                    Zn2+
BoNT/A             FSPDFTFGFEESLEVDTNPLLGAGKFARDPAVTLAHELIHAGHRLYGIAI   237
BoNT/A (inf)       FSPDFTFGFEESLEVDTNPLLGAGKFARDPAVTLAHELIHAEHRLYGIAI   237
BoNT/B (GpI)       FCPEYVSVFNNVQENKGASIFNRRGYFSDPALILMHELIHVLHGLYGIKV   244
BoNT/B (GpII)      FCPEYVSVFNNVQENKGASIFNRRGYFSDPALILMHELIHVLHGLYGIKV   244
BoNT/C             ISPRFMLTYSNATNDVGEGRFSKSEFCMDPILILMHELNHAMHNLYGIAI   243
BoNT/D             VAPEFLLTFSDVTSNQSSAVLGKSIFCMDPVIALMHALTHSLHQLYGINI   243
BoNT/CD            ISPRFMLTYSNATNNVGEGRFSKSEFCMDPILILMHELNHTMHNLYGIAI   243
BoNT/E             FSPEYSFRFNDN---------SMNEFIQDPALTLMHELIHSLHGLYGAKG   226
BoNT/E (buty)      FSPEYSFRFKDN---------SMNEFIQDPALTLMHELIHSLHGLYGAKG   226
BoNT/F (GpI)       FSPEYEYTFNDI---SGGYNSSTESFIADPAISLAHELIHALHGLYGARG   241
BoNT/F (GpII)      FSPEYEYTFNDI---SGGHNSSTESFIADPAISLAHELIHALHGLYGARG   241
BoNT/F (bara)      FSTEYEYAFND----------NTDLFIADPAISLAHELIHVLHGLYGAKG   233
BoNT/G             FCPSCLNVFNNVQENKDTSIFSRRAYFADPALTLMHELIHVLHGLYGIKI   244
TeNT               FCPEYVPTFDNVIENITSLTIGKSKYFQDPALLLMHELIHVLHGLYGMQV   247

                         250
BoNT/A             NPNR-VKFVNTNAYYEMSGLEVSFEELRTFGGHDAKFIDSLQENEFRLYY   286
BoNT/A (inf)       NPNR-VKFVNTNAYYEMSGLEVSFEELRTFGGHDAKFIDSLQENEFRLYY   286
BoNT/B (GpI)       -DDLPIVPNEKKFF-MQSTDAIQAEELYTFGGQDPSIITPSTDKSIYDKV   292
BoNT/B (GpII)      -DDLPIVPNEKKFF-MQSTDTIQAEELYTFGGQDPSIISPSTDKSIYDKV   292
BoNT/C             PNDQTISSVTSNIFYSQYNVKLEYAEIYAFGGPTIDLIPKSARKYFEEKA   293
BoNT/D             PSDKRIRPQVSEGFFSQDGPNVQFEELYTFGGLDVEIIPQIERSQLREKA   293
BoNT/CD            PNDQRISSVTSNIFYSQYKVKLEYAEIYAFGGPTIDLIPKSGRKYFEEKA   293
BoNT/E             ITTKYTITQKQNPLITNIRG-TNIEEFLTFGGTDLNIITSAQSNDIYTNL   275
BoNT/E (buty)      ITTKYTITQKQNPLITNIRG-TNIEEFLTFGGTDLNIITSAQSNDIYTNL   275
BoNT/F (GpI)       VTYKETIKVKQAPLMIAEKP-IRLEEFLTFGGQDLNIITSAMKEKIYNNL   290
BoNT/F (GpII)      VTYEETIEVKQAPLMIAEKP-IRLEEFLTFGGQDLNIITSAMKEKIYNNL   290
BoNT/F (bara)      VTNKKVIEVDQGALMAAEKD-IKIEEFITFGGQDLNIITNSTNQKIYVIL   282
BoNT/G             -SNLPITPNTKEFF-MQHSDPVQAEELYTFGGHDPSVISPSTDMNIYNKA   292
TeNT               -SSHEIIPSKQEIY-MQHTYPISAEELFTFGGQDANLISIDIKNDLYEKT   295

                              300
BoNT/A             YNKFKDIASTLNKAKSIVGTT--ASLQYMKNVFKEKYLLSEDTSGKFSVD   334
BoNT/A (inf)       YNKFKDVASTLNKAKSIIGTT--ASLQYMKNVFKEKYLLSEDTSGKFSVD   334
BoNT/B (GpI)       LQNFRGIVDRLNKVLVCISDPNIN-INIYKNKFKDKYKFVEDSEGKYSID   341
BoNT/B (GpII)      LQNFRGIVDRLNKVLVCISDPNIN-INIYKNKFKDKYKFVEDSEGKYSID   341
BoNT/C             LDYYRSIAKRLNSITTANPSSFNKYIGEYKQKLIRKYRFVVESSGEVTVN   343
BoNT/D             LGHYKDIAKRLNNINKTIPSSWISNIDKYKKIFSEKYNFDKDNTGNFVVN   343
BoNT/CD            LDYYRSIAKRLNSITTANPSSFNKYIGEYKQKLIRKYRFVVESSGEVAVD   343
BoNT/E             LADYKKIASKLSKVQ--VSNPLLNP---YKDVFEAKYGLDKDASGIYSVN   320
BoNT/E (buty)      LADYKKIASKLSKVQ--VSNPLLNP---YKDVFEAKYGLDKDASGIYSVN   320
BoNT/F (GpI)       LANYEKIATRLSRVNSA---PPEYDINEYKDYFQWKYGLDKNADGSYTVN   337
BoNT/F (GpII)      LANYEKIATRLSEVNSA---PPEYDINEYKDYFQWKYGLDKNADGSYTVN   337
BoNT/F (bara)      LSNYTAIASRLSQVNRN---NSALNTTYYKNFFQWKYGLDQDSNGNYTVN   329
BoNT/G             LQNFQDIANRLN-IVSSAQGSGID-ISLYKQIYKNKYDFVEDPNGKYSVD   340
TeNT               LNDYKAIANKLSQVTSC-NDPNID-IDSYKQIYQQKYQFDKDSNGQYIVN   343

                                 350
BoNT/A             KLKFDKMYKMLTEIYTEDNFVKFFKVNLRKTYLNFDK-AVFKINIVPKVN   383
BoNT/A (inf)       KLKFDKMYKMLTEIYTEDNFVNFFKVNIRKTYLNFDK-AVFRINIVPDEN   383
BoNT/B (GpI)       VESFDKLYKSLMFGFTETNIAENYKIKTRASYFSDSLPPVKIKNLLDNEI   391
BoNT/B (GpII)      VESFNKLYKSLMLGFTEINIAENYKIKTRASYFSDSLPPVKIKNLLDNEI   391
BoNT/C             RNKFVELYNELLTQIFTEFNYAKIYNVQNRKIYLSNVYTPVTA-NILDDNV   392
BoNT/D             IDKFNSLYSDLLTNVMSEVVYSSQYNVKNRTHYFSRHYLPVFA-NILDDNI   392
BoNT/CD            RNKFAELYKELLTQIFTEFNYAKIYNVQNRKIYLSNVYTPVTA-NILDDNV   392
BoNT/E             INKFNDIFKKLL-YSFTEFDLATKFQVKCRQTYIGQYK-YFKLSNLLNDSI   368
BoNT/E (buty)      INKFNDIFKKLL-YSFTEFDLATKFQVKCRQTYIGQYK-YFKLSNLLNDSI   368
BoNT/F (GpI)       ENKFNEIYKKLL-YSFTEIDLANKFKVKCRNTYFIKYG-FLKVPNLLDDDI   385
BoNT/F (GpII)      ENKFNEIYKKLL-YSFTESDLANKFKVKCRNTYFIKYE-FLKVPNLLDDDI   385
BoNT/F (bara)      ISKFNAIYKKLL-FSFTECDLAQKFQVKNRSTYLFHFK-PFRLLDLLDDNI   377
BoNT/G             KDKFDKLYKALMFGFTETNLAGEYGIKTRYSYFSEYLPPIKTEKLLDNTI   390
TeNT               EDKFQILYNSIMYGFTEIELGKKFNIKTRLSYFSMNHDPVKIPNLLDDTI   393
```

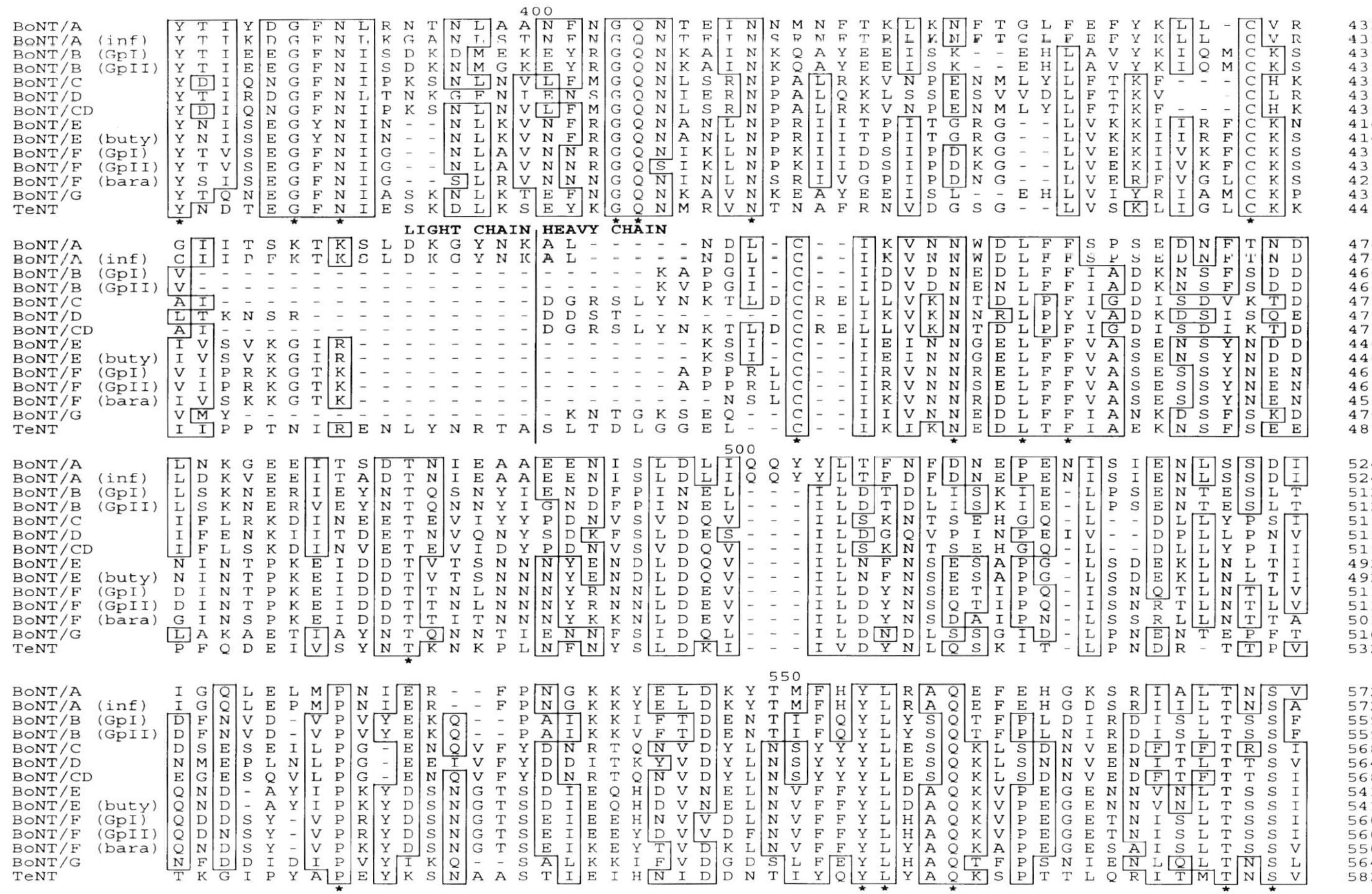

Figure 17.2 Continued.

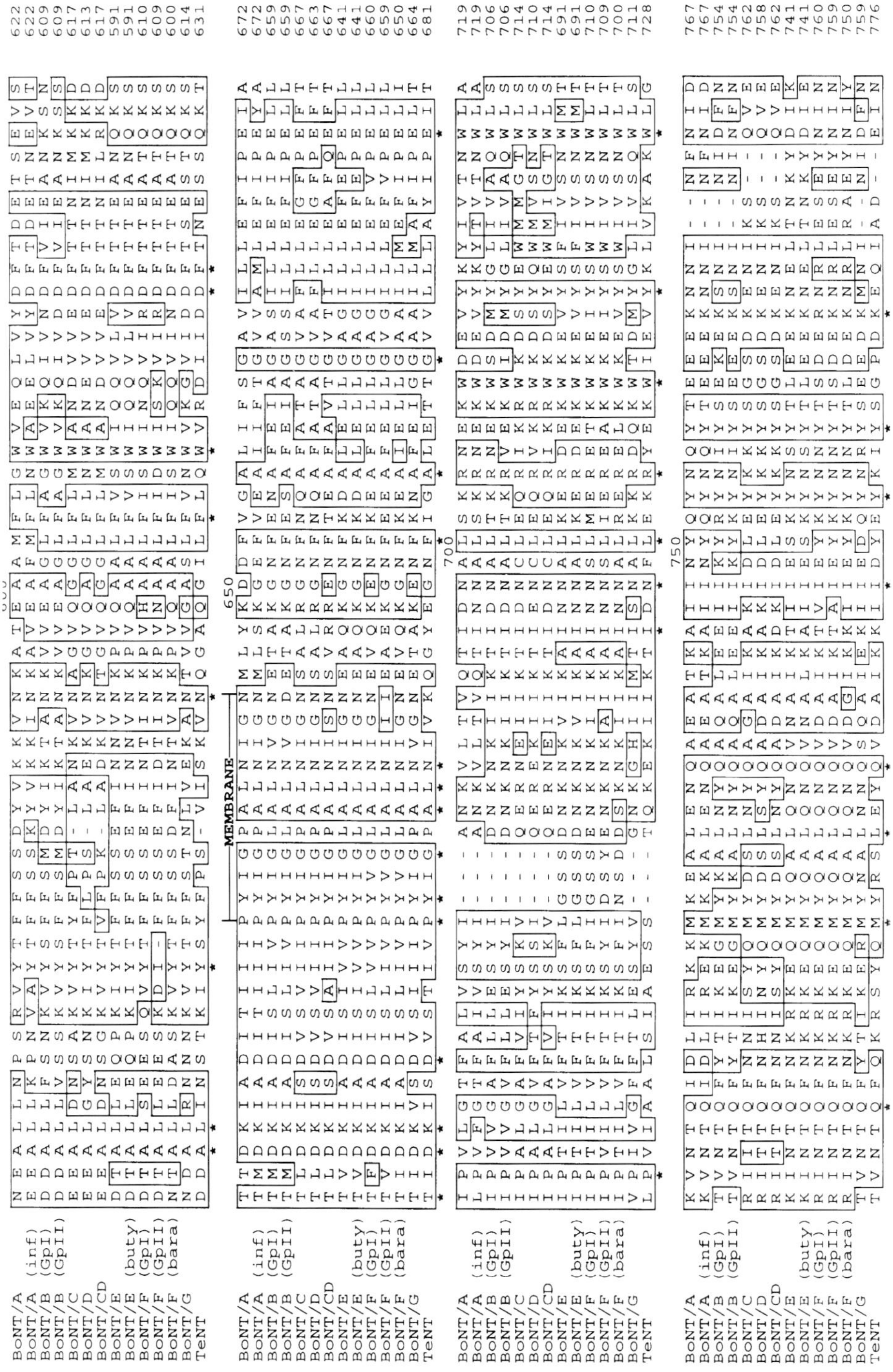

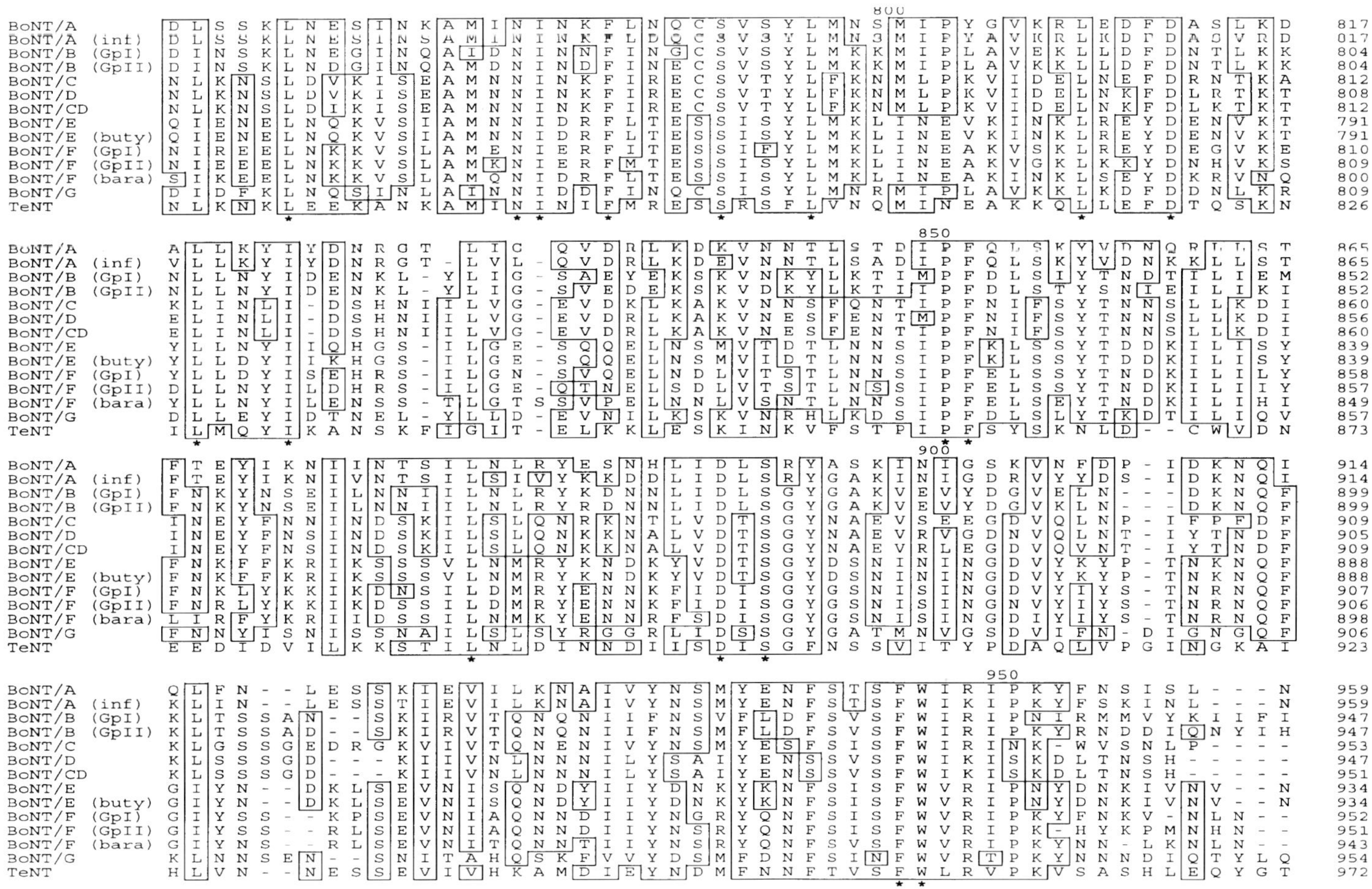

Figure 17.2 Continued.

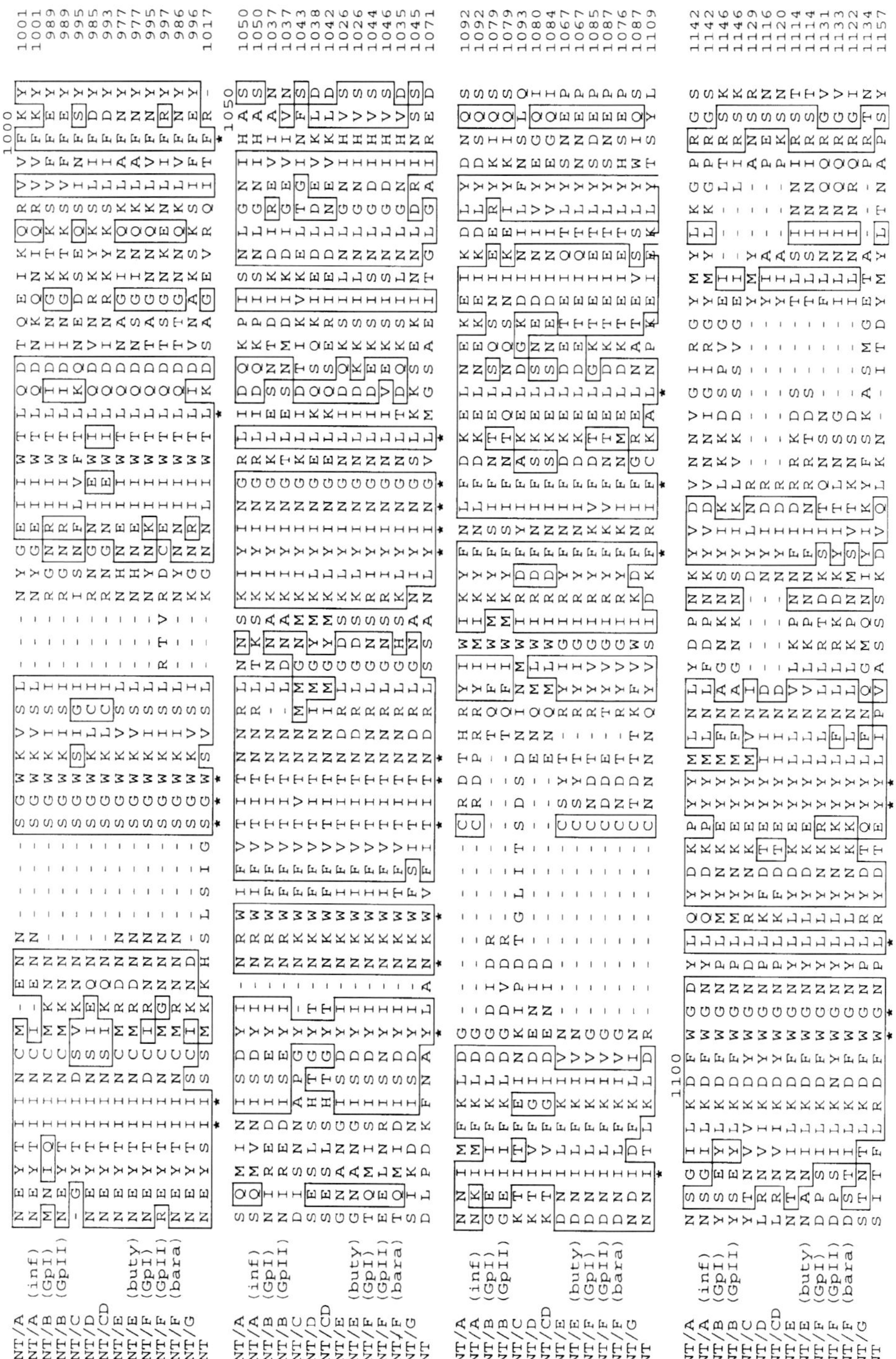
BoNT/A (inf)
BoNT/A
BoNT/B (GpI)
BoNT/B (GpII)
BoNT/C
BoNT/D
BoNT/CD
BoNT/E (buty)
BoNT/E
BoNT/F (GpI)
BoNT/F (GpII)
BoNT/F (bara)
BoNT/G
TeNT

Figure 17.2 Continued.

indicates, for example, that transfer of the well-conserved BoNT/B gene between group I and II *C. botulinum* (93% similarity) took place relatively recently on an evolutionary scale. In comparison, the BoNT/F genes of *C. botulinum* and *C. barati* (less than 87% similarity) transferred at a much earlier point in evolutionary time.[25]

It is apparent that the degree of similarity shared by toxins varies depending on which region of the protein is compared. Thus, while the light chains of all three BoNT/F variants exhibit an equivalent degree of sequence divergence, the heavy chains exhibit marked differences (79% for I/II and 73.8% for II/VI). This indicates that, following lateral transfer of genes into the different physiological groupings, the different domains of the protein chains evolve at different rates. One process that may speed up such a process might be recombination between different toxin genes to form hybrid toxins, such as the BoNT/C:D mosaic toxin.[22]

III. THE RELATIONSHIP BETWEEN BoNT GENES AND MOBILE GENETIC ELEMENTS

Dissemination of the neurotoxin genes between the physiological groups implies that the genes are, or have at one time been encoded on mobile genetic elements such as transposons, plasmids and bacteriophage genomes.

A. The role of bacteriophage-mediated transfer of BoNT genes

Bacteriophages that infect clostridial strains expressing BoNT's were identified in the late 1960s and early 1970s.[50–54] Those isolated from group I and II *C. botulinum* strains were shown to exist in a lysogenic state, with the bacteriophage DNA maintained as a chromosomally integrated prophage.[55] As a result, bacteriophage-sensitive strains from these groups are difficult to isolate. For the few strains in which this has been possible, no change in toxigenicity was detected.

In group III strains expressing BoNT/C and BoNT/D, toxigenicity could be lost by curing of bacteriophage, and then regained by reinfection with the appropriate phage. Once infected, the strains demonstrated superinfection immunity to other potential reinfecting bacteriophages.[56] Atoxinogenic strains, that previously expressed either of these two neurotoxin serotypes, were not only indistinguishable from each other, but also from *C. novyi* strains cured of a bacteriophage encoding α-toxin.[57] This indicated that the presence of toxinogenic bacteriophages were solely responsible for the subclassification of these clostridial strains. It was also observed that curing of the strains with such reagents as Acridine Orange, or with u.v. light, occurred at high frequency. This led to the suggestion that the strains had not been lysogenized (group I and II strains), but that the phage DNA was present in a so-called state of 'pseudolysogeny', in which it existed 'free' within the host cytoplasm. This interpretation was subsequently experimentally proven for BoNT/C and BoNT/D through the use of DNA probes.[19,21]

Evidence for the participation of bacteriophages in the transfer of other BoNT genes is more tentative. Work in our laboratory with the group I *C. botulinum* strain NCTC 2916 has shown that approximately 1 kb downstream of the BoNT/A gene is a gene, *lycA*, encoding a protein which has similarity to various types of lysozymes. Of particular significance is the homology with the lytic enzymes of the *Lactobacillus bulgaricus* bacteriophage ϕMV-1[58] and the *Streptococcus pneumoniae* phage ϕCP-1.[59] Members of the Charalopsis group of lysozymes, these enzymes share conserved aspartate and glycine residues at their active sites[60] and possess C-terminal repeat sequences responsible for substrate recognition, that are common to enzymes that utilize complex polymeric substrates.[61] Expression of *lycA* in *Escherichia coli*, under the control of the *lpp* promoter, has yielded a product that is active against clostridial cell wall preparations.[62]

Genes similar to *lycA* have been detected by PCR in other group I *C. botulinum* strains at least one of which, *lycF* from the group I type F *C. botulinum* strain Langeland, is found at a similar distance from the stop codon of the BoNT gene. Nucleotide sequencing of the appropriate region from strain Langeland (unpublished data) has shown that this *lycF* sequence shares 96% identity with the NCTC 2916 *lycA*, and that this homology continues for 116 bp upstream of its start codon. Thereafter the two sequences exhibit complete divergence. Given the intimate involvement of lytic enzymes in the life cycles of bacteriophages, the possibility exists that these *lyc* genes, and the associated BoNT genes, may form part of an integrated prophage in certain group I *C. botulinum* strains. In this light, it is interesting to note that release of botulinum toxin is intimately linked with autolysis.[63]

Bacteriophages have also been implicated in the transfer of the BoNT/E gene between *C. butyricum* and the E-like strain of *C. botulinum* S-5. Experiments have been described[64] in which a nontoxinogenic strain of *C. botulinum* (S-5) acquired the ability to produce BoNT/E following its growth in the presence of bacteriophage obtained by Mitomycin C treatment of a *C. butyricum* type E strain (ATCC 5839). Transfer would, however, only occur if strain S-5 and the bacteriophage particles were grown in a culture filtrate derived from a third, nontoxinogenic *C. butyricum* strain (ATCC 19398). It was suggested that the BoNT/E gene of ATCC 5839 resides on a defective prophage which requires the assistance of an unknown factor from the helper strain ATCC 19398 in order to transfer to the nontoxinogenic strain S-5.[64] DNA–DNA hybridization studies demonstrated that the BoNT/E gene and the defective bacteriophage DNA hybridized to apparently similar restriction enzyme fragments of the ATCC 5839 genome. Attempts to prove that the BoNT/E gene was carried by the bacteriophage were, however, inconclusive. Thus, a fragment of the expected size could be generated using isolated bacteriophage DNA as a template in a PCR and appropriate BoNT/E gene-specific primers. In contrast, negative results were obtained in a dot–blot analysis of phage DNA using a radiolabeled BoNT/E gene fragment as a probe. It is clear from these results that the BoNT/E gene is only present on a minor fraction of the phage genomes present. Assuming that all of the isolated DNA is derived from a single phage type, the subpopulation carrying BoNT/E may represent defective prophages which have both become defective and acquired the BoNT/E gene as a result of imperfect excision from the clostridial chromosome. This explanation would require that the bacteriophage responsible has an attachment site in the chromosome adjacent to the BoNT/E gene.

B. Plasmid-encoded neurotoxins

The first clostridial neurotoxin shown to be plasmid encoded was TeNT. Such a location was suggested as early as 1977,[65] but more direct evidence was not obtained until a few years later when loss of toxinogenicity was shown to correlate with curing of a large 75 kb plasmid.[66] Thereafter, an oligonucleotide probe based on the N-terminal sequence of TeNT was used to confirm that this plasmid (pCL1) did indeed carry the TeNT gene.[67] Plasmids have also been detected in many of the clostridial strains which elaborate BoNT. Those found in *C. botulinum* strains which produce BoNT/A and BoNT/E range in size from around 2 MDa up to 61 MDa.[68–70] No correlation with toxinogenicity has yet been reported.

Evidence for a plasmid location of the BoNT/E gene in strains of *C. butyricum* has also been presented. In these experiments, the subjective observation was made that greater quantities of a BoNT/E-specific DNA fragment could be PCR-amplified from a ‘plasmid’ DNA preparation than from a ‘chromosome’.[71] More substantive data are required (e.g., DNA–DNA hybridization of probes to a defined plasmid species) before the location of the BoNT/E gene in this strain may be confidently assigned as being on an extrachromosomal element. Such definitive experiments have been undertaken to demonstrate clearly that BoNT/G is carried by a 114 kb plasmid.[72]

C. Do transposons have a role in the dissemination of BoNT genes?

The observation that neurotoxin genes can have many different locations (chromosome, plasmid or bacteriophage) suggests that they themselves may be mobile through the agency of transposable genetic elements. Although there is no direct evidence for this supposition, a number of observations are worthy of note. It is, for instance, apparent in a number of cases, that nucleotide sequence conservation between toxin genes ends abruptly at defined points past their 3′-ends. This is most obvious between BoNT/A genes of the type A *C. botulinum* strains NCTC 2916 and 62A, where any semblance of sequence homology ends at a point 100 bp downstream of their translational stop codons. The opposite is evident in comparison between strain NCTC 2916 and the type F *C. botulinum* strain Langeland. In this case, the neurotoxin structural genes, and the first 600 bp following their stop codons, share no significant homology, before becoming 96% identical, owing to the presence of highly conserved *lyc* genes. In this case, it would appear that these two toxin genes have 'transposed' to the same location in the genome of *C. botulinum*, Group I.

Evidence for the involvement of transposons has also emerged from the analysis of the genome of bacteriophage C-st and a mutant derivative which had lost a 21.5 kb fragment of DNA.[73,74] Nucleotide sequence analysis demonstrated that the fragment in question was flanked by a six-nucleotide motif (5′-AAGGAG-3′) typical of the Tn*554* family of site-specific transposons, and that only a single copy of this motif remained in the CN deletion variant. The suggestion that this 21.5 kb fragment is a transposable element is supported by the fact that an essentially identical fragment is present on the genome of the type D phage D-1873. Furthermore, the nucleotide sequences of the regions of the phage genome surrounding this element are completely dissimilar from the equivalent regions of the C-st phage. However, although the point at which the sequences of the two phages diverge exactly corresponds to the 6 bp core motif at one end of the 21.5 kb element, the two phages remain homologous for 61 bp past this motif at the opposite end. The involvement of the 6 bp sequence, therefore, remains open to question.

IV. THE ORGANIZATION OF THE BOTULINUM PROGENITOR TOXIN GENES

Unlike TeNT, BoNTs are always found in association with a varying number of non-toxic proteins. The complexes formed are known as progenitor toxins. Three forms are recognized: LL toxin (900 kDa), L toxin (500 kDa) and M toxin (300 kDa). M toxin comprises just two components, the 150 kDa BoNT and an equivalent sized nontoxic, nonhaemagglutinating protein (NTNH). The larger

L and LL forms also contain NTNH, but in addition they incorporate an undefined number of smaller nontoxic proteins which possess haemagglutinating activity (HA). The nature of the progenitor toxin produced varies according to toxin type and host strain (Table 17.1). In type E and F strains, only type M progenitor toxin is produced. Type B, C and D BoNT is usually produced as either the M or L form, while BoNT/G is composed exclusively of the L complex. All three forms are found in type A *C. botulinum* cultures.

Progenitor toxins exhibit far greater stability to pH extremes, temperature and proteases than the BoNT alone. Botulism is principally caused by the ingestion of preformed toxin arising from growth of a clostridial contaminant in foodstuffs. This has led to the suggestion that the function of the ancillary proteins of the toxin complex is to maintain the toxin in a stable form during its exposure to the strongly acidic and protease-rich environment encountered in the stomach.[75,76] Consistent with this view, it is generally the case that the larger the complex, the greater the toxicity when administered by the oral route.[77]

A. Location of a genes encoding nontoxic components

From an early stage it was apparent that the genes coding for at least some of the non-toxic components of the toxin complex were in some way linked to the gene responsible for production of the neurotoxin. Thus, it was shown that atoxinogenic strains of *C. botulinum* which were reinfected with BoNT/C encoding bacteriophage acquired the ability to produce haemagglutinating (HA) activity, in addition to the reacquisition of the ability to synthesize neurotoxin.[78] This fact was exploited to clone a gene expressing HA activity by constructing a gene library of the bacteriophage C-st genome in *E. coli*, and screening it with anti-HA sera. Through nucleotide sequence analysis, the gene was subsequently shown to be on the opposite strand to, and 4.3 kb upstream of, the BoNT/C1 neurotoxin gene.[79] The translated sequence specified a protein with a predicted molecular mass of 33 kDa. Its N-terminal sequence proved to be identical to a sequence derived experimentally from an equivalent sized protein (HA33/C) subsequently purified from the type C toxin complex.

Shortly after the identification of HA33/C, Somers and DasGupta[80] published the results of their detailed analysis of the constituents of Group I *C. botulinum* type A and type B toxin complexes. They were able to show that, in addition to the NTNH proteins, the haemagglutinating activity of both complexes could be separated into five distinct proteins with molecular masses of 17, 21.5, 35, 57 and 92 kDa. Each protein was shown to have a unique N-terminal sequence, but the corresponding subunits from the two botulinum serotypes were extremely homologous. In the case of the 35 kDa protein, substantial homology (84% identity) with the N-terminus of HA33/C was evident.

Using the determined[80] N-terminal amino acid sequence of the 17 kDa protein, we have been able to design degenerate PCR primers which, in combination with primers based on the BoNT/A gene,[14] amplify a 4 kb region of DNA residing immediately upstream of the BoNT/A gene. Thereafter, the availability of this fragment allowed the identification and subsequent cloning of genomic fragments from the *C. botulinum* strains NCTC 2916 which encompassed an 8.0 kb region of DNA from immediately upstream of the BoNT/A gene. This has been sequenced[81] and shown to contain the gene encoding NTNH/A and sufficient genes to bring about the production of all five HA proteins identified by Somers and DasGupta.[80] The equivalent region was also cloned from upstream of the BoNT/C1 gene of the C468 phage genome[82] and shown to carry the same genes as those found upstream of BoNT/A. Some of these genes have also been detected upstream of other neurotoxins. The current extent of knowledge regarding the organization of the genes of the various toxin loci is presented in Figure 17.3, the individual components of which are discussed in the following sections.

B. The NTNH proteins

The first NTNH gene to be sequenced was that of the bacteriophage C-st.[83] It was shown to be located between the previously identified HA33/C[79] and BoNT/C1[20] genes, residing immediately upstream of the latter. Its identity was unequivocally established by the equivalence of the predicted size of the encoded protein (138 kDa) and purified NTNH/C, and the fact the translated N-terminal amino acid sequence was identical to that experimentally determined for NTNH purified from type C toxin complex. Similar genes have subsequently been cloned and sequenced from *C. botulinum* strains expressing BoNT/A,[81] BoNT/C,[82] BoNT/E[84] and BoNT/F.[85] The NTNH genes associated with the other toxin serotypes, including those of *C. butyricum* and *C. barati*, have been either inferred through translation of sequences residing immediately 5′ to the beginning of the neurotoxin genes, or have been detected by PCR. More specifically, NTNH genes have been shown to be located immediately upstream of toxin serotypes by PCR using primers targeted at conserved regions of NTNH and BoNT genes. No NTNH gene has been detected in any *C. tetani* strain.

NTNH protein sequences are highly conserved and show greater overall sequence similarity than their neurotoxin counterparts. Thus, for instance, NTNH/A shares 65%, 65% and 71% sequence similarity with NTNH/C, NTNH/E and NTNH/F, respectively. The equivalent respective homology

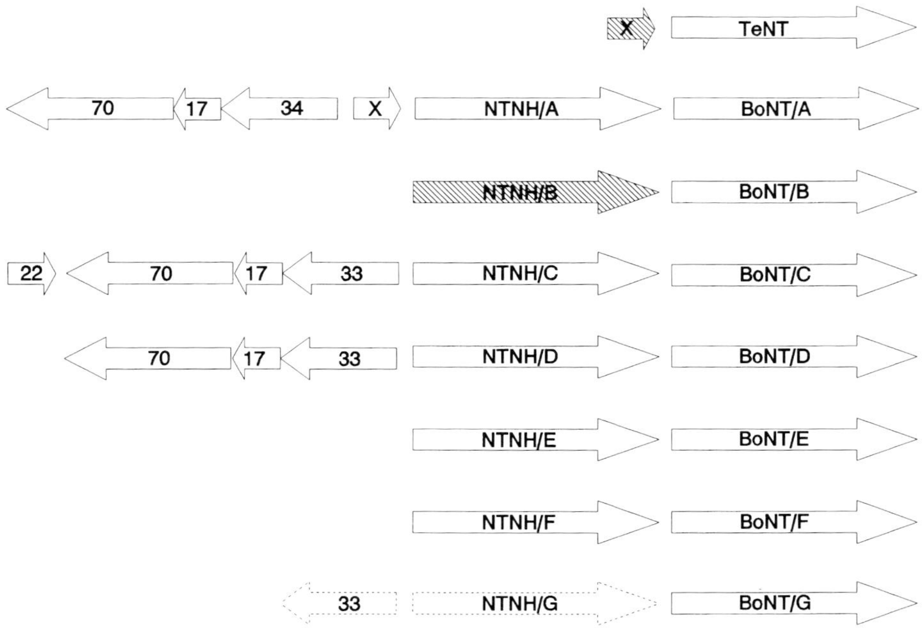

Figure 17.3 Organization of BoNT and TeNT gene loci. Open arrows indicate genes whose nucleotide sequences have been determined. Shaded arrows indicate open reading frames for which only partial sequences are available, and arrows with a dotted outline are those for which their existence is only inferred through DNA/DNA hybridization studies or through the amplification of DNA fragments of the expected size in PCR experiments. The numbers 17, 33, 34 and 70 refer to the predicted size (in kDa) of the encoded haemagglutinating protein (HA). X and 22 are OrfX and Orf22, the putative transcriptional factors.

between neurotoxins is 29%, 37% and 41%. The greater sequence conservation exhibited by NTNH proteins is, perhaps, an unexpected finding, given the simplistic 'protective' role of this class of protein compared with the sophisticated function of neurotoxins. Because the two encoding genes are always found in association with one another (e.g., on the same plasmid or bacteriophage), both must be subjected to the same mutational influences. Despite this, a greater degree of divergence is evident in BoNT sequences. It follows, therefore, that a greater evolutionary pressure must exist to conserve NTNH structure than for BoNTs. It seems reasonable to assume that the factors involved relate to NTNH function. It is intriguing to note, that the extreme N-terminal regions of NTNHs share significant amino acid homology with the equivalent domain of BoNTs.[44] The functional significance of these shared sequence motifs are unclear, but may indicate that the two proteins interact with a common component.

Evidence for the specific interaction of NTNH with other toxin complex components is now beginning to emerge. In a recent study with the M and L toxins of a type D *C. botulinum* strain, it was shown that the NTNH present in the two progenitor forms was different.[86] Thus, amino acid sequencing of the NTNH component isolated from the M form toxin demonstrated that 140 amino acids were absent from its N-terminal end. These amino acids had presumably been removed by proteolytic cleavage between Thr-140 and Ser-141 by an unspecified protease. The evidence obtained, however, suggested that this processing took the form of proteolytic 'nicking', rather than removal, as the resultant 15 kDa polypeptide was apparently still in direct contact with the toxin complex. It was concluded that an intact N-terminal portion of NTNH protein is required for the binding of the HA proteins, allowing the L complex to form. Similar observations have also been made in complexes of type C and type A *C. botulinum* strains, where the nick site corresponds to a point between Pro-144 and Phe-145.[87] These amino acid residues lie in a 33 amino acid insert common to NTNH/A, NTNH/C and NTNH/D, but which is absent from NTNH/E and NTNH/F. Type E and F botulinum toxin does not, of course, occur in a form larger than M.

The silent B gene encoded by the *C. botulinum* strain NCTC 2916 also has a NTNH protein associated with it, but it has an unusual structure (see Figure 17.4). Sequencing of the region upstream of the BoNT/B gene (unpublished

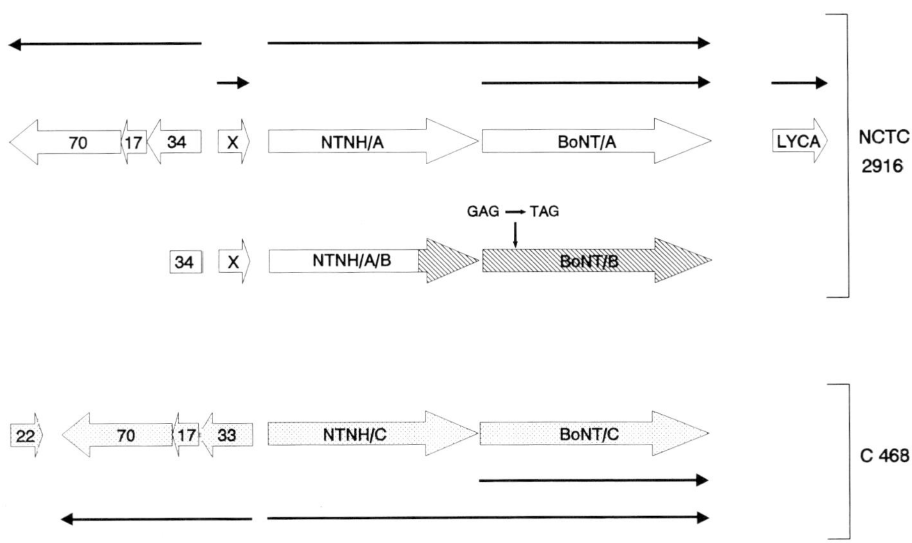

Figure 17.4 Gene structure at the neurotoxin loci in the *C. botulinum* strains NCTC 2916 and C468. Open and striped arrows indicate the genes encoded at the BoNT/A and silent BoNT/B loci of NCTC 2916,[45,81] and stippled arrows those at the BoNT/C locus of C468.[82] Solid arrows identify those genes and operons for which transcriptional initiation sites have been mapped.

data) has revealed an open reading frame whose encoded protein is a hybrid composed of the N-terminus of the NCTC 2916 NTNH/A, and the C-terminus of NTNH/B from the type B *C. botulinum* strain Danish.[17] A DNA sequence alignment of NTNH/A, NTNH/AB hybrid and NTNH/B Danish revealed a sequence of 94 nucleotides that were absolutely identical between all three sequences. The DNA sequence of the hybrid prior to this was identical to that of NTNH/A, and after to NTNH/B Danish. This hybrid gene has no doubt arisen through a recombinational event between a NTNH/A and a NTNH/B gene, driven by the homologous stretch of 94 bases shared by the two sequences.

C. The HA proteins

Homologous equivalents to HA33/C and HA34/A have been detected in other *C. botulinum* strains[88] through the cloning and sequencing of the encoding genes from the *C. botulinum* strains NCTC 7272 (group I, type A), NCTC 7273 (group I, type B) and ATCC 25765 (Eklund 17B; group II, type B). As predicted by the analysis of Somers and DasGupta,[80] the proteins of type A and B complexes exhibit a high degree of similarity, whether the type B toxin is derived from a group I (83% similarity) or a group II (90%) *C. botulinum* strain. The genes encoding the 17 kDa peptide identified by Somers and DasGupta[80] have only been isolated and characterized at the sequence level for *C. botulinum* NCTC 2916, C468 and C-st. In all cases the encoding genes reside immediately downstream of the HA34/A and HA33/C genes. The two type C sequences (HA17/C) are identical, while HA17/A from the NCTC 2916 exhibits 66% identity with its type C counterparts. Neither HA17/A or HA17/C exhibit any similarity to any protein currently deposited in the Genbank/EMBL database.

Following the gene encoding HA17, is a third ORF, designated HA70. Amino acid position 16–25 of the deduced sequence of the HA70/A exactly corresponds to the N-terminal sequence of the 21.5 kDa protein purified from the type A progenitor toxin complex,[80] while residues 204–213 are identical to the N-terminal sequence of the 57 kDa toxin complex component. A similar homology occurs at the equivalent position in HA70/C. This suggests that HA70 initially is produced by the clostridial host as a single protein, and subsequently becomes processed into these smaller polypeptides. Interestingly, the smaller of the two polypeptides exhibits some similarity (20% identity in a 202 amino acid overlap) with the cytotoxic/receptor binding domain of the enterotoxin of *C. perfringens*.[89]

V. REGULATION OF GENE EXPRESSION AT THE NEUROTOXIN GENE LOCUS

In bacterial pathogens, tightly managed regulatory systems have evolved precisely to control the circumstances under which virulence factors are produced. Such control is exerted through modulation of gene expression, and frequently involves complex regulatory cascades, acting predominantly at the level of transcription. Although such mechanisms have yet to be defined in the neuropathogenic clostridia, their presence can be predicted from a number of factors. These include the observed modular organization of the various toxin complex genes, and the observation that toxin production is significantly affected by exogenous nitrogen levels,[90–93] an environmental signal frequently sensed and acted upon through so-called two-component signal transduction systems.[94]

A. Putative regulatory proteins

A consistent feature which has emerged from the analysis of the region of DNA upstream of BoNT genes has been the presence of a small open reading frame (ORF) encoding a protein which exhibits a number of features characteristic of regulatory proteins. In the case of NCTC 2916 it has been designated *orfX*,[81] while the C468 homolog has been called *orf22*.[82] The encoded proteins are both 22 kDa, possess putative helix-turn-helix (HTH) motifs at their C-termini and have predicted pI values of approximately 10.4. These properties are highly suggestive of a DNA-binding function. This assumption is further supported by the fact that they share amino acid similarity with known regulatory proteins MsmR and UviA (Figure 17.5). The MsmR protein is involved in the regulation of expression of genes encoding proteins involved in the sugar transport machinery of *Streptomyces mutans*.[95,96] UviA regulates bacteriocin production from the *C. perfringens* plasmid pIP404,[97] and shares homology with the −35 binding domain of σ factors, such as σ^{70} of *E. coli* and σ^{D} of *Bacillus subtilis*.[98] More significantly, it has become apparent that OrfX/Orf22 possess sequence similarities with Orftxe1, the product of a gene recently shown to be located immediately upstream of, and on the same strand as, the toxin B gene of *C. difficile*.[99]

Genes encoding proteins equivalent to OrfX and Orf22 have also been cloned and sequenced from *C. botulinum* NCTC 7272 (type A), *C. botulinum* NCTC 7273 (type B, group I) and *C. botulinum* Eklund 17B (type B, group II).[88] It is also apparent, from a closer examination of the upstream region of the TeNT gene sequence, that an equivalent gene resides immediately upstream of this neurotoxin gene. It appears likely that this family of proteins is in some way involved in the expression of all clostridial neurotoxin genes. Unfortunately, to date, no similar regulatory sequences have been detected in strains that produce type E and F neurotoxins. In these strains, the NTNH genes are preceded by ORFs of as yet unknown function.

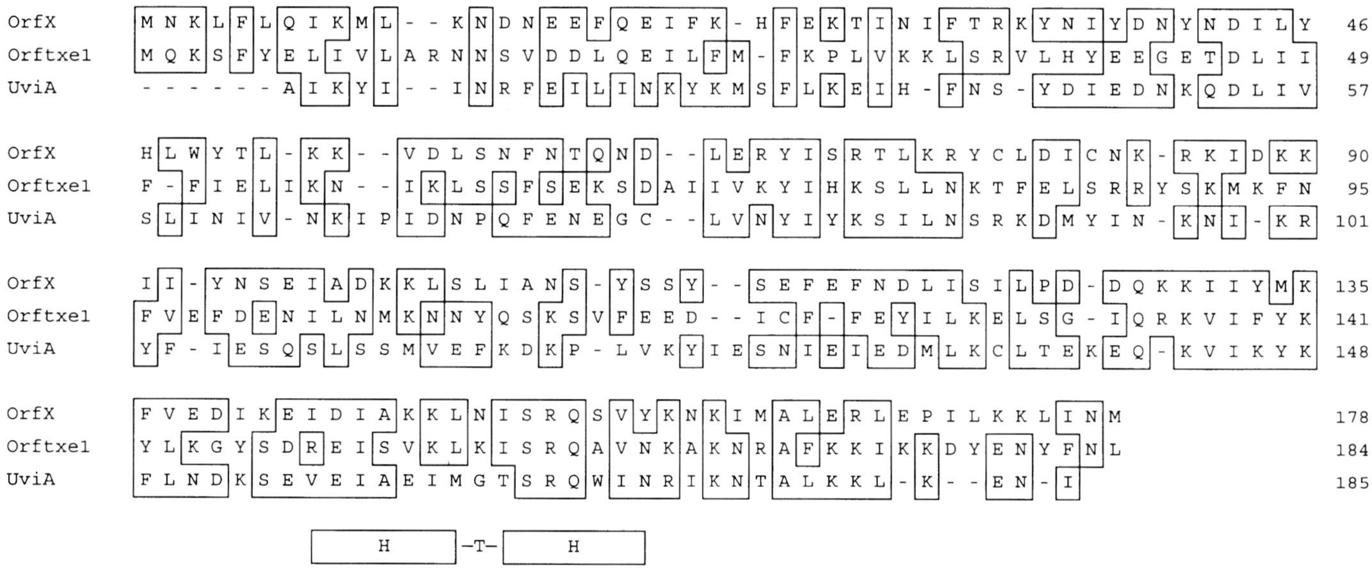

Figure 17.5 Comparison of the amino acid sequences of OrfX with Orftxel from *Clostridium difficile* VPI 10463 and UviA from *Clostridium perfringens.* Boxed sequences represent identical/similar amino acids. The putative helix-turn-helix motif (HTH) is indicated below the *C*-terminal portion of these proteins.

B. Transcriptional analysis of neurotoxin gene loci

It is immediately apparent that the toxin complex genes of type C and type A *C. botulinum* strains are organized in two distinct, divergent units, namely, the NTNH-BoNT genes and a unit composed of the HA-encoding genes. The suspicion that these two units may form two distinct operons has been confirmed in two separate studies. In our own laboratory, Northern blots of RNA isolated from the type A strain NCTC 2916, and appropriate DNA probes, have indicated the presence of a single 3.2 kb transcript encompassing all three HA genes, a 7.5 kb transcript specific to the genes encoding NTNH/A and BoNT/A, and a 4.0 kb transcript specific to BoNT/A alone.[81] From this it would appear that the HA genes are translated from a single polycistronic message, whereas BoNT/A is transcribed as either a monocistronic message or as a bicistronic transcript which also specifies the NTNH/A gene (Figure 17.4). The existence of the last two transcripts was confirmed in primer extension studies, when it was shown that transcriptional initiation points mapped 5′ to both the NTNH/A and BoNT/A gene.[81] Essentially similar results were previously obtained for the corresponding genes of the *C. botulinum* type C strain C468.[82] In this case the presence of transcripts encoding more than one gene was confirmed using reverse transcription (RT)PCR and appropriate gene-specific primers. The reason why the BoNT/A gene is transcribed as either a bi- or mono-cistronic message remains unclear.

No detailed analysis of other botulinum gene clusters has yet been undertaken. The similar spatial organization of the genes, however, suggests that the same transcript pattern will be found. It should be noted, however, that other workers[15] mapped the start point of the type A *C. botulinum* strain 62A as being 118 nucleotides 5′ to the BoNT/A structural gene. We determined the equivalent start point in NCTC 2916 to 20 nucleotides 5′ to the ATG colon. Although this discrepancy could be attributable to strain differences, it is notable that the upstream regions of both genes are identical for at least 340 bp. The primer used in the 62A study,[15] however, spanned the BoNT/A gene ATG start codon. Small transcript products beginning 20 nucleotides upstream may, therefore, not have been detected.

C. Putative regulatory DNA sequences

An alignment of the promoter region of the HA-operon of NCTC 2916 and the promoter region of its NTNH/A-BoNT/A operon has enabled the identification of a highly conserved sequence motif (Figure 17.6). This motif is composed of sequence boxes (box 1, 5′-ATTTTagGTTTACAAAA-3′; box 2, 5′-ATGTTA-

```
                                 BOX 1                                        BOX 2

                                 "-35"                                        "-10"
                                                                                              ↓
           TTATATTTTTAAAAA  ATTTTAGGTTTACAAAA  AATAGTGTGGCT  ATGTTATATaTAA  ATG--84nt--AGGAGA---6nt-ATG-NTNH/A
                                                                                             ↓
           ATTTGATATATTTAA  ATTTTAGGTTTACAAAA  AATAACTTGATT  ATGTTATATGTtA  TAT--21nt--AGGAGGA--6nt-ATG-HA34
                                                             ↓
           TTTATATATTTTTAT  ATTTTAGGTTTACAAAA  GACGATGAAGAA  ATGTTATATaTAA  GTG--82nt--AGGAGA--10nt-ATG-NTNH/C1

           AATAGATATATTATA  ATTTTAGGTTTACAAAA  TTTGGCTTAAAT  ATGTTATATGTAA  GTG--20nt--GGGAGGA--6nt-ATG-HA33
                                                                            ↓
           TTTTTAAAAATTTAA  ATTTTcaGTTTACAAAA  AATAACCTGATT  ATGTTATATGTAA  TTG-110nt--AGGAGA---8nt-ATG-TeNT

Consensus                   ATTTTagGTTTACAAAA  RatRRYNtRRNt  ATGTTATATgTaA

(a)                                TTTACA--------21nt-----------tatgtt
(b)                                TTTACA--------18nt--------ttatat
(c)                                TTTACA--------16nt------tgttat
Gram +ve
vegetative                         TTGACA--------------------TATAAT
promoter
                                                                                             ↓
           GATATGTCAAAGTATTTGTATTTATGGTCATTTAAATAATTAATAATTTAATTAATTTTAAAT----5nt--AAGAGG---8nt-ATG-BoNT/A
                                                                                              ↓
           TTAGTATGTTTACATAATTGATTATGGATATTTCGTAAAAATGGCTTATTAAAAATTTAAAGG---46nt--GGAGGG--12nt-ATG-OrfX
                                                                                                        SD
```

Figure 17.6 Transcriptional analysis of the genes at the BoNT/A locus and putative regulatory sites in the promoters of the two major operons at the BoNT/A, BoNT/C1 and TeNT gene loci. Arrows indicate transcriptional initiation sites.[29,81,82] Boxes 1 and 2 represent conserved sequences in the promoter regions indicated, the consensus of which is presented below the boxed DNA sequences, as are the putative −35 and −10 promoter sites of boxes 1 and 2 (a–c). The transcriptional start sites for the BoNT/A and OrfX genes are also presented.

TATgTaA-3′), separated by a 12 nucleotide gap, which span the −35 and −10 promoter regions of both operons. Furthermore, this same sequence motif is almost exactly conserved upstream of the equivalent operons of the genome of the C468 phage. In NCTC 2916, the two motifs are separated by *orfX*. In C468, *orf22* is located elsewhere, so the two motifs exist as contiguous inverted repeats. This led to the probably incorrect previous assumption[82] that, in C468 at least, they play a role in the formation of a stem–loop structure. It appears more likely that these sequences are operator sites at which as yet unidentified transcriptional factors interact. It is tempting to speculate that the OrfX/Orf22 polypeptides could fulfill this hypothetical role. Interestingly, the same sequence motif is found immediately upstream of the TeNT gene.

VI. ACKNOWLEDGMENT

We acknowledge the financial support of the UK Department of Health and Medical Research Council (Grant No. G9207831CB).

VII. REFERENCES

1. Hatheway C.L. (1989) Bacterial sources of clostridial neurotoxins. In: Simpson L.L. (ed.), *Botulinum Neurotoxin and Tetanus Toxin*, pp. 3–24. Academic Press, San Diego, CA.
2. Smith L.D.S. (1975) *The Pathogenic Anaerobic Bacteria.* CC Thomas, Springfield, IL.
3. Popoff M.R. (1995) Ecology of neurotoxigenic strains of clostridia. *Curr Top Microbiol* **95**: 1–29.
4. Hutson R.A., Thompson D.E. and Collins M.D. (1993) Genetic interrelationships of saccharolytic *Clostridium botulinum* types B, E and F and related clostridia as revealed by small-subunit rRNA gene sequences. *FEMS Microbiol Lett* **108**: 103–110.
5. Hutson R.A., Thompson D.E., Lawson P.A., Schocken-Itturino R.P., Bottger E.C. and Collins M.D. (1983) Genetic interrelationships of proteolytic *Clostridium botulinum* types A, B and F and other members of the *Clostridium botulinum* complex as revealed by small subunit rRNA gene sequences. *Ant Leewenhoek* **64**: 278–283.
6. McCroskey L.M., Hatheway C.L., Fenicia L., Pasolini B. and Aureli P. (1986) Characterisation of an organism that produces type E botulinual toxin but which resembles *Clostridium butyricum* from the faeces of an infant with type E botulism. *J Clin Microbiol* **23**: 201–202.
7. Aureli P., Fenicia L., Pasolini B., Gianfranceschi M., McCroskey L.M. and Hatheway C.L. (1986) Two cases of type E infant botulism caused by neurotoxigenic *Clostridium butyricum* in Italy. *J Inf Dis* **154**: 207–211.
8. Hall J.D., McCroskey L.M., Pincomb B.J. and Hatheway C.L. (1985) Isolation of an organism resembling *Clostridium barati* which produces type F botulinal toxin from an infant with botulism. *J Clin Microbiol* **21**: 654–655.
9. McCroskey L.M., Hatheway C.L., Woodruff B.A., Greenberg J.A. and Jurgenson P. (1991) Type F botulism due to neurotoxinogenic *Clostridium barati* from an unknown source in an adult. *J Clin Microbiol* **29**: 2618–2620.
10. Hatheway C.L. (1990) Toxinogenic bacteria, *Clin Microbiol Rev* **3**: 66–98.

11. Suen J.C., Hatheway C.L., Steigerwalt A.G. and Brenner D.J. (1988) *Clostridium argentinense* sp.nov.: a genetically homogeneous group composed of all strains of *Clostridium botulinum* toxin type G and some nontoxigenic strains previously identified as *Clostridium subterminale* or *Clostridium hastiforme*. *Int J System Bacteriol* **38**: 375–381.
12. Cato E.P., George W.L. and Finegold S.M. (1986) Genus *Clostridium*. In: Sneath P.H.A., Mair N.S., Sharpe M.E. and Holt J.G. (eds), *Bergey's Manual of Systematic Bacteriology*, pp. 1141–1200. Williams and Wilkins, Baltimore, MD.
13. Sugiyama H. (1980) *Clostridium botulinum* neurotoxin. *Microbiol Rev* **44**: 419–448.
14. Thompson D.E., Brehm J.K., Oultram J.D., Swinfield T.J., Shone C.C., Atkinson T., Melling J. and Minton N.P. (1990) The complete amino acid sequence of the *Clostridium botulinum* type A neurotoxin, deduced by nucleotide sequence analysis of the encoding gene. *Eur J Biochem* **189**: 73–81.
15. Binz T., Kurazono H., Wille M., Frevert J., Wernars K. and Niemann H. (1990) The complete sequence of botulinum neurotoxin type A and comparison with other clostridial neurotoxins. *J Biol Chem* **265**: 9153–9158.
16. Willems A., East A.K., Lawson P.A. and Collins M.D. (1993) Sequence of the gene coding for the neurotoxin of *Clostridium botulinum* type A associated with infant botulism: comparison with other clostridial neurotoxins. *Res Microbiol* **144**: 547–556.
17. Whelan S.M., Elmore M.J., Bodsworth N.J., Brehm J.K., Atkinson T. and Minton N.P. (1992) Molecular cloning of the *Clostridium botulinum* structural gene encoding the type B neurotoxin and determination of its entire nucleotide sequence. *Appl Environ Microbiol* **58**: 2345–2354.
18. Hutson R.A., Collins M.D., East A.K. and Thompson D.E. (1993) Nucleotide sequence of the gene coding for non-proteolytic *Clostridium botulinum* type B neurotoxin: comparison with other clostridial neurotoxins. *Curr Microbiol* **28**: 101–110.
19. Hauser D., Eklund M.W., Kurazono H., Binz T., Niemann H., Gill D.M., Boquet P. and Popoff M.R. (1990) Nucleotide sequence of *Clostridium botulinum* C1 neurotoxin *Nucl Acids Res* **18**: 4924.
20. Kimura K., Fujii N., Tsuzuki K., Murakami T., Indoh T., Yokosawa N., Takeshi K., Syuto B. and Oguma K. (1990) The complete nucleotide sequence of the gene coding for botulinum type C1 toxin in the C-ST phage genome. *Biochem Biophys Res Comm* **171**: 1304–1311.
21. Binz T., Kurazono H., Popoff M.R., Eklund M.W., Sakaguchi G., Kozaki S., Krieglstein K., Henschen A., Gill D.M. and Niemann H. (1990) Nucleotide sequence of the gene encoding *Clostridium botulinum* neurotoxin type D. *Nucl Acids Res* **18**: 5556.
22. Moriishi K., Koura M., Fujii N., Fujinaga Y., Inoue K., Syuto B. and Oguma K. (1996) Molecular cloning of the gene encoding the mosaic neurotoxin, composed of parts of botulinum neurotoxin types C1 and D, and PCR detection of this gene from *Clostridium botulinum* type C organisms. *Appl Environ Microbiol* **62**: 662–667.
23. Whelan S.M., Elmore M.J., Bodsworth N.J., Atkinson T. and Minton N.P. (1992) Complete amino acid sequence of the *Clostridium botulinum* type E neurotoxin deduced by nucleotide sequence analysis of the encoding gene. *Eur J Biochem* **204**: 657–667.
24. Poulet S., Hauser D., Quanz M., Niemann H. and Popoff M.R. (1992) Sequences of the botulinal neurotoxin E derived from *Clostridium botulinum* type E (strain Beluga) and *Clostridium butyricum* (strains ATCC 43181 and ATCC 43755). *Biochem Biophys Res Commun* **183**: 107–113.
25. Elmore M.J., Hutson R.A., Collins M.D., Bodsworth N.J., Whelan S.M. and Minton N.P. (1995) Nucleotide sequence of the gene coding for proteolytic (group I) *Clostridium botulinum* type F neurotoxin: genealogical comparison with other clostridial neurotoxins. *Syst Appl Microbiol* **18**: 23–31.
26. East A.K., Richardson P.T., Allaway D., Collins M.D., Roberts T.A. and Thompson D.E. (1992) Sequence of the gene encoding type F neurotoxin of *Clostridium botulinum*. *FEMS Microbiol Lett* **96**: 225–230.
27. Thompson D.E., Hutson R.A., East A.K., Allaway D., Collins M.D. and Richardson P.T. (1993)

Nucleotide sequence of the gene coding for *Clostridium barati* type F neurotoxin: comparison with other clostridial neurotoxins. *FEMS Microbiol Lett* **108**: 175–182.

28. Campbell K.D., Collins M.D. and East A.K. (1993) Nucleotide sequence of the gene coding for *Clostridium botulinum* (*Clostridium argentinense*) type G neurotoxin: genealogical comparison with other clostridial neurotoxins. *Biochim Biophys Acta* **1216**: 487–491.
29. Eisel U., Jarausch W., Goretzki K., Henschen A., Engels J., Weller U., Hudel M., Habermann E. and Niemann H. (1986) Tetanus toxin: primary structure, expression in *E. coli* and homology with botulinum toxins. *EMBO J* **5**: 2495–2502.
30. Fairweather N.F. and Lyness V.A. (1986) The complete nucleotide sequence of tetanus toxin. *Nucl Acid Res* **14**: 7809–7812.
31. Young M., Staudenbauer W.L. and Minton N.P. (1989) Genetics of *Clostridium*. In Minton N.P. and Clarke D.J. (eds), *Clostridia* pp. 63–103. Plenum Publishing Co, New York.
32. Minton N.P., Maule A., Luton P. and Outram J.D. (1992) Obligate anaerobes and their biotechnological potential. In: Herbert R.A. and Sharp R.J. (eds), *Molecular Biology and Biotechnology of Extremophiles*, pp. 281–320. Chapman and Hall, New York.
33. Szabo E.A., Pemberton J.M. and Desmarchelier P.M. (1993) Detection of the genes encoding botulinum neurotoxin types A to E by the polymerase chain reaction. *Appl Environ Microbiol* **59**: 3011–3020.
34. Campbell K.D., Collins M.D. and East A.K. (1993) Gene probes for identification of botulinal the neurotoxin gene and specific identification of neurotoxin types B, E and F. *J Clin Microbiol* **31**: 2255–2262
35. Franciosa G., Ferreira J.L. and Hatheway C.L. (1994) Detection of type A, B and E botulism neurotoxin genes in *Clostridium botulinum* and other *Clostridium* species by PCR: evidence of unexpressed type B toxin genes in type A toxigenic organisms. *J Clin Microbiol* **32**: 1911–1917.
36. Fach P., Hauser D., Guillou J.P. and Popoff M.R. (1993) Polymerase chain reaction for the rapid identification of *Clostridium botulinum* type A strains and detection in food samples. *J Appl Bacteriol* **75**: 234–238.
37. Gimenez D.F. and Ciccarelli A.S. (1970) Studies on strain 84 of *Clostridium botulinum*. *Zbl Bakt, I Abt Orig* **215**: 212–220.
38. Hatheway C.L., McCroskey L.M., Lombard G.L. and Dowell V.R. (1981) Atypical toxin variant of *Clostridium botulinum* type B associated with infant botulism. *J Clin Microbiol* **14**: 607–611.
39. Poumeyrol M., Billon J., DeLille F., Haas C., Marmonier A. and Sebald M. (1983) Intoxication botulique mortelle due a une souche de *Clostridium botulinum* de type AB. *Med Malad Infect* **13**: 750–754.
40. Hatheway C.L. and McCroskey L.M. (1987) Examination of feces and serum for diagnosis of infant botulism in 336 patients. *J Clin Microbiol* **25**: 2334–2338.
41. Sugiyama H., Mizutani K. and Yang K.H. (1972) Basis of type A and F toxicities of *Clostridium botulinum* strain 84. *Proc Soc Exp Biol Med* **141**: 1063–1067.
42. Gimenez D.F. (1984) *Clostridium botulinum* subtype Ba. *Zbl Bakt Parasit Infekt Hyg Abt I Orig Reihe A* ***257**: 68–72.*
43. Cordoba J.J., Collins M.D. and East A.K. (1995) Studies on the genes encoding botulinum neurotoxin type A of *Clostridium botulinum* from a variety of sources. *Syst Appl Microbiol* **18**: 13–22.
44. Minton N.P. (1995) Molecular genetics of clostridial neurotoxins. *Curr Top Microbiol* **95**: 161–194.
45. Davis T.O., Henderson I., Whelan S.M. and Minton N.P. (1995) Molecular characterisation of a 'silent' gene encoding a type B botulinum neurotoxin carried by the type A *Clostridium botulinum* strain NCTC 2916. *Movement Disorders* **10**: 374.
46. Ochanda J.O., Syuto B., Oguma K., Iida H. and Kubo S. (1984) Comparison of antigenicity of toxins produced by *Clostridium botulinum* type C and D strains. *Appl Environ Microbiol* **47**: 1319–1322.
47. Oguma K., Murayama S., Syuto B., Iida H. and Kubo S. (1984) Analysis of antigenicity of

Clostridium botulinum type C1 and D toxins by polyclonal and monoclonal antibodies. *Infect Immun* **43**: 584–588.

48. Terajima J., Syuto B., Ochanda J.O. and Kubo S. (1985) Purification and characterisation of neurotoxin produced by *Clostridium botulinum* type C 6813. *Infect Immun* **48**: 312–317.
49. Higgins D.G. and Sharp P.M. (1988) CLUSTAL: A package for performing multiple sequence alignment on a microcomputer. *Gene* **73**: 237–244.
50. Vinet G., Berthiaume L. and Fredette V. (1968) Un bacteriophage dans une culture de *C. botulinum* C. *Rev Can Biol* **27**: 73–74.
51. Inoue K. and Iida H. (1968) Bacteriophages of *Clostridium botulinum*. *J Virol* **2**: 537–540.
52. Inoue K. and Iida H. (1970) Conversion to toxinogenicity in *Clostridium botulinum* type C. *Jap J Microbiol* **14**: 87–89.
53. Eklund M.W., Poysky F.T. and Boatman E.S.. (1969) Bacteriophages of *Clostridium botulinum* types A, B, E and F and nontoxigenic strains resembling type E. *J Virol* **3**: 270–274.
54. Dolman C.E. and Chang E. (1972) Bacteriophages of *Clostridium botulinum*. *Can J Microbiol* **18**: 67–76.
55. Eklund M.W., Poysky F.T. and Habig W.H. (1989) Bacteriophages and plasmids in *Clostridium botulinum* and *Clostridium tetani* and their relationships to production of toxins. In: Simpson L.L. (ed.), *Botulinum Neurotoxin and Tetanus Toxin*, pp. 26–52. Academic Press, San Diego, CA.
56. Eklund M.W. and Poysky F.T. (1974) Interconversion of type C and D strains of *Clostridium botulinum* by specific bacteriophages. *Appl Microbiol* **27**: 251–258.
57. Eklund M.W., Poysky F.T., Peterson M.E. and Meyers J.A. (1976) Relationship of bacteriophage to alpha toxin production in *Clostridium novyi* types A and B. *Infect Immun* **14**: 798–803.
58. Boizet B., Lahbib-Mansais Y., Dupont L., Ritzenthaler P. and Mata M. (1990) Cloning, expression and sequence analysis of an endolysin-encoding gene of *Lactobacillus bulgaricus* bacteriophage mv1. *Gene* **94**: 61–67.
59. Garcia E., Garcia J.L., Garcia P., Arraras A., Sanchez-Puelles J.M. and Lopez R. (1988) Molecular evolution of lytic enzymes of *Streptococcus pneumoniae* and its bacteriophages. *Proc Natl Acad Sci USA* **85**: 914–918.
60. Fouche P.B. and Hash J.H. (1978) The *N-O*-diacetylmuramidase of *Charalopsis* species. Identification of aspartyl and glutamyl residues of the active site. *J Biol Chem* **253**: 6787–6793.
61. Croux C. and Garcia J.L. (1991) Sequence of the lyc gene encoding the autolytic lysozyme of *Clostridium acetobutylicum* ATCC 824: comparison with other lytic enzymes. *Gene* **104**: 25–31.
62. Whelan S.M., Garcia J.L., Elmore M.J. and Minton N.P. (1994) The botulinum neurotoxin gene of the type A *Clostridium botulinum* strain NCTC 2916 is followed by a gene (*lycA*) encoding a lysozyme. In: Freer J., Aitken R. and Alouf J.E. (eds), *Bacterial Protein Toxins, Zbl Bakt Suppl 24*, pp. 162–163. Gustav Fischer, Stuttgart.
63. Bonventre P.F. and Kempe L.L. (1960) Physiology of toxin production of *Clostridium botulinum* types A and B. I. Growth, autolysis and toxin production. *J Bacteriol* **79**: 18–23.
64. Zhou Y., Sugiyama H. and Johnson E.A. (1993) Transfer of neurotoxigenicity from *Clostridium butyricum* to a nontoxinogenic *Clostridium botulinum* type E-like strain. *Appl Environ Microbiol* **59**: 3825–3831.
65. Hara T., Matsuda M. and Yoneda M. (1977) Isolation and some properties of nontoxigenic derivatives of a strain of *Clostridium tetani*. *Biken J* **20**: 105–115.
66. Laird W.J., Aaronson W., Silver R.P., Habig W.H. and Hardegree M.C. (1980) Plasmid-associated toxigenicity in *Clostridium tetani*. *J Inf Dis* **142**: 623.
67. Finn C.W. Jr, Silver R.P., Habig W.H. and Hardegree M.C. (1984) The structural gene for tetanus neurotoxin is on a plasmid. *Science* **224**: 881–884.
68. Scott V.N. and Duncan C.L. (1978). Cryptic plasmids in *Clostridium botulinum* and *C. botulinum*-like organisms. *FEMS Microbiol Lett* **4**: 55–58.

69. Weikert M.J., Chambliss G.H. and Sugiyama H. (1986). Production of toxin by *Clostridium botulinum* type A strains cured by plasmids. *Appl Environ Microbiol* **51**: 52–56.
70. Strom M.S., Eklund M.W. and Poysky F.T. (1984) Plasmids in *Clostridium botulinum* and related *Clostridium* species *Appl Environ Microbiol* **48**: 956–963.
71. Hauser, D., Gibert M., Boquet P. and Popoff M.R. (1992) Plasmid localisation of a type E botulinal neurotoxin gene homologue in toxigenic *Clostridium butyricum* strains, and absence of this gene in non-toxigenic *C. butyricum* strains. *FEMS Microbiol Lett* **99**: 251–256.
72. Zhou Y., Sugiyama H., Nakano H. and Johnson E.A. (1995) The genes for the *Clostridium botulinum* type G toxin complex are on a plasmid. *Infect Immun* **63**: 2087–2091.
73. Hauser D., Gibert M., Eklund M.W., Boquet P. and Popoff M.R. (1993) Comparative analysis of C3 and botulinal neurotoxin genes and their environment in *Clostridium botulinum* types C and D. *J Bacteriol* **175**: 7260–7268.
74. Hauser D., Gibert M., Marvaud J.C., Eklund M.W. and Popoff M.R. (1995) Botulinal neurotoxin C1 complex genes, clostridial neurotoxin homology and genetic transfer in *Clostridium botulinum*. *Toxicon* **33**: 515–526.
75. Ohishi I., Sughii S. and Sakaguchi G. (1977) Oral toxicities of *Clostridium botulinum* toxins in response to molecular size. *Infect Immun* **16**: 107–109.
76. Ohishi I. and Sakaguchi G. (1980) Oral toxicities of *Clostridium botulinum* type C and D toxins of different molecular size. *Infect Immun* **28**: 303–309.
77. Sugii S. and Sakaguchi J. (1977) Botulinogenic properties of vegetables with special reference to the molecular size of the toxin in them. *J Food Safety* **1**: 53–65.
78. Oguma K., Iida H. and Shiozaki M. (1976) Phage conversion to haemagglutinin production in *Clostridium botulinum* types C and D. *Infect Immun* **14**: 597–602.
79. Tsuzuki K., Kimura K., Fujii N., Yokosawa N., Indoh T., Murakami T. and Oguma K. (1990) Cloning and complete nucleotide sequence of the gene coding for the main component of hemagglutinin produced by *Clostridium botulinum* type C. *Infect Immun* **58**: 3173–3177.
80. Somers E. and DasGupta B.R. (1991) *Clostridium botulinum* types A, B, C1 and E produce proteins with or without haemagglutinating activity: do they share common amino acid sequences and genes. *J Pro Chem* **10**: 415–425.
81. Henderson I., Davis T.O., Whelan S.M. and Minton N.P. (1996) Genetic characterisation of the botulinum toxin complex of *Clostridium botulinum* strain NCTC 2916. *FEMS Microbiol Lett* **140**: 151–158.
82. Hauser D., Eklund M.W., Boquet P. and Popoff M.R. (1994) Organisation of the botulinum neurotoxin C1 gene and its associated non-toxic protein genes in *Clostridium botulinum* C468. *Mol Gen Genet* **243**: 631–640.
83. Tsuzuki K., Kimura K., Fujii N., Yokosawa N. and Oguma K. (1992) The complete nucleotide sequence of the gene coding for the nontoxic-nonhemagglutinin component of *Clostridium botulinum* type C progenitor toxin. *Biochem Biophys Res Commun* **183**: 1273–1279.
84. Fujii N., Kimura K., Murakami T., Yokosawa N., Yashiki T., Tsuzuki K. and Oguma K. (1993) The complete nucleotide sequence of the gene encoding the nontoxic component of *Clostridium botulinum* type E progenitor toxin. *J Gen Microbiol* **139**: 79–86.
85. East A.K. and Collins M.D. (1994) Conserved structure of genes encoding components of botulinum neurotoxin complex M and the sequence of the gene encoding for the non-toxic component in nonproteolytic *Clostridium botulinum* type F. *Curr Microbiol* **29**: 69–77.
86. Ohyama T., Watanabe T., Fujinaga Y., Inoue K., Hiroyuki S., Fujii N., Inoue K. and Oguma K. (1995) Characterisation of nontoxic-nonhemagglutinin component of the two types of progenitor toxin (M and L) produced by *Clostridium botulinum* type D CB-16. *Microbiol Immunol* **39**: 457–465.
87. Fujita R., Fujinaga Y., Inoue K., Nakajima H., Kumon H. and Oguma K. (1995) Molecular characterisation of two forms of nontoxic-nonhemagglutinin components of *Clostridium botulinum* type A progenitor toxins. *FEBS Lett* **376**: 41–44.

88. East A.K., Stacey J.M. and Collins M.D. (1995) Cloning and sequencing of a hemagglutinin component of the botulinum neurotoxin complex encoded by *Clostridium botulinum* types A and B. *Syst Appl Microbiol* **17**: 306–312.
89. Hanna P.C., Wieckowski E.U., Meitzner T.A. and McClane B.A. (1992) Mapping of functional regions of *Clostridium perfringens* type A enterotoxin. *Infect Immun* **60**: 2110–2114.
90. Perkins W.F. and Tsuji K. (1962) Sporulation of *Clostridium botulinum.* II Effect of arginine and its degradation products on sporulation in synthetic medium. *J Bacteriol* **84**: 86–94.
91. Bowers L.E. and Williams O.B. (1963) Effect of arginine on the growth and lysis of *Clostridium botulinum. J Bacteriol* **85**: 1175–1176.
92. Patterson Curtis S.I. and Johnson E.A. (1989) Regulation of neurotoxin and protease formation in *Clostridium botulinum* Okra B and Hall A by arginine. *Appl Environ Microbiol* **55**: 1544–1548.
93. Schantz E.J. and Johnson E.A. (1992) Properties and use of botulinum toxin and other microbial neurotoxins in medicine. *Microbiol Revs* **56**: 80–99.
94. Parkinson J.S. and Kofoid E.C. (1992) Communication modules in bacterial signaling proteins. *Annu Rev Genet* **26**: 71–112.
95. Russell R.R.B., Aduse-Opoku J., Sutcliffe I.C., Tao L. and Ferretti J.J. (1992) A binding protein-dependent transport system in *Streptococcus mutans* responsible for multiple sugar metabolism. *J Biol Chem* **267**: 4631–4637.
96. Tao L., Sutcliffe I.C., Russell R.R.B. and Ferretti J.J. (1993) Cloning and expression of the multiple sugar metabolism (*msm*) operon of *Streptococcus mutans* in heterologous streptococcal hosts. *Infect Immun* **61**: 1121–1125.
97. Garnier T. and Cole S.T. (1988) Studies of UV-inducible promoters from *Clostridium perfringens in vivo* and *in vitro. Mol Microbiol* **2**: 607–614.
98. Kahn D. and Ditta G. (1991) Modular structure of FixJ: homology of the transcriptional activator domain with the −35 binding domain of sigma factors. *Mol Microbiol* **5**: 987–997.
99. Hammond G.A. and Johnson J.L. (1995) The toxigenic element of *Clostridium difficile* VPI 10463. *Microb Path* **19**: 203–213.

Chapter 18

The Structure and Mode of Action of Botulinum and Tetanus Toxins

GIAMPIETRO SCHIAVO AND CESARE MONTECUCCO*

*Imperial Cancer Research Fund, PO Box 123, 44 Lincoln's Inn Fields, London WC2A 3PX, UK and *Centro CNR Biomembrane and Dipartimento di Scienze Biomediche Università di Padova, Via Trieste 75, 35121 Padova Italy*

I. INTRODUCTION

Tetanus is a painful and often lethal syndrome, characterized by a spastic paralysis. Death follows bodily exhaustion and occurs by respiratory failure or circulatory collapse (see Chapter 7). Tetanus was recognized as a neurological disease for 24 centuries before the identification of *Clostridium tetani*, the bacterium that causes tetanus via the release of a protein toxin, named tetanus neurotoxin (TeNT).[1–4]

In contrast, botulism is characterized by a generalized muscular weakness that starts at the eye and throat muscles and extends gradually to all skeletal muscles. In its severe form, a generalized flaccid paralysis associated with the impairment of respiration and autonomic functions becomes evident in the victim and death results from respiratory failure.[5] The causative agents of botulism are the neurotoxigenic strains of *Clostridium botulinum,*[6] *Clostridium barati* and *Clostridium butyricum.*[7,8] At present, seven different serotypes of botulinum neurotoxin (BoNT), called A to G, have been identified (see Chapter 17).

The Clostridia: Molecular Biology and Pathogenesis
ISBN 0-12-595020-9

Because a single protein is responsible for all the clinical symptoms of either tetanus or botulism, these diseases can be completely prevented by administration of antitoxin-specific antibodies.[9] Toxin-neutralizing antibodies can be acquired passively by treatment with antibodies isolated from immunized donors or actively as a result of immunization with toxoids. Toxoids are chemically or genetically transformed toxins that remain immunogenic, but lose their toxicity. Together with the diphtheria toxoid, tetanus toxoid is the major product of the biotechnology industry. Because of the rarity of botulism in developed countries, only people involved in manipulation of BoNTs routinely receive immunization against botulism.

TeNT and BoNTs are the most potent toxins known. For example, the 50% lethal dose (LD_{50}) in mice, human and horses varies between 0.1 ng and 1 ng of toxin/kg of body weight. Interestingly different animal species show a great range of sensitivity to TeNT and BoNTs. While mice are exquisitely sensitive to TeNT, rats and birds are quite resistant and turtles are completely insensitive to TeNT.[10]

The extreme toxicity of clostridial neurotoxins derives from their absolute neurospecificity and also from catalytic activity. TeNT and BoNTs first bind specifically to the neuromuscular junction (NMJ) of motor neurons. The identity of the receptor(s) on the presynaptic membrane is unknown, but the binding affinity to the cognate receptor must be very high because deadly doses in the mouse are, after dilution in the body fluids, well below the picomolar range. Receptor-bound BoNT is internalized at the presynaptic membrane of the NMJ and gains access to the neuronal cytosol. Here it blocks the release of acetylcholine (ACh), causing a flaccid paralysis.[11] TeNT also binds to the presynaptic membrane of the motor neuron, but its action is limited to the level of the central nervous system. TeNT undergoes retrograde transport inside the motor neuron to the spinal cord,[12,13] where it migrates trans-synaptically into inhibitory interneurons.[14] In this type of neuron, TeNT blocks the release of inhibitory neurotransmitters.[15] During trans-synaptic migration, TeNT is exposed to the extracellular medium and can be neutralized by antitoxin antibodies injected in the cerebrospinal fluid.[16] The amount of toxin that reaches the CNS is an important parameter that determines the severity of tetanus and may account for the varying degree of TeNT toxicity observed in different vertebrates.[10] The blockade of inhibitory synapses at the spinal cord caused by TeNT impairs the neuronal circuitry that controls voluntary muscle contraction, thereby causing the spastic paralysis characteristic of tetanus.[11]

The opposite clinical symptoms of tetanus and botulism result from the sites of action of TeNT and BoNTs within the nervous system, rather than a different pathophysiological mechanism. TeNT acts on the central nervous system, while BoNTs exert their action on the peripheral nervous system. The only known physiological activity of TeNT and BoNTs is a persistent inhibition of neurotransmitter release, including glycine, GABA, ACh, D-Asp, glutamate and norepinephrine. The release of opioids, oxytocin and vasopressin is also inhibited by clostridial neurotoxins.[17] The action of tetanus and botulinum neurotoxins can be extended to different exocytotic events in a wide range of non-neuronal cells if the neurospecific binding and uptake processes are circumvented.[18–30]

Electrophysiological evidence indicates that TeNT and BoNTs greatly decrease the pool of synaptic vesicles (SSV) capable of fusion, but do not alter the neurotransmitter content of SSV or interfere with the processes of neurotransmitter synthesis, uptake and storage.[31] Similarly, TeNT and BoNTs interfere neither with the propagation of electrical nerve impulses nor calcium entry into neurons, nor calcium disposal at the synaptic terminals.[32] Careful morphological examinations of toxin-poisoned NMJs reveal no apparent structural changes, except for an increase in the number of SSVs present at the active zones of synaptic terminals.[33,34] Although their final physiological effects are virtually identical, the eight clostridial neurotoxins show remarkable qualitative differences in their presynaptic action. For example, BoNT/A and BoNT/E poisoning of the NMJ does not alter the synchronization of the quantal release elicited, while BoNT/B, BoNT/D and BoNT/F cause massive desynchronization of this process.[11,17,31,32] This electrophysiological observation suggests that the neurotoxins have different targets within the presynaptic terminal and this conclusion has been fully substantiated by the identification of the molecular targets of each clostridial neurotoxin (see below).

Clostridial neurotoxins are now used in the treatment of human diseases characterized by a hyperactivity of the NMJ, such as dystonias and strabismus.[35] Injection of minute amounts of BoNT into the hyperactive muscle(s) leads to a depression of the symptoms for a few months. BoNT/A is generally used, but other BoNT types are currently under clinical trial. Neurotoxins are now recognized as the best available treatment for dystonias and for certain types of strabismus. The treatment can be repeated several times, without major side effects such as the development of an immune response. If the patient develops an immune response to BoNT/A, the treatment can be continued with another BoNT serotype.[35]

II. STRUCTURE OF CLOSTRIDIAL NEUROTOXINS

Clostridial neurotoxins share not only a similar mechanism of action but also a common structural organization. As depicted in Figure 18.1, they are produced as inactive polypeptide chains of 150 kDa. Upon bacterial lysis, TeNT and BoNTs are released and cleaved by endogenous or exogenous proteases at an

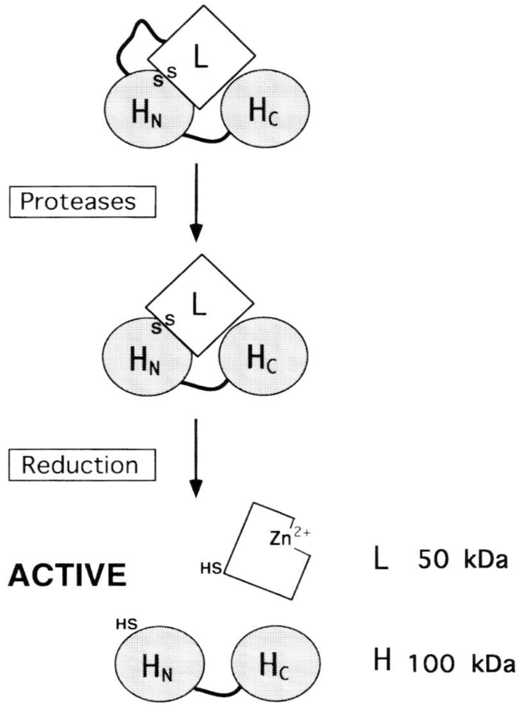

Figure 18.1 Scheme of the structure and mechanism of activation of tetanus and botulinum neurotoxins. The toxins are produced as an inactive single polypeptide chain of 150 kDa, composed of three 50 kDa domains, connected by protease-sensitive loops. The toxins are activated upon selective proteolytic cleavage which generates two disulfide-linked chains: L (50 kDa) and H (100 kDa). The three domains play different functional roles in cell penetration: H_C is responsible for cell binding and H_N for cell penetration. Reduction takes place inside the nerve cells and liberates the activity of the L chain in the cytosol. L is a zinc-endopeptidase specific for protein components of the neuroexocytotic apparatus.

exposed loop. An active di-chain neurotoxin is thus generated (Figure 18.1).[36] The heavy chain (H, 100 kDa) and the light chain (L, 50 kDa) are bridged by a single interchain disulfide bridge. This interchain bond is essential for their neurotoxicity when the toxin is applied to the extracellular space.[37,38] BoNT/A and the carboxy-terminal domains of TeNT have recently been crystallized,[39,40] but no high-resolution structures are presently available. Analysis of their secondary structure by biophysical methods revealed a composition of α-helix and β-sheet compatible with an overall globular structure.[41,42] Biochemical studies and low-resolution electron microscopic analysis led to the hypothesis that TeNT and BoNTs are folded into three distinct 50 kDa domains, each with a different set of physiological functions (Figure 18.1).[11,17,31] The mechanism of neurotoxin cell intoxication consists of four distinct steps: (1) binding to the neuronal membrane, (2) internalization into an endosomal-like compartment, (3) membrane translocation and (4) enzymatic target modification.[43] The L chain is responsible for the intracellular catalytic activity of clostridial neurotoxins. The amino-terminal 50 kDa domain of the H chain (H_N) is implicated in membrane translocation, while the carboxy-terminal part (H_C) is mainly responsible for the neurospecific binding.[11,17,31]

The amino acid sequence of all eight clostridial neurotoxins has been derived from their corresponding genes (see Chapter 17).[44] The L chains and H chains are composed, on average, of 439 and 843 residues, respectively. Both chains contain homologous domains separated by regions of very little similarity (Figure 18.2). The most conserved portions of L chains are the amino-terminal and central regions. The latter region (residues 216–244, numbering of TeNT) contains the His-Glu-Xaa-Xaa-His binding motif of zinc-endopeptidases (Figure

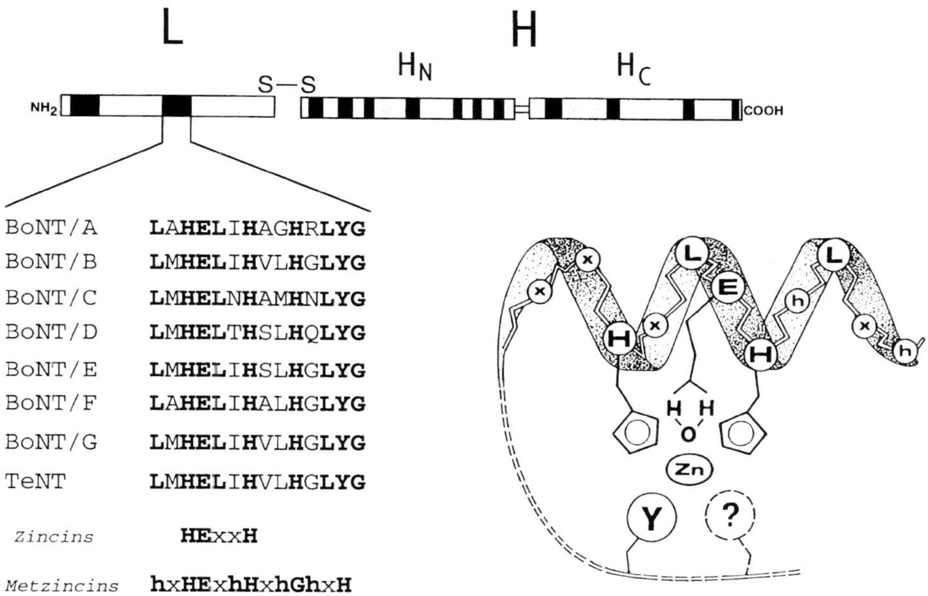

Figure 18.2 Structure and zinc coordination of clostridial neurotoxins. The upper panel reports the sequence of the L and H chain of clostridial neurotoxins. The segments that show greater homology between the different serotypes are black.[44] The highest homology is presented by a short segment corresponding to amino acid residues 216–244 in TeNT. This segment contains the zinc-binding motif of metallo-proteinases (zincins) and it is dissimilar from the consensus sequence of the metzincins metallo-proteinase family (left lower panel). The right lower panel show the mode of zinc coordination at the active site of clostridial neurotoxins, as suggested by available data. Whereas the involvement of the two imidazole groups and the water molecule is well established, the involvement of the conserved Tyr present in the segment and of a possible additional aromatic residue is still equivocal. The Tyr likely to be the fourth ligand of clostridial neurotoxins occupies the same positions as the His fourth zinc ligand of metzincins.

18.2).[45,46] This observation led to the demonstration that clostridial neurotoxins are zinc-containing proteins.[47–53] The zinc content of purified preparations of clostridial neurotoxins has been determined directly on all but BoNT/D and BoNT/G. One atom of zinc is bound to the L chain of TeNT, BoNT/A, BoNT/B, BoNT/E and BoNT/F. These neurotoxins show a single zinc-binding site with a dissociation constant (K_d) of 50–100 nM, a binding affinity that is in the same range as most metallo-proteases. Multiple divalent cation-binding sites with a lower affinity are also present.[48,52,54] BoNT/C binds two atoms of zinc,[51] similar to the neutrophil collagenase, whose three-dimensional structure has been recently determined.[55] One atom of zinc is present at the active site of collagenase and is exchangeable, while the second one is bound very strongly and is thought to play a structural role. Zinc can be removed from the clostridial neurotoxins with heavy metal chelators. Activity can be restored upon incubation of the metal-depleted neurotoxin in zinc-containing buffers.[47–51,56,57]

The zinc atom of zinc-endopeptidases is coordinated by either two or three histidine residues[46] in all zinc-endopeptidases studied to date. In thermolysin-like enzymes, the zinc atom is coordinated by the imidazole rings of two histidines present within the consensus sequence for zinc-endopeptidases, along with a water molecule bound to the glutamic acid of the same motif and the carboxylate side-chain of another glutamate reside. Astacin, a crayfish metallo-protease, penta-coordinates zinc via three histidine, one tyrosine residues and one water molecule. Adamalysin, the alkaline protease of *Pseudomonas aeruginosa* and collagenase adopt a tetrahedral zinc coordination via three histidines and a glutamate-bound water molecule. The active site of zinc-endopeptidases resides in a cleft with the zinc atom in its center and the residues of the zinc-binding motif forming an α-helix. The orientation and volume of the amino acid side-chains at the active site determine the peptide bond specificity of the toxin cleavage. The water molecule bound to the glutamate residue of the motif is involved in peptide bond hydrolysis in a mechanism that has been studied in detail only for thermolysin.[58]

In order to determine the number of histidine residues involved in zinc coordination, the L chains of TeNT and BoNT/A, BoNT/B and BoNT/E were modified with diethylpyrocarbonate (DEPC), a reagent that specifically modifies histidine residues and has been used to determine the critical role of these residues in the activity of BoNT/A and BoNT/E.[59] In each case, two additional histidines were modified in the zinc-depleted neurotoxin with respect to those in the zinc-containing neurotoxin.[47,48] This finding was also observed with thermolysin, but not with astacin and adamalysin.[60] Mutation of Glu-234 of TeNT led to complete loss of proteolytic activity.[53,61] These results indicate that the zinc atom of clostridial neurotoxins is coordinated via two histidines and a Glu-bound water molecule, as in thermolysin. Mutations at the two histidines of the motif inactivate TeNT and suppress its ability to bind radiolabeled $^{65}Zn^{2+}$.[53] In addition, mutations of the conserved Glu-271 and Glu-272 of TeNT, predicted to be in a α-helical segment,[62] result in decreased zinc binding and loss of activity. Based on these experimental results, it has been suggested that clostridial neurotoxins are thermolysin-like proteases and that one of these two Glu residues represents the fourth zinc ligand.[53] A mutational approach to study the active site of a given protein in the absence of its structural basis is always biased by the fact that many amino acid residues are required for proper folding and structural stability of a protein. Mutation of such residues leads to the same experimental result: loss of catalytic activity, although the mutated residues are not directly involved in catalysis, but rather in protein folding. This argument is supported by the fact that the two TeNT-corresponding Glu residues of collagenase are located in a segment distant from the active site.

The metal atom of TeNT can be exchanged with other transition metals of similar size with preservation of toxic activity, as shown for thermolysin and astacin.[60,63] A comparison of the extended X-ray absorption fine structure (EXAFS) spectra of TeNT, astacin, alkaline protease and thermolysin shows a close similarlity of TeNT with astacin and alkaline protease.[64] This result indicates that one (or possibly two) additional aromatic residue(s) is present around the zinc atom of TeNT, in addition to the two histidines required for cation coordination. Sequence comparison (Figure 18.2) indicates that a Tyr residue (Tyr-243 in TeNT), conserved among all clostridial neurotoxins, is located in the same position as the third histidine in astacin and astacin-like proteinases. In accordance with this model, chemical modification experiments demonstrated that Tyr residues are essential for the activity of BoNT/A, BoNT/B and BoNT/E.[65] Moreover, the replacement of Tyr-243 with Phe in TeNT leads to a great loss of toxicity, while replacement of the third conserved histidine residue preserves its activity.[53] These results suggest that in TeNT the polypeptide chain bends after the first two histidines of the zinc-binding motif to bring the following conserved Tyr close to the active site zinc, as depicted in Figure 18.2. If this model holds true, it would represent a novel manner of zinc coordination among metallo-proteinases. In addition, studies on the denaturant sensitivity of the holo- and apo-L chain indicate that the zinc atom does not contribute significantly to the structural stability of the clostridial neurotoxins.[42]

The H chains are less conserved that the L chains, with the carboxyl-terminal part (H_C) the most dissimilar region of the toxin (Figure 18.2).[44] This is consistent with the belief that the H_C domain is involved in binding to the nerve terminals and that different neurotoxins bind to different cognate receptors. On this basis it may be suggested that the receptor binding regions of TeNT and BoNTs are located mainly within the 180 carboxy-terminal residues of the H chain. The H_N domain and the first half of the H_C domain have stretches predicted to fold into amphipathic α-helices.[62,66] These stretches could be involved in the process of membrane translocation of the L domain in the neuron cytosol (see below).

Nucleotide and amino acid sequence comparisons of the clostridial neurotoxins clearly indicate that they derive from a common ancestral gene. In this respect, it is significant that several clostridial neurotoxin genes are located on mobile genetic elements (see Chapter 17).[44] Bacteriophages, conjugative plasmids, and transposons may have spread these genes among *Clostridium* bacteria. Mutations in clostridial neurotoxin genes are apparent from the discovery of variants of the seven BoNTs among the same serotypes; these have been detected in different clostridia by using modern molecular genetics methods.[44] Moreover, strains that harbor more than one BoNT gene have been identified.[5]

III. CELL BINDING AND TARGETING

After diffusion into the body fluids from the site of production or absorption, BoNTs and TeNT bind to the presynaptic membrane at the NMJ of α-motor neurons. TeNT is also capable of binding to sympathetic and adrenergic neurons.[17,67] Clostridial neurotoxins bind to a variety of non-neuronal cells *in vitro*, but at concentrations far exceeding those of clinical relevance. The rat NMJ contains a few hundred binding sites for BoNT/A and BoNT/B per square Ångstrom.[68] The number of TeNT receptors in a neuroblastoma-glioma cell line is around 450/Å^2.[69]

Many studies have been devoted to assessing the role of polysialogangliosides in neurotoxin binding.[17,67,70] The outer layer of the presynaptic membranes contain a large proportion of polysialogangliosides. This provides a large number of acceptors sites for clostiridal neurotoxins that are known to absorb on negatively charged membrane surfaces.[17,67,70] Incubation of cultured chromaffin cells with polysialogangliosides increases their sensitivity to TeNT and BoNT/A. Preincubation of these neurotoxins with ganglioside mixtures prior to neuronal exposure reduces TeNT toxicity and intra-axonal toxin transport in neurons.[17,67] In addition, treatment of membranes with neuraminidase, which removes the negatively charged sialic acid residues, decreases toxin binding.[17,67,70] Although these studies have clearly established that polysialogangliosides interact with the various clostridial neurotoxins, no clear demonstration of their direct involvement in the neurospecific binding of these toxins *in vivo* has been provided. As discussed in detail elsewhere,[17,67,70] it is very unlikely that polysialogangliosides are the sole receptors of the clostridial neurotoxins at synaptic terminals.

Parallel experiments with cells in culture indicate that cell surface proteins may be involved in toxin binding.[71–73] In a NGF-differentiated subclone of PC-12 cells, a protein with an apparent molecular mass of 20 kDa was shown to interact with the H chain of TeNT.[74] A 'double receptor' model for clostridial neurotoxin binding to neuronal cells has emerged to account for the action of polysialogangliosides and the putative receptor protein.[70] In this model, the neurotoxin first comes into contact with the negatively charged lipids of the presynaptic membrane via a high-capacity interaction(s). This model of binding would constitute a large trapping device for the minute amounts of neurotoxin known to cause disease. The toxin then moves laterally on the presynaptic membrane surface to bind a protein receptor that is reponsible for neurospecificity and uptake at the NMJ. Specific interaction of each neurotoxin with its protein receptor is greatly favored by the bi-dimensionality of the cell surface membrane, which greatly reduces the reaction volume and increases the rate of neurotoxin binding. Nishiki and colleagues recently have experimentally validated this model. They demonstrated that the high-affinity receptor for BoNT/B is composed of polysialogangliosides and of synaptotagmin, a transmembrane protein of SSVs.[75,76] The protein receptor of the other clostridial neurotoxins is not yet known.

To reach its final site of action, TeNT has to enter two different neurons: a peripheral neuron and an inhibitory interneuron of the spinal cord. There is evidence that the mode of TeNT binding to peripheral and central presynaptic terminals is different.[17,67,70] It is likely that TeNT reaches high concentrations in the motor neuron-inhibitory interneuron synaptic cleft and hence polysialogangliosides can mediate TeNT entry into CNS neurons. Acidic lipids may be sufficient to bind and guide the internalization of TeNT into the inhibitory interneurons, when the toxin is present in large amounts. Experiments with lipid monolayers have clearly documented the ability of TeNT to interact with acidic lipids at concentrations as low as 10^{-8}M;[77] similar concentrations are routinely used with cells in culture.[17] In clinical tetanus and botulism, however, peripheral clostridial neurotoxin concentrations are in the subpicomolar range, and therefore a high-affinity protein receptor must be involved to increase toxin affinity and provide binding specificity to the neuron. The TeNT protein receptor may also be responsible for localizing the toxin to a vesicle that moves in a retrograde direction, while the BoNTs protein receptors drive these neurotoxins inside small synaptic vesicles.[77a] The vesicle carrying TeNT may reach the Golgi or trans-Golgi network or another membrane compartment in the neuronal soma, where the toxin may separate from its protein receptor. TeNT exits the neuron inside a vesicle that moves to dendritic terminals and releases the toxin into the space between the pre- and post-synaptic membranes.

IV. INTERNALIZATION

Because the L chains of TeNT and BoNTs are responsible for the cytosolic activity of the clostridial neurotoxins, this domain of the toxin molecule, at least, must reach the cytosol. Pharmacological and morphologic evidence indicates that the clostridial neurotoxins enter the cell by endocytosis and reach an acidic compartment. Electron microscopic investigations demonstrate that the clostridial neurotoxins enter the lumen of vesicular structures in a temperature- and energy-dependent process.[64,78–81] Although the precise nature of the toxin-containing intracellular compartment remains unclear, there is considerable evidence that clostridial neurotoxins must pass through an acidic compartment for intoxication to occur. Furthermore, agents that interfere with intracellular vesicular acidification inhibit the toxicity of clostridial neurotoxins.[82,83]

Nerve stimulation facilitates intoxication,[17] so we suggest that TeNT and BoNTs are taken up at the NMJ inside SSVs or other vesicles whose recycling is linked to synaptic activity. According to this hypothesis the clostridial neurotoxins bind the intraluminal portion of a vesicle protein while the lumen is in contact with the external medium. The uptake of antibodies, which are similar in mass to neurotoxins, by endocytosis in hippocampal neurons has been clearly documented.[84] This observation is supported by the finding that BoNT/B binds to synaptotagmin in the presence of polysialogangliosides.[75,76] However, if both TeNT and BoNT are inside SSVs at peripheral synaptic terminals, it has yet to be explained how BoNT escapes from the vesicle lumen into the cytosol to block ACh release, while TeNT remains inside the vesicle and is delivered to the CNS.

V. TRANSLOCATION INTO THE NEURONAL CYTOSOL

To reach the cytosol, the clostridial neurotoxins contained inside vesicles of the presynaptic terminal must cross the hydrophobic barrier of the vesicle membrane. Indirect, but compelling evidence suggests that TeNT and BoNTs have to pass through a low pH step for neuron intoxication to occur.[77a,82,83] It is also clear that acidic pH does not activate the toxin directly via a structural change, since the direct introduction of the L chain in the neutral pH environment of the cytosol is sufficient to block exocytosis.[18,21–23,85–87] Hence, low pH is necessary for the process of L chain membrane translocation from the vesicle lumen into the cytosol. With regard to membrane translocation, clostridial neurotoxins resemble the other bacterial protein toxins that have a three-domain structure, such as diphtheria toxin.[43]

The interaction of clostridial neurotoxins with membrane bilayers has been studied mainly with model membrane systems and limited data have been obtained *in vivo*. Evidence indicates that at low pH TeNT and BoNTs undergo a conformational change from a water-soluble 'neutral' form to an 'acidic' form, the latter characterized by the exposure of hydrophobic segments. This increase in hydrophobicity enables penetration of both the H and L chains into the hydrocarbon core of the lipid bilayer.[31] Following this low pH-induced membrane insertion, TeNT and BoNTs form ion channels in planar lipid bilayers.[31,88] These channels are cation-selective, have few tens of picoSiemens (pS) conductance and are permeable to molecules smaller than 700 Da. The H_N domain of clostridial neurotoxins includes several segments which may form amphipathic α-helices and thus may be candidate regions for channel formation after toxin oligomerization.[62,66,89,90] Three-dimensional image reconstruction of the channel formed by BoNT/B in phospholipid bilayers[88] is consistent with this model.

Single channels formed by TeNT in spinal cord neurons have been recently recorded.[66] These channels appear with high frequency at pH 5.0, but not at neutral pH, have a single channel conductance of 45 pS and are nonselective for a particular ion. Like other three-domain bacterial protein toxins, TeNT forms ion-conducting channels in cell membranes at acidic pH values.[43] A general consensus exists that these toxin channels are related to the translocation process of the enzymatic domain across the vesicle membrane into the neuronal cytosol. The most likely possibility is that the L chain translocates into the cytosol at the lipid–protein boundary, rather than inside a proteinaceous pore.[43,91] This cleft model proposes that the two toxin subunits change conformation at low pH, exposing hydrophobic surfaces and contacting the hydrophobic core of the lipid bilayer. In its 'acidic' conformation, the H chain is suggested to form a transmembrane hydrophilic cleft that allows the passage of the partly unfolded acidic form of the L chain through the membrane. The neutral pH environment found in the cytosol induces the L chain to refold and to regain its water-soluble 'neutral' conformation. While the L chain is leaving the vesicle membrane, the transmembrane hydrophilic cleft of the H chain 'tightens up' to reduce the amount of hydrophilic protein surface exposed to the hydrophobic core of the lipid bilayer. However, this process leaves a peculiarly shaped channel in the membrane, with two rigid protein walls and a small mobile lipid seal on one side. This is proposed to be the structure responsible for the ion-conducting properties of TeNT and BoNTs. In the cleft model, the ion channel is a consequence of membrane translocation, rather than a prerequisite for this process.

VI. INTRACELLULAR ZINC-ENDOPEPTIDASE ACTIVITY OF CLOSTRIDIAL NEUROTOXINS AND THEIR TARGETS

One of the more conserved segments of the L chain of clostridial neurotoxins is a central region that contains a His-Glu-Xaa-Xaa-His zinc-binding motif characteristic of zinc-endopeptidases. The presence of this motif suggested that TeNT and the BoNTs may inhibit neuroexocytosis via a zinc-endopeptidase activity. This hypothesis was confirmed by two experimental approaches in *Aplysia* neurons. First, the lack of toxicity of apo-TeNT L chain demonstrated the essential role of the metal atom in toxin activity.[47,92] Second, phosphoramidon, a very specific inhibitor of zinc-endopeptidases, was shown to inhibit TeNT-induced blockade of ACh release.[47] These results were the first evidence that the L chain of TeNT acted inside neurons via a metallo-protease activity.

Because neurotransmitters at the presynaptic terminals are contained mainly inside SSVs, these organelles were the first candidate targets of the neurotoxin proteolytic activity. TeNT and BoNT/B, BoNT/D, BoNT/F and BoNT/G, free of contaminant proteases, cleave a membrane protein of SSVs called VAMP or synaptobrevin.[47,49,53,92–94] Conversely, BoNT/A, BoNT/C and BoNT/E act on proteins associated with the presynaptic membrane: BoNT/A and BoNT/E cut SNAP-25,[93,95] while serotype C cleaves syntaxin, in addition to SNAP-25.[51,96–99] Figure 18.3 shows schematically the cleavage sites of the various clostridial neurotoxins on VAMP, SNAP-25 and syntaxin. Recombinant VAMP, SNAP-25 and syntaxin are cleaved at the identical peptide bonds and with the same rate as the corresponding cellular proteins. This indicates that no additional endogenous factors are involved in neurotoxin proteolytic activity. The peptide bonds hydrolyzed by each neurotoxin has been identified (Table 18.1). With the exception of TeNT and BoNT/B, every clostridial neurotoxin catalyses the hydrolysis of a different peptide bond. The amino acid residues flanking the cleavage sites differ in terms of charge, polarity and size. Hence, the active sites of these metallo-proteinases must all differ in their spatial organization in order to accommodate and cleave different peptide bonds with disparate flanking sequences. The finding that VAMP, SNAP-25 and syntaxin are the only known substrates of these neurotoxins clearly demonstrates that these three proteins play a central role in neuroexocytosis.

A. VAMP/synaptobrevin

VAMP (vesicle associated membrane protein) or synaptobrevin gene was first cloned from a *Torpedo* electric organ library, and then from different animal species including human and yeast.[100,101] VAMP is about 120 residues long and

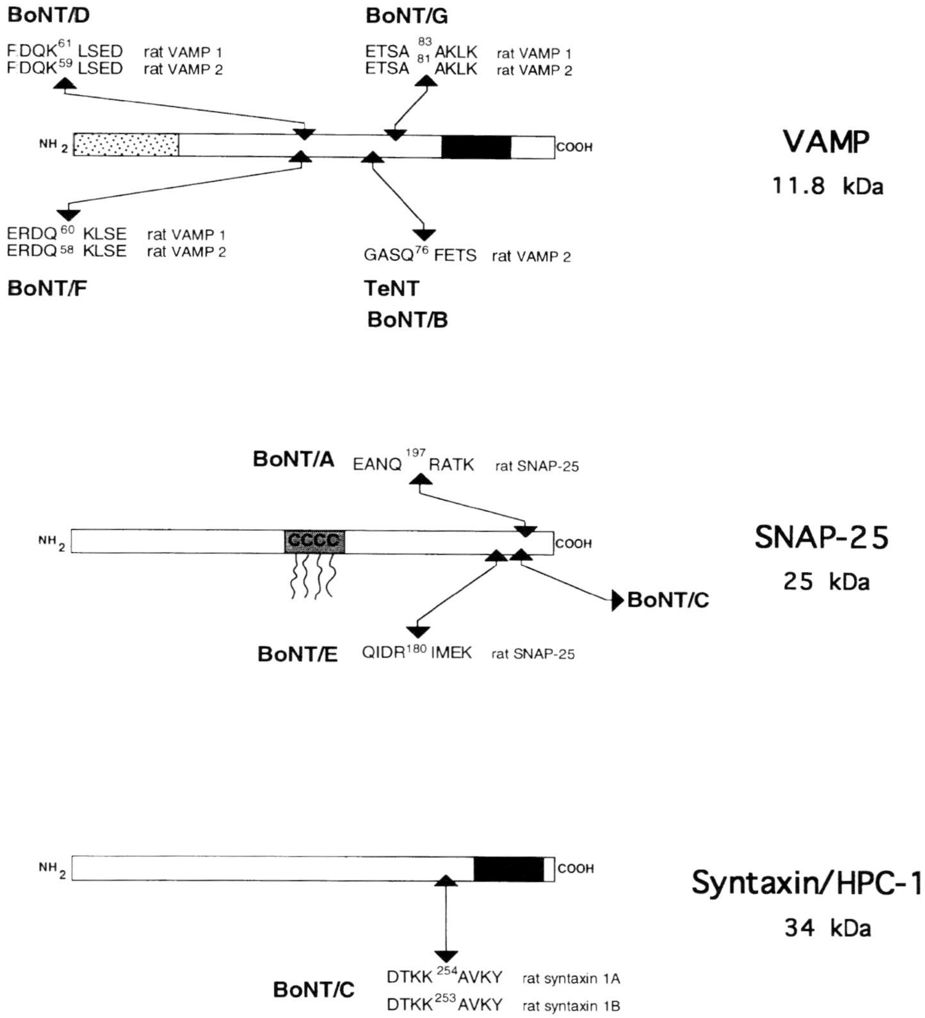

Figure 18.3 Specificity and sites of cleavage of the clostridial neurotoxins. VAMP is bound to the synaptic vesicle membrane through a single transmembrane domain (black box), with most of the protein exposed to the cytoplasm. SNAP-25 and syntaxin are bound to the target membrane, via palmitoylation of a cluster of conserved cysteines (SNAP-25) or via a single transmembrane domain. TeNT and BoNT/B, BoNT/D, BoNT/F or BoNT/G act on the conserved central portion of VAMP and release the amino-terminal part of VAMP in the cytosol. The sequences report the peptides bonds cleaved by clostridial neurotoxins on rat VAMP-1 and VAMP-s. BoNT/A and BoNT/E cleave SNAP-25 at the carboxy terminus with the release of 9 and 26 residues, respectively. BoNT/C also cleaves SNAP-25 at the carboxy terminus. BoNT/C cleaves syntaxin at a single site near the cytosolic membrane surface. The action of TeNT and BoNT/B, BoNT/C, BoNT/D, BoNT/F and BoNT/G causes the release of a large portion of the cytosolic domain of VAMP and syntaxin. Conversely, only a small segment of SNAP-25 is released by the selective proteolysis of BoNT/A, BoNT/C and BoNT/E.

Table 18.1. Tetanus and botulism neurotoxins target and peptide bond specificity

Toxin type	Target	Peptide bond cleaved p_3-p_2-p_1–p_1'-p_2'-p_3'
TeNT	VAMP	A-S-Q–F-E-T
BoNT/A	SNAP-25	A-N-Q-R-A-T
BoNT/B	VAMP	A-S-Q–F-E-T
BoNT/C	Syntaxin	T-K-K–A-V-K
BoNT/C	SNAP-25	N-Q-R–A-T-K[a]
BoNT/D	VAMP	D-Q-K–L-S-E
BoNT/E	SNAP-25	I-D-R–I-M-E
BoNT/F	VAMP	R-D-Q–K-L-S
BoNT/G	VAMP	T-S-A–A-K-L

[a] SNAP-25 was also cleaved in neurons exposed to BoNT/C.[96–98]

the exact number of residues depends on its source and isotype. As shown in Figure 18.3, VAMP contains a short carboxyl-terminal segment located inside the vesicle lumen, while the majority of the molecule is exposed to the cytosol. The 30 residue long amino-terminus diverges considerably among species and isoforms and has a high proline content. The highly conserved central part of VAMP (residues 30–96), rich in charged and hydrophilic residues, includes the known cleavage sites of the neurotoxins. On the SSV membrane, VAMP is associated with synaptophysin,[102–104] a major membrane component of SSV. Synaptophysin is capable of forming oligomers endowed with channel properties, that may be directly involved in neurotransmitter release.[100,101] VAP-33, another integral membrane protein of SSVs, interacts functionally with VAMP in *Aplysia* neurons.[105]

Three isoforms of VAMP have been identified: VAMP-1, VAMP-2 and cellubrevin.[100,101,106] VAMP appears to be present in all vertebrate tissues, but the distribution of VAMP-1 and VAMP-2 varies among different cell types.[107] In contrast, cellubrevin has an ubiquitous distribution.[106,107] The reason for this differential distribution of VAMP isotypes is unknown, but it suggests that VAMP plays a role in many events of vesicle fusion in addition to neuroexocytosis. Recent findings indicate that the neurospecific action of TeNT and BoNTs results from the availability of cell surface receptors able to mediate their cellular uptake, and not the lack of targets in non-neuronal cells.[30,108–111]

Rat VAMP-1 is not cleaved by clostridial neurotoxins.[92] These VAMP isotypes contain a Val residue in place of the Gln residue at the cleavage site of TeNT and BoNT/B in human and mouse VAMP-1. This single amino acid replacement at the cleavage site has been proposed to be associated with the resistance of rats and chickens to tetanus and type B botulism.[112] The differential susceptibility to TeNT and BoNT/B proteolysis of rat VAMP-1 and VAMP-2 has been used to identify the VAMP isotypes.[108,113] In contrast, BoNT/D, BoNT/F and BoNT/G cleave both VAMP-1 and VAMP-2 with similar rates.

B. SNAP-25

SNAP-25 (25 kDA synaptosomal-associated protein) is a 206-residue protein that lacks a classical transmembrane segment, but is bound to the cytosolic surface of the neuronal plasmalemma, probably via palmitoylation of four cysteine residues located near the center of the molecule (Figure 18.3).[100,101] SNAP-25 is required for axonal growth during development and nerve terminal plasticity in the mature nervous system.[114] This protein is highly conserved in organisms from *Drosophila* to mammals. In humans, there are two isoforms which are differentially expressed during development. Isotype *a* is constitutively expressed at the beginning of the embryonic stage, whereas isotype *b* appears at birth and is the predominant form in the adult.[115] The tissue distribution of SNAP-25 is less well characterized than that of VAMP, however, its presence in pancreatic cells[29,116] may indicate that it is also expressed outside the nervous system. A SNAP-25 related protein required for post-Golgi transport has been cloned from yeast.[117]

C. Syntaxin

Syntaxin 1 is located on the cytosolic surface of the neuronal plasmalemma.[100,101] It is a typical type II membrane protein with most of its molecular mass exposed to the cytosol (Figure 18.3). Syntaxin 1 is associated with calcium channels at the active zones of the presynaptic membrane, where neurotransmitter release takes place. Syntaxin also interacts with synaptotagmin, the putative high-affinity calcium sensor on SSV.[100,101] Syntaxin, together with SNAP-25, undergoes a complex recycling process in organelles indistinguishable from SSVs.[118] Two isoforms of slightly different length have been identified in neurons (isotypes 1A and 1B) and a vast polymorphism is present in other tissues.[119] Syntaxin-related proteins have been identified in yeast and in plants.[120] BoNT/C cleaves syntaxin only when syntaxin is present in lipid bilayers containing acidic lipids. The various syntaxin isoforms show different sensitivity to BoNT/C: isoforms 1A, 1B, 2, 3 are sensitive,[51,96] while isoforms 4 and 5 are resistant to toxin cleavage.

D. Are VAMP, SNAP-25 and syntaxin the real clostridial neurotoxins targets?

Evidence is rapidly accumulating that TeNT and BoNTs block neuroexocytosis via the specific cleavage of VAMP, SNAP-25 and syntaxin. The following experimental observations suggest strongly that these membrane proteins are true targets of the metallo-protease activity of clostridial neurotoxins:

(1) The intracellular activity of the toxins is inhibited by specific inhibitors of zinc-endopeptidases such as phosphoramidon and captopril;[27,47,121,122]
(2) VAMP, SNAP-25 and syntaxin are cleaved in synaptosomes and PC12 cells intoxicated with TeNT or the BoNTs;[56,89,90,123,124]
(3) Peptides spanning the cleavage site of VAMP inhibit TeNT and BoNT/B action in *Aplysia* neurons or chromaffin cells;[27,92]
(4) An antibody highly specific for VAMP prevents the inhibition of neurotransmitter release in *Aplysia* neurons induced by TeNT and BoNT/B, but not that caused by BoNT/A;[125]
(5) An antibody against the zinc-binding motif of TeNT inhibits the activity of TeNT and BoNT/A in chromaffin cells;[126]
(6) Mutations in the histidine-rich motif of clostridial neurotoxins block the neurotoxin-derived inhibition of neurotransmitter release and the proteolysis of the substrate simultaneously;[127]
(7) The inhibition of neurotransmitter release in cultured cortical neurons and in a variety of secreting cells and is accompanied by the proteolysis of VAMP, SNAP-25 or syntaxin.[56,97–99,110]

These results do not exclude the possibility that clostridial neurotoxins may act on other substrates, but clearly indicate that VAMP, SNAP-25 and syntaxin are major targets of the proteolytic activity of clostridial neurotoxins. Clostridial neurotoxins are therefore increasingly used to probe the involvement of these three proteins in a variety of cell events.[27–30,108,110,111,113,124]

VII. TARGET RECOGNITION

More than 100 peptides have been tested *in vitro* as possible proteolytic substrates to determine the specificity of the various clostridial neurotoxins. Some were tested on the basis of the known protein sequence flanking the cleavage sites of the various toxins. Invariably, none of these peptides were cleaved by clostridial neurotoxins.[47,94,128–131] In contrast, peptides encompassing the cleavage sites do bind to the respective toxins and inhibit their proteolytic activity against endogenous substrates.[50,92] BoNT/B proteolysis of VAMP peptides requires a 40 residue peptide and maximal cleavage rates were observed with a VAMP segment spanning residues 33–96 of the native protein.[130] Interestingly, Yamasaki and colleagues[94] found that for cleavage BoNT/F and BoNT/D appear to require peptides more extended at their amino-terminus than those hydrolyzed by TeNT and BoNT/B. Another peculiarity of these metallo-proteinases is that they hydrolyze only one out of several identical peptide bonds present in the target protein sequence. These features suggest that TeNT and BoNTs recognize the tertiary structure of their targets and that the cleaved peptide bond and flanking residues is not the sole determinant for VAMP, SNAP-25 and syntaxin site recognition and cleavage.[132]

As mentioned above, the L chains of clostridial neurotoxins show a strong sequence similarity, thus suggesting that they derive from a common ancestral metallo-protease. This favors the possibility that the targets share a common region involved in toxin binding. Close inspection of the sequences of VAMP, SNAP-25 and syntaxin, reveals that all three proteins contain a short motif predicted to adopt an α-helical conformation.[132] Helical wheel projections of this motif show a face with three negative charges next to a hydrophobic region (Figure 18.4). This motif is present in all VAMP, SNAP-25 and syntaxin isoforms sensitive to the neurotoxins. Variations are present in VAMP and syntaxin of *Drosophila* and yeast, species which are resistant to these neurotoxins, and in syntaxin isoforms known not to be involved in exocytosis. There are two copies of this motif in VAMP, four copies in SNAP-25 and two copies in syntaxin

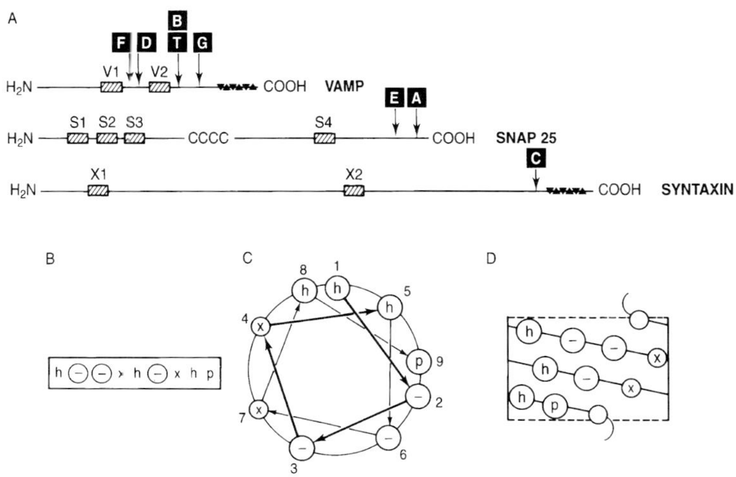

Figure 18.4 Neurotoxin recognition motifs of VAMP, SNAP-25 and syntaxin. (a) Hatched boxes indicate the presence and positions of a motif common to the three targets of clostridial neurotoxins; (B) the motif composed of h (hydrophobic residue), – (Asp or Glu residue), p (polar residue) and x (any residue). The motif is included in the three proteins within segments predicted to adopt an α-helical conformation. (C) and (D) show a top and a lateral view of the motif, respectively.

(Figure 18.4A). Peptides corresponding to the specific sequences of this motif from VAMP, SNAP-25 and syntaxin inhibit neurotoxin activity *in vitro* and *in vivo*, irrespective of their origin and toxin isoform. Antibodies raised against these peptides crossreact among the three target proteins.[132,133] These results indicate that the motif is exposed on the protein surface and adopts a similar configuration in each of the three neurotoxin targets. Consequently, the VAMP-specific, SNAP-25-specific, and syntaxin-specific neurotoxins inhibit one another because they compete for the same binding site. However, they only cleave their specific target protein.[132,133] In VAMP this toxin-binding region is essential for the recycling of SSVs after their fusion with the presynaptic membrane.[134] In summary, clostridial neurotoxins recognize their protein substrates via two distinct regions: (1) a segment that includes the peptide bond to be cleaved and (2) another region conserved in VAMP, SNAP-25 and syntaxin, which accounts for antibody crossreactivity and cross-inhibition of the different neurotoxin types. The relative contribution of these segments to the specificity and strength of neurotoxin binding has not been determined.

VIII. CLOSTRIDIAL NEUROTOXINS AND NEUROEXOCYTOTIC APPARATUS

Recently, Söllner and colleagues[135–137] have shown that VAMP, SNAP-25 and syntaxin, together with a group of cytosolic proteins (NSF, α- and γ-SNAP), form a 20S protein complex involved in the docking and fusion of SSV with the presynaptic membrane. Synaptotagmin, the likely calcium sensor in exocytosis,[100] interacts with this core complex.[136] In the presence of the neuronal isoform β-SNAP, synaptotagmin induces the formation of a large docking and fusion particle, likely to be involved in neurotransmission.[138] In addition to exocytosis, vesicle targeting and fusion are the basis of many events of cell membrane trafficking, including endocytosis, protein maturation and export and formation and control of cell organelles.[101,137,139] The specific delivery of 'cargo' within each vesicle type is ensured by the recognition between a vesicle receptor proteins (termed v-SNARE) and a complementary receptor proteins located on the target membrane (t-SNARE). In neuroexocytosis, VAMP and synaptotagmin are likely v-SNAREs and syntaxin and SNAP-25 are the t-SNAREs.[135,138] Specific VAMP and syntaxin isoforms couple and control each vesicle docking event within the cell.[135,137] Experimental evidence supporting of the SNARE hypothesis is rapidly accumulating and suggests that the SNAREs functionally define the borders of intracellular transport units.[137,139]

SNAP-25 and syntaxin form a stoichiometric complex.[140] This complex can bind one molecule of VAMP with high affinity. This trimeric SNARE complex is stable in sodium dodecylsulfate.[141–143] In the process of neuroexocytosis,

SNARE complex formation precedes the recruitment of cytosolic and membrane protein components required for fusion of the lipid bilayers.[135,139] It is likely that NSF-mediated hydrolysis of ATP provides energy for priming the neuroexocytotic apparatus. The primed system is then ready to trigger exocytosis upon calcium influx into the synapse. It has not yet been established whether the last step of neurotransmitter release takes place via a fusion pore or via a complete membrane fusion with lipid intermixing.[144,145]

BoNT/A removes only nine residues from the SNAP-25 carboxy terminus and yet this cleavage is sufficient to impair neuroexocytosis. This indicates that the extreme carboxy-terminal portion of the molecule is essential for SNAP-25 folding or for correct interaction with other SNAREs. The functional intoxication with BoNT/A can be reversed by α-latratoxin or by increasing the intracellular calcium concentration. This rescue process is not possible when SNAP-25 is cleaved with BoNT/E, a process that removes an additional 18 amino acid residues from its carboxy terminus, or by intoxication of the synapse with other clostridial neurotoxins. These results suggest that SNAP-25 may have an additional modulatory role, which is strictly related to the calcium-sensitive step.

Cleavage of VAMP and syntaxin by TeNT, BoNT/B, BoNT/D, BoNT/F, BoNT/G and BoNT/C leads to the release of a large part of the target molecule into the cytosol. These fragments, and also fragments of SNAP-25 generated by BoNT/A and BoNT/E, can still form a SNARE complex capable of recruiting SNAPs and NSF. In the presence of ATP, the 20S particle is able to dissociate with a kinetic rate similar to that of the native complex.[142,143] This suggests that clostridial neurotoxins do not directly impair the docking process of vesicles to their target membranes, but more likely a late step in exocytosis. Ultrastructural analysis of poisoned and electrically silent synapses show a similar number of docked vesicles as control synapses.[34,145] This suggests that the docking step is multiprotein-mediated process involving additional components. However, these docked vesicles are unable to fuse with the plasma membrane, even after prolonged stimulation, which indicates that VAMP, SNAP-25 and syntaxin may also be involved in the fusion process and/or in the vesicle re-uptake, in addition to their proposed role in docking.

IX. OTHER ACTIVITIES OF CLOSTRIDIAL NEUROTOXINS

Very recently, several experimental approaches have provided findings difficult to reconcile with a single enzymatic activity for clostridial neurotoxins. Among others, the results of Ashton and colleagues indicate that nonconservative mutations in the active site of tetanus toxin do not abolish its inhibitory activity on neuroexocytosis.[147] The fact that the L chain of clostridial neurotoxins may possess more than one biological activity is not unlikely in the light of the recent findings that a protein with a similar molecular weight, the leukotriene A4 hydrolase, also is a zinc endopeptidase.[148] It has been shown that TeNT is an activator of transglutaminase and that synapsin is a good transglutaminase substrate.[149–151] On this basis Facchiano *et al.*[151,152] proposed that an activity of TeNT may be the activation of the transglutaminase associated with SSV and crosslinking of synapsin. The SSV would then be anchored to actin bundles, thus preventing their subsequent activation for docking and fusion. On this basis it is possible to envisage a rapid 'acute' step in TeNT action with cleavage of VAMP and a long-term phase consisting of the fixation of SSV to the cytoskeletal network. A critical discussion of this model is presented elsewhere.[153]

An alternative mechanism of clostridial neurotoxins action is based on the alteration of the arachidonic acid cycle.[154] In this model, the primary inhibitory effect caused by BoNT/A is the reduction of the release of arachidonic acid from synaptic membrane stores. Based on the facilitatory effect of arachidonic acid on secretion, this primary effect may account for the neurotoxin-induced inhibition of neurotransmitter release. Future experiments are necessary to map at the molecular level the active site(s) of these putative activities and to establish their importance for the *in vivo* action of clostridial neurotoxins.

X. CONCLUDING REMARKS

The clostridial neurotoxins responsible for tetanus and botulism form a new group of zinc-endopeptidases endowed with peculiar properties. They are produced as inactive precursors which are activated by specific proteolysis followed by intracellular reduction of a single disulfide bond. The amino acid sequence around the zinc binding motif is different from that of any group of zinc-endopeptidases. They act in the neuronal cytosol and are very specific in terms of their protein target and the peptide bond hydrolyzed. Their exquisite specificity for proteins of the neuroexocytotic apparatus makes them valuable tools in modern neuroscience research.

Important discoveries in the future will be the identification of the neuron-specific receptors of clostridial neurotoxins and of the mode of internalization and membrane translocation of the neurotoxins. Another important line of research is aimed at finding specific inhibitors of these metallo-proteinases. Inhibitors which can cross the neuronal plasmalemma into the cytosol would be potential therapeutic agents in the treatment of tetanus and botulism. The modification of BoNTs to prolong their life inside the NMJ is an important research goal to improve the treatment of dystonias. The determination of the three-dimensional structure of these neurotoxins will greatly accelerate the research on these fronts.

XI. ACKNOWLEDGMENTS

We thank the referees and T. Hohl, J. McNew, T.H. Söllner for critical reading of the manuscript and for suggestions. Work in the author's laboratory (C.M.) is supported by CNR-Biotecnologie, Telethon-Italia grant 473 and MURST.

XII. REFERENCES

1. Faber K. (1890) Die Pathogenie des Tetanus. *Berl Klin Wochenschr* **27**: 717–720.
2. Kitasato S. (1891) Experimentelle Untersuchungen über das Tetanusgift. *Zeitschr Hyg Infektkrh* **10**: 267–305.
3. Tizzoni G. and Cattani G. (1890) Über das Tetanusgift. *Zentralbl Bakt* **8**: 69–73.
4. Tizzoni G. and Cattani G. (1890) Untersuchungen über das Tetanusgift. *Arch exp Pathol Pharmakol* **27**: 432–450.
5. Hatheway C.L. (1995) Botulism: the present status of the disease. *Curr Top Microbiol Immunol* **195**: 55–75.
6. van Ermengem E. (1897) Über ein neuen anaeroben Bacillus und seine Beziehungen zum Botulismus. *Zeitschr Hyg Infektkrh* **26**: 1–56.
7. Aureli P., Fenicia L., Pasolini B., Gianfranceschi M., McCroskey L.M. and Hatheway C.O. (1986) Two cases of type E infant botulism in Italy caused by neurotoxigenic *Clostridium butyricum*. *J Inf Dis* **154**: 207–211.
8. Hall J.D., McCroskey L.M., Pincomb B.J. and Hatheway C.O. (1985) Isolation of an organism resembling *Clostridium barati* which produces type F toxin from an infant with botulism. *J Clin Microbiol* **21**: 654–655.
9. Middlebrook J.L. and Brown J.E. (1995) Immunodiagnosis and immunotherapy of tetanus and botulinum neurotoxins. *Curr Top Microbiol Immunol* **195**: 89 –122.
10. Payling-Wright G. (1955) The neurotoxins of *Clostridium botulinum* and *Clostridium tetani*. *Pharmacol Rev* **7**: 413 –465.
11. Simpson L.L. (ed.) (1989) *Botulinum Neurotoxin and Tetanus Toxin*. Academic Press, San Diego, CA.
12. Bruschettini A. (1892) Sulla diffusione del veleno del tetano nell'organismo. *Rif Med* **8**: 270 –273.
13. Vallee R.B. and Bloom G.S. (1991) Mechanisms of fast and slow axonal transport. *Annu Rev Neurosci* **14**: 59–92.
14. Schwab M.E., Suda K. and Thoenen H. (1979) Selective retrograde trans-synaptic transfer of a protein, tetanus toxin, subsequent to its retrograde axonal transport. *J Cell Biol* **82**: 798–810.
15. Curtis D.R., Game C.J.A., Lodge D. and McCulloch R.M. (1976) A pharmacologic study of Renshaw cell inhibition. *J Physiol* **258**: 227–242.
16. Erdmann G., Wiegand H. and Wellhoner H. (1975) Intraxonal and extraaxonal transport of ^{125}I-tetanus toxin in early local tetanus. *Naunyn Schmiedeberg Arch Pharmacol* **290**: 357–373.
17. Wellhoner H.H. (1992) Tetanus and botulinum neurotoxins. In Herken H. and Hucho F. (eds), *Handbook of Experimental Pharmacology*, Vol. 102, pp. 357–417. Springer-Verlag, Berlin.

18. Penner R., Neher E. and Dreyer F. (1986) Intracellularly injected tetanus toxin inhibits exocytosis in bovine adrenal chromaffin cells. *Nature* **324**: 76–77.
19. Bittner M.A. and Holz R.W. (1988) Effects of tetanus toxin on catecholamine release from intact and digitonin-permeabilized chromaffin cells. *J Neurochem* **51**: 451–456.
20. Anhert-Hilger G., Bader M.F., Bhakdi S. and Gratzl M. (1989) Introduction of macromolecules into bovine adrenal medullary chromaffin cells and rat pheochromocytoma cells (PC12) by permeabilization with streptolysin O – inhibitory effect of tetanus toxin on catecholamine secretion. *J Neurochem* **52**: 1751–1758.
21. Anhert-Hilger G., Weller U., Dauzenroth M.E., Habermann E. and Gratzl M. (1989) The tetanus toxin light chain inhibits exocytosis. *FEBS Lett* **242**: 245–248.
22. Bittner M.A., DasGupta B.R. and Holz R.W. (1989) Isolated light chains of botulinum neurotoxins inhibit exocytosis. Studies in digitonin-permeabilized chromaffin cells. *J Biol Chem* **264**: 10354–10360.
23. Bittner M.A., Habig W.H. and Holz R.W. (1989) Isolated light chain of tetanus toxin inhibits exocytosis: studies in digitonin-permeabilized cells. *J Neurochem* **53**: 966–968.
24. Stecher B., Weller U., Habermann E., Gratzl M. and Anhert-Hilger G. (1989) The light chain but not the heavy chain of botulinum A toxin inhibits exocytosis from permeabilized adrenal chromaffin cells. *FEBS Lett* **255**: 391–394.
25. McInnes C. and Dolly J.O. (1990) Ca^{2+}-dependent noradrenaline release from permeabilised PC12 cells is blocked by botulinum neurotoxin A or its light chain. *FEBS Lett* **261**: 323–326.
26. Dayanithi G., Weller U., Anhert-Hilger G., Link H., Nordmann J.J. and Gratzl M. (1992) The light chain of tetanus toxin inhibits calcium dependent vasopressin release from permeabilized nerve endings. *Neuroscience* **46**: 489–493.
27. Dayanithi G., Stecher B., Höhne-Zell B., Yamasaki S., Binz T., Weller U., Niemann H. and Gratzl M. (1994) Exploring the functional domain and the target of the tetanus toxin light chain in neurohypophysial terminals. *Neuroscience* **58**: 423–431.
28. Ikonen E., Tagaya M., Ullrich O., Montecucco C. and Simons K. (1995) Different requirements for NSF, SNAP and Rab proteins in apical and basolateral transport in MDCK cells. *Cell* **81**: 571–580.
29. Sadoul K., Lang J., Motecucco C., Weller U., Catsicas S., Wollheim C. and Halban P. (1995) SNAP-25 is expressed in islets of Langerhans and is involved in insulin release. *J Cell Biol* **128**: 1019–1028.
30. Regazzi R., Wolheim C., Lang J., Theler J.M., Rossetto O., Montecucco C., Sadoul K., Weller U., Palmer U. and Thorens B. (1995) Vamp-2 and cellubrevin are expressed in pancreatic β-cells and are essential for Ca^{2+} but not GTPγS-induced secretion. *EMBO J* **14**: 2723–2730.
31. Montecucco C. and Schiavo G. (1995) Structure and function of botulinum neurotoxins. *Quarterly Rev Biophys* **28**: 423–472.
32. Dreyer F., Mallart A. and Brigant J.L. (1983) Botulinum A toxin and tetanus toxin do not affect presynaptic membrane currents in mammalian motor nerve endings. *Brain Res* **270**: 373–375.
33. Pecot-Dechavassine M., Molgo J. and Thesleff S. (1991) Ultrastructure of botulinum type A poisoned frog motor nerve terminals after enhanced quantal transmitter release caused by carbonyl cyanide m-chlorophenylhydrazone. *Neurosci Lett* **130**: 5–8.
34. Hunt J.M., Bommert K., Charlton M.P., Kistner A., Habermann E., Augustine G.J. and Betz H. (1994) A post-docking role for synaptobrevin in synaptic vesicle fusion. *Neuron* **12**: 1269–1279.
35. Jankovic J. and Hallett M. (eds) (1994) *Therapy with Botulinum Toxin.* Marcel Dekker, New York.
36. DasGupta B.R. (1994) Structures of botulinum neurotoxin, its functional domains, and perspectives on the crystalline type A toxin. In Jankovic J. and Hallett M. (eds), *Therapy with Botulinum Toxin*, pp. 15–39. Marcel Dekker, New York.
37. Schiavo G., Papini E., Genna G. and Montecucco C. (1990) An intact interchain disulfide bond is required for the neurotoxicity of tetanus toxin. *Infect Immun* **58**: 4136–4141.
38. de Paiva A., Poulain B., Lawrence G.W., Shone C.C., Tauc L. and Dolly J.O. (1993) A role for the interchain disulfide or its participating thiols in the internalization of botulinum neurotoxin A

revealed by a toxin derivative that binds to ecto-acceptors and inhibits transmitter release intracellularly. *J Biol Chem* **268**: 20838–20844.

39. Stevens R.C., Evenson M.L., Tepp W. and DasGupta B.R. (1991) Crystallization and preliminary X-ray analysis of botulinum neurotoxin type A. *J Mol Biol* **222**: 877–880.
40. Anderson M.D., Fairweather N., Charles I.G., Emsley P., Isaacs N.W. and MacDermott G. (1993) Crystallographic characterization of tetanus toxin fragment C. *J Mol Biol* **230**: 673–674.
41. Singh B.R., Wasacz F.M., Strand S., Jakobsen R.J. and DasGupta B.R. (1990) Structural analysis of botulinum neurotoxin types A and E in aqueous and non polar solvents by Fourier transform infrared, second derivative UV absorption and circular dichroic spectroscopies. *J Protein Chem* **9**: 705–713.
42. De Filippis V., Vangelista L., Schiavo G., Tonello F. and Montecucco C. (1995) Structural studies on the zinc-endopeptidase light chain of tetanus neurotoxin. *Eur J Biochem* **229**: 61–69.
43. Montecucco C., Papini E. and Schiavo G. (1994) Bacterial protein toxins penetrate cells via a four-step mechanism. *FEBS Lett* **346**: 92–98.
44. Minton N. (1995) Molecular genetics of clostridial neurotoxins. *Curr Top Microbiol Immunol* **195**: 161–194.
45. Jongeneel C.V., Bouvier J. and Bairoch A. (1989) A unique signature identifies a family of zinc-dependent metallopeptidases. *FEBS Lett* **242**: 211–214.
46. Jiang W. and Bond J.S. (1992) Families of metalloendopeptidases and their relationships. *FEBS Lett* **312**: 110–114.
47. Schiavo G., Poulain, B., Rossetto O., Benfenati F., Tauc L. and Montecucco C. (1992) Tetanus toxin is a zinc protein and its inhibition of neurotransmitter release and protease activity depend on zinc. *EMBO J* **11**: 3577–3583.
48. Schiavo G., Rossetto O., Santucci A., DasGupta B.R. and Montecucco C. (1992) Botulinum neurotoxins are zinc proteins. *J Biol Chem* **267**: 23479–23483.
49. Schiavo G., Shone C.C., Rossetto O., Alexandre F.C.G. and Montecucco C. (1993) Botulinum neurotoxin serotype F is a zinc endopeptidase specific for VAMP/synaptobrevin. *J Biol Chem* **268**: 11516–11519.
50. Schiavo G., Malizio C., Trimble W.S., Polverino de Laureto P., Milan G., Sugiyama H., Johnson E.A. and Montecucco C. (1994) Botulinum G neurotoxin cleaves VAMP/synaptobrevin at a single Ala/Ala peptide bond. *J Biol Chem* **269**: 20213–20216.
51. Schiavo G., Shone C.C., Bennett M.K., Scheller R.H. and Montecucco C. (1995) Botulinum neurotoxin type C cleaves a single Lys-Ala bond within the carboxyl-terminal region of syntaxins. *J Biol Chem* **270**: 10566–10570.
52. Wright J.F., Pernollet M., Reboul A., Aude C. and Colomb M. (1992) Identification and partial characterization of a low affinity metal-binding site in the light chain of tetanus toxin. *J Biol Chem* **267**: 9053–9058.
53. Yamasaki S., Hu Y., Binz T., Kalkuhl A., Kurazono H., Tamura T., Jahn R., Kandel E. and Niemann H. (1994) Synaptobrevin/VAMP of *Aplysia californica*: structure and proteolysis by tetanus and botulinal neurotoxins type D and F. *Proc Natl Acad Sci USA* **91**: 4688–4692.
54. Bhattacharyya S.D. and Sugiyama H. (1989) Inactivation of botulinum and tetanus toxins by chelators. *Infect Immun* **57**: 3053–3057.
55. Lovejoy B., Cleasby A., Hassell A.M., Longley K., Luther M.A., Weigl D., McGeehan G., McElroy A.B., Drewry D., Lambert M.H. and Jordan S.R. (1994) Structure of the catalytic domain of fibroblast collagenase complexed with an inhibitor. *Science* **263**: 375–377.
56. Simpson L.L., Coffield J.A. and Bakry N. (1993) Chelation of zinc antagonizes the neuromuscular blocking properties of the seven serotypes of botulinum neurotoxin as well as tetanus toxin. *J Pharmacol Exp Ther* **267**: 720–727.
57. Höhne-Zell B., Ecker A., Weller U. and Gratzl M. (1994) Synaptobrevin cleavage by tetanus toxin light chain is linked to inhibition of exocytosis in chromaffin cells. *FEBS Lett* **355**: 131–134.

58. Matthews B.W. (1988) Structural basis of the action of thermolysin and related zinc peptidases. *Acc Chem Res* **21**: 333–340.
59. Dasgupta B.R. and Rasmussen S. (1984) Effect of diethylpyrocarbonate on the biological activities of botulinum neurotoxin types A and E. *Arch Biochem Biophys* **232**: 172–178.
60. Tonello F., Schiavo G. and Montecucco C. (1997) Metal substitution of tetanus neurotoxin. *Biochem J* (in press).
61. Li Y., Foran P., Fairweather N., de Paiva A., Weller U., Dougan G. and Dolly O. (1994) A single mutation in the recombinant light chain of tetanus toxin abolishes its proteolytic activity and removes its toxicity seen after reconstitution with native heavy chain. *Biochemistry* **33**: 7014–7020.
62. Lebeda F.J. and Olson M.A. (1994) Secondary structural predictions for the clostridial neurotoxins. *Proteins* **20**: 293–300.
63. Höhne-Zell B., Stecher B. and Gratzl M. (1993) Functional characterization of the catalytic site of the tetanus toxin light chain using permeabilized adrenal chromaffin cells. *FEBS Lett* **336**: 175–180.
64. Morante S., Furenlid L., Schiavo G., Tonello F., Zwilling R. and Montecucco C. (1996) An X-ray absorption spectroscopy study of zinc coordination in tetanus neurotoxin, astacin, thermolysin and alkaline protease. *Eur J Biochem* **235**: 606–612.
65. Woody M. and DasGupta B.R. (1989) Effect of tetranitromethane on the biological activities of botulinum neurotoxin types A, B and E. *Mol Cell Biochem* **85**: 159–169.
66. Beise J., Hahnen J., Andersen-Beckh B. and Dreyer F. (1994) Pore formation by tetanus toxin, its chain and fragments in neuronal membranes and evaluation of the underlying motifs in the structure of the toxin molecule. *Naunyn Schmiedebergs Arch Pharmacol* **349**: 66–73.
67. Halpern J.L. and Neale E.A. (1995) Neurospecific binding, internalization, and retrograde axonal transport. *Curr Top Microbiol Immunol* **195**: 221–241.
68. Black J.D. and Dolly J.O. (1986) Interaction of ^{125}I-labeled neurotoxins with nerve terminals. I. Ultrastructure autoradiographic localization and quantitation of distinct membrane acceptors for types A and B on motor nerves. *J Cell Biol* **103**: 521–534.
69. Wellhoner H.H. and Neville D. Jr (1987) Tetanus toxin binds with high affinity to neuroblastoma × glioma hybrid cells NG 108–15 and impairs their stimulated acetylcholine release. *J Biol Chem* **262**: 17374–13738.
70. Montecucco C. (1986) How do tetanus and botulinum neurotoxins bind to neuronal membranes? *Trends Biochem Sci* **11**: 314–317.
71. Pierce E.J., Davison M.D., Parton R.G., Habig W.H. and Critchley D.R. (1986) Characterization of tetanus toxin binding to rat brain membranes. *Biochem J* **236**: 845–852.
72. Parton R.G., Ockleford C.D. and Critchley D.R. (1988) Tetanus toxin binding to mouse spinal cord cells: an evaluation of the role of gangliosides in toxin internalization. *Brain Res* **475**: 118–127.
73. Yavin E. and Nathan A. (1986) Tetanus toxin receptors on nerve cells contain a trypsin sensitive component. *Eur J Biochem* **154**: 403–407.
74. Schiavo G., Rossetto O., Ferrari G. and Montecucco C. (1991) Tetanus toxin receptor. Specific cross-linking of tetanus toxin to a protein of NGF-differentiated PC12 cells. *FEBS Lett* **290**: 227–230.
75. Nishiki T., Kamata Y., Nemoto Y., Omori A., Ito T., Takahashi M. and Kozaki S. (1994) Identification of protein receptor for *Clostridium botulinum* type B neurotoxin in rat brain synaptosomes. *J Biol Chem* **269**: 10498–10503.
76. Nishiki T., Tokuyama Y., Kamata Y., Nemoto Y., Yoshida A., Sato K., Sekiguchi M., Takahashi M. and Kozaki S. (1996) The high-affinity binding of *Clostridium botulinum* type B neurotoxin to synaptotagmin II associated with gangliosides GT1b/GD1a. *FEBS Lett* **378**: 253–257.
77. Schiavo G., Demel R. and Montecucco C. (1991) On the role of polysialogangliosides as tetanus toxin receptors: a study with lipid monolayers. *Eur J Biochem* **199**: 705–711.
77a. Matteoli M., Verderio C., Rossetto O., Iezzi N., Coco S., Schiavo G. and Montecucco C. (1996) Synaptic vesicle endocytosis mediates the entry of tetanus toxin into hippocampal neurons. *Proc Natl Acad Sci USA* **93**: 13310–13315.
78. Montesano R., Roth J., Robert A. and Orci L. (1982) Non coated invaginations are involved in binding and internalization of cholera and tetanus toxin. *Nature* **296**: 651–653.

79. Dolly J.O., Black J., Williams R.S. and Melling J. (1984) Acceptors for botulinum neurotoxin reside on motor nerve terminals and mediate its internalization. *Nature* **307**: 457–460.
80. Black J.D. and Dolly J.O. (1986) Interaction of ^{125}I-labeled neurotoxins with nerve terminals. II. Autoradiographic evidence for its uptake into motor nerves by acceptor-mediated endocytosis. *J Cell Biol* **103**: 535–544.
81. Parton R.G., Ockleford C.D. and Critchley D.R. (1987) A study of the mechanism of internalisation of tetanus toxin by primary mouse spinal cord cultures. *J Neurochem* **49**: 1057–1068.
82. Simpson L.L., Coffield J.A. and Bakry N. (1994) Inhibition of vacuolar adenosine triphosphatase antagonizes the effects of clostridial neurotoxins but not phospholipase A2 neurotoxins. *J Pharmacol Exp Ther* **269**: 256–262.
83. Williamson L.C. and Neale E.A. (1992) Bafilomicin A1 inhibits the action of tetanus toxin in spinal cord neurons in cell culture. *J Neurochem* **63**: 2342–2345.
84. Matteoli M., Takei K., Perin M.S., Südhof T.C. and De Camilli P. (1992) Exo-endocytic recycling of synaptic vesicles in developing processes of cultured hippocampal neurons. *J Cell Biol* **117**: 849–861.
85. Poulain B., Tauc L., Maisey E.A., Wadsworth J.D.F., Mohan P.M. and Dolly J.O. (1988) Neurotransmitter release is blocked intracellularly by botulinum neurotoxins, and requires uptake of both toxin polypeptides by a process mediated by the larger chain. *Proc Natl Acad Sci USA* **85**: 4090–4094.
86. Mochida S., Poulain B., Weller U., Habermann E. and Tauc L. (1989) Light chain of tetanus toxin intracellularly inhibits acethylcholine release at neuro-neuronal synapses, and its internalization is mediated by heavy chain. *FEBS Lett* **253**: 47–51.
87. Weller U., Dauzenroth M.-E., Gansel M. and Dreyer F. (1991) Cooperative action of the light chain of tetanus toxin and the heavy chain of botulinum toxin type A on the transmitter release of mammalian motor endplates. *Neurosci Lett* **122**: 132–134.
88. Schmid M.F., Robinson J.P. and DasGupta B.R. (1993) Direct visualization of botulinum neurotoxin-induced channels in phospholipid vesicles. *Nature* **364**: 827–830.
89. Montal M.S., Blewitt R., Tomich J.M. and Montal M. (1992) Identification of an ion channel-forming motif in the primary structure of tetanus and botulinum neurotoxins. *FEBS Lett* **313**: 12–18.
90. Oblatt-Montal M., Yamazaki M., Nelson R. and Montal M. (1995) Formation of ion channels in lipid bilayers by a peptide with the predicted transmembrane sequence of botulinum neurotoxin A. *Protein Sci* **4**: 1490–1497.
91. Montecucco C., Papini E. and Schiavo G. (1991) Molecular models of toxin membrane translocation. In Alouf J.E. and Freer J.H. (eds), *Sourcebook of Bacterial Protein Toxins*, pp. 45–56. London, Academic Press.
92. Schiavo G., Benfenati F., Poulain B., Rossetto O., Polverino de Laureto P., DasGupta B.R. and Montecucco C. (1992) Tetanus and botulinum-B neurotoxins block neurotransmitter release by a proteolytic cleavage of synaptobrevin. *Nature* **359**: 832–835.
93. Schiavo G., Rossetto O., Catsicas S., Polverino de Laureto P., DasGupta B.R., Benfenati F. and Montecucco C. (1993) Identification of the nerve-terminal targets of botulinum neurotoxins serotypes A, D and E. *J Biol Chem* **268**: 23784–23787.
94. Yamasaki S., Baumeister A., Binz T., Blasi J., Link E., Cornille F., Roques B., Fykse E.M. Südhof T.C., Jahn R. and Niemann H. (1994) Cleavage of members of the synaptobrevin/VAMP family by types D and F botulinal neurotoxins and tetanus toxin. *J Biol Chem* **269**: 12764–12772.
95. Blasi J., Chapman E.R., Link E., Binz T., Yamasaki S., DeCamilli P., Südhof T.C., Niemann H. and Jahn R. (1993) Botulinum neurotoxin A selectively cleaves the synaptic protein SNAP-25. *Nature* **365**: 160–163.
96. Blasi J., Chapman E.R., Yamasaki S., Binz T., Niemann H. and Jahn R. (1993) Botulinum neurotoxin C blocks neurotransmitter release by means of cleaving HPC-1/syntaxin. *EMBO J* **12**: 4821–4828.

97. Williamson L.C., Halpern J.L., Montecucco C., Brown J.E. and Neale E.A. (1996) Clostridial neurotoxins and substrate proteolysis in intact neurons: botulinum neurotoxin C acts on synaptosomal-associated protein of 25 kDa. *J Biol Chem* **271**: 7694–7699.
98. Osen-Sand A., Staple J.K., Naldi E., Schiavo G., Rossetto O., Petit-Pierre S., Malgaroli A., Catsicas S. and Montecucco C. (1996) Common and distinct fusion-proteins in axonal growth and transmitter release. *J Comp Neurol* **367**: 222–234.
99. Foran P., Lawrence G.W., Shone C.C., Foster K.A. and Dolly J.O. (1996) Botulinum neurotoxin C1 cleaves both syntaxin and SNAP-25 in intact and permeabilized chromaffin cells: correlation with its blockade of catecholamine release. *Biochemistry* **35**: 2630–2636.
100. Bennett M.K. and Scheller R.H. (1994) A molecular description of synaptic vesicle membrane trafficking. *Annu Rev Biochem* **63**: 63–100.
101. Südhof T.C. (1995) The synaptic vesicle cycle: a cascade of protein–protein interactions. *Nature* **375**: 645–653.
102. Calakos N. and Scheller R.H. (1994) Vesicle-associated membrane protein and synaptophysin are associated on the synaptic vesicle. *J Biol Chem* **269**: 24534–24537.
103. Edelmann L., Hanson P.I., Chapman E.R. and Jahn R. (1995) Synaptobrevin binding to synaptophysin: a potential mechanism for controlling the exocytotic fusion machine. *EMBO J* **16**: 224–231.
104. Washbourne P., Schiavo G. and Montecucco C. (1995) VAMP-2 forms a complex with synaptophysin. *Biochem J* **305**: 721–724.
105. Skehel P.A., Martin K.C., Kandel E.R. and Bartsch D. (1995) A VAMP-binding protein from *Aplysia* required for neurotransmitter release. *Science* **269**: 1580–1583.
106. McMahon H.T., Ushkaryov Y.A., Edelmann L., Link E., Binz T., Niemann H., Jahn R. and Südhof T.C. (1993) Cellubrevin is a ubiquitous tetanus-toxin substrate homologous to a putative synaptic vesicle fusion protein. *Nature* **364**: 346–349.
107. Rossetto O., Gorza L., Schiavo G., Schiavo N., Scheller R.H. and Montecucco C. (1996) VAMP/synaptobrevin isoforms 1 and 2 are widely and differentially expressed in non-neuronal tissues. *J Cell Biol* **132**: 167–179.
108. Braun J.E.A., Fritz B.A., Wong S.M.E. and Lowe A.W. (1994) Identification of a vesicle-associated membrane protein (VAMP-like) membrane protein in zymogen granules of the rat exocryne pancreas. *J. Biol Chem* **269**: 5328–5335.
109. Galli T., Chilcote T., Mundigl O., Binz T., Niemann H. and De Camilli P. (1994) Tetanus toxin-mediated cleavage of cellubrevin impairs exocytosis of transferrin receptor-containing vesicles in CHO cells. *J Cell Biol* **125**: 1015–1024.
110. Gaisano H.Y., Sheu L., Foskett J.K and Trimble W.S. (1994) Tetanus toxin light chain cleaves a vesicle-associated membrane protein (VAMP) isoform 2 in rat pancreatic zymogen granules and inhibits enzyme secretion. *J Biol Chem* **269**: 17062–17066.
111. Tamori Y., Hashiramoto M., Araki S., Kamata Y., Takahashi M., Kozaki S. and Kasuga M. (1996) Cleavage of vesicle-associated membrane protein (VAMP)-2 and cellubrevin on GLUT4-containing vesicles inhibits the translocation of GLUT4 in 3T3-L1 adipocytes. *Biochem Biophys Res Commun* **220**: 740–745.
112. Patarnello T., Bargelloni L., Rossetto O., Schiavo G. and Montecucco C. (1993) Neurotransmission and secretion. *Nature* **364**: 581–582.
113. Steinhardt R.A., Bi G. and Alderton J.M. (1994) Cell membrane resealing by a vesicular mechanism similar to neurotransmitter release. *Science* **263**: 390–393.
114. Osen-Sand A., Catsicas M., Staple J.K., Jones K.A., Ayala G., Knowles J., Grenningloh G. and Catsicas S. (1993) Inhibition of axonal growth by SNAP-25 antisense oligonucleotides *in vitro* and *in vivo*. *Nature* **364**: 445–448.
115. Bark C. and Wilson M.C. (1994) Human cDNA clones encoding two different isoforms of the nerve terminal protein SNAP-25. *Gene* **139**: 291–292.

116. Jacobsson G., Bean A.J., Scheller R.H., Juntti-Berggren L., Beeney J.T., Berggren P.O. and Meister B. (1994) Identification of synaptic proteins and their isoform mRNA in compartments of pancreatic endocrine cells. *Proc Natl Acad Sci USA* **91**: 12487–12491.
117. Brennwald P., Kearns B., Champion K., Keranen S., Bankaitis V. and Novick P. (1994) Sec9 is a SNAP-25-like component of a yeast SNARE complex that may be the effector of Sec4 function in excytosis. *Cell* **79**: 245–258.
118. Walch-Solimena C., Blasi J., Edelmann L., Chapman E.R., von Mollard G.F. and Jahn R. (1995) The t-SNAREs syntaxin 1 and SNAP-25 are present on organelles that participate in synaptic vesicle recycling. *J Cell Biol* **128**: 637–645.
119. Bennett, M.K., Garcia-Arras J.E., Elferink L.A., Peterson K., Fleming A.M., Hazuka C.D. and Scheller R.H. (1993) The syntaxin family of vesicular transport receptors. *Cell* **74**: 863–873.
120. Bassham D.C., Gal S., da Silva Conceicao A. and Raikhel N.V. (1995) An *Arabidopsis* syntaxin homologue isolated by functional complementation of a yeast pep12 mutant. *Proc Natl Acad Sci USA* **92**: 7262–7266.
121. de Paiva A., Ashton A.C., Foran P., Schiavo G., Montecucco C. and Dolly J.O. (1993) Botulinum A like type B and tetanus toxin fulfils criteria for being a zinc-dependent protease. *J Neurochem* **61**: 2338–2341.
122. Deshpande S.S., Sheridan R.E. and Adler M. (1995) A study of zinc-dependent metallo-endopeptidase inhibitors as pharmacological antagonists in botulinum neurtoxin poisoning. *Toxicon* **33**: 551–557.
123. Link E., Edelmann L., Chou J.H., Binz T., Yamasaki S., Eisel U., Baumert M., Südhof T.C., Niemann H. and Jahn R. (1992) Tetanus toxin action: inhibition of neurotransmitter release linked to synaptobrevin. *Biochem Biophys Res Commun* **189**: 1017–1023.
124. Papini E., Rossetto O. and Cutler D. (1995) Vesicle-associated membrane protein (VAMP/synaptobrevin-2) is associated with dense core secretory granules in PC12 neuroendocryne cells. *J Biol Chem* **270**: 1332–1336.
125. Poulain B., Rossetto O., Deloye F., Schiavo G., Tauc L. and Montecucco C. (1993) Antibodies against rat brain vesicle-associated membrane protein (synaptobrevin) prevent inhibition of acetylcholine release by tetanus toxin of botulinum neurotoxin type B. *J Neurochem* **61**: 1175–1178.
126. Bartels F., Bergel H., Bigalke H., Frevert J., Halpern J. and Middlebrook J. (1994) Specific antibodies against the Zn-binding domain of clostridial neurotoxins restore exocytosis in chromaffin cells treated with tetanus or botulinum A neurotoxin. *J Biol Chem* **269**: 8122–8127.
127. Zhou L., de Paiva A., Liu D., Aoki R. and Dolly J.O. (1995) Expression and purification of the light chain of botulinum neurotoxin A: a single mutation abolishes its cleavage of SNAP-25 and neurotoxicity after reconstitution with the heavy chain. *Biochemistry* **21**: 15175–15181.
128. Shone C.C., Quinn C.P., Wait R., Hallis B., Fooks S.G. and Hambleton P. (1993) Proteolytic cleavage of synthetic fragments of vesicle-associated membrane protein, isoform-2 by botulinum type B neurotoxin. *Eur J Biochem* **217**: 965–971.
129. Shone C.C. and Roberts A.K. (1994) Peptide substrate specificity and properties of the zinc-endopeptidase activity of botulinum type B neurotoxin. *Eur J Biochem* **225**: 263–270.
130. Cornille F., Goudreau N., Ficheux D., Niemann H. and Roques B.P. (1994) Solid-phase synthesis, conformational analysis and in vitro cleavage of synthetic human synaptobrevin II 1–93 by tetanus toxin L chain. *Eur J Biochem* **222**: 173–181.
131. Foran P., Shone C.C. and Dolly J.O. (1995) Differences in the protease activities of tetanus and botulinum B toxins revealed by the cleavage of vesicle-associated membrane protein and various sized fragments. *Biochemistry* **33**: 15365–15374.
132. Rossetto O., Schiavo G., Montecucco C., Poulain B., Deloye F., Lozzi L. and Shone C.C. (1994) SNARE motif and neurotoxin recognition. *Nature* **372**: 415–416.
133. Pellizzari R., Rossetto O., Lozzi L., Gioved-EC S., Johnson E., Shone C.C. and Montecucco C. (1996)

Structural determinants of the specificity for synaptic vesicle-associated membrane-protein/ synaptobrevin of tetanus and botulinum type B and G neurotoxins. *J Biol Chem* **271**: 2035–2038.
134. Grote E., Hao J.C., Bennett M.K. and Kelly R.B. A targeting signal in VAMP regulating transport to synaptic vesicles. *Cell* **81**: 581–589.
135. Söllner T., Whiteheart S.W., Brunner M., Erdjument-Bromage H., Geromanos S., Tempst P. and Rothman J.E. (1993) SNAP receptors implicated in vesicle targeting and fusion. *Nature* **362**: 318–324.
136. Söllner T., Bennett M., Whiteheart S.W., Scheller R.H. and Rothman J.E. (1993) A protein assembly-disassembly pathway in vitro that may correspond to sequential steps of synaptic vesicle docking, activation and fusion. *Cell* **75**: 409–418.
137. Rothman J.E. (1994) Mechanisms of intracellular protein transport. *Nature* **372**: 55–63.
138. Schiavo G., Gmachl M.J., Stenbeck G., Söllner T.H. and Rothman J.E. (1995) A possible docking and fusion particle for synaptic transmission. *Nature* **378**: 733–736.
139. Söllner T.H. (1995) SNAREs and targeted membrane fusion. *FEBS Lett* **369**: 80–83.
140. Chapman E.R., An S., Barton N. and Jahn R. (1994) SNAP-25, a t-SNARE which binds to both syntaxin and synaptobrevin via domains that may form coiled coils. *J Biol Chem* **269**: 27427–27432.
141. Hayashi T., McMahon H., Yamasaki S., Binz T., Hata Y., Südhof T.C. and Niemann H. (1994) Synaptic vesicles membrane fusion complex: action of clostridial neurotoxins on assembly. *EMBO J* **13**: 5051–5061.
142. Hayashi T., Yamasaki S., Nauenburg S., Binz T. and Niemann H. (1995) Disassembly of the reconstituted synaptic vesicle membrane fusion complex *in vitro*. *EMBO J* **14**: 2317–2325.
143. Pellegrini L.L., O'Connor V., Lottspeich F. and Betz H. (1995) Clostridial neurotoxins compromise the stability of a low energy SNARE complex mediating NSF activation of synaptic vesicle fusion. *EMBO J* **14**: 4705–4713.
144. Burgoyne R.D. and Morgan A. (1993) Regulated exocytosis. *Biochem J* **293**: 305–316.
145. Monk J.R. and Fernandez J.M. (1994) The exocytotic fusion pore and neurotransmitter release. *Neuron* **12**: 707–716.
146. Broadie K., Prokop A., Bellen H.J., O'Kane C.J., Schulze K.L. and Sweeney S.T. (1995) Syntaxin and synaptobrevin function downstream of vesicle docking in *Drosophila*. *Neuron* **15**: 663–673.
147. Ashton A.C., Li Y., Doussau F., Weller U., Dougan G., Poulain B. and Dolly J.O. (1995) Tetanus toxin inhibits neuroexocytosis even when its Zn^{2+}-dependent protease activity is removed. *J Biol Chem* **270**: 31386–31390.
148. Wetterholm A., Medina J.F., Radmark O., Shapiro R., Haeggstrom J.Z., Vallee B.L. and Samuelsson B. (1992) Leukotriene A4 hydrolase: abrogation of the peptidase activity by mutation of glutamic acid-296. *Proc Natl Acad Sci USA* **89**: 9141–9145.
149. Facchiano F. and Luini A. (1992) Tetanus toxin potently stimulates tissue transglutaminase. A possible mechanism of neurotoxicity. *J Biol Chem* **267**: 13267–13271.
150. Facchiano F., Benfenati F., Valtorta F. and Luini A. (1993) Covalent modification of synapsin I by a tetanus toxin-activated transglutaminase. *J Biol Chem* **268**: 4588–4591.
151. Facchiano F., Di Tullio M. and Luini A. (1994) Evidence that transglutiminase and synapsin I are involved in the neuroparalytic action of tetanus toxin. *Ann NY Acad Sci* **710**: 107–119.
152. Facchiano F., Valtorta F., Benfenati F. and Luini A. (1993) The transglutaminase hypothesis for the action of tetanus toxin. *Trends Biochem Sci* **18**: 327–329.
153. Coffield J.A., Considine R.V., Jeyapaul J., Maksymowych A.B., Zhang R.D. and Simpson L.L. (1994) The role of transglutaminase in the mechanism of action of tetanus toxin. *J Biol Chem* **269**: 24454–24458.
154. Ray P., Berman J.D., Middleton W. and Brendle J. (1993) Botulinum toxin inhibits arachidonic acid release associated with acetylcholine release from PC12 cells. *J Biol Chem* **268**: 11057–11064.

G: Enterotoxins

Chapter 19

The *Clostridium perfringens* Enterotoxin

JOHN F. KOKAI-KUN AND BRUCE A. McCLANE
E1240 Biomedical Science Tower, Department of Molecular Genetics and Biochemistry, University of Pittsburgh School of Medicine, Pittsburgh, PA 15261, USA

I. INTRODUCTION

One of the most common and important human gastrointestinal illnesses caused by *Clostridium perfringens* is *C. perfringens* type A food poisoning (see Chapter 8).[1–3] In the USA and Europe, this food poisoning accounts for hundreds of thousands of cases and millions of dollars in economic costs each year.[4–6] Diarrheal and cramping symptoms associated with *C. perfringens* type A food poisoning are believed to be caused by a protein toxin, named *C. perfringens* enterotoxin (CPE).[1] Evidence for the involvement of CPE in *C. perfringens* type A food poisoning includes: (1) 80–100% of feces from individuals ill with *C. perfringens* type A food poisoning contain CPE,[7,8] while feces from healthy humans rarely, or never, contain detectable levels of this toxin;[9] (2) CPE is often present in the feces of these *C. perfringens* type A poisoning victims at levels[2,7,8] known to cause serious gastrointestinal effects in animal models;[10] (3) human volunteers fed highly purified CPE develop all symptoms typically associated with *C. perfringens* type A food poisoning;[11] (4) CPE-positive *C. perfringens* isolates[12] produce significantly more fluid accumulation in rabbit ileal loops or diarrhea in human volunteers than do CPE-negative *C. perfringens* isolates[13] and

The Clostridia: Molecular Biology and Pathogenesis
ISBN 0-12-595020-9

(5) antisera raised against CPE neutralizes the ability of CPE-positive *C. perfringens* to induce rabbit ileal loop fluid accumulation.[14]

Clostridium perfringens type A food poisoning develops when an individual consumes food contaminated with large numbers of vegetative cells of enterotoxigenic *C. perfringens* (see Chapter 8).[1] Many of these bacteria die when exposed to the acidity of the stomach.[15] However, if the ingested food was sufficiently contaminated, some *C. perfringens* cells survive and passage from the stomach into the intestine, where they multiply and then sporulate. It is during sporulation in the intestine that CPE is produced (see Chapter 27).[1–3] Therefore, since *C. perfringens* type A food poisoning involves *in vivo* toxin production, this disease should be considered an infection rather than a classical clostridial intoxication (such as food-borne botulism).[16] CPE, which accumulates intracellularly in sporulating *C. perfringens* cells, is released into the intestine at the completion of sporulation, as the sporulating cells lyse to release their endospore.[1]

Animal studies[3] indicate that the small intestine is probably the site of action of this released CPE (see Section III.A); it is thought[3] that after its release into the lumen of the small intestine, enterotoxin rapidly binds to the brush border membranes of small intestinal epithelial cells, where it exerts its characteristic cytotoxic action (described in Section IV). This action, in turn, is believed[17] to lead to the development of the diarrhea and cramping symptoms characteristic of *C. perfringens* type A food poisoning. Because *C. perfringens* cells must complete sporulation before any CPE can be released into the intestine and this sporulation process typically takes 8–15 h, the first symptoms of *C. perfringens* type A food poisoning do not typically start until around 12 h after ingestion of contaminated foods.[1–3] These symptoms persist for 12–24 h, and then spontaneously resolve (see Chapter 8).[1–3] Somewhat contrary to its mild reputation, *C. perfringens* type A food poisoning ranks as the fifth leading cause of death from food poisoning in the USA, with most deaths occurring in the elderly or in debilitated individuals.[1,4–6]

In addition to its traditional association with *C. perfringens* type A food poisoning, CPE has now also been linked to a number of other human and veterinary gastrointestinal (GI) illnesses (Table 19.1). While the pathogenesis of many of these CPE-associated GI illnesses remains poorly understood, it is noteworthy that they often have a longer duration and/or involve more serious symptoms (e.g., more severe diarrhea) than are typical of most cases of the more common *C. perfringens* type A food poisoning.[1,18–20] It is interesting that CPE has also been recently suggested to be involved in some cases of sudden infant death syndrome (SIDS).[21,22]

Table 19.1. Human and veterinary diseases linked to CPE

Clostridium perfringens type A food poisoning of humans
Antibiotic-associated diarrhea of humans
Chronic non-foodborne diarrhea
Acute non-foodborne diarrhea
Sudden infant death syndrome
Veterinary diarrhea of pigs, dogs, llamas, etc.

See McClane[17] for original citations.

II. BIOCHEMICAL ATTRIBUTES OF THE CPE MOLECULE

CPE was first purified and characterized in the early 1970s;[23] additional information about this protein was later supplied by amino acid and nucleotide sequencing studies.[24–28] Collectively, these studies indicate that the CPE made by *C. perfringens* NCTC 8239 is a single, 319 amino acid polypeptide with a molecular weight of 35 317 Da[24] and an isoelectric point of 4.3.[1–3] Nucleotide sequence analysis indicates that CPEs made by different *C. perfringens* strains share extensive amino acid homology, but may also have some limited microsequence variations.[24,28,29] CPE does not share sequence homology with any known proteins,[24] except for Antp 70/C1, one of the nonneurotoxic proteins found in the botulinum C1 neurotoxin gene complex.[30] The significance, if any, of the limited homology between CPE and the Antp 70/C1 protein is unclear (see Chapter 17).

Physical studies suggest that the secondary structure of CPE consists of approximately 80% β-pleated sheet and 20% random coil.[17] The tertiary structure of CPE currently remains undetermined owing to difficulties in crystallizing the toxin for X-ray diffraction analysis.

CPE's biological activity is both heat-labile, i.e., heating the toxin to 60°C for 5 min causes an irreversible inactivation[17] and pH labile, i.e., CPE's biological activity is lost at pH <5 or >10.[17] However, this toxin's activity is resistant to treatment with some proteases; in fact, limited treatment of the purified enterotoxin with trypsin or chymotrypsin causes a two- to three-fold activation in CPE biological activity.[31,32] This activation effect has prompted suggestions that intestinal proteases may activate CPE during food poisoning.[31]

III. *IN VIVO* EFFECTS OF *C. PERFRINGENS* ENTEROTOXIN

A. CPE effects on the gastrointestinal system

The diarrheal symptoms associated with *C. perfringens* type A food poisoning (and CPE-linked nonfoodborne diarrhea) provided the first indication that CPE was active on the GI tract.[1–3] Subsequent studies with purified CPE have shown that this enterotoxin causes a rapid fluid and electrolyte loss from the GI tract in several different animal models.[1–3,33–35] These effects can develop within as little as 15–30 min of exposure to CPE.[33,34] All regions of the small intestine respond to CPE treatment, although the ileum is particularly CPE sensitive.[3,35] Interestingly, while rabbit colonic cells bind CPE at high levels, the rabbit colon does not show significant effects when challenged with enterotoxin;[36] it remains unclear whether CPE affects the human large intestine.

The intestinal effects of CPE in animal models may be distinguished from those of cholera enterotoxin in several respects. First, unlike cholera toxin, CPE does not affect small intestinal cAMP levels,[3] but does inhibit small intestinal glucose absorption.[3] Second, while treatment with cholera toxin does not cause visible tissue damage to the gastrointestinal tract,[37] CPE produces extensive histopathological damage to the small intestine,[3,10,33,34] i.e., unlike cholera toxin, CPE belongs to the class of 'cytotoxic' enterotoxins. CPE can be distinguished from other cytotoxic enterotoxins, such as the *Clostridium difficile* toxins, on the basis of how quickly CPE-induced intestinal tissue damage develops; in animal models, histopathological damage is detectable within as little as 15–30 minutes of treatment with purified CPE,[34] while damage from other cytotoxic enterotoxins typically develops more slowly in these models.[38]

In animal models, the initial CPE-induced histopathological damage is limited to the villi tips of the small intestine, but with time the entire small intestinal villus of CPE-treated animals eventually suffers extensive desquamation.[10,33,34] Strong circumstantial evidence suggests that this histopathological damage plays an important role in producing the fluid and electrolyte losses that are associated with CPE treatment of the small intestine.[34] For example, the onsets of both fluid and electrolyte transport alterations have been shown to closely coincide with the development of histopathological damage in CPE-treated rabbit ileal loops.[34] Further, only those CPE doses able to cause histopathologic damage in rabbit ileal loops can also produce fluid and electrolyte loss in this model.[10]

B. CPE lethality

In addition to these intestinal effects, highly purified CPE rapidly kills animals when administered intravenously (the LD_{50} of CPE in mice is 50 μg/kg).[39] The lethal action of CPE may now be assuming clinical relevance, given recent reports of detection of CPE in the serum of some infants dying from SIDS.[21,22]

How does CPE kill when it becomes systemically distributed? Animal studies have identified at least three pathophysiologic mechanisms that could potentially be involved in lethal effect; (1) CPE may cause hyperpotassemia through its effects on the liver – this hyperpotassemia may, in turn, lead to sudden cardiac failure;[39] (2) CPE may induce an endotoxin-like shock response (possibly due to CPE effects on immune cells), triggering cardiorespiratory failure,[40] and/or (3) CPE may interfere with neuromuscular transmission by inducing increased exocytosis in neurosecretory cells.[41,42] *In vivo*, these different mechanisms are likely to act synergistically, i.e., when CPE becomes systemically distributed there may not be a single lethal effect of this toxin.

IV. CPE CYTOTOXICITY

A. An overview

Detailed studies conducted over the past 15–20 years indicate that the action of CPE on the small intestine appears to involve the overall sequence of events shown in Figure 19.1. Because CPE-induced histopathologic damage to the small intestine is believed to result directly from the lethal effects of CPE on mammalian cells,[17] considerable effort has been devoted recently to understanding the cytotoxic action of CPE at the cellular and molecular levels. As will be elaborated (see sections IV.B.4 and IV.C), results from these studies indicate that CPE should be considered a membrane-active toxin whose primary cytotoxic effect is to induce small molecule membrane permeability alterations in mammalian cells. These small molecule permeability changes collapse the colloid–osmotic equilibrium normally present in mammalian cells,[17] setting off a cascade of events that includes inhibition of macromolecular synthesis, morphologic damage and cell lysis.[17] Thus, mammalian cells should be considered nonviable after the onset of CPE-induced small molecule permeability changes (i.e., after the fourth event shown on the Figure 19.1 time-line),[17] because there is no evidence, to date, that mammalian cells are capable of reversing CPE-induced small molecule permeability changes.[17] Continuing this explanation, the events shown in the Figure 19.1 timeline occurring after the induction of small molecule membrane permeability alterations appear to be secondary consequences of CPE treatment. It is also worth mentioning that it was once thought that CPE cytotoxicity might require the presence of extracellular Ca^{2+}, however, as shown in Figure 19.1, it is now clear[17] that extracellular Ca^{2+} is only required for the development of some secondary CPE-associated events, i.e., CPE-sensitive mammalian cells are irreversibly damaged whether treated with enterotoxin in Ca^{2+}-containing or Ca^{2+}-free media because the development of small molecule permeability alterations (CPE's primary cytotoxic effect) is a Ca^{2+}-independent process.[17,46]

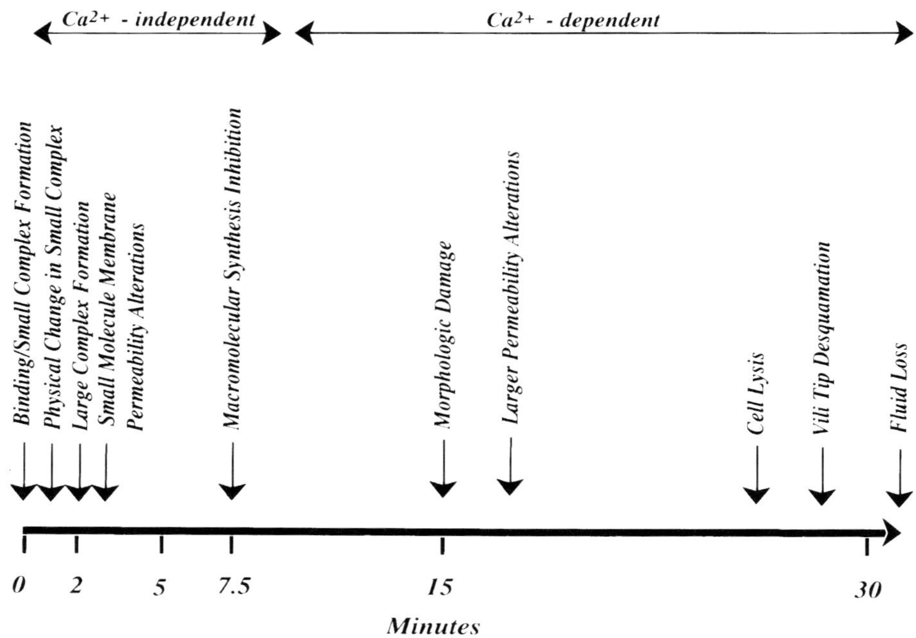

Figure 19.1 Time line of events in CPE action. This figure graphically depicts the onset and sequence of known events in CPE action, up to the induction of pathophysiological intestinal effects. CPE is added at time zero in this figure. See text for discussion of these events.

Placing CPE-induced cytotoxicity in its most common pathophysiologic context, it appears that as intestinal epithelial cells die during food poisoning (and, presumably, during other CPE-linked G.I. diseases) from the consequences of CPE-induced small molecule membrane permeability alterations, these dead cells slough from the villi of the small intestine. This desquamation produces histopathological damage (as mentioned in Section III.A) that disrupts the integrity of the small intestinal villus,[33,34] thereby leading to the fluid and electrolyte losses from the CPE-treated small intestine[10,33,34] that are clinically manifested as the diarrheal symptoms associated with *C. perfringens* type A food poisoning.[1–3]

B. Early events

Because of the apparent importance of early events in CPE-induced cytotoxicity, each early event will be discussed individually, with particular emphasis placed on recent findings.

1. Binding

CPE binds rapidly to many, but not all, cell types from a number of mammalian species (discussion below).[17,43–45] CPE binding is specific (the binding of ^{125}I-CPE can be significantly reduced by the presence of excess unlabeled CPE) and saturable (with estimates that around 10^6 CPE molecules bind per mammalian cell), fulfilling the criteria expected of a receptor-mediated process.[17,44,46] Several lines of evidence indicate that the specific binding of CPE to its receptor(s) is an essential first step for initiating CPE-induced cytotoxicity. For example, cells that do not specifically bind CPE are completely insensitive to CPE-induced cytotoxicity (see discussion below).[43,45,47] A monoclonal antibody that blocks the specific binding of CPE to its receptor on CPE-sensitive cells also completely abrogates CPE-induced cytotoxicity.[48] Finally, a CPE deletion mutant that has lost the ability to bind specifically to the CPE receptor is completely noncytotoxic for CPE-sensitive mammalian cells (see section V).[49]

What is the mammalian membrane receptor that mediates CPE-specific binding? Because protease pretreatment of mammalian cells or isolated intestinal brush border membranes (BBMs) destroys their subsequent ability to specifically bind CPE,[44,46] the CPE receptor must be at least partially proteinacious. In the mid-1980s, affinity chromatography experiments[50,51] using immobilized CPE demonstrated that two different rabbit BBM proteins are able to bind to CPE. When BBMs were extracted with the detergent NP-40 prior to chromatography, a 50 kDa BBM protein bound to the immobilized CPE column,[50] while a 70 kDa BBM protein bound to CPE affinity columns if the detergent CHAPS was used.[51] For a number of years it remained unclear whether these results were indicating the existence of two kinds of CPE receptors, or, instead, were indicating a more complicated situation, where only one of the two CPE-binding proteins is actually the CPE receptor and the second protein interacts with CPE after the toxin is already bound to its receptor.

Recent studies now appear to have resolved this confusion.[45] When Triton X-100 extracts of intact cells or isolated BBMs treated with CPE at 4°C were analyzed using a nondenaturing electrophoresis system, a single *c.* 90 kDa species containing specifically bound CPE was detected (Figure 19.2).[45] This CPE-containing species, now called 'small complex', forms in membranes (or cell extracts) of several cell types (including, intestinal epithelial cells, Vero

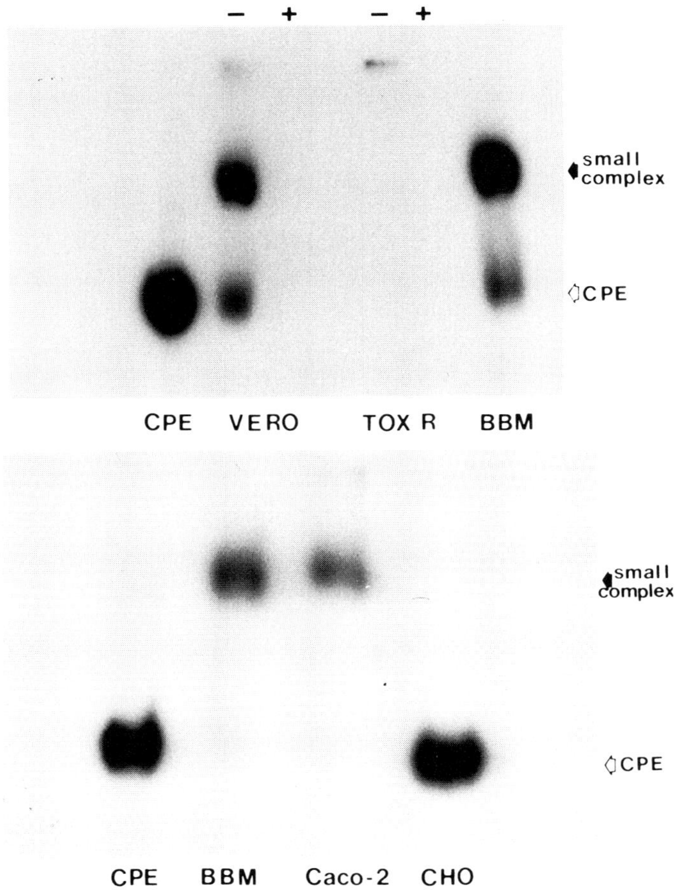

Figure 19.2 CPE small complex formation correlates with CPE binding and sensitivity. Top panel. CPE, [^{125}I]CPE alone; Vero, [^{125}I]CPE incubated with intact Vero cells at 4°C; ToxR, [^{125}I] CPE incubated with intact ToxR cells at 4°C, and BBMs, isolated intestinal brush border membranes incubated with [^{125}I]CPE at 4°C; + denotes incubation in the presence of 50-fold excess unlabeled CPE, while – has no excess CPE added. After incubation with [^{125}I]CPE, cells or membranes containing bound [^{125}I]CPE were treated with Triton X-100 and the resultant Triton X-100 extracts were subjected to nondenaturing electrophoresis.[45]

Bottom panel. CPE, [^{125}I]CPE incubated in the presence of Triton X-100 alone; BBMs, Triton X-100 extracts of BBMs incubated with [^{125}I]CPE; Caco2, Triton X-100 extracts of Caco2 cells incubated with [^{125}I]CPE; and CHO, Triton X-100 extracts of CHO cells incubated with [^{125}I]CPE. After incubation, extracts were subjected to nondenaturing electrophoresis. Caco2 and Vero cells are CPE-sensitive cells that bind CPE at high levels, while ToxR and CHO cells are CPE-insensitive cells that do not bind CPE.[45] BBMs are isolated from CPE-sensitive small intestinal epithelial cells that bind high levels of CPE.[44] Reprinted with permission of the American Society for Biochemistry and Molecular Biology from Wieckowski *et al.* (1994).[45]

cells and CaCO-2 cells; Figure 19.2) that can specifically bind large amounts of CPE and are also highly CPE-sensitive.[43,45] However, this small complex is absent from membranes of cells (or cell extracts) such as CHO cells and ToxR cells that cannot specifically bind significant amounts of CPE and are, therefore, CPE-insensitive (Figure 19.2).[45] This correlation strongly suggests that small complex represents CPE bound to its 'functional' receptor, i.e., to the receptor that conveys the cytotoxic response to CPE challenge.

When the composition of small complex was investigated using immunoprecipitation analysis of cells treated with CPE at 4°C, evidence was obtained that small complex represents CPE associated with a 50 kDa CPE-binding membrane protein, i.e., the 90 kDa small complex appears to contain one 35 kDa CPE molecule and one 50 kDa CPE-binding protein.[45] Thus, considerable evidence now points to a 50 kDa protein as the major (if not only) functional CPE receptor (the role of the 70 kDa CPE-binding protein identified by affinity chromatography is discussed in section IV.B.3). A major challenge for the future is to identify the normal physiological role of this 50 kDa receptor protein. For example, might this CPE receptor also function as a receptor for other ligands, such as hormones? The widespread distribution of the 50 kDa protein among widely divergent cell types may suggest that this protein plays an important normal physiological role for many mammalian cells;[45] however, the 50 kDa CPE receptor cannot be an essential molecule for all mammalian cells because there are mammalian cells (e.g., CHO cells) that do not bind the enterotoxin.[45]

While the specific binding of CPE to its receptor is the essential first step for CPE-induced cytotoxicity, several lines of studies have clearly established that mere occupancy of the CPE receptor is not sufficient to trigger a cytotoxic response. For example, several CPE mutants are able to specifically bind to the CPE receptor, yet are completely noncytotoxic (see section V).[25,49] Further, while rabbit colonic cells can specifically bind high levels of CPE, the rabbit colon is not affected by this toxin.[36] Collectively, these results indicate that CPE-induced cytotoxicity must be a multistep process and that the post-binding early events described below are also required for the enterotoxin's action.

2. *A post-binding physical change in CPE small complex*

The second early step in CPE action appears to involve a post-binding physical change to CPE-containing small complex. When CPE first binds to a mammalian cell, the toxin can be easily released by treating that cell with Pronase® or trypsin.[44,46] However, with increasing time after binding, specifically bound CPE becomes progressively more resistant to protease-induced release.[46,52,53] The progressive development of this resistance argues that this effect cannot be explained simply as CPE binding to its receptor to form a protease-resistant structure, but instead implies that some type of physical change must occur after CPE becomes bound to its receptor, i.e., after CPE becomes sequestered in small complex.[17] This post-binding physical change to CPE small complex appears to be an essential step in CPE action, since CPE deletion mutants that are specifically blocked for their ability to perform this physical change are also not cytotoxic (see section V).[49,54]

What type of physical change to small complex could explain this protease-resistance phenomenon? While internalization of CPE into the cytoplasm could theoretically lead to a progressively developing protease resistance for bound CPE, internalization cannot be the actual explanation for this physical change in small complex because (1) the same protease resistance phenomenon is detectable using isolated BBMs[44,46,53], as well as intact cells[46,53] and (2) other studies have clearly established that CPE remains localized to the plasma membrane of mammalian cells throughout its action.[55] After eliminating internalization, at least two other physical processes remain as viable explanations for the post-binding physical change to small complex. First, after binding, CPE (possibly together with its 50 kDa receptor) could mimic many other membrane-active toxins (e.g., staphylococcal α-toxin)[56] and insert itself into the lipid bilayer of plasma membranes, where it would be protected from external proteases. Some evidence indirectly supports the possibility of CPE insertion into membranes: (1) specifically bound CPE does not dissociate from intact cells or isolated membranes,[44,46] an effect that could be explained by CPE becoming entrapped within the lipid bilayer after binding; (2) kinetic analysis of CPE specific binding indicates a biphasic pattern of association that would be consistent with CPE action involving a specific binding step that is rapidly followed by a second step, such as insertion,[44,46] and (3) specifically bound CPE is not released from membranes by treatment with EDTA or chaotropic salts,[44,46] substances that are known to release peripherally bound membrane proteins,[56–58] but not inserted protein toxins.[56–58]

Recent studies have further evaluated whether CPE action might involve insertion by asking whether this enterotoxin can exhibit the amphiphilic characteristics expected of an insertion-capable toxin.[45] Using Triton X-114 phase-separation analysis (see Figure 19.3) and charge shift analysis as assays for amphiphilic behavior, it was shown that purified CPE, when alone in solution, displays only hydrophilic properties, even when characterized over a broad pH range (pH 3, 5, 7, or 9).[45] However, these studies showed that when CPE binds to the 50 kDa protein to form small complex, this small complex does exhibit the amphiphilic characteristics considered necessary for protein insertion into lipid bilayers.[45] Upon disruption of small complex, however, CPE reverts to its hydrophilic nature,[45] thus indicating that CPE amphiphilicity does not result from covalent modifications when the toxin is exposed to membranes, implying

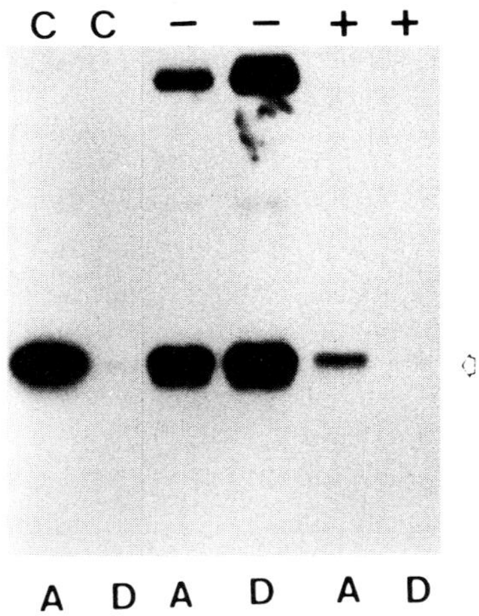

Figure 19.3 CPE small complex demonstrates amphiphilic behavior following Triton X-114 phase separation. Samples include: C lanes, [^{125}I]CPE alone; – lanes, BBMs incubated with [^{125}I]CPE at 4°C prior to extraction with Triton X-114 and phase separation; and + lanes, which are the same as – lanes except samples were treated with [^{125}I]CPE in the presence of a 50-fold excess of unlabeled CPE prior to extraction with Triton X-114 phase separation analysis. A lanes, show SDS–PAGE analysis of a sample from the aqueous phase after Triton X-114 phase separation while D lanes show SDS–PAGE analysis of a sample from the detergent phase after Triton X-114 phase separation. The higher molecular weight species on this gel is CPE in small complex, while the open arrow shows where free CPE migrates in this gel system. Reprinted with permission of the American Society for Biochemistry and Molecular Biology from Wieckowski *et al.* (1994).[45]

that CPE only acquires amphiphilic characteristics when it is sequestered in small complex. To our knowledge, this represents the first report of a membrane-active toxin acquiring amphiphilic characteristics as a result of its association with a mammalian protein; other insertion-capable, membrane-active toxins appear to develop amphiphilic behavior spontaneously.[56,57,59]

While there is now a considerable amount of circumstantial evidence consistent with CPE 'insertion' into membranes, this evidence does not yet rule out the alternative explanation for the post-binding physical change to CPE small complex, i.e., the possibility that small complex undergoes a progressively-developing, post-binding conformational change that results in the development of a protease-resistant species. One approach to help distinguish between the 'insertion' versus 'conformational change' hypothesis is to determine the topology of CPE in membranes. In initial studies aimed at addressing the topology of CPE in membranes,[52] we have recently shown that ^{125}I-labeled anti-CPE antibodies are able to specifically react with intact BBMs containing CPE small complex.[52] It is noteworthy that the CPE-treated BBMs used in these studies were allowed sufficient time for their small complex to complete the 'post-binding physical change' before any antibody probes were added. Immunoprecipitation studies have confirmed that the CPE antibodies react specifically with the small complex present in these BBMs (Figure 19.4).[52]

The antibody probe studies described above indicate that at least a portion of the CPE molecule sequestered in the small complex must remain surface-exposed, even after the completion of the post-binding physical change. While, this result by itself does not necessarily exclude the possibility of CPE insertion (it is possible that, after the development of the physical change to the small complex, as little as one epitope on a small portion of the toxin molecule remains surface exposed), by demonstrating that the entire CPE molecule cannot be inserted into the lipid bilayer these studies do increase interest in the conformational change hypothesis as a possible explanation for the post-binding physical change to small complex. It should be mentioned that virtually all the evidence consistent with CPE 'insertion' appears equally applicable to explain the conformational change hypothesis. For example, if a conformational change to the

---

Figure 19.4 (opposite) CPE sequestered in small complex and large complex can be immunoprecipitated by anti-CPE antibodies applied to intact BBMs containing these species. Top panel. Triton X-100 native gel electrophoretic analysis of immunoprecipitation supernatants from BBMs containing only small complex. [^{125}I]CPE, in the presence (+ lanes) or absence (− lanes) of 50-fold excess unlabeled CPE, was bound and allowed to undergo the post-binding physical change in BBMs at 4°C in order to form CPE small complex (note that large complex does not form at 4°C, see Figure 19.5). These intact BBMs were then treated with IgG purified from either rabbit polyclonal anti-CPE serum (RPC-IgG) or normal rabbit serum (NRS-IgG) and then extracted with 1% Triton X-100. After incubation of extracts with protein A beads, samples were microcentrifuged to remove immunoprecipitated materials and 40 μl of each remaining supernatant was analyzed by Triton X-100 electrophoresis to identify CPE small complex.[45] CPE, free [^{125}I]CPE (i.e., no BBMs) in PBS containing 1% Triton X-100 was included as a control in all experimental sets described below. Samples shown are; Control, a set of microcentrifugation supernatants from [^{125}I]CPE-treated BBMs samples that were extracted with 1% Triton X-100, and then directly microcentrifuged; +RPC-IgG and +NRS-IgG, microcentrifugation supernatants from two sets of [^{125}I]CPE-treated BBMs samples that were incubated with RPC-IgG or NRS-IgG, respectively, prior to Triton X-100 extraction, incubation with protein A beads and microcentrifugation. See Kokai-Kun and McClane (1996)[52] for a discussion of the other control lanes. The top arrow depicts the migration of small complex on this gel system, while the bottom arrow indicates the location of free [^{125}I]CPE.

Bottom panel. SDS–PAGE analysis of immunoprecipitation supernatants from BBMs containing large complex. [^{125}I]CPE was added to BBMs in the presence (+ lanes) or absence (− lanes) of 50-fold excess unlabeled CPE at 22°C in order to allow the radiolabeled toxin to specifically bind, undergo the post-binding physical change and then form CPE large complex. Samples are the same on the bottom panel as described on the top panel. However, these samples are analyzed by SDS–PAGE (without sample boiling) to identify CPE large complex.[45] The top arrow depicts the location of large complex on these gels, while the bottom arrow indicates the migration of free [^{125}I]CPE (note that small complex is not visible using this gel system because small complex dissociates in the presence of SDS). Reprinted with permission of the American Society for Microbiology from Kokai-Kun and McClane (1996).[52]

small complex 'traps' CPE on the cell surface, this would explain why CPE does not dissociate after binding, is not released by EDTA or chaotropic salts, and shows a two-step kinetics profile (i.e., in this case, binding would rapidly be followed by a conformational change).

3. Large complex formation

Under physiological conditions, 'physically changed' small complex is thought[45] to rapidly interact with a 70 kDa membrane protein to form a CPE-containing complex of *c.* 160 kDa (see Figure 19.5).[51] Presumably, the 70 kDa protein

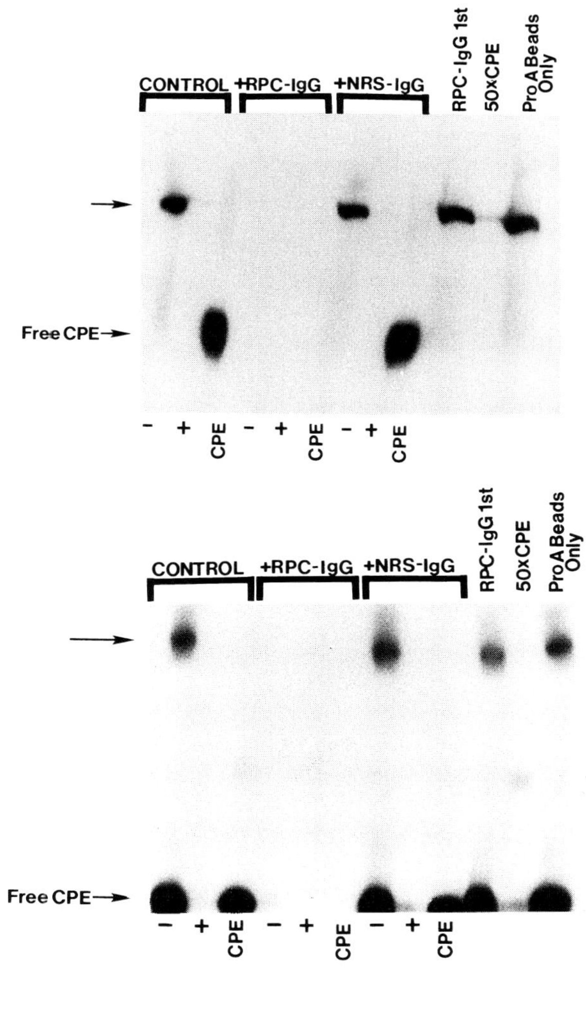

Figure 19.4

associated with this 'large' CPE complex is the same 70 kDa protein that was observed in earlier studies (see section IV.B.1) to bind to CPE affinity columns.[51] If so, this would imply that CPE directly interacts with one 50 kDa and one 70 kDa mammalian membrane protein when this toxin is sequestered in large complex. Large complex clearly represents a much more stable membrane species than small complex; for example, large complex remains intact in the presence of sodium dodecyl sulfate (SDS),[53] while small complex falls apart after exposure to SDS.[45] However, large complex can be at least partly dissociated by boiling.[51]

Large complex formation is blocked at low temperatures[53] (Figure 19.5). This observation suggests that large complex formation requires a fluid lipid bilayer (as found in membranes held at higher temperatures), possibly indicating that large complex formation involves the diffusion of small complex and/or the 70 kDa protein through membranes. Large complex formation appears to be an important step in the pathophysiologic pathway leading to CPE-induced cytotoxicity since, at 4°C, CPE-treated cells are able to perform the first two steps (i.e., binding and the physical change to small complex) in the cytotoxic pathway, but are blocked for both large complex formation and the subsequent development of the small molecule membrane permeability alterations that represent CPEs primary cytotoxic effect (see section III.B.4).[53]

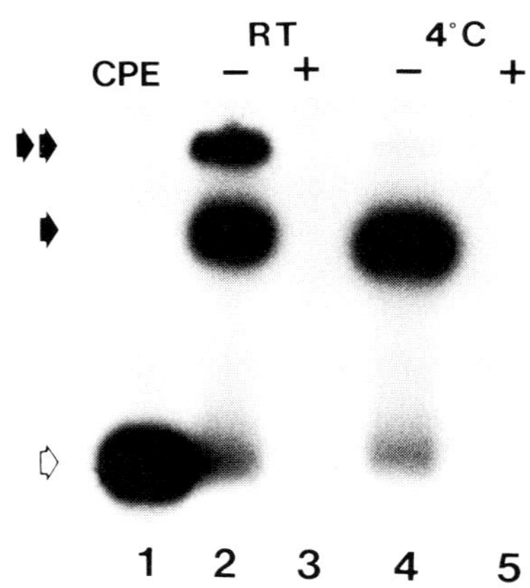

Figure 19.5 Large complex formation does not occur at low temperature. [^{125}I]CPE incubated with BBMs at 4°C or room temperature (RT) in the presence (+), or absence (–), of 50-fold excess unlabeled CPE, prior to extraction with Triton X-100 and analysis by nondenaturing electrophoresis. CPE represents free [^{125}I]CPE dissolved in Triton X-100 that was analyzed by nondenaturing electrophoresis. The double arrow depicts the migration of large complex. The single closed arrow depicts the migration of small complex, and the open arrow depicts the migration of free CPE in this gel system. Reprinted with permission of the American Society for Biochemistry and Molecular Biology from Wieckowski *et al.* (1994).[45]

When antibody probing experiments, similar to those described in the previous section, were conducted with BBMs containing large complex to help determine the topology of CPE when it is sequestered in large complex,[52] it was found that, (1) ^{125}I-labeled anti-CPE antibodies react specifically with BBMs containing large complex, and (2) anti-CPE antibodies can specifically immunoprecipitate CPE large complex (Figure 19.4). When these results are combined with similar topology studies of BBMs containing small complex, it is now apparent that at least a portion of the CPE molecule remains surface-exposed throughout most, if not all, of this toxin's cytotoxic action.

4. Small molecule membrane permeability alterations

When CPE-sensitive mammalian cells are treated with CPE, their plasma membranes rapidly develop massive permeability changes.[17,43,60,61] As shown in Figure 19.6, these permeability alterations are initially restricted to small molecules of less than 200 Da,[17,43,60–62] suggesting that the initial CPE lesion is 55–91 Å^2.[63] This initial CPE-induced membrane lesion can develop within as little as 5 min of CPE treatment time and is permeable to a wide range of small molecules, including, cations, anions and amino acids.[17,43,60,61]

What causes these CPE-induced small molecule permeability alterations to develop? Some insight into this important question was gained when cells were pretreated with CPE at 4°C and then rapidly warmed to 37°C. It was observed that, under these conditions, CPE-induced small molecule permeability alterations develop concurrently with the formation of large complex.[53] Since there would not appear to be sufficient time for extensive intermediary steps to occur between large complex formation and the onset of small molecule membrane permeability alterations under these experimental conditions, these results strongly suggest that large complex is directly responsible for the small molecule membrane permeability changes that occur in CPE-treated mammalian cells.

How could large complex formation disrupt normal membrane permeabiliby properties? This important question remains one of the major unanswered questions in CPE research. The simplest explanation would appear to be that large complex has pore-like properties, causing 'functional holes' to develop in CPE-treated membranes. However, there are other possible explanations for large complex-mediated disruption of membrane permeability. For example, large complex formation could affect the activation/inactivation of existing ion channels. Learning the identities and normal physiologic role(s) of the 50 kDa and 70 kDa CPE-binding proteins may provide further insights into the unique mechanism by which CPE disrupts normal plasma membrane function.

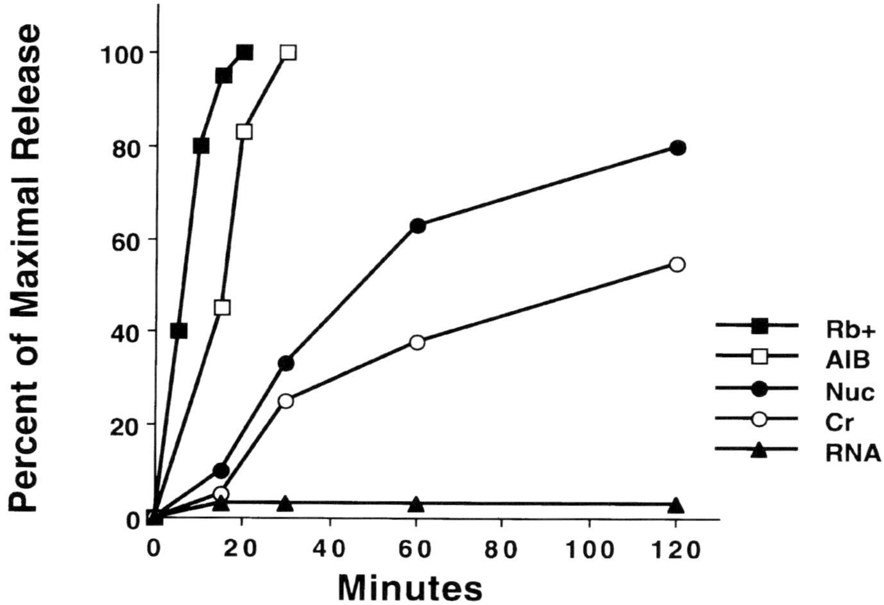

Figure 19.6 CPE induces small molecule membrane permeability alterations in affected cells. The figure represents cumulative results from several studies measuring CPE effects on Vero cell permeability using CPE-induced release of intracellular radioactive markers of defined sizes. Markers included; ^{86}Rb (Rb+, M_r ~100), [^{14}C]aminoisobutyric acid (AIB, M_r ~100), ^{3}H-nucleotides (Nuc, M_r <1000), ^{51}Cr (Cr, M_r ~3000) and ^{3}H-RNA (RNA, M_r > 25 000). Note the rapid release of small molecules (i.e., M_r ~100) following CPE treatment. Reprinted with permission of the American Society for Microbiology from McClane (1997).[71]

C. A model summarizing the early steps in CPE action

Figure 19.7 diagrammatically summarizes the current understanding of the early events in CPE action that have been discussed in detail in section IV.B. At physiological temperatures, this model predicts that: (1) CPE binds to a 50 kDa eukaryotic membrane receptor protein to form small complex; (2) a physical change develops in this CPE small complex (this change could involve a partial insertion of CPE into the lipid bilayer or a conformational change to small

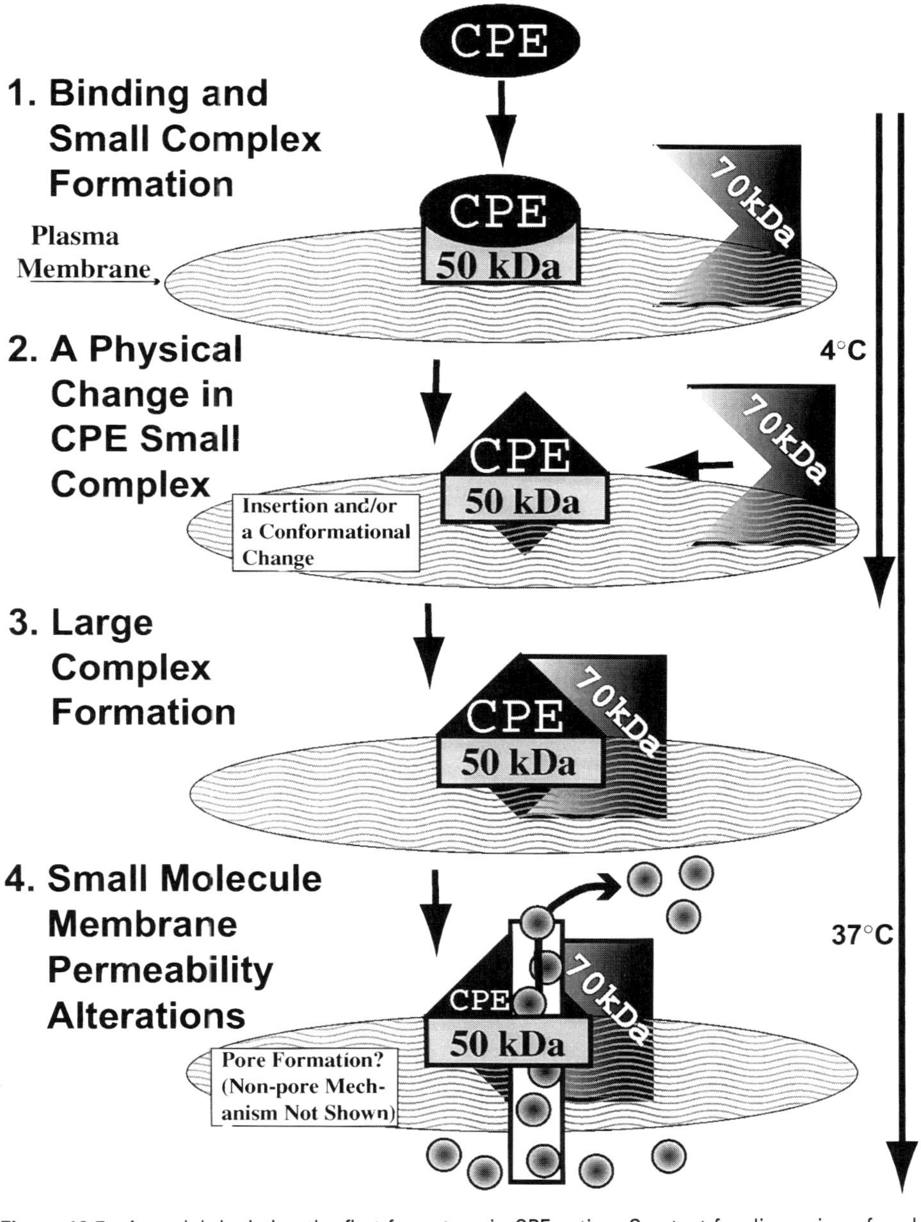

Figure 19.7 A model depicting the first four steps in CPE action. See text for discussion of each step in this model.

complex); (3) a large complex forms between CPE small complex and a 70 kDa membrane protein; and (4) this large complex causes small molecule membrane permeability alterations to develop; this effect may involve large complex serving directly as a pore or may involve some as yet undefined nonpore mechanism.

D. Later (secondary) events in CPE-induced cytotoxicity

What are the cytopathologic consequences of the CPE-induced small molecule permeability alterations described in Section IV.B.4? It has become well established that, like CPE, a number of other membrane-active bacterial toxins also cause small molecule membrane permeability alterations in mammalian membranes (although the CPE mechanism for causing this effect appears to be unique) and that these permeability changes result in the collapse of the colloid–osmotic equilibrium of the mammalian cell.[62,64] To understand this effect it is important to appreciate that healthy mammalian cells normally maintain a lower net intracellular ion concentration relative to the external medium in order to balance osmotically the large numbers of macromolecules that are present in their cytoplasm.[62,64,65] When a toxin changes the normal permeability properties of the plasma membrane for small molecules, there is a net influx of ions (and consequently water) into the cytoplasm of the toxin-treated cell.[62,64,65] These changes produce morphological damage (including membrame 'blebbing') and permeability alterations for larger molecules as the plasma membrane stretches owing to water influx.[17,33,60,62,64,65] Together these changes may eventually cause the toxin-treated cell to lyse.[17]

CPE treatment clearly leads to a collapse of the cellular colloid–osmotic equilibrium.[62,64] CPE-treated cells develop morphologic damage, including bleb formation (Figure 19.8), that is consistent with water influx into these cells. With time, these cells also become permeable to increasingly larger molecules (up to *c.* 5 kDa),[62,64] which is another effect consistent with water influx into CPE-treated cells. However the best proof that the later effects of CPE action include a disruption of the colloid–osmotic equilibrium comes from studies using 'osmotic stabilizers', which are molecules larger than the size (55–91 Å^2)[63] of the initial CPE lesion. Because osmotic stabilizers cannot enter cells through the CPE lesion because of their large size, they increase the osmolarity of the extracellular medium and thereby prevent water (but not ion) influx into CPE-treated cells. As would be expected if CPE is damaging the colloid–osmotic equilibrium, the presence of osmotic stabilizers in the medium of CPE-treated cells has been shown to inhibit significantly the

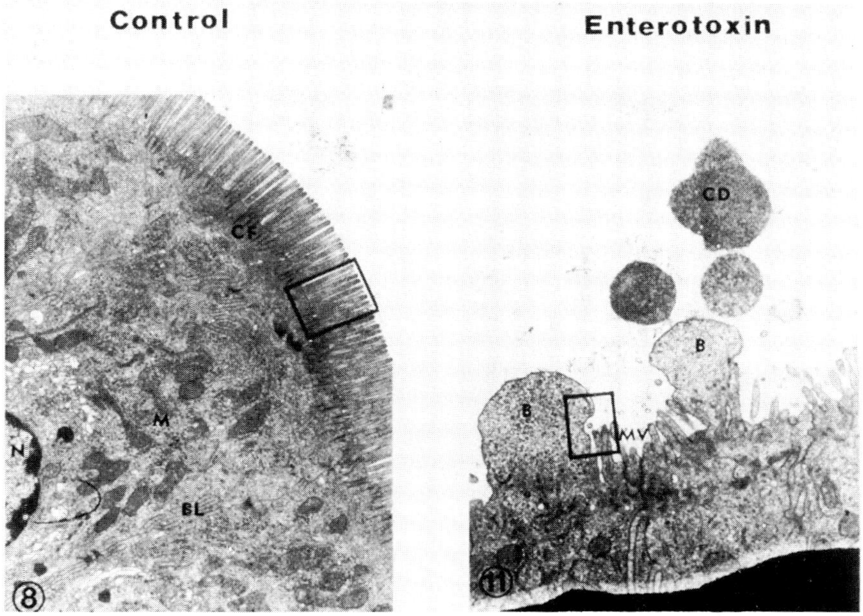

Figure 19.8 CPE-induced morphologic damage to rabbit small intestinal cells. The control specimen (left) shows a normal brush border membrane in a rabbit intestinal epithelial cell (box). CPE-treatment of rabbit intestinal cells (right) produces damage to brush border membranes (B). Reprinted with permission of the copyright holder, Williams and Wilkins, from McDonel *et al.* (1978) The effects of *Clostridium perfringens* enterotoxin on rat and rabbit ileum: an electron microscopy study. *Laboratory Investigations* 39: 210–218.

development of both CPE-induced morphologic damage and larger molecule membrane permeability alterations.[62,64]

It should be noted that preventing CPE-induced morphologic damage and larger molecule membrane permeability damage does not mean that cells treated with CPE in the presence of osmotic stabilizers are protected against the lethal effects of CPE.[62,64] While osmotic stabilizers prevent water influx in CPE-treated cells, they do not affect CPE-induced small molecule permeability changes in these cells.[62,64] In addition to damaging the colloid–osmotic equilibrium and thereby inducing water influx (as described above), these small-molecule membrane permeability alterations can also lead to shut down of important metabolic processes in the CPE-treated mammalian cell.[65] For example, it has been shown that the inhibition of macromolecular synthesis noted in CPE-treated mammalian cells involves small molecule permeability changes through at least two mechanisms: (1) the permeability changes alter intracellular ion levels, an effect that can directly shut-off some macromolecular synthetic pathways,[65] and (2) the permeability changes cause loss of some intracellular precursors (such as amino acids) that are necessary for macromolecular synthesis.[66] It is also likely that CPE-induced small molecule permeability changes affect other metabolic pathways (such as energy production) that are necessary for macromolecular synthesis.[65]

Since available evidence[62,64,65] indicates that both colloid–osmotic and metabolic disruptions occur simultaneously in cells treated with CPE in the absence of osmotic stabilizers (i.e., under normal physiologic conditions), both of these secondary effects are likely to contribute to the death of CPE-treated cells.

V. STRUCTURE–FUNCTION RELATIONSHIP STUDIES

Understanding a toxin molecule's structure–function relationship usually proves to be an invaluable aid for understanding how that toxin acts. CPE presents researchers interested in mapping its structure–function relationship with certain advantages, but also with some challenges. Problems encountered with mapping the CPE structure–function relationship include: (1) since CPE does not display significant sequence homology with other toxins,[24] it is not possible to make inferences about the CPE structure–function relationship (or CPE action) based upon information gleaned from other toxins; (2) because attempts to crystallize the CPE molecule have been unsuccessful to date, it is not yet possible to interpret results from structure–function relationship in a three-dimensional context; and (3) the close interactions that exist between CPE and eukaryotic membrane proteins throughout CPE cytotoxicity (as discussed in section IV.B) complicates assignment of certain cytotoxic functions to specific regions of the CPE molecule, e.g., does the CPE molecule directly form a pore or channel in membranes when it is sequestered in large complex, or does the CPE molecule cause the 50 kDa and/or 70 kDa CPE-binding proteins to alter membrane permeability? The advantages of working with CPE for structure–function studies are: (1) CPE is a relatively small molecule (35 kDa)[24] compared with many bacterial toxins (including the clostridial neurotoxins and *C. difficile* toxins),[16, 38] making it easier to probe the enterotoxin molecule for its defined functional regions; (2) CPE is a single continuous polypeptide i.e., it is not composed of fragments or multiple-subunits,[24] so all functional activities associated with CPE cytotoxicity obviously must be localized to the CPE 319 amino acid single polypeptide; and (3) it is possible to express full-length, cytotoxic CPE in *E. coli*,[24] allowing defined alterations of the *cpe* gene to be made using the many powerful *E. coli* genetic and expression systems that are available.

To date, the CPE structure–function relationship has been mapped primarily using defined CPE fragments. In the earliest mapping studies, these CPE fragments were produced using chemical- or protease-induced cleavage of native CPE.[31,67] However, since the cloning of the first *cpe* gene fragments in 1989,[25,26,28] most CPE structure–function studies have used recombinant DNA approaches to express defined CPE fragments.[25,49,54,68,69] Recombinant approaches offer a number of significant advantages over chemical or protease based cleavage methods for generating CPE fragments for mapping studies including: (1) it is easier to ascertain the precise identity of CPE fragments generated by recombinant approaches than it is to identify fragments generated by proteolytic or chemical cleavage of native enterotoxin (i.e., nucleotide sequencing is much easier to perform than protein sequencing); (2) the use of recombinant CPE fragments produced by *E. coli* eliminates the possibility, inherent to proteolytic/chemical cleavage techniques, that fragment preparations may still contain small amounts of uncleaved native CPE capable of producing false positive conclusions about the CPE activities retained by that fragment; and (3) the use of recombinant DNA techniques allow the construction/expression of virtually any defined CPE fragment that may be desired for study, unlike specific protease and chemical cleavage techniques which are often severely limited by cleavage site availability in the native CPE molecule.

A. Structure–function studies using proteolytically and chemically derived CPE fragments

In the earliest CPE structure–function study, Granum *et al.*[31] showed that limited trypsin treatment of native CPE removes the first 25 *N*-terminal amino acids from the enterotoxin, creating a CPE_{26-319} fragment. This trypsin-generated fragment displays approximately a two- to three-fold increase in biological activity compared with native CPE, as measured by its erythemal activity in guinea pigs[31] (this is an older assay for CPE biologic activity which is based upon the ability of CPE to affect capillary permeability when injected intra-dermally).[23] Follow-up studies from our laboratory later demonstrated that, on a molar basis, CPE_{26-319} also causes a two- to three-fold increase in the primary cytotoxic effect of CPE, i.e., it causes a two- to three-fold increase in CPE-induced small molecule permeability alterations.[69] Our studies also showed that the increased cytotoxicity possessed by the CPE_{26-319} fragment is not due to enhanced receptor binding ability,[69] implying that removal of the 25 *N*-terminal amino acids of CPE must affect a post-binding event in CPE cytotoxicity. More recently, Granum and Richardson[32] extended their original findings by reporting that, on a molar basis, a CPE_{37-319} fragment produced by limited chymotrypsin digestion of CPE also possesses two- to three-fold more cytotoxic activity than native enterotoxin. It is certainly noteworthy that both chymotrypsin and trypsin are present in the small intestine where CPE is thought to be active during food poisoning; as mentioned earlier, this may suggest that during food poisoning intestinal proteases may serve to increase CPE's activity *in vivo*.[31]

The first information suggesting a location for receptor binding region(s) on the CPE molecule was produced by Sakaguchi's group[67] using 2-nitro-5-thio-cyanobenzoic acid (NTCB)-induced cleavage of native CPE to generate CPE fragments. The principle of this approach is that since NTCB cleaves proteins specifically at cysteine residues and CPE contains only one cysteine (located at residue 186, i.e., approximately in the middle of the enterotoxin's 319 amino acids[24]), NTCB cleavage of enterotoxin should produce only two, approximately equal size, toxin fragments.[67] It was convincingly demonstrated that NTCB-induced cleavage of native CPE produced a preparation of 'purified' NTCB fragment that possessed receptor binding but not cytotoxic activity.[67] Based upon amino acid composition analysis of this fragment preparation, it was concluded that the preparation containing this binding fragment more closely resembles the amino acid composition expected of the *C*-terminal, rather than the *N*-terminal, half of CPE,[67] thus providing a suggestion that the *C*-terminal half of CPE molecule possesses receptor-binding activity.

However, there were concerns over the definitiveness of this assignment because: (1) it was not clear that the binding fragment preparation used in these studies contained only a single CPE fragment (the purity of the binding fragment preparation was demonstrated by SDS–PAGE,[67] which may not show multiple bands if two CPE fragments of almost equal size are present in the same preparation); (2) the amino composition analysis presented in this study did not clearly establish that the binding fragment preparation contained a *C*-terminal CPE fragment (while the amino acid composition of the binding fragment preparation more closely matched the composition of the *C*-terminal versus the *N*-terminal half of the enterotoxin,[67] the compositional analysis of this preparation showed significant deviation from the expected composition of the *C*-terminal CPE fragment);[67] and (3) it remained conceivable that in addition to possible contamination with the *N*-terminal CPE fragment produced by NTCB cleavage, the binding fragment preparation might also have contained some contaminating uncleaved enterotoxin that could be contributing to (or causing) the binding activity that was associated with the binding fragment preparation. Because of these concerns, it was considered important to evaluate further the location of the CPE receptor binding region(s) using more precise recombinant DNA approaches.

B. Structure–function studies performed with CPE fragments generated using recombinant DNA techniques

Early attempts in our laboratory to clone the *cpe* gene in *E. coli* involved the use of a λgt11 vector.[25] Cloning studies with this vector often produce fusion proteins, which result from the cloning of a fragment of the gene of interest into the β-galactosidase (β-gal) gene of the vector. During our early *cpe* cloning studies, an *E. coli* clone expressing a CPE-β-gal fusion protein of *c.* 135 kDa was identified.[25] This fusion protein, which was shown to contain 119 kDa of β-gal sequence and 19 kDa of enterotoxin sequence, did not possess either CPE-like cytotoxic or receptor-binding activity.[25] However, since it was reasoned that the presence of the large β-gal moiety in the fusion protein might cause steric hindrance that would block any activity possessed by the 19 kDa CPE fragment, the portion of the *cpe* gene encoding this 19 kDa CPE fragment was subcloned into an *E. coli* expression vector to permit expression of the fragment without the addition of significant amounts of vector-encoded amino acids. When the 19 kDa CPE fragment was expressed, without any β-gal sequences, this fragment was still not cytotoxic for Vero cells but it did possess a strong ability to compete against [^{125}I]CPE for binding to BBMs.[25] When the *cpe* gene fragment encoding the 19 kDa CPE fragment was sequenced, it was determined that this 19 kDa fragment corresponds to the *C*-terminal half of native enterotoxin,[25] i.e., it is a $_r$CPE$_{171-319}$ fragment. Further studies then directly demonstrated that $_r$CPE$_{171-319}$ can bind specifically to BBMs in a saturable and specific manner[25] and that the neutralizing monoclonal antibody (MAb 3C9), which blocks the binding of native enterotoxin to its receptor[48], is also able to abolish the binding of $_r$CPE$_{171-319}$ to BBMs,[25] as would be expected if $_r$CPE$_{171-319}$ contains the receptor-binding region of native enterotoxin.

These studies produced three important conclusions. First, they definitively localized a receptor-binding region to the *C*-terminal half of CPE. Second, as mentioned earlier, these studies confirmed that simple occupancy of the CPE receptor is not sufficient to trigger a CPE-like cytotoxic response, implying that post-binding events are essential for CPE-induced cytotoxicity. Third, they also demonstrated that some sequences present in the *N*-terminal half of CPE must be involved in CPE cytotoxic activity.

Studies were then conducted to evaluate why $_r$CPE$_{171-319}$ is noncytotoxic.[54] These studies determined that following its binding to BBMs, $_r$CPE$_{171-319}$ remains sensitive to protease induced release from these membranes,[54] suggesting that the nontoxic $_r$CPE$_{171-319}$ fragment is specifically unable to perform the post-binding physical change that is the second step in CPE cytotoxicity. In addition to supporting the importance of the post-binding physical change in

CPE cytotoxicity, this result also indicates that some *N*-terminal CPE sequences must be involved in mediating this second step of CPE action.

The location of the receptor binding region within the *C*-terminal half of the CPE molecule was further defined in follow-up studies by subcloning the *cpe* gene fragment encoding $_{r}CPE_{171-319}$ to produce three smaller *C*-terminal CPE fragments ($_{r}CPE_{171-238}$, $_{r}CPE_{239-289}$, and $_{r}CPE_{290-319}$).[68] When *E. coli* lysates containing these fragments were tested for their abilities to compete against [^{125}I]CPE for binding to BBMs, only the lysate containing the $_{r}CPE_{290-319}$ fragment was found to exhibit competitive binding activity (Figure 19.9),[68] a result that localizes a CPE receptor binding region to the extreme *C*-terminal 30 amino acids of the enterotoxin molecule. A synthetic peptide corresponding to the $_{r}CPE_{290-319}$ sequence was then shown to inhibit the binding of [^{125}I]CPE to BBMs at equimolar amounts as native CPE (Figure 19.9),[68] suggesting that most, if not all, of the receptor binding activity of the native enterotoxin molecule is localized to the $CPE_{290-319}$ region.

When epitope mapping experiments were conducted with various CPE fragments,[69] it was found that MAb 3C9 (which, as mentioned earlier, neutralizes the cytotoxicity of native CPE by blocking its binding to the CPE receptor)[48] reacts with the $CPE_{290-319}$ synthetic peptide,[69] indicating that this region contains at least one CPE-neutralizing epitope. Since $_{r}CPE_{290-319}$ is noncytotoxic,[68] yet contains a neutralizing epitope,[69] this fragment appears to be potentially useful as a CPE vaccine (the fact that antibodies to this neutralizing epitope result in a blockage of CPE binding can be regarded as particularly promising for vaccine development since many of the most effective antitoxin vaccines induce antibodies capable of blocking toxin binding). Therefore, the $CPE_{290-319}$ synthetic peptide was conjugated to a thyroglobulin carrier and this CPE conjugate was administered to mice by intraperitoneal injection.[70] These conjugate-immunized mice developed a strong serum antibody response that was capable of neutralizing the cytotoxicity of native CPE for Vero cells;[70] as would be predicted from our structure–function mapping studies, this neutralization resulted from an antibody-induced abrogation of CPE binding.[70] While these CPE vaccine results remain preliminary and many challenges lie ahead, these initial studies appear particularly promising regarding the possibility of developing vaccine protection

---

Figure 19.9 (opposite) The $CPE_{290-319}$ fragment displays CPE binding activity. Top panel. $_{r}CPE_{290-319}$ lysates inhibit binding of [^{125}I]CPE to BBM receptors while lysates containing other *C*-terminal CPE fragments or control lysates do not. Bottom panel. $CPE_{290-319}$ synthetic peptide inhibits the binding of [^{125}I]CPE to BBM receptors at equimolar ratios as native CPE, while a negative control peptide has no effect. Reprinted with permission of the American Society for Biochemistry and Molecular Biology from Hanna *et al.* (1991).[68]

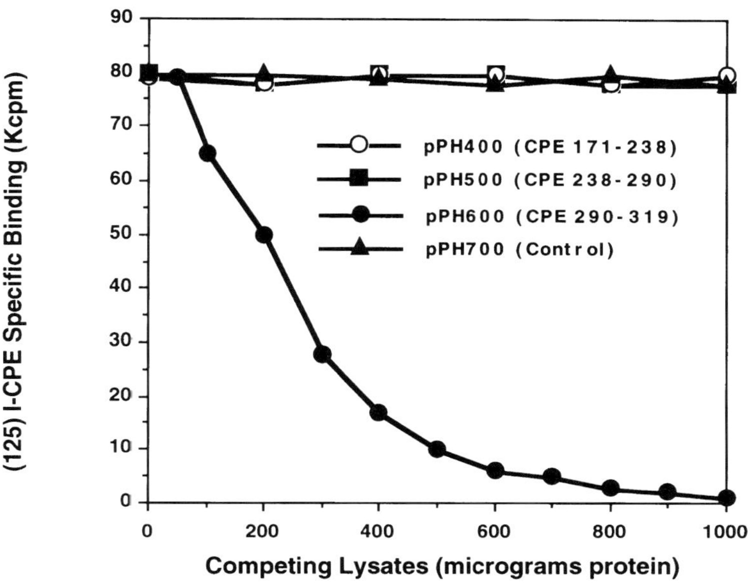
90
80
70
60
50
40
30
20
10
0
(125) I-CPE Specific Binding (Kcpm)
pPH400 (CPE 171-238)
pPH500 (CPE 238-290)
pPH600 (CPE 290-319)
pPH700 (Control)
0
200
400
600
800
1000
Competing Lysates (micrograms protein)

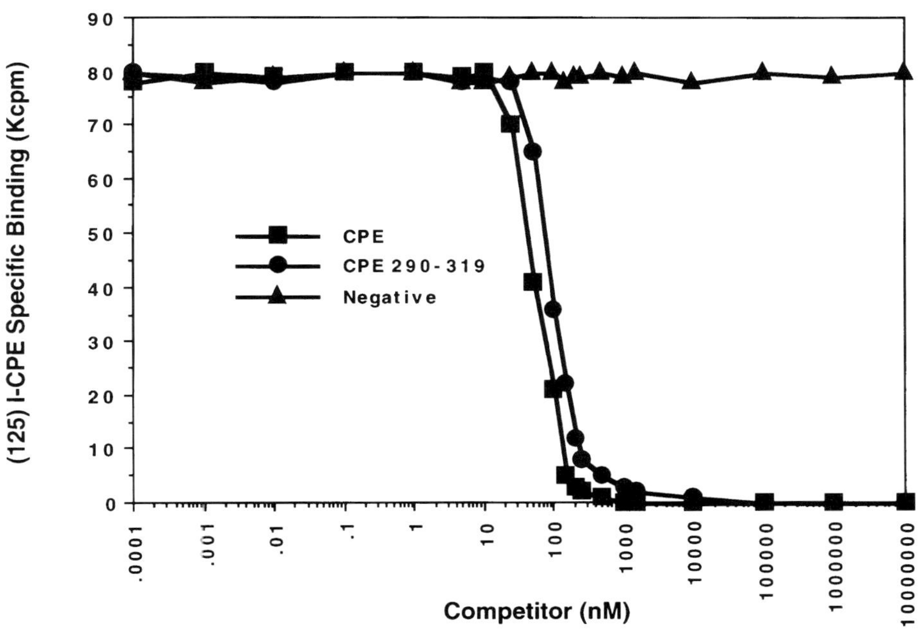
90
80
70
60
50
40
30
20
10
0
(125) I-CPE Specific Binding (Kcpm)
CPE
CPE 290-319
Negative
.0001
.001
.01
.1
1
10
100
1000
10000
100000
1000000
10000000
Competitor (nM)

for diseases purportedly involving systematically distributed CPE (e.g., SIDS).[21,22]

An additional series of CPE *N*- and *C*-terminal deletion fragments (see Figure 19.10) has recently been prepared to map further the CPE structure–function relationship.[49] This deletion series includes; $_{r}CPE_{1-319}$, a full-length CPE construct used as a positive control for CPE expressed in *E. coli*; $_{r}CPE_{37-319}$, a recombinant CPE fragment corresponding to the fragment produced by limited chymotrypsin digestion of native CPE (as discussed in section V.A);[32] several *N*-terminal deletion mutants that have progressively larger portions of their *N*-terminal CPE sequences removed and $_{r}CPE_{1-289}$, a fragment that lacks the putative $CPE_{290-319}$ receptor-binding region of native enterotoxin.[68] These fragments have now been systematically surveyed for their ability to perform the four early steps known to be involved in CPE cytotoxicity (e.g., binding, post-binding physical change, large complex formation and small molecule membrane permeability alterations), as discussed briefly below.

1. Binding

The binding characteristics of these CPE fragments were measured by their ability to competitively inhibit the binding of [^{125}I]CPE to BBMs. In this assay, $_{r}CPE_{1-319}$ and all of the *N*-terminal deletion mutants were found to exhibit approximately equivalent binding activity to native enterotoxin,[49] as might be predicted since they retain the putative $CPE_{290-319}$ receptor-binding region identified in our earlier study.[68] Importantly, the $_{r}CPE_{1-289}$ *C*-terminal deletion fragment did not show any receptor-binding activity.[49] This result provides strong evidence that the $CPE_{290-319}$ region of the enterotoxin is absolutely required for receptor-binding and is consistent with this region being the sole receptor-binding region present in native enterotoxin.

2. Post-binding physical change

The *N*-terminal deletion fragments capable of binding were then tested for their sensitivity to protease-induce release from BBMs. It was observed that, like native enterotoxin, $_{r}CPE_{1-319}$, $_{r}CPE_{37-319}$ and $_{r}CPE_{45-319}$ all develop resistance to protease-induced release;[49] however, deletion of any further *N*-terminal sequences beyond residue 45 of native enterotoxin was found to block the ability of these CPE fragments to undergo the post-binding physical change step of CPE action, i.e., these fragments remained susceptible to protease-induced release from membranes.[49] This result indicates that *N*-terminal amino acid residues between 45 and 53 must be involved in mediating the second step in CPE action.

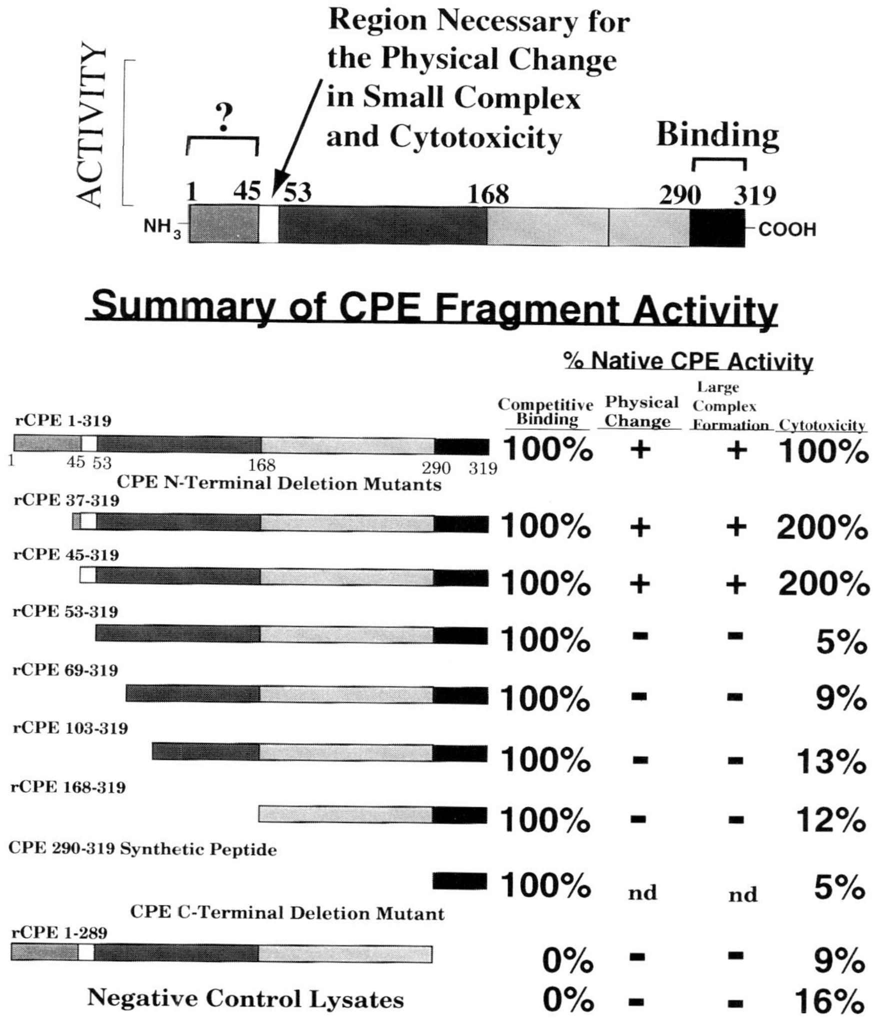

Figure 19.10 A current map of the structure–function relationship of the CPE molecule. The top portion of the figure depicts the current map for the CPE structure–function relationship (see text for discussion). The bottom portion of the figure depicts the various CPE fragments used to determine this relationship and each fragment's ability to perform CPE activities. The figures represent portions of CPE present in each fragment; + denotes that a fragment retains that CPE-activity, while – denotes that a fragment is unable to perform the specified CPE-activity; n.d. denotes that the ability to perform that CPE-activity was not determined for a fragment.

3. Large complex formation

The ability of binding-capable *N*-terminal deletion fragments to form large complex was assayed by Western blot analysis of SDS-extracted samples from BBMs treated with either native CPE or a specific CPE fragment at temperatures permissive for large complex formation.[53] Using this approach, it was found that only the same three CPE constructs capable of developing resistance to protease-induced release (i.e., capable of undergoing the post binding physical change) are also able to form large complex.[49] Since no fragments have been identified that are capable of forming large complex but are unable to undergo the post-binding change to small complex, these results independently support previous conclusions, drawn from low temperature studies,[53] suggesting that large complex formation occurs after the development of the physical change in small complex during CPE action. In the current large-complex studies, it was also noted that $_{r}CPE_{37-319}$ and $_{r}CPE_{45-319}$ are consistently able to form approximately twofold more large complex than native CPE or $_{r}CPE_{1-319}$,[49] a result that suggests the increased ^{86}Rb-release activity of $_{r}CPE_{37-319}$ and $_{r}CPE_{45-319}$ (see below) may result from their enhanced ability to form large complex. This result further supports the direct involvement of large complex formation in CPE-induced small molecule membrane permeability alterations.

4. Small molecule membrane permeability alterations

Turning to the final early event in CPE cytotoxicity, the ability of each CPE fragment to induce small molecule permeability alterations was measured using an assay based on enterotoxin-induced ^{86}Rb release from Vero cells.[53] These experiments indicate that recombinant CPE_{1-319} is equally as capable as native CPE at inducing small molecule membrane permeability alterations.[49] However, on a molar basis, $_{r}CPE_{37-319}$ is able to cause about a twofold increase in, small molecule membrane permeability alterations compared with native CPE,[49] a result consistent with Granum and Richardson's study demonstrating that a chymotrypsin-generated CPE_{37-319} fragment also possesses enhanced cytotoxicity.[32] Interestingly, the current study also indicated that $_{r}CPE_{45-319}$ displays enhanced cytotoxicity.[49] This result also demonstrates that at least the first 45 *N*-terminal CPE amino acids are not necessary for CPE cytotoxicity.

C. An overview of the current understanding of CPE structure–function relationships

Overall, these CPE structure–function mapping studies indicate that the enterotoxin, like many other bacterial toxins, segregates its regions responsible for binding and post binding activities (Figure 19.10). The CPE molecule appears to contain a single-receptor binding region, located at the extreme *C*-terminus of the enterotoxin[49,68,69] (amino acids 290–319). *N*-terminal sequences lying between residues 45 and 53 appear to be required for CPE sequestered in small complex to undergo the post-binding physical change that is required for the development of CPE-induced cytotoxicity.[49] When these *N*-terminal residues are removed to create $_r$CPE$_{53-319}$, this fragment lacks cytotoxic properties, and becomes unable to develop the post-binding physical change that is the second step in CPE action.[49] However, it is premature to conclude that amino acids between residues 45 and 53 play a direct role in the physical change that occurs to CPE small complex, since removal of these residues may simply alter the conformation of $_r$CPE$_{53-319}$ enough to disrupt this step. Residues 1–44 of the enterotoxin apparently serve no known function for enterotoxin action, and appear to interfere with large complex formation and, ultimately, cytotoxicity.[49] Combining this conclusion with the observation that the *C*-terminal deletion mutant ($_r$CPE$_{1-289}$) is nontoxic,[49] it can now be concluded that the minimum size CPE fragment that retains cytotoxicity is about CPE$_{45-319}$.[49]

CPE fragment studies have thus far been unable to identify a CPE region(s) specifically involved in large complex formation or in the induction of small molecule permeability alteration. However, these regions may soon be identified using a series of CPE point mutants that are now being produced and characterized in our laboratory.

VI. CONCLUDING REMARKS

Due in large part to the use of recombinantly derived CPE fragments for both structure–function and mechanism of action studies, the past 5 or 6 years have seen unprecedented progress in our understanding of how CPE affects mammalian cells. In particular, these studies have now clearly established that the molecular action of CPE involves a unique series of at least four early events, each step of which includes close participation between the enterotoxin and eukaryotic membrane proteins.

Many exciting challenges remain for future CPE researchers. Answers must be found to such questions as, what is the identity/normal physiological role of the 50 kDa and 70 kDa membrane proteins that appear to play such important roles in CPE action? Does the post-binding physical change to small-complex represent CPE small complex insertion or a conformational change in small complex? How does large complex formation disrupt membrane permeability properties for small molecules? What regions of CPE mediate large complex formation and the onset of these small molecule permeability changes? How are CPE functional regions situated within the three-dimensional context of the CPE molecule?

Studies are currently underway to answer many of these questions; results from these studies are likely to provide exciting and surprising new insights into this unique toxin.

VII. ACKNOWLEDGMENTS

This work is supported by Public Health Service grant AI-19844–13 from the National Institute of Allergy and Infectious Diseases.

VIII. REFERENCES

1. McClane B.A. (1992) *Clostridium perfringens* enterotoxin: structure, action and detection. *J Food Safety* **12**: 237–252.
2. Labbe R.G. (1989) *Clostridium perfringens*. In Doyle M.P. (ed.), *Foodborne Bacterial Pathogens*, pp. 192–234. Marcel Decker, New York.
3. McDonel J.L. (1986) Toxins of *Clostridium perfringens* types A, B, C, D, and E. In Dorner F. and Drews H. (eds), *Pharmacology of Bacterial Toxins*, pp. 477–517. Pergamon Press, Oxford.
4. Bean N.H. and Griffen P.M. (1990) Foodborne disease outbreaks in the United States, 1973–1987; pathogens, vehicles and trends. *J Food Protect* **53**: 804–817.
5. Todd E.C.D. (1989) Cost of acute bacterial foodborne disease in Canada and the United States. *Int J Food Microbiol* **9**: 313–326.
6. Todd E.C.D. (1989) Preliminary estimates of costs of foodborne disease in the United States. *J Food Prot* **52**: 595–601.
7. Batholomew B.A., Stringer M.F., Watson G.N. and Gilbert R.J. (1985) Development and application of an enzyme-linked immunosorbent assay for *Clostridium perfringens* type A enterotoxin. *J Clin Pathol* **38**: 222–228.
8. Birkhead G., Vogt R.L., Heun E.M., Snyder J.T. and McClane B.A. (1988) Characterization of an outbreak of *Clostridium perfringens* food poisoning by quantitative fecal culture and fecal enterotoxin measurement. *J Clin Microbiol* **26**: 471–474.
9. Saito M. (1990) Production of enterotoxin by *Clostridium perfringens* derived from humans, animals and the natural environment in Japan. *J Food Protect* **53**: 115–118.
10. McDonel J.L. and Duncan C.L. (1975) Histopathological effect of *Clostridium perfringens* enterotoxin in the rabbit ileum. *Infect Immun* **12**: 1214–1218.
11. Skjelkvale R. and Uemura T. (1977) Experimental diarrhea in human volunteers following oral administration of *Clostridium perfringens* enterotoxin. *J Appl Bacteriol* **46**: 281–286.
12. Kokai-Kun J.F., Songer J.G., Czeczulin J.R., Chen F. and McClane B.A. (1994) Comparison of Western immunoblots and gene detection assays for identification of potentially enterotoxigenic isolates of *Clostridium perfringens*. *J Clin Microbiol* **32**: 2533–2539.
13. Strong D.H., Duncan C.L. and Perna G. (1971) *Clostridium perfringens* type A food poisoning. II. Response of the rabbit ileum as an indication of enteropathogenicity of strains of *Clostridium perfringens*. *Infect Immun* **3**: 171–178.

14. Hauschild A.H., Niilo L. and Dorward W.J. (1971) The role of enterotoxin in *Clostridium perfringens* type A enteritis. *Can J Microbiol* **17**: 987–991.
15. Hobbs B.C. (1979) *Clostridium perfringens* gastroenteritis. In Riemann H. and Bryan F.L. (eds), *Food-Borne Infections and Intoxications*, 2nd edn, pp. 131–167. Academic Press, New York.
16. Niemann H. (1991) Molecular biology of clostridial neurotoxins. In Alouf J. and Freer J. (eds), *Sourcebook of Bacterial Protein Toxins*, pp. 303–348. Academic Press, London.
17. McClane B.A. (1994) *Clostridium perfringens* enterotoxin acts by producing small molecule permeability alterations in plasma membranes. *Toxicology* **87**: 43–67.
18. Brett M.M., Rodhouse J.C., Donovan T.J., Tebbut G.M. and Hutchinson D.N. (1992) Detection of *Clostridium perfringens* and its enterotoxin in cases of sporadic diarrhea. *J Clin Pathol* **45**: 609–611.
19. Collins J.E., Bergeland M.E., Bouley D., Duocommun A.L., Francis D.H. and Yeske P. (1989) Diarrhea associated with *Clostridium perfringens* type A enterotoxin in neonatal pigs. *J Vet Diagn Invest* **1**: 351–353.
20. Jackson S., Yip-Chuck D., Clark J. and Brodsky M. (1986) Diagnostic importance of *Clostridium perfringens* enterotoxin analysis in recurring enteritis among the elderly, chronic care psychiatric patients. *J Clin Microbiol* **23**: 748–751.
21. Lindsay J., Mach A., Wilkinson M., Martin L., Wallace L., Keller F. and Wojciechowski M. (1993) *Clostridium perfringens* type A cytotoxic- enterotoxin(s) as triggers for death in sudden infant death syndrome: development of a toxico-infection hypothesis. *Curr Microbiol* **27**: 51–59.
22. Murrell W.G., Stewart B.J., O'Neill C.O., Siarakas S. and Kariks S. (1993) Enterotoxigenic bacteria in the sudden infant death syndrome. *J Med Microbiol* **39**: 114–127.
23. McDonel J.L. and McClane B.A. (1988) Production, purification and assay of *Clostridium perfringens* enterotoxin. *Method Enzymol* **165**: 94–103.
24. Czeczulin J.R., Hanna P.C. and McClane B.A. (1993) Cloning, nucleotide sequencing, and expression of the *Clostridium perfringens* enterotoxin gene in *Escherichia coli*. *Infect Immun* **61**: 3429–3439.
25. Hanna P.C., Wnek A.P. and McClane B.A. (1989) Molecular cloning of the 3' half of the *Clostridium perfringens* enterotoxin gene and demonstration that this region encodes receptor-binding activity. *J Bacteriol* **171**: 6815–6820.
26. Iwanejko L.A., Routledge M.N. and Stewart G.S.A.B. (1989) Cloning in *Escherichia coli* of the enterotoxin gene from *Clostridium perfringens* type A. *J Gen Microbiol* **135**: 903–909.
27. Richardson M. and Granum P.E. (1985) The amino acid sequence of the enterotoxin from *Clostridium perfringens* type A. *FEBS Lett* **182**: 479–484.
28. van Damme-Jongsten M., Wernars K. and Notermans S. (1989) Cloning and sequencing of the *Clostridium perfringens* enterotoxin gene. *Ant Leeuwenhoek J Microbiol* **56**: 181–190.
29. Cornillot E., Saint-Joanis B., Daube G., Katayama S., Granum P.E., Canard B. and Cole S.T. (1995) The enterotoxin gene (*cpe*) of *Clostridium perfringens* can be chromosomal or plasmid-borne. *Mol Microbiol* **15**: 639–647.
30. Hauser D., Eklund M.W., Boquet P. and Popoff M.R. (1994) Organization of the botulinum neurotoxin C1 gene and its non-toxic protein genes in *Clostridium botulinum* C 468. *Mol Gen Genet* **243**: 631–640.
31. Granum P.E., Whitaker J.R. and Skjelkvale R. (1981) Trypsin activation of enterotoxin from *Clostridium perfringens* type A. *Biochim Biophys Acta* **668**: 325–332.
32. Granum P.E. and Richardson M. (1991) Chymotrypsin treatment increases the activity of *Clostridium perfringens* enterotoxin. *Toxicon* **29**: 445–453.
33. McDonel J.L., Chang L.W., Pounds J.L. and Duncan C.L. (1978) The effects of *Clostridium perfringens* enterotoxin on rat and rabbit ileum: an electron microscopy study. *Lab Invest* **39**: 210–218.
34. Sherman S., Klein E. and McClane B.A. (1994) *Clostridium perfringens* type A enterotoxin induces

concurrent development of tissue damage and fluid accumulation in the rabbit ileum. *J Diarrheal Dis Res* **12**: 200–207.

35. McDonel J.L. and Duncan C.L. (1977) Regional localization of activity of *Clostridium perfringens* type A enterotoxin in Vero cells. *J Inf Dis* **136**: 661–666.
36. McDonel J.L. and Demers G.W. (1982) *In vivo* effects of enterotoxin from *Clostridium perfringens* type A in rabbit colon: binding vs. biologic activity. *J Inf Dis* **145**: 490–494.
37. Finkelstein R.A. and Dorner F. (1986) Cholera enterotoxin (choleragen). In Dorner F. and Drews J. (eds), *Pharmacology of Bacterial Toxins*, pp. 161–172. Pergamon Press, Oxford.
38. Wilkins T.O., Krivan H., Stiles B., Carman R.J. and Lyerly D.M. (1985) Clostridial toxins active locally in the gastrointestinal tract. In Evered D. and Whelan J. (eds), *Microbial Toxins and Diarrheal Diseases*, pp. 230–241. Pitman, London.
39. Sugimoto N., Chen Y., Lee S., Matsuda M. and Lee C. (1991) Pathodynamics of intoxication in rats and mice by enterotoxin of *Clostridium perfringens* type A. *Toxicon* **29**: 751–759.
40. Siarakas S., Damas E. and Murrell W. (1995) Is cardiorespiratory failure induced by bacterial toxins the cause of sudden infant death syndrome? Studies with an animal model (the rabbit). *Toxicon* **33**: 635–649.
41. Matsuda M., Okabe T., Sugimoto N., Senda T. and Fujita H. (1994) Tetanus toxin and *Clostridium perfringens* enterotoxin as tools for the study of exocytosis. *Ann N Y Acad Sci* **710**: 94–106.
42. Sugimoto N., Miyamoto A., Horiguchi Y., Okabe T. and Matsuda M. (1992) Inhibition of neuromuscular transmission in isolated mouse phrenic nerve-diaphragm by the enterotoxin of *Clostridium perfringens* type A. *Toxicon* **30**: 825–834.
43 McDonel J.L. and McClane B.A. (1979) Binding vs. biological activity of *Clostridium perfringens* enterotoxin in Vero cells. *Biochem Biophys Res Commun* **87**: 497–504.
44. McDonel J.L. (1980) Binding of *Clostridium perfringens* ^{125}I-enterotoxin to rabbit intestinal cells. *Biochem* **21**: 4801–4807.
45. Wieckowski E.U., Wnek A.P. and McClane B.A. (1994) Evidence that an ~50 kDa mammalian plasma membrane protein with receptor-like properties mediates the amphiphilicity of specifically-bound *Clostridium perfringens* enterotoxin. *J Biol Chem* **269**: 10838–10848.
46. McClane B.A., Wnek A.P., Hulkower K.I. and Hanna P.C. (1988) Divalent cation involvement in the action of *Clostridium perfringens* type A enterotoxin. *J Biol Chem* **263**: 2423–2435.
47. Horiguchi Y., Uemura T., Kozaki S. and Sakaguchi G. (1985) The relationship between cytotoxic effects and binding to mammalian cultures cells of *Clostridium perfringens* enterotoxin. *FEMS Microbiol Lett* **28**: 131–135.
48. Wnek A.P., Strouse R.J. and McClane B.A. (1985) Production and characterization of monoclonal antibodies against *Clostridium perfringens* type A enterotoxin. *Infect Immun* **50**: 442–448.
49. Kokai-Kun J.F. and McClane B.A. (1997) Deletion analysis of the *Clostridium perfringens* enterotoxin. *Infect Immun* **65**. In press.
50. Wnek A. and McClane B. (1983) Identification of a 50 000 M_r protein from rabbit brush boarder membranes that binds *Clostridium perfringens* enterotoxin. *Biochem Biophys Res Com* **112**: 1099–1105.
51. Wnek A.P. and McClane B.A. (1989) Preliminary evidence that *Clostridium perfringens* type A enterotoxin is present in a 160 000-M_r complex in mammalian membranes. *Infect Immun* **57**: 574–581.
52. Kokai-Kun J.F. and McClane B.A. (1996) Evidence that region(s) of the *Clostridium perfringens* enterotoxin molecule remain exposed on the external surface of the mammalian plasma membrane when the toxin is sequestered in small or large complex. *Infect Immun* **64**: 1020–1025.
53. McClane B.A. and Wnek A.P. (1990) Studies of *Clostridium perfringens* enterotoxin action at different temperatures demonstrate a correlation between complex formation and cytotoxicity. *Infect Immun* **58**: 3109–3115.
54. Hanna P.C. and McClane B.A. (1991) A recombinant *C*-terminal toxin fragment provides evidence

that membrane insertion is important for *Clostridium perfringens* enterotoxin cytotoxicity. *Mol Microbiol* **5**: 225–230.

55. Tolleshaug H., Skjelkvale R. and Berg T. (1982) Quantitation of binding and subcellular distribution of *Clostridium perfringens* enterotoxin in rat liver cells. *Infect Immun* **37**: 486–491.
56. Bhakdi S. and Tranum-Jensen J. (1991) Alpha-toxin of *Staphylococcus aureus. Microbiol Rev* **55**: 733–751.
57. Bhakdi S. and Tranum-Jensen J. (1986) Membrane damage by pore-forming bacterial cytolysins. *Microb Pathog* **1**: 5–14.
58. Bhakdi S. and Tranum-Jensen J. (1987) Damage to mammalian cells by proteins that form transmembrane pores. *Rev Physiol Biochem Pharmacol* **107**: 147–222.
59. Bhakdi S., Menestrina G., Hugo F., Seeger W. and Tranum-Jensen J. (1988) Pore-forming bacterial cytolysins. In Fehrenbach A. (ed.), *Bacterial Protein Toxins, Zbl. Bakt. Suppl 17*, pp. 71–77. Gustav Fischer, Stuttgart.
60. McClane B.A. and McDonel J.L. (1980) Characterization of membrane permeability alterations induced in Vero cells by *Clostridium perfringens* enterotoxin. *Biochim Biophys Acta* **600**: 974–985.
61. Matsuda M., Ozutsumi K., Iwashi H. and Sugimoto N. (1986) Primary action of *Clostridium perfringens* type A enterotoxin on HeLa and Vero cells in the absence of extracellular calcium: rapid and characteristic changes in membrane permeability. *Biochem Biophys Res Commun* **141**: 704–710.
62. McClane B.A. and McDonel J.L. (1981) Protective effects of osmotic stabilizers on morphological and permeability alterations induced in Vero cells by *Clostridium perfringens* enterotoxin. *Biochim Biophys Acta* **641**: 401–409.
63. Menestrina G. (1991) Electrophysiologic methods for the study of toxin-membrane interaction. In Alouf J. and Freer J. (eds), *Sourcebook of Bacterial Protein Toxins*, pp. 215–242. Academic Press, London.
64. McClane B.A. (1984) Osmotic stabilizers differentially inhibit permeability alterations induced in Vero cells by *Clostridium perfringens* enterotoxin. *Biochim Biophys Acta* **777**: 99–106.
65. Hulkower K.I., Wnek A.P. and McClane B.A. (1989) Evidence that alterations in small molecule permeability are involved in the *Clostridium perfringens* type-A enterotoxin-induced inhibition of macromolecular synthesis in Vero cells. *J Cell Physiol* **140**: 498–504.
66. Hulkower K.I. and McClane B.A. (1988) The effects of *Clostridium perfringens* enterotoxin on intracellular levels or transport of uridine, thymidine and leucine do not fully explain enterotoxin-induced inhibition of macromolecular synthesis. *Biochem Biophys Res Commun* **153**: 699–707.
67. Horiguchi Y., Akai T. and Sakaguchi G. (1987) Isolation and function of a *Clostridium perfringens* enterotoxin fragment. *Infect Immun* **55**: 2912–2915.
68. Hanna P.C., Mietzner T.A., Schoolnik G.K. and McClane B.A. (1991) Localization of the receptor-binding region of *Clostridium perfringens* enterotoxin utilizing cloned toxin fragments and synthetic peptides. The 30 C-terminal amino acids define a functional binding region. *J Biol Chem* **266**: 11037–11043.
69. Hanna P.C., Wieckowski E.U., Mietzner T.A. and McClane B.A. (1992) Mapping functional regions of *Clostridium perfringens* type A enterotoxin. *Infect Immun* **60**: 2110–2114.
70. Mietzner T.A., Kokai-Kun J.F., Hanna P.C. and McClane B.A. (1992) A conjugated synthetic peptide corresponding to the C-terminal region of *Clostridium perfringens* type A enterotoxin elicits an enterotoxin-neutralizing antibody response in mice. *Infect Immun* **60**: 3947–3951.
71. McClane B.A. (1997) *Clostridium perfringens.* In Doyle M.P., Beuchat L. and Montville T.J. (eds), *Food Microbiology: Fundamentals and Frontiers*, pp. 305–326. ASM Press, Washington D.C.

Chapter 20

Binary Toxins from *Clostridium spiroforme* and *Clostridium perfringens*

ROBERT J. CARMAN, SYLVIE PERELLE*
AND MICHEL R. POPOFF*
*TechLab, Inc., 1861 Pratt Drive, Blacksburg, VA 24060-6364, USA and *Laboratoire des Toxines, Institut Pasteur, 28 Rue du Docteur Roux, 75724 Paris Cedex 15, France*

I. INTRODUCTION

Both *Clostridium spiroforme* and *Clostridium perfringens* type E produce ι-toxin, a binary toxin which causes diarrhea in mammals. The clinical conditions that each form of the toxin cause have been thoroughly reviewed.[1] Consequently, we have concentrated in this chapter on information which has become available over the last decade.

Based on 16S rRNA and cellular fatty acid analysis (R.J. Carman, unpublished data), *C. spiroforme* and *C. perfringens* are not closely related species (see Chapter 1). According to Collins *et al.*[2] the former is a member of cluster XVIII of the clostridia, a taxonomic grouping which also includes *Clostridium ramosum*[3] and *Clostridium cocleatum* (R.J. Carman, unpublished data), another helically coiled species. Furthermore, based on their 16S rRNA and cellular fatty acid profiles, ι-toxigenic strains of *C. spiroforme*, which to date have been isolated only from rabbits, constitute a species distinct from the nontoxigenic strains of true *C. spiroforme*, which come from humans and chickens. For

The Clostridia: Molecular Biology and Pathogenesis
ISBN 0-12-595020-9

simplicity we will continue to refer to the ι-toxigenic strains as *C. spiroforme* throughout this chapter. *Clostridium perfringens* falls into cluster I with *Clostridium butyricum*, the type species of the genus *Clostridium* (see Chapter 1).[3] ι-Toxigenic *C. perfringens* and *C. spiroforme* are infectious agents causing diarrhea in hosts with compromised intestinal floras. The most common predisposing factors are use of antibiotics and weaning; both result in a reduction in the host's ability to resist colonization. Bacteria that successfully negotiate gastric acidity and the anaerobic environment of the large bowel can exploit exposed ecological niches and colonize. Freely sporulating clostridia are eminently suited to the task and, if they produce a toxin, diarrhea may follow.

First described in the 1940s, *C. perfringens* type E is only occasionally a cause of diarrhea of animals, notably domesticated livestock. There is a single report of a fatal diarrhea in adult guinea pigs, occurring when germ-free animals were removed from positive-pressure isolators.[4]

Since the early 1980s *C. spiroforme*, so-called because of its spirally coiled chains of curved rods, has become a well-documented cause of an often fatal diarrhea of laboratory and farmed rabbits. There is a single report of a similarly shaped organism being isolated from a leukemic human with diarrhea.[5] However, this particular strain has a cellular fatty acid profile which is distinct from the pathogenic rabbit isolates and which more closely resembles that of *C. cocleatum* (R.J. Carman, unpublished data). Several similar strains have been isolated from non-*C. difficile* antibiotic-associated diarrhea; but none have been shown to produce ι or any other toxin (R.J. Carman, unpublished data).

Clostridium spiroforme and *C. perfringens* type E have occasionally been isolated from the feces of apparently healthy animals. These animals may be only transiently colonized with spores which pass directly through their intestinal tracts without germinating or they may represent an asymptomatic carrier state without true colonization. To distinguish between the two it would be necessary to compare the total count per gram (vegetative cells and spores) with the spore count. This has not yet been done. As *C. perfringens* type E has been isolated from animal fodder, an asymptomatic carrier state might be explained by ingestion followed by sporulation but not by subsequent growth.

II. ι-TOXIN

ι-Toxin is a binary toxin, consisting of an active component (s_a and i_a for *C. spiroforme* and *C. perfringens*, respectively) and a binding protein (s_b and i_b, respectively). Carman *et al.*[2] showed that *in vitro* production of ι-toxin by *C. spiroforme* requires divalent cations, whereas production by *C. perfringens* appears to be free from this constraint. The biological activity of ι-toxin requires the presence of both the binding and active components. Combinations of i_a and s_b, and vice versa, possess all the activities of the homologous protein pair. The toxin activity is neutralized by antitoxin raised against formalized culture filtrates of either bacterium. More specifically, antibodies to a single component neutralizes all *in vivo* activities.

A. Biological activity and effects

Sterile filtrates of ι-toxin are dermonecrotic when injected, intradermally, into the flank of a guinea pig, and kill mice when given intraperitoneally. The specific neutralization of these activities forms the basis of diagnostic assays for ι-toxin in clinical samples and culture fluids. Protocols for both assays, and illustrations of the former, are included in Sterne and Batty.[6] A more humane, cheaper, and potentially quantitative assay for the biological activity of ι-toxin is the specific neutralization of its cytotoxic effect, which is most readily seen as the rounding of Vero cells.[7] Compared with other cytotoxic clostridia, e.g. *Clostridium difficile*, the levels of activity from clinical samples may be low. A dilution of 10^{-2} frequently may be sufficient to make activity in diarrheic feces inapparent. This may be one reason that ι-toxin has never been detected from the feces of humans with antibiotic-associated diarrhea.

The ι-toxin toxin of *C. spiroforme* causes fluid accumulation when injected into ligated sections of rabbit ileum, the amount of fluid accumulation, degree of inflammation and villus tip degeneration are all proportional to the amount of toxin injected into the loop. ι-Toxin from *C. perfringens* has not yet been shown to have the same effects, which is surprising and deserves further study.

In weanling rabbits with diarrhea caused by *C. spiroforme*, the principal clinical observations are peracute diarrhea and death, which is often so sudden that no external signs of scouring, such as perianal soiling with feces, are seen. At autopsy, petechiation of the cecal serosa is commonly seen and the cecum is invariably filled with a fetid, green to dark brown watery content. Thickening of the submucosa, which is visible to the eye, is frequent and is caused by an inflammatory exudate. Changes to the mucosal epithelium range from the loss of brush borders to varying degrees of necrosis, which in turn lead to multifocal ulcerations and even diffuse mucosal denudation. There is evidence of rapid epithelial turnover and, when suppurative exudation is mixed with fibrin, pseudomembranes can be seen. Changes to the small bowel are very similar to those in the cecum. A more extensive review of the pathology of ι-enterotoxemia can be found in Carman and Evans.[8]

B. Molecular basis of activity

The active and binding components of ι-toxin (ι_a and ι_b from *C. perfringens* and s_a and s_b from *C. spiroforme*) are not linked by covalent or noncovalent bonds. The larger component, ι_b, which is about 100 kDa, is activated by proteolysis releasing a peptide of about 20 kDa. Although their identities have not been established, proteases able to activate ι_b and s_b are produced by *C. perfringens* and *C. spiroforme*, respectively. Commercial preparations of trypsin are also able to activate these. The activated binding components recognize a cell surface receptor which has not yet been identified. The receptor appears to be ubiquitous since both forms of the toxin interact with a large variety of cell lines.[9,10] The activated binding component linked to its receptor appears to induce a binding site for the smaller component, ι_a, which is about 45 kDa and has enzymatic properties.[9] The whole toxin is internalized, probably by receptor-mediated endocytosis. The binding component is probably involved in the translocation of the enzymatic component across the cytoplasmic membrane.[11]

In the cytosol, ι_a and s_a catalyze the monoADP-ribosylation of actin. ι-Toxin ADP-ribosylates both nonmuscular and muscular actin.[12] The NAD molecule is cleaved and the ADP-ribose moiety is transferred to Arg-177 of monomeric G-actin.[13–15] In the cells, the F-actin filaments are continuously polymerized at one end and depolymerized at the other. The ADP-ribosylated G-actin inhibits the polymerization of actin filaments resulting in a dissociation of the actin cytoskeleton. The intoxicated cells show morphological changes (rounding) and the permeability of cell monolayers is modified. The impairment of the barrier function of endothelium and epithelium tissues accounts for the pharmacological effects of these toxins, such as increase in vascular permeability, dramatic decrease of blood pressure, lung edema and intestinal hemorrhages.[16] Aktories[16] has reviewed these and other clostridial ADP-ribosylating toxins.

III. MOLECULAR ANALYSIS OF ι-TOXIN

The active site of the ADP-ribosylating toxins consists of a cavity formed by a β-strand and α-helix able to bind NAD, and flanked by two catalytic residues (Arg or His, and Glu).[17] Sequence comparison show that this structure is conserved in ι_a and that Arg-295 and Glu-380 (numbering of the mature proteins) can be aligned with catalytic residues of cholera toxin and pertussis toxin. Site-directed mutagenesis of these two residues induces a drastic decrease of the ADP-ribosylation activity indicating that Arg-295 and Glu-380 are involved in the active site of ι_a (S. Perrelle and M.R. Popoff, unpublished data). The ι-toxin genes have been characterized for *C. perfringens* type E strain NCIB10748 in which they are

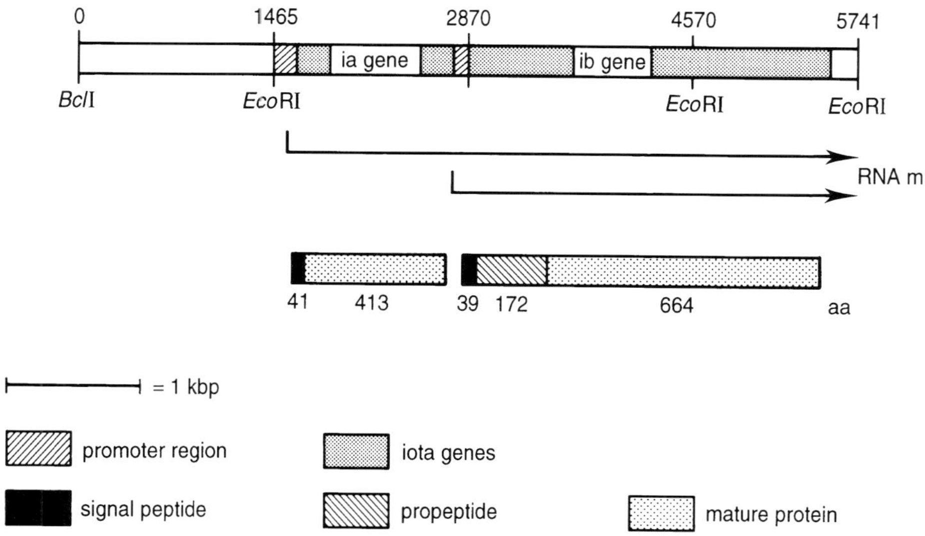

Figure 20.1 ι-Toxin genes, RNAm and deduced polypeptides.

localized on a large plasmid (about 140 kb). Components ι_a and ι_b are encoded by two genes transcribed in the same orientation and separated by 40 noncoding nucleotides.[11] The transcription start site has been identified 135 nucleotides upstream from the AUG codon of the ι_a gene. The ι_b gene can be transcribed from a promoter localized in the intergenic region but less efficiently than from the ι_a promoter. This result suggests that the ι_a and ι_b genes form a bi-cistronic operon under the control of the promoter upstream of the ι_b gene. We have also found that a protein related to the Hpr transition state regulator of *Bacillus subtilis* is involved in the ι gene regulation (S. Perelle and M.R. Popoff, unpublished data). Figure 20.1 shows the ι-toxin genes and their predicted polypeptides.

A. ι_a

The ι_a gene encodes a 454 amino acid protein. The *N*-terminal 41 amino acids display the characteristics of a signal peptide (charged *N*-terminal residues with a long hydrophobic core). Mature ι_a consists of 413 amino acids with a predicted molecular mass of 47 588 Da, which is very similar to native ι_a (47.5 kDa) and s_a from *C. spiroforme* (44 kDa).

The *C*-terminal part of ι_a has 23.4% identity with the C3 ADP-ribosyltransferase, but ι_a is not significantly related to other ADP-ribosylating toxins. It has been found that the NAD^+ binding and catalytic site of ADP-ribosylating toxins consists of a conserved tertiary structure showing low homology in amino acid sequences.[17] Using site-directed mutagenesis, Glu-421 has been found to be essential for the enzymatic activity of ι_a (S. Perelle and M.R. Popoff, unpublished data).

B. ι_b

Clostridium perfringens ι_b consists of 875 amino acids and contains three domains. The 39 *N*-terminal residues form an hydrophobic region which probably constitutes a signal peptide. The next 171 residues (position 40 to 211, 19.8 kDa) and 664 residues (position 212 to 875, 80.8 kDa) correspond to the propeptide and mature protein, respectively. This conclusion is confirmed by the protein *N*-terminal sequence of naturally activated s_b which matches residues 212–219 of the *C. perfringens* sequence. The presence of a propeptide located between the signal peptide and the mature protein is a common feature of most bacterial extracellular proteases[18] and of some other toxins such as *C. perfringens* ε-toxin. Unlike the propeptide of ε-toxin, which is short (13 residues) and highly basic[19] the ι_b propeptide is much longer and its release does not induce a significant change in the pI of the protein. ι_b has 33.9% homology with the protective antigen (PA) of anthrax toxin. PA has a similar structure (signal peptide of 29 residues, propeptide of 19.2 kDa and mature protein of 63 kDa). By analogy with PA, the *C*-terminal part of mature ι_b could be involved in the recognition of a cell surface receptor, and the *N*-terminal domain could be the translocation domain. However, PA acts in a slightly different manner than ι_b. The PA precursor binds to the cell surface receptor and is cleaved by a cell surface protease, releasing the propeptide and mature PA which remains bound to the cell, and is able to bind and translocate either edema factor or lethal factor.[20] Interestingly, the translocation domains of ι_b and PA are the most conserved regions (45% identity) and contain a hydrophobic sequence (Leu-292 to Ser-309 in ι_b sequence), which is predicted to form a transmembrane segment. This suggests that ι_b and PA could share a common mechanism of translocation of the enzymatic component across the cell membrane.

IV. DIVERSITY OF *CLOSTRIDIUM* BINARY TOXINS

A. Variant strains deficient in binding component

Certain clostridial strains produce only the enzymatic component and not the binding component, but still exhibit cryptic binding component genes as shown by DNA–DNA hybridization. One *C. difficile* strain (CD196) has been found to produce an actin-specific ADP-ribosyltransferase (CDT), which is different from *C. difficile* toxins A and B.[14] CDT is immunologically closely related to ι_a and s_a. This component is ineffective on Vero cells or in mice and no functional binding component for CDT has been found in culture supernatant of CD196, although the ι_b gene used as a probe hybridizes with CD196 DNA. However, combinations of CDT with ι_b or s_b are cytotoxic and lethal. The concentration of CDT needed for these activities is similar to the level of either ι_a and s_a required for activity.

A *C. perfringens* strain determined as type A by the standard method has been recognized to harbor ι_a and ι_b genes by DNA–DNA hybridization. But, the restriction patterns of ι_a and ι_b genes of this strain and type E strain NCIB10748 are different. This indicates a divergent evolution of ι_a and ι_b genes in these two strains. The culture supernatant of the defective strain displays actin-specific ADP-ribosyltransferase activity but not functional binding component, and is cytotoxic for Vero cells in combination with extraneous ι_b or s_b. This suggests that the enzymatic and binding component genes of the clostridial binary toxins probably derive from common ancestors, respectively. The enzymatic and binding component genes have probably been transferred together between *Clostridium* species and have evolved independently from each other.

B. Other clostridial binary toxins

Two families of clostridial binary toxins can be defined on the basis of the immunological relatedness, and functional complementation between the enzymatic and binding components. One family consists of ι-toxin from *C. perfringens* type E, *C. spiroforme* (ι) toxin, and the enzymatic components produced by certain defective strains of *C. perfringens* type E and *C. difficile* (ι-like family), and the other family of C2 toxin of *C. botulinum* types C and D (C2 toxin family). Beyond their binary nature, the C2 toxins are unrelated to the ι-toxins, either genetically or immunologically.

V. SUMMARY

The ι-toxins produced by *C. perfringens* type E and *C. spiroforme* species cause diarrhea in several species of animal. Both toxins consist of a binding and active, or enzymatic, component. There are only slight differences between the physical properties and amino acid sequences of the related components. They also exhibit a high degree of cross-immunological reactivity. Both components are needed for biological activity, and both homologous and heterologous reconstituted toxins have virtually identical biological activities.

VI. REFERENCES

1. Borriello S.P. and Carman R.J. (1985) Clostridial diseases in the gastrointestinal tract in animals. In Borriello S.P. (ed.), *Clostridia in Gastrointestinal Disease*, pp. 195–221. CRC Press, Boca Raton, FL.
2. Carman R.J., van Tassell R.L., and Wilkins T.D. (1987) Production of iota toxin by *Clostridium spiroforme*: a requirement for divalent cations. *Vet Microbiol*, **15**: 115–120.
3. Collins M.D., Lawson P.A., Willems A., Cordoba J.J., Fernandez-Garayzabal J., Garcia P., Cai J., Hippe H. and Farrow J.A.E. (1994) The phylogeny of the genus *Clostridium*: proposal of five new genera and eleven new species combinations. *Int J Syst Bacteriol* **44**: 812–826.
4. Madden D.L., Horton R.E. and McCullough N.B. (1970) Spontaneous infection in ex-germfree guinea pigs due to *Clostridium perfringens. Lab Anim Care*, **20**: 454–455.
5. Baudieri S., Borriello S.P., Pantosti A., Luzzi I., Testore G.P. and Panichi G. (1986) Diarrhoea associated with toxigenic *Clostridium spiroforme. J Infect* **12**: 278–279.
6. Sterne M. and Batty I. (1975) *Pathogenic Clostridia*. Butterworths, London.
7. Butt M.T., Papendick R.E., Carbonne L.G. and Quimby F.W. (1994) A cytotoxicity assay for *Clostridium spiroforme* enterotoxin in cecal fluid of rabbits. *Lab Anim Sci* **44**: 52–54.
8. Carman R.J. and Evans R.L. (1984) Experimental and spontaneous clostridial enteropathies of laboratory and free living lagomorphs. *Lab Anim Sci* **34**: 443–452.
9. Considine R.V. and Simpson L.L. (1991) Cellular and molecular actions of binary toxins possessing ADP-ribosyltransferase activity. *Toxicon* **29**: 913–936.
10. Fritz G., Schroeder P. and Aktories K. (1995) Isolation and characterization of a *Clostridium botulinum* C2 toxin-resistant cell line: evidence for possible involvement of the cellular C2II receptor in growth regulation. *Infect Immun* **63**: 2334–2340.
11. Perelle S., Gibert M., Boquet P. and Popoff M.R. (1993) Characterization of *Clostridium perfringens* iota-toxin genes and expression in *Escherichia coli. Infect Immun* **61**: 5147–5156.
12. Mauss S., Chaponnier C., Just I., Aktories K. and Gabbiani G. (1990) ADP-ribosylation of actin isoforms by *Clostridium botulinum* C2 toxin and *Clostridium perfringens* iota toxin. *Eur J Biochem* **194**: 237–241.
13. Aktories K., Barmann M., Ohishi I., Tsuyama S., Jakobs K.H. and Habermann E. (1986) Botulinum C2 toxin ADP-ribosylates actin. *Nature* **322**: 390–392.
14. Popoff, M.R. and Boquet P. (1988) *Clostridium spiroforme* toxin is a binary toxin which ADP-ribosylates cellular actin. *Biochem Biophys Res Commun* **152**: 1361–1368.
15. Simpson L.L., Stiles B.G., Zepeda H.H. and Wilkins T.D. (1987) Molecular basis for the pathological actions of *Clostridium perfringens* iota toxin. *Infect Immun* **55**: 118–122.
16. Aktories K. (1994) Clostridial ADP-ribosylating toxins; effects on ATP and GTP-binding proteins. *Mol Cell Biochem* **138**: 167–176.
17. Domenighini M., Magagnoli C., Pizza M. and Rappuoli R. (1994) Common features of the NAD-binding and catalytic site of ADP-ribosylating toxins. *Mol Microbiol* **14**: 41–50.
18. Vandersman C. (1989) Secretion, processing and activation of bacterial extracellular proteases. *Mol Microbiol* **3**: 1825–1831.
19. Hunter S.E., Clarke I.N., Kelly D.C. and Titball R.W. (1992) Cloning and nucleotide sequencing of the

Clostridium perfringens epsilon-toxin gene and its expression in *Escherichia coli*. *Infect Immun* **60**: 102–110.

20. Leppla S.H. (1991) The anthrax toxin complex. In Alouf J.E. and Freer J. (eds), *Sourcebook of Bacterial Protein Toxins*, pp. 277–302. Academic Press, New York.

Chapter 21

Molecular Biology of the *Clostridium difficile* Toxins

J. SCOTT MONCRIEF, DAVID M. LYERLY
AND TRACY D. WILKINS
Department of Biochemistry and Anaerobic Microbiology, Virginia Polytechnic Institute and State University, Blacksburg, Virginia, USA

I. INTRODUCTION

At the time it was first implicated as the cause of pseudomembranous colitis (PMC), *Clostridium difficile* was almost an unknown species of bacteria. First isolated from healthy newborn infants by Hall and O'Toole in 1935, it was named *Bacillus difficilis* after the apparent difficulty the researchers encountered in isolating it.[1] The organism was shown to produce a toxic culture filtrate which killed animals upon injection. Snyder[2] confirmed and extended these findings and also showed that the toxic activity was neutralized by antisera to *C. difficile*. Because the organism did not appear to cause disease, few studies followed its initial discovery.

Over 40 years later, in 1977, *C. difficile* was implicated as the major causative agent of PMC. It was first suspected that the disease was caused by viruses. In a major breakthrough, however, it was discovered that the cytotoxic activity in diarrheal specimens from PMC patients was neutralized by antiserum consisting of a mixture of clostridial antitoxins. The specific neutralizing antitoxin in the

The Clostridia: Molecular Biology and Pathogenesis
ISBN 0-12-595020-9

mixture was identified as *Clostridium sordellii* antitoxin. Curiously, investigators were unable to isolate *C. sordellii* from the patients' stools. They soon showed, however, that *C. difficile* could be isolated consistently from the cytotoxic fecal specimens and that cytotoxic activity from *C. difficile* was neutralized by *C. sordellii* antitoxin. Furthermore, they showed that a similar disease in Golden Syrian hamsters was caused by the toxigenic *C. difficile* isolates from PMC patients. Thus, the fortuitous cross neutralizing activity of *C. sordellii* antitoxin led to the discovery of *C. difficile* as the cause of PMC (see Chapter 8).[3–16]

A number of investigators began work on purification and characterization of the cytotoxic factor produced by *C. difficile*. It was initially assumed that a single cytotoxin was responsible for the activity. It was soon demonstrated, however, that the organism produced at least two toxins which could be separated by anion-exchange chromatography.[17,18] The toxin which eluted first and had weak cytotoxic activity was designated toxin A. The second toxin, designated toxin B, was comparatively more cytotoxic. It is responsible for the cytotoxic activity observed in fecal specimens from PMC patients and masks the cytotoxic activity of toxin A. Toxin A also possesses enterotoxic activity, and is responsible for the initial damage to the intestine during the course of disease.

The disease itself was first described by Finney in 1893, following intestinal surgery on a 22-year-old woman.[19] The woman died 15 days after surgery and the autopsy showed 'diphtheritic colitis'. Since the initial discovery of *C. difficile* as the major etiological agent of PMC, extensive research has established it as a major enteric pathogen. In addition to being the cause of most cases of PMC, it also causes a less severe, but more common, form of the disease termed antibiotic associated diarrhea (ADD). *Clostridium difficile* is now recognized as the primary cause of ADD in the United States.

Further research established the toxins as the major virulence factors in *C. difficile* disease. Protection against both toxins are consistently found in fecal specimens from patients and experimental animals with the disease.[20] Moreover, the intestinal pathology and death caused by the toxins in experimental animals mimics that of the disease itself.[21] Furthermore, effective vaccination of experimental animals against the disease requires protection against both toxins:[21,22] vaccination against either toxin alone is insufficient. As a result, much of the research on *C. difficile* disease has focused on purification and characterization of the molecular and biological properties of toxins A and B.

Toxins A and B are the largest known bacterial toxins, with M_r of 308 000 and 270 000, respectively. This led to some confusion in initial efforts to purify and characterize the toxins because many researchers had a difficult time accepting the idea that the toxins could be this large. Compounding the problem was the fact that toxin B is an extremely potent cytotoxin. In some studies, protein fractions attributed to 'purified toxin' erroneously showed lower molecule weight proteins when analyzed by gel electrophoresis. Cloning, sequencing and expression of the toxin genes in *Escherichia coli* established that toxins A and B are indeed the largest known bacterial toxins and are each composed of a single polypeptide.[23–26] This chapter focuses on the recent advances in reseach on these unusual and interesting toxins, with an emphasis on molecular biology.

II. PURIFICATION AND CHARACTERIZATION OF TOXINS A AND B

A. Toxin production

Numerous media have been used for the growth of *C. difficile* and production of the toxins. Taylor and Bartlett [27] purified toxin B from cultures grown anaerobically in brain–heart infusion broth. Rolfe and Finegold[28] reported that chopped meat–glucose broth, brain–heart infusion broth, and a basal synthetic medium supplemented with proteose peptone gave the highest yields of cytotoxic activity. Synthetic media have also been described for the production of the toxins, however, the density of growth achievable is considerably lower than that in rich media.[29,30] The highest levels of toxin can be produced in dialysis sac cultures. The method was originally developed by Sterne and Wentzel for the production of botulinum toxin and works quite well for the production of toxins A and B.[31] In this method the organisms are grown in a dialysis sac suspended in a rich media such as brain–heart infusion broth. The organisms grow slowly and to a very high density within the dialysis sac. The toxins are released within the sac and, thus, are concentrated in a small volume. Further, the membrane of the dialysis sac excludes any high molecular weight material from the culture filtrate, eliminating any interference during purification. This method has been used routinely to produce culture filtrate for purification of the toxins. More recently, Kazamias *et al.*[32] produced unusually high cytotoxicity titers by growth in alkaline trypticase–yeast extract medium supplemented with mannitol. The level of toxin A in this medium was not reported. The toxins are produced and released post-exponentially, near the beginning of the decline phase of the bacterial growth cycle.[33] Although the toxins are produced during stationary phase, there is no correlation of toxin production with sporulation as occurs with *C. perfringens* enterotoxin.[34]

Toxin-producing strains almost invariably produce both toxins. However, a strain that produces only one toxin, an unusual form of toxin B, has been described recently,[35–37] but this is a rare exception. The toxins appear to be coordinately regulated, since toxigenic strains always produce either high or low levels of each toxin The level of toxins produced can vary as much as 10 000-

fold among toxigenic *C. difficile*. As a result, the isolation and maintenance of a high-producing toxigenic strain was an important step in research on the toxins. The most extensively studied toxigenic strain has been VPI 10463 (also ATCC 43255), first characterized by Sullivan *et al.* in 1982.[18] This strain consistently produces high levels of both toxins and has been used for most studies on the biological and molecular properties of toxins A and B.

B. Purification and physicochemical properties

A number of methods have been developed for the purification of toxins A and B from *C. difficile* culture filtrate.[18,38–40] In one method the toxins are separated by anion exchange chromatography on DEAE-Sepharose CL-6B. Toxin A does not bind under high salt conditions and, therefore, in most media toxin A will pass through the column while toxin B binds to Sepharose CL-6B. The column is then washed with Tris-HCl pH 7.5 buffer containing 0.25 M NaCl, followed by elution of toxin B with Tris-HCl 7.5 buffer containing 0.5 M NaCl. Toxin A is then purified from the media from which toxin B was removed. In the first step toxin A is precipitated by ammonium sulfate at 40% saturation. The precipitant is dialyzed against Tris-HCl pH 7.5 buffer containing 0.15 M NaCl. The solution is then passed through a column containing Sepharose CL-6B which removes contaminants while toxin A passes through. Toxin A is then further purified by acid precipitation in dialysis tubing with acetate buffer at pH 5.6. Other classical methods of protein purification have been employed to purify the toxins. In general, the methods require careful attention to separation of toxins A and B owing to the similar size of the toxins. In the case of toxin B it is also particularly difficult to separate the enolase enzyme. Enolase of *C. difficile* is a protein of *c.* M_r 50 000 which has a similar elution profile to that of toxin B. It also forms a large molecular weight multimeric protein under native conditions. Thus, toxin B preparations are often contaminated with enolase. This caused some confusion in previous work on toxin B, which was erroneously reported to contain enolase activity.[41,42]

Toxin A can also be purified by thermal affinity chromatography based on its ability to bind to immobilized bovine thyroglobulin.[40] Bovine thyroglobulin contains large amounts of the toxin A carbohydrate receptor Galα1–3Galβ1–4GlcNAc. Toxin A binds the carbohydrate tightly at 4°C, but binding is markedly decreased at 37°C. An affinity matrix can be prepared by covalently coupling thyroglobulin to agarose. Toxin A binds tightly to the affinity column at 4°C. Contaminating proteins are removed by washing at 4°C. Toxin A is then eluted at 37°C. Elution is gentle and yields highly purified toxin A with full biological activity.

Toxin A purified by Banno *et al.*[17,38] and our research group[18,39] exhibited similar physicochemical properties. These researchers all reported the toxin to be unusually large. Under denaturing conditions the toxin had an $M_r > 250\,000$ and did not dissociate into subunits. There was controversy about the large molecular weight of toxin A, as others reported that purified toxin had an estimated M_r of 52 000.[43] Cloning and sequencing of the toxin A gene, however, has established that the toxin is extremely large with an M_r of 308 000 calculated from the deduced amino acid sequence.[24]

Toxin B purified by Banno *et al.*,[17,38] Taylor *et al.*,[27] and our group[18,39] was also found to be an extremely large toxin. Like toxin A, under denaturing conditions the toxin did not dissociate and has an M_r of at least 250 000. It has also been erroneously reported that toxin B is composed of lower molecular weight subunits estimated at M_r 50 000.[44] As with toxin A, however, the large size of toxin B has been established by cloning and sequencing of the toxin B gene.[23] The calculated M_r from the deduced amino acid sequence of toxin B is 270 000.

Toxin B is more susceptible to extremes of pH and proteases than toxin A.[18,39] Both toxins are inactivated by oxidizing agents and can be protected by reducing agents.[18,39] The toxins contain few cysteine residues, and the fact that they are not inactivated by reducing agents indicates that disulfide bonds are probably not involved in toxin conformation. Site-directed mutagenesis of the toxin B gene, however, indicates that conserved cysteine residues of the toxins are important for cytotoxic activity (see below). It has been determined that both toxins contain zinc and iron. Moreover, the toxins are inhibited by the zinc-specific chelator 1,10-phenanthroline (J.S. Moncrief, D.M. Lyerly and T.D. Wilkins, unpublished data). A number of other enzymes are known to require free sulfhydryls for zinc-binding.[45] Free sulfhydryls, therefore, may play a similar role in toxins A and B. However, this has not been determined.

C. Biological properties

Toxin A is an extremely potent enterotoxin.[46,47] On a molar basis it is as active as cholera toxin in the rabbit ileal loop assay for enterotoxic activity. Its mechanism of action, however, is quite different. Unlike cholera toxin, toxin A causes extensive damage to the intestine. The villus tips of the epithelium are initially disrupted, followed by damage to the brush border membrane. The mucosa eventually becomes eroded. Damage to the intestinal mucosa is accompanied by extensive infiltration with inflammatory neutrophils, and this probably plays an important role in the extensive damage caused. The fluid response presumably results from the tissue damage. It has been observed, however, that the cytotoxic activity of toxin A results in disruption of the tight junctions and this alone may play a role in the initial fluid loss.[48] For example, the disruption of tight junctions is responsible for the enterotoxicity of the zonula occludens toxin of *Vibrio cholera*.[49]

Despite the fact that toxin B is a much more potent cytotoxin than toxin A, it does not cause a fluid response in animal models.[46,47] The lack of enterotoxicity of toxin B may be due to an inability to bind tightly to a receptor on the brush border membrane cells that line the intestine. Thus, toxin A, which is able to bind to specific carbohydrate receptors on the surface of intestinal cells, initiates damage to the intestine. Toxin B then gains access to the underlying tissues and contributes to the extensive damage during the course of disease. This hypothesis is supported by experiments on the effects of the toxins in the hamster intestine.[50] In these studies, we found that toxin B given intragastrically to hamsters does not cause a response. However, when given along with sublethal amounts of toxin A, the animals died with no observable intestinal pathology. Similarly, if the intestine is physically disrupted, toxin B gains access to the systemic circulation and causes death.

It has recently been observed that toxin B is more potent than toxin A in damaging human colonic epithelium *in vitro*.[51] In this study, human colonic mucosal sheets in Ussing chambers were treated with the toxins. Both toxins caused damage to the mucosa and resulted in decreased electrical resistance and increased permeability. The effects of both toxins were dose dependent, but toxin B was approximately 10 times more potent on a molar basis. Toxin B is cytotoxic *in vitro* to all tissue culture cells and the relative importance of each of the toxins in human disease requires more research.

Both toxins cause a similar type of rounding of tissue culture cells, and are active against every cell line which has been tested. Toxin A is far less cytotoxic than toxin B on most cell lines; however, when tested on certain cell lines, particularly those of intestinal origin, its activity approaches that of toxin B.[52] This is presumably due to an increased density of toxin A carbohydrate receptors on these cell lines. Rounding of intoxicated cells is accompanied by disassembly of the actin microfilaments, whereas microtubules and intermediate filaments are only affected secondarily.[53] Although the microfilament cytoskeleton is preferentially affected, actin itself does not appear to be the target.[54,55] Rearrangement of actin, and also the actin-binding proteins vinculin and talin, has been reported.[53] Some recent research has revealed the molecular mode of action by which these toxins cause disassembly of the actin cytoskeleton (see below).

Both toxins are lethal when injected intraperitoneally or subcutaneously in small amounts.[46] Contrary to the cytotoxic activity of the toxins, the lethal activity of toxin A is at least as potent as toxin B, with both toxins having a minimum lethal dose of 50 ng. The mechanism by which the toxins cause death when administered systemically is unknown. In mice injected intraperitoneally there is some evidence of liver damage.[56] Arnon *et al.*[57] examined the effects of the toxins in rhesus monkeys and noted that the animals apparently died from cessation of breathing. Death is not accompanied by the floppy paralysis characteristic of the botulinum and tetanus neurotoxins.

D. Toxin receptors

Both toxins bind to the cell surface and cluster in coated pits, indicating that receptor-mediated endocytosis is involved in toxin uptake.[58,59] Each of the toxins contains a large series of repeating units at its carboxyl end. In the case of toxin A, the repeating units are responsible for binding to specific cell surface carbohydrate receptors.[26] Toxin A is multivalent as shown by its ability to agglutinate rabbit erythrocytes. Agglutination occurs at 4°C, but not at 37°C, and is specific for rabbit erythrocytes which contain the carbohydrate Galα1–3Galβ1–4GlcNAc. This property of the toxin led to the discovery and characterization of the carbohydrate portion of the receptors.[60] We found that toxin A binds to hamster intestinal brush border cells which also contain Galα1–3Galβ1–4GlcNAc. This led to speculation that this was the functional carbohydrate receptor in humans. However, this trisaccharide is not expressed by normal human cells, and therefore, cannot be the receptor for toxin A in the human intestine. We later determined, however, that toxin A also binds to the carbohydrate antigens I, X, and Y, all of which have a conformationally similar type 2 core structure.[61] Moreover, binding occurs at 37°C and these antigens are all expressed on the surface of the human intestinal epithelium suggesting they may function as receptors for the toxin. It should be noted that the X antigen is also present in large amounts on the surface of human neutrophils, indicating that toxin A may also target these cells. A 163 kDa glycoprotein receptor for toxin A in hamsters has been purified from brush border cells from the small intestine of infant hamsters.[62] Binding was inhibited by lectins specific for galactose and *N*-acetylglucosamine and by nonimmunoglobulin components of human milk.[63] Glycoprotein receptors from human intestinal epithelial cells have not been identified.

It is interesting that toxin A does not bind to rabbit erythrocytes at 37°C. This is probably due to the low density of receptors on the erythrocyte surface. The affinity of the repeating units for a single carbohydrate receptor appears to be low at 37°C. Thus, effective binding at elevated temperatures requires a high density of receptors. Toxin A thus appears to have evolved an interesting strategy to target specific cells: by binding tightly only to surfaces with high receptor density, it is able to find its target.

Table 21.1. Physicochemical and biological properties of toxins A and B

Property	Toxin A	Toxin B
M_r[a]	308 000	270 000
pI[a]	5.3	4.1
Cytotoxic dose[b]	10 ng	1 pg
Enterotoxic dose[b]	1 μg	Negative
Lethal dose[b]	50 ng	50 ng
Hemagglutination[c]	Yes	No
Receptors	Galα1–3Galβ1–4GlcNAc I, X, and Y antigens	Unknown
Metal ion content	Zn, Fe	Zn, Fe

[a] Based on the deduced amino acid sequence of the toxin genes.
[b] Minimum dose as reported by Lyerly *et al.*[46]
[c] Agglutination of rabbit erythrocytes at 4°C.

A receptor(s) for toxin B has not been determined. Wheat germ agglutinin, concanavalin A, and ricin have been reported to delay the cytotoxic effect of toxin B, suggesting that a carbohydrate receptor may be present on tissue culture cells.[64] It is assumed that the repeating units of toxin B also function in binding to the cell surface, however, there is no formal proof of this assumption. Deletion mutants of toxin B which lack the repeating units are tenfold less cytotoxic than the native toxin (see below). This suggests that this portion of the molecule functions in binding, however, the toxin can enter tissue culture cells in the absence of receptor-mediated endocytosis. Similar results have been observed with deletion mutants of toxin A which lack the repeating units (Wilkins, unpublished data). Thus, both toxins can enter tissue culture cells by nonspecific uptake. The physicochemical and biological properties of toxins A and B are summarized in Table 21.1.

III. MOLECULAR GENETICS OF THE TOXINS

A. Cloning, sequencing and expression of the toxin A gene

In 1987, Muldrow *et al.*[65] reported the cloning and expression of a 0.3 kb fragment of the toxin A gene from *C. difficile* by screening of a λ gt11 library with toxin A antisera. Wren *et al.*[66] also described λ gt11 clones containing a toxin A gene fragment. The clone expressed a protein that caused elongation of Chinese hamster ovary cells and had an estimated molecular weight of 235 000. In both cases, however, the clones were unstable. In 1987, we reported cloning of the carbohydrate binding portion of toxin A in pBR322.[26] The cloned 4.7 kb *Pst*I restriction fragment expressed antigen that reacted with affinity-purified monospecific antibodies against toxin A. Recombinant lysates of the clone also agglutinated rabbit erythrocytes and reacted with monoclonal antibody to toxin A. However, no cytotoxic activity was observed. In further studies, the clone (designated pCD11), was used for chromosomal walking to clone flanking regions of the *C. difficile* VPI 10463 chromosome.[24,25] It was found that the genes of toxins A and B were located in close proximity and this work led to the cloning and sequencing of both toxins (Figure 21.1).

Dove *et al.*[24] sequenced the entire toxin A gene and confirmed that the toxin is indeed, as indicated by protein chemistry studies, an unusually large protein. The structural gene is 8130 nucleotides long and encodes a protein of 2710 amino acids with a deduced M_r of 308 000. The toxin A gene contains a 26.9% G+C content, consistent with the overall G+C content of 28 mol% of the *C. difficile* chromosome.

Previous findings that toxin A agglutinates rabbit erythrocytes and is precipitated by monoclonal antibody PCG-4, demonstrated its multivalent properties and suggested the possibility that the toxin contains repeats. This was also confirmed by the sequencing of the toxin A gene. Nearly one-third of the gene at the 3′ end encodes a series of 38 contiguous repeating units composed of 855 amino acids. The repeating units were grouped into classes I and II, based on length and low levels of sequence similarity. The repeating units contain seven class I repeating units, each 30 amino acids long. The additional 31 class II repeats, were further subdivided into four groups, A, B, C, and D. Group II repeats contain 20 or 21 amino acids each. The repeating units form the domain responsible for the binding and agglutination properties of toxin A. The repetitive domains of the toxins are probably the site of the size-variant toxins which have been observed.[67] They are also the immunodominant portion of the molecule,[67] and contain at least two epitopes for the monoclonal antibody PCG-4.[68]

The deduced amino acid sequence of toxin A does not contain any sequence consistent with a signal peptide at its amino-terminal end. This is in agreement with lack of post-translational processing of the *N*-terminal end of the protein. Toxin A contains a small hydrophobic C-terminal sequence following the repeating units. Therefore, if toxin A is secreted it may be by an atypical pathway, such as those mediated by a C-terminal secretion signal.[73] *C*-terminal signals are known to mediate the export of a number of bacterial proteins.[69] Whether toxin A is secreted, or simply released upon lysis requires further research.

An additional open reading frame was identified immediately 5′ to the toxin A gene, with transcription in the same direction as the toxin A gene.[24] It encodes a small hydrophilic protein of *c.* M_r 19 000 or *c.* M_r 16 000, depending on the actual start codon. The smaller protein contains a potential signal sequence similar to those of prokaryotic lipoproteins. The function of the protein is unknown.

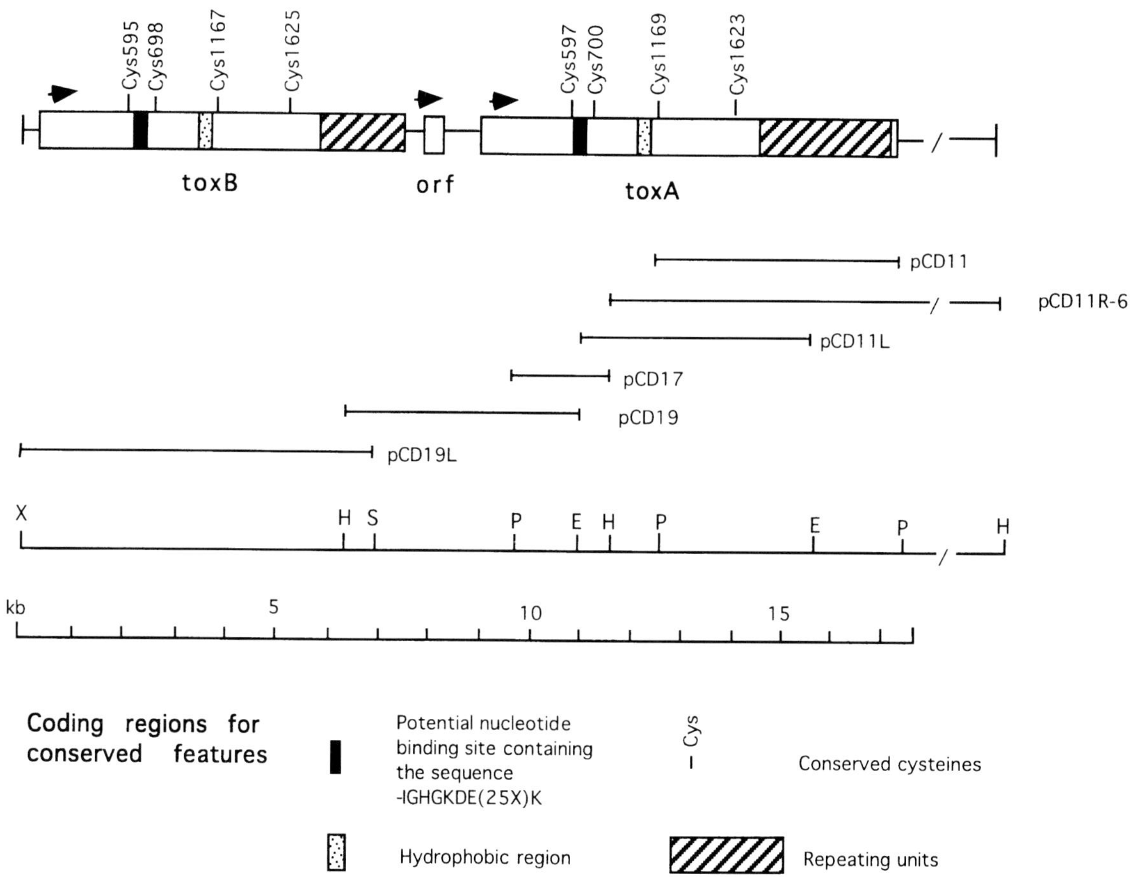

Figure 21.1 Organization of the toxin A and B genes. The toxin genes are separated by 1350 bp of DNA which contains an open reading frame (orf) of unknown function. Arrows indicate the direction of transcription of each gene. Clones pCD11, pCD11R-6, pCD11L, pCD17, pCD19, and pCD19L used to sequence and reconstruct the toxin genes are shown below the genetic map along with a partial restriction map of the region. Restriction enzymes are: E, *Eco*RI; H, *Hin*dIII; P, *Pst*I; S, *Sph*I. The toxins share a potential nucleotide-binding site, a hydrophobic region of about 50 amino acids, four conserved cysteines, and a series of contiguous repeating units at the carboxy terminus. Coding regions for conserved features of the toxins are illustrated as indicated below the genetic map.

The entire toxin A gene was reconstructed from cloned fragments of the toxin A gene and expressed in *E. coli*. Lysates of the recombinant *E. coli* expressing the toxin A gene contained the cytotoxic activity, enterotoxic activity and lethal properties of the toxins.[70] Moreover, the activity of the lysates was inhibited by antibody to toxin A. This work provided definitive proof that the toxic activity is attributable solely to the large protein produced by the toxin A gene.

B. Cloning, sequencing and expression of the toxin B gene

The 3′ end of the toxin B gene is located 1350 bp upstream of the toxin A initiation codon. Chromosomal walking experiments used for cloning the toxin A gene thus led directly to cloning of the toxin B gene.[25] Clone pCD19, which contains the 5′-end of the toxin A gene and a small open reading frame, was found to contain 1.2 kb of DNA which, when subcloned expressed a nontoxic peptide that reacted with toxin B antibodies. A 6.8 kb fragment, pCD19L, that overlapped the pCD19 clone by 0.8 was found to contain the remainder of the toxin B gene. The two overlapping fragments were used to reconstruct the full-length toxin gene. The protein expressed by the complete gene showed complete immunological identity with toxin B and was cytotoxic and lethal. Immunoblot analysis of the expressed full-length recombinant toxin showed that toxin B had an M_r of 240 000 or greater. One surprising finding of this study was that the cloned toxin B gene was expressed quite well in *E. coli* under control of its own promoter. This made it possible to proceed with mutagenesis studies on the toxin B gene in *E. coli* (see below).

The toxin B gene has been sequenced.[23] The structural gene is 7098 nucleotides long and has a deduced amino acid sequence of 2366 (M_r 270 000). Once again, the gene sequence confirmed earlier findings that toxin B, like toxin A, is an extremely large protein expressed as a single polypeptide.

The toxin A and B genes have also been cloned from VPI 10463 by Eichel-Steiber *et al.*[71] The sequencing data are identical to that reported by Dove *et al.*[24] and Barroso *et al.*[23]

C. Sequence identity of toxins A and B

Comparison of the deduced amino acid sequences of the toxin A and B genes confirmed their relatedness at the structural level. A comparative sequence analysis has been reported by Eichel-Streiber *et al.*[71] Alignment of the amino acid sequences showed extensive sequence identity; the toxins share 44.8%

identical amino acids and are 63.1% similar if conservative substitutions are considered.

The sequence identity is striking in some areas. One of these areas in the center of the toxins consists of a large number of hydrophobic amino acids. Other bacterial toxins have hydrophobic regions near the center of the molecule and these are thought to function as membrane-spanning regions necessary for toxin transport. Whether the hydrophobic center of toxins A and B has a similar function has not been determined. A putative nucleotide binding site is located within the first one-third of each of the molecules, between the beginning of the toxins and the hydrophobic region. In other proteins this consensus motif has been reported to function in binding to ATP.[72] Additional clusters with exceptional sequence identity are also present within the region between the hydrophobic area and the beginning of the repeating units.[71]

The toxins share four conserved cysteines. Site-directed mutagenesis indicates that at least two of these conserved cysteines are important in toxin function (see below). Reducing agents do not affect the toxins, and therefore, the cysteines are apparently not involved in formation of disulfide bridges critical to toxin function. Instead, the cysteine residues may function as free sulfhydryls involved in ligand binding. Alternatively, the conserved cysteines may be strategically located near processing points essential for toxin activation.

The extensive sequence identity and similar structure and function of the toxins, combined with the fact that the genes are in close proximity on the chromosome, indicate that the genes may be the result of gene duplication.[71] Furthermore, considering the homology and close proximity of the toxin genes it would not be surprising to find altered toxins arising from recombination events. A toxin variant has recently been described in some detail.[35–37]

Putative promoters and transcriptional terminators of the toxin A and B genes have been proposed by Eichel-Streiber.[71] The proposed promoters have unusual distances between the transcriptional start site and the ATG start codons, i.e. 169 nucleotides (nt) for toxin A and 239 nt for toxin B. Similar spacing is found in the promoter regions of tetanus and botulinum toxins. Whether these are the functional promoters of the toxin genes has not been determined. Stem loop structures beginning 69 nt downstream of the toxin A stop codon and 100 nt downstream of the toxin B stop codon were proposed as likely transcriptional terminators.

D. The repeating units

Toxins A and B both contain a series of repeating units at the carboxy terminus that comprise nearly one-third of the molecule. The repeating units of the toxins share homology with the carbohydrate binding region of streptococcal glycosyl transferases (GTF).[73] The GTFs and repeating units of toxins A and B are rich in aromatic amino acids and, with rare exceptions, each unit contains three consecutive aromatic amino acids, YYF. All the repeats have the dipeptide YF and this feature has been used to align the sequences for classification into subgroups based on additional sequence similarity.[24,73] It has been proposed that the repeats have a modular design in which a stretch of aromatic amino acids function in primary protein–carbohydrate interaction.[74] The affinity for binding is amplified by the repetition of the sequence and binding specificity is determined by a second, as yet undefined, domain characteristic of the arrangement and sequence of the repetitive structure. In this manner the proteins have evolved an ability to bind different carbohydrate structures based on a similar fundamental unit.

IV. MECHANISM OF ACTION

Because of the similarity of toxins A and B, both structurally and in their effect on cells, they probably have a similar mechanism of action. Although toxin B is significantly more potent as a cytotoxin, the effects of each toxin on tissue culture cells is qualitatively very similar. Intoxicated cells exhibit a retraction of cell processes and rounding of the cell body.[58,59] The cytopathic effect results from disassembly of filamentous actin, accompanied by a decrease in F-actin and an increase in G-actin prior to the onset of cell rounding. The fact that the toxins are active in very small amounts supports the idea that the mechanism is dependent on enzymatic activity within the target cell. Since the toxins cause cell rounding by disruption of the actin cytoskeleton, proteins regulating this process are likely targets for the toxins.

The Rho/Rac family of low molecular weight GTP-binding proteins are therefore ideal candidates for modification by the toxins. This Ras-related family of proteins, which includes RhoA, RhoB, RhoC, Rac 1, Rac 2, and CDC42, has recently been implicated in the regulation of actin microfilament assembly (see Chapter 22).[75] Additionally, C3 exoenzyme of *C. botulinum* and the newly discovered exoenzyme of *C. limosum* catalyze ADP-ribosylation of Rho and Rac proteins.[76,77] Thus, a precedence exists for targeting of these proteins by clostridial enzymes.

Two research groups have now independently reported that toxins A and B do, in fact, modify Rho proteins.[78–80] The finding that Rho from cells pretreated with toxin B was resistant to modification by C3 provided the first evidence that the toxins modify Rho. Just *et al.*[78] first demonstrated that treatment of cells with toxin B resulted in a time- and concentration-dependent inhibition of C3-catalysed ADP-ribosylation of Rho in cell lysates. Furthermore, toxin B caused an inhibition of C3 activity on Rho, not only in intact cells, but also in cell lysates. RhoA–glutathione S-transferase (RhoA–GST) fusion protein added to the cell lysate showed a decrease in ADP-ribosylation after treatment of the lysate with toxin B. Immunoblot analysis revealed no obvious decrease in the amount or apparent molecular weight of RhoA–GST fusion protein in the toxin B treated cell lysates. However, ADP-ribosylation of RhoA–GST protein purified from toxin-B-treated cell lysate was inhibited, indicating a modification of Rho itself. Moreover, transfection of cells with the *rhoA* gene, under the control of a strong promoter, conferred protection from toxin B. The investigators concluded that toxin B acts directly or indirectly by modification of Rho, and that the cytopathic effect involves dysfunction of Rho.

A similar study by Just *et al.*[79] revealed that toxin A also inhibits C3 modification of Rho. This provided the first evidence that, as suspected, the toxins have a similar molecular mode of action. They also demonstrated that the action of toxin A is not due to proteolytic degradation of Rho or an intrinsic ADP-ribosyl transferase activity. As with toxin B, the activity occurs in cell lysates and also in intact cells treated with toxin A. The modification of Rho by the toxins was determined to be dependent on a heat-stable low molecular mass cytosolic factor.

Dillon *et al.*[80] later confirmed that modification of Rho was involved in the mechanism of action of toxins A and B. They extended these findings to show that Rho is covalently modified in cell lysates treated with the toxins and that the covalent modification prevents ADP-ribosylation of Rho by the C3 exoenzyme. An epitope tag from the c-Myc protein was used to follow the effects of toxin treatment on Rho in cells transfected with the Myc–RhoA fusion. Immunoblots of toxin-treated lysates compared with untreated samples showed a small (< 1 kDa) but detectable increase in the molecular weight of the Myc–RhoA fusion. Therefore, covalent modification of Rho in toxin-treated cell lysates was apparently due to the addition a very small molecule. Dillon *et al.*[80] also showed that microinjection of recombinant RhoA protected over 90% of the cells from toxin-induced rounding. Microinjection of the closely related low-molecular weight GTP-binding protein Ras did not confer protection, indicating that the molecular mode of action might be specific for the Rho family of regulatory proteins.

Conflicting results on the stability of Rho upon toxin treatment were observed in these studies. Just *et al.*[79] did not observe any apparent decrease in detectable RhoA levels by immunoblot analysis of toxin-treated lysates with anti-RhoA. In contrast, Dillon *et al.*[80] noted a marked decrease in detection of Rho by immunoblot analysis with anti-RhoA. In a separate but related experiment they used the Rho-specific probe GS–GDP dissociation inhibitor (GST–GDI) coupled to agarose beads to precipitate Rho from treated and untreated cell lysates. Consistent with the results from direct immunoblotting, Rho was detectable only in untreated lysates, further indicating that Rho was degraded upon toxin treatment. Of note here is that Just *et al.*[79] used RBL cells for the immunoblotting experiments, while Dillon *et al.*[81,82] examined lysates of toxin-treated NIH 3T3 cells. RBL cells express more immunoreactive Rho than NIH 3T3 cells, and this may explain the difference in observations on Rho stability.

Perhaps most interestingly, both groups of investigators determined that the activity of toxins A and B was not due to ADP-ribosylation of Rho. Many bacteria produce ADP-ribosyl transferases specific for Rho including *Clostridium* strains, *Bacillus cereus*, and *Staphylococcus aureus*.[81,82] Toxins A and B, however, covalently modify Rho by a novel mechanism.

The molecular mode of action has recently been determined. Both toxins modify members of the Rho family of proteins by UDP-glucose dependent glucosyltransferase activity.[83,84] Using electrospray mass spectrophotometry Aktories *et al.* determined that the toxins were increased by 162 Da, consistent with the incorporation of a hexose into Rho. They further showed that the covalent modification was dependent on UDP-glucose, since preincubation of recombinant Rho in the presence of toxin A inhibited ADP-ribosylation with C3 only when UDP-glucose, and not other forms of activated hexoses, were present. Using proteolytic peptide mapping the acceptor amino acid of glucosylation was identified as Thr-37 of RhoA. This is in close proximity to Asn-41 of Rho, which is ADP-ribosylated by C3, explaining why toxins A and B inhibit modification of Rho by C3. (A more detailed discussion of recent progress in understanding the precise mechanism by which toxins A and B modify Rho is presented in the following chapter.)

Cytotoxic necrotizing factor produced by virulent *E. coli* also modifies Rho by an unknown mechanism.[85] The *E. coli* toxin does not have ADP-ribosyltransferase activity and leads to an enrichment of actin filaments rather than disassembly. Thus, a number of bacteria have evolved enzymes that attack the Rho proteins, however, the enzymatic mechanism and end result are not always the same.

V. MUTAGENESIS OF THE TOXIN B GENE

Toxins A and B share several distinct structural features, including four conserved cysteines, a potential nucleotide binding site, a central hydrophobic region, and a series of contiguous repeating units at the carboxyl terminus (Figure 21.1). We developed a series of deletion and site-directed mutants of the toxin B gene altered in these conserved features and studied the effect of the mutation on cytotoxicity.[86] The results are summarized in Figure 21.2. Three deletion mutants of the carboxy-terminal end were developed to investigate the

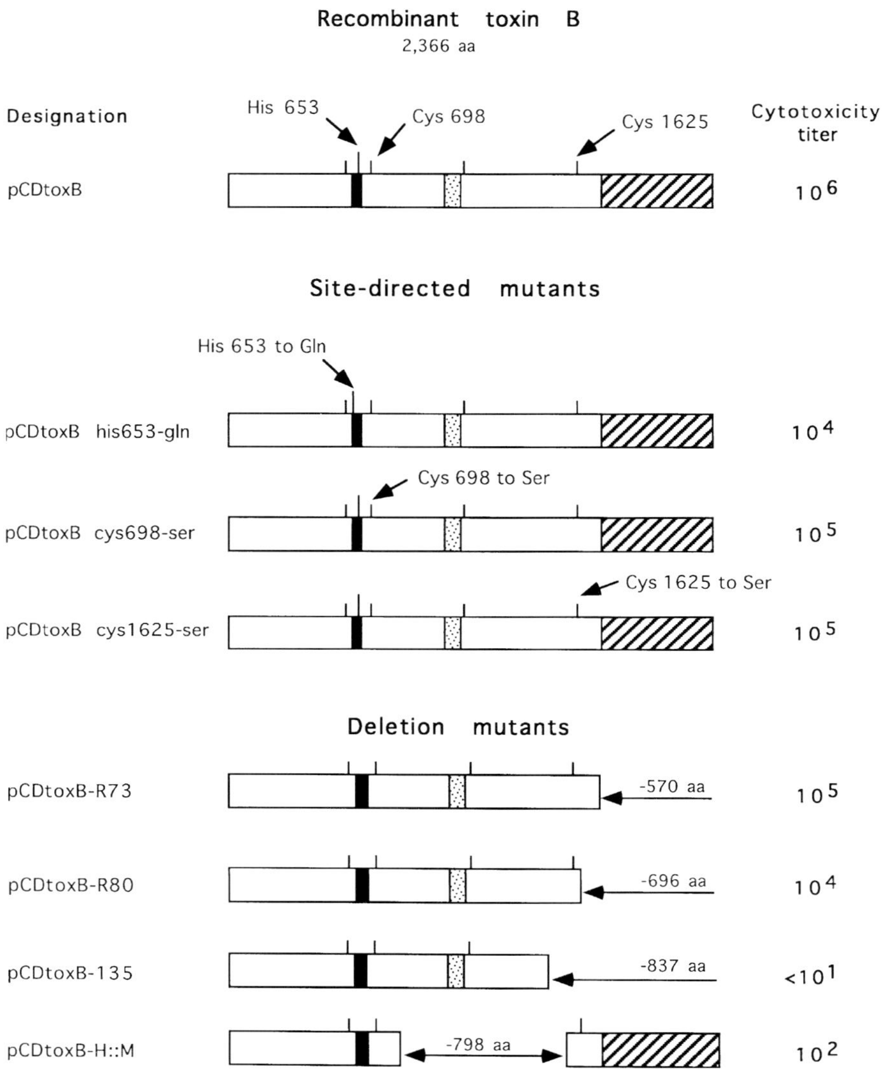

Figure 21.2 Mutagenesis of the toxin B gene. Features conserved with toxin A are illustrated as shown in Figure 21.1. The cytotoxicity titer of the recombinant toxin B lysate is compared with the altered recombinant toxins in the column on the right. Amino acid residues altered by site-directed mutagenesis and regions deleted from the toxin are indicated by arrows. The resulting number of amino acids (aa) removed by each deletion is indicated above the arrows.

role of the repeating units and fourth cysteine. The toxin B gene was deleted from the 3′ end by exonuclease digestions. Removal of the repeating units of toxin B (pCDtoxB-R73) resulted in a tenfold reduction in activity. Surprisingly, however, the mutant toxin still had significant activity, since the recombinant lysate had a cytotoxicity titer of 10^5 compared with 10^6 for the unaltered toxin. Removal of a further 125 amino acids (pCDtoxB-R80) decreased the cytotoxicity another tenfold, however, a significant amount of activity (10^4) remained. In contrast, deletion of an additional 141 amino acid region containing the fourth conserved cysteine (pCDtoxB-135), resulted in complete inactivation of the toxin. An additional mutant was constructed by removing a large central portion of the toxin gene which contained the hydrophobic region and the third conserved cysteine. The resulting deletion mutant (pCDtoxB-H::M) contained the first one-third of toxin B, including the putative nucleotide-binding site, fused in-frame with the carboxy-terminal one-third of the toxin, including the fourth conserved cysteine and repeating units. Surprisingly, this mutant, which was deleted for nearly one-third of the toxin molecule, still retained some cytotoxic activity.

Another series of mutants were developed by site-directed mutagenesis. To examine the possible role of the putative nucleotide binding site, a positively charged histidine residue within the nucleotide binding site was changed to an uncharged glutamine (pCDtoxB His653-Gln). The mutant altered in the nucleotide-binding site was reduced in activity by 99%. When serine was substituted for the second conserved cysteine (pCDtoxB Cys698-Ser), located downstream of the proposed nucleotide binding site, a tenfold decrease in activity was observed. Substitution of the fourth cysteine with serine (pCDtoxB cys1625-ser) also resulted in a tenfold decrease in activity.

These results provide evidence that the enzymatic activity of the toxins is contained in a domain within the first one-third of the molecule. Moreover, a single substitution within the putative nucleotide binding site, located within this region of the toxin eliminated 99% of the activity, suggesting that it may be the enzymatic site of the toxin.

VI. GENE TRANSFER IN *CLOSTRIDIUM DIFFICILE*

The genetic analysis of *C. difficile* virulence factors has been hampered to some extent by the lack of a system for gene transfer. However, some recent success has been reported by Mullany *et al.*[87–89] on conjugative gene transfer in *C. difficile*. A pBR322-based vector, pCI195, containing a 4.2 kb region of the conjugative transposon Tn*919* was used for site-specific integration into the *C. difficile* chromosome. Conjugative transfer was accomplished by first transforming pCI195 into a *B. subtilis* strain that contained the related transposon Tn*916*Δ*E*. The transposon::plasmid structure could be transferred to *C. difficile* by filter mating at a frequency of 10^{-8}. They further demonstrated that pCI195 could be used for gene cloning into *C. difficile* by transfer of a 1.1 kb fragment of the toxin B gene into a nontoxigenic strain of *C. difficile*. As in *B. subtilis*, the transposon::plasmid structure entered the genome at a specific site. The ability to mobilize genes into *C. difficile* means that it may now be possible to construct isogenic mutants that express only toxin A, only toxin B, or neither toxin. Comparative studies of toxin-deleted isogenic mutants would provide important information on the relative importance of each toxin in the course of *C. difficile* pathogenesis.

VII. RELATED TOXINS OF *CLOSTRIDIUM SORDELLII* AND *CLOSTRIDIUM NOVYI*

Cross-neutralization of toxins A and B by *C. sordellii* antitoxin led to the discovery of *C. difficile* as the primary cause of PMC. This suggested that *C. sordellii* produced two toxins immunologically related to the toxins of *C. difficile*. The production of two toxins with distinctive biological activities by *C. sordellii* was first described in 1969.[90] Intradermal injection of cell-free extracts into guinea pigs caused edema and hemorrhage. The activities were shown to be due to separate toxins, with the toxin that caused edema being the more lethal toxin (LT) than the hemorrhagic toxin (HT). The toxins have been separated by ion-exchange chromatography and, like toxins A and B, are unusually large, with molecular weights estimated at 300 000 for HT and 260 000 for LT.[91–95]

HT has biological activities and immunological properties remarkably similar to toxin A.[92] It causes a hemorrhagic fluid response in rabbit ileal loops, similar to the type of reaction caused by toxin A. Furthermore, monoclonal antibody PCG-4, which recognizes epitopes in the repeating units of toxin A, neutralizes the enterotoxicity of HT and toxin A. HT is also cytotoxic and lethal with minimum doses similar to those reported for toxin A. The *N*-terminal amino acid sequence of toxins A and HT also share close identity.

The lethal toxin of *C. sordellii* is immunologically related to toxin B, and has similar physicochemical properties.[93,94] However, toxin B is 10 000 times more cytotoxic than LT. In contrast, LT is tenfold more lethal than toxin B. The gene for *C. sordellii* LT has recently been cloned and sequenced.[96] The structural gene is similar in size to the *C. difficile* toxin B gene, with a corresponding molecular weight of 270 000. The LT deduced amino acid sequence shares 76% identity with toxin B and 47% identity with toxin A, confirming that it is more closely related to toxin B.

Despite the relatedness of the *C. difficile* and *C. sordellii* toxins, recent studies have shown that the cytopathic effect and mechanism of action of *C. sordellii* LT toxin is very different from those described for toxins A and B.[97] LT causes an increase in cortical actin, with a concomitant decrease in the number of actin stress fibers. Numerous microvilli composed of the actin-binding and microfilament-bundling protein fimbrin appear on the surface of toxin treated cells. Moreover, Rho isolated from LT treated cells is not resistant to ADP-ribosylation by C3, as is observed when cells are treated with toxins A and B. Overexpression of Rho in transfected cells protects them from toxins A and B, but not from LT. Thus, LT, despite its similarity to toxin B, does not exert its activity by modifying Rho.

It has been suggested that α-toxin of *C. novyi*, due to its large molecular weight, may be a member of a class of large molecular weight toxins that include the *C. difficile* and *C. sordellii* toxins.[98] α-Toxin, like the toxins of *C. difficile* and *C. sordellii* has lethal, cytotoxic, and edema-producing activity.[98–100] The presumed similarity to the large molecular weight toxins of *C. difficile* and *C. sordellii* was confirmed by sequencing and analysis of the α-toxin gene.[101] The deduced amino acid sequence shows 48% identity with toxins A and B. Moreover, like toxins A and B, it contains a series of repeat motifs at the carboxy-terminal portion of the molecule and also a central hydrophobic region with potential membrane-spanning segments.

VIII. CONCLUSIONS AND FUTURE RESEARCH

Approximately the final one-third of each toxin functions in binding. If, as recent studies suggest, the enzymatic site responsible for cytotoxicity is contained within the first one-third of the toxin molecules, this leaves a very large part of the molecules for which the function is unknown. In fact, the region between the hydrophobic center of the molecules and the beginning of the repeating units is nearly M_r 100 000. This alone is larger than most bacterial proteins, including most toxins. The function of this portion of the molecule will require further analysis.

Recent breakthroughs have clearly established that the toxins covalently modify low molecular weight GTP-binding proteins that regulate the actin microfilaments. A number of other clostridial enzymes modify the Rho/Rac family of proteins by ADP-ribosylation. The mode of action of toxins A and B however, does not involve ADP-ribosylation. Instead, the toxins modify Rho by UDP-glucose-dependent glucosylation. *C. sordellii* toxin LT, although closely related to toxin B, also disrupts the cytoskeleton, but by a different mechanism and with different cytopathic effects. Thus, a large family of clostridial enzymes has evolved which are capable of modifying the architecture of eukaryotic cells. The mode of action, however, varies. Actin plays a central role in the architecture and function of all eukaryotic cells, including cell division, motility, and intercellular communication. Undoubtedly further studies on the molecular biology of this group of clostridial enzymes will provide important insights into cellular function.

IX. REFERENCES

1. Hall J. and O'Toole E. (1935) Intestinal flora in newborn infants with description of a new pathogenic organism, *Bacillus difficilis. Am J Dis Child* **135**: 390–402.
2. Snyder M.L. (1937) Further studies of *Bacillus difficilis* (Hall and O'Toole). *J Inf Dis* **60**: 223–231.
3. Bartlett J.G. and Gorbach S.L. (1977) Pseudomembranous enterocolitis (antibiotic-related colitis). *Adv Intern Med* **22**: 455–476.
4. Bartlett J.G., Onderdonk A.B., Cisneros R.L. and Kasper D.L. (1977) Clindamycin-associated colitis due to a toxin-producing species of *Clostridium* in hamsters. *J Inf Dis* **136**: 701–705.
5. Bartlett J.G., Chang T.W., Gurwith M., Gorbach S.L. and Onderdonk A.B. (1978) Antibiotic associated pseudomembranous colitis due to a toxin-producing clostridia. *New Engl J Med* **298**: 531–534.
6. Bartlett J.G., Moon N., Chang T.-W. and Taylor N. (1978) Role of *Clostridium difficile* in antibiotic-associated pseudomembranous colitis. *Gastroenterology* **75**: 778–782.
7. Chang T.-W., Bartlett J.G., Gorbach S.L. and Onderdonk A.B. (1978) Clindamycin-induced enterocolitis in hamsters as a model of pseudomembranous colitis in patients. *Infect Immun* **20**: 526–529.
8. Chang T.-W., Gorbach S.L. and Bartlett J.G. (1978) Neutralization of *Clostridium difficile* toxin by *Clostridium sordellii* antitoxins. *Infect Immun* **22**: 418–422.
9. Larson H.E. and Price A.B. (1977) Pseudomembranous colitis: Presence of clostridial toxin. *Lancet* **ii**: 1312–1314.
10. Larson H.E., Parry J.V., Price A.B., Davies D.R., Dolby J. and Terryl D.A.J. (1977) Undescribed toxin in pseudomembranous colitis. *Br Med J* **1**: 1246–1248.
11. Larson H.E., Price A.B., Honour P. and Borriello S.P. (1978) *Clostridium difficile* and the aetiology of pseudomembranous colitis. *Lancet* **i**: 1063–1066.
12. Allo M., Silva J., Fekety R., Rifkin G.D. and Waskin H. (1979) Prevention of clindamycin-induced colitis in hamsters by *Clostridium sordellii* antitoxin. *Gastroenterology* **76**: 351–355.
13. Rehg J.E. (1980) Cecal toxin(s) from guinea pigs with clindamycin-associated colitis, neutralized by *Clostridium sordellii* antitoxin. *Infect Immun* **27**: 387–390.
14. Willey S. and Bartlett J.G. (1979) Cultures for *Clostridium difficile* in stools containing a cytotoxin neutralized by *Clostridium sordellii* antitoxin. *J Clin Microbiol* **10**: 880–884.
15. George R.H., Simonds J.M., Dimock F., Brown J.D., Arabi Y., Shinagawa N., Keighly M.R.B., Alexander-Williams L. and Burdon D.W. (1978) Identification of *Clostridium difficile* as a cause of pseudomembranous colitis. *Br Med J* : **1**: 695.
16. George W.L., Sutter V.L., Goldstein E.J.C., Ludwig S.L. and Finegold S.M. (1978) Aetiology of antimicrobial-agent-associated colitis. *Lancet* **i**: 802–803.
17. Banno Y., Kobayashi T., Watantabe K., Ueno K. and Nozawa Y. (1981) Two toxins (D-1 and D-2) and *Clostridium difficile* causing antibiotic associated colitis: purification and some biological characterization. *Biochem Int* **2**: 269–635.
18. Sullivan N.M., Pellet S. and Wilkins T.D. (1982) Purification and characterization of toxins A and B of *Clostridium difficile. Infect Immun* **35**: 1032–1040.

19. Finney J.M.T. (1893) Gastroenterostomy for cicatrizing ulcer of the pyloris. *Bull Johns Hopkins Hosp* **4**: 53–55.
20. Lyerly D.M., Sullivan N.M. and Wilkins T.D. (1983) Enzyme-linked immunosorbent assay for *Clostridium difficile* toxin A. *J Clin Microbiol* **17**: 72–78.
21. Libby J.M., Jortner B.S. and Wilkins T.D. (1982) Effects of the two toxins of *Clostridium difficile* in antibiotic-associated cecitis in hamsters. *Infect Immun* **36**: 822–829.
22. Fernie D.S., Thomson R.O., Batty I. and Walker P.D. (1982) Active and passive immunization to protect against antibiotic associated caecitis in hamsters. *Dev Biol Stand* **53**: 325–332.
23. Barrosso L.A., Wang S.Z., Phelps C.J., Johnson J.L. and Wilkins T.D. (1990) Nucleotide sequence of the *Clostridium difficile* toxin B gene. *Nucl Acids Res* **18**: 4004.
24. Dove C.H., Wang S.-Z., Price S.B., Phelps C.J., Lyerly D.M., Wilkins T.D. and Johnson J.L. (1990) Molecular characterization of the *Clostridium difficile* toxin A gene. *Infect Immun* **58**: 480–488.
25. Johnson J.L., Phelps C.J., Barroso L.A., Roberts M.D., Lyerly D.M. and Wilkins T.D. (1990) Cloning and expression of the toxin B gene of *Clostridium difficile. Curr Microbiol* **20**: 397–401.
26. Price S.B., Phelps C.J., Wilkins T.D. and Johnson J.L. (1987) Cloning of the carbohydrate binding portion of the toxin A gene of *Clostridium difficile. Curr Microbiol* **16**: 55–60.
27. Taylor N.S. and Bartlett J.G. (1979) Partial purification and characterization of a cytotoxin from *Clostridium difficile. Rev Infect Dis* **1**: 379–385.
28. Rolfe R.D. and Finegold S.M. (1979) Purification and characterization of *Clostridium difficile* toxin. *Infect Immun* **25**: 191–201.
29. Haslam S.C., Ketley J.M., Mitchell T.J., Stephen J., Burdon D.W. and Candy D.C.A. (1986) Growth of *Clostridium difficile* and production of toxins A and B in complex and defined media. *J Med Microbiol* **21**: 293–297.
30. Seddon S.V. and Borriello S.P. (1989) A chemically defined and minimal media for *Clostridium difficile. Lett Appl Microbiol* **9**: 237–239.
31. Sterne M. and Wentzel L.M. (1950) A new method for the large scale production of high titre botulinum Formal-toxoid types C and D. *J Immunol* **65**: 175–183.
32. Kazamias M.T. and Sperry J.F. (1995) Enhanced fermentation of mannitol and release of cytotoxin by *Clostridium difficile* in alkaline culture media. *Appl Environ Microbiol* **61**: 2425–2427.
33. Ketly J.M., Haslam S.C., Mitchell T.J., Stephen J., Candy D.C.A. and Burdon D.W. (1984) Production and release of toxins A and B by *Clostridium difficile. J Med Microbiol* **18**: 385–391.
34. Ketly J.M., Mitchell T.J., Haslam S.C., Stephen J., Candy D.C.A. and Burdon D.W. (1986) Sporogenesis and toxin A production by *Clostridium difficile. J Med Microbiol* **22**: 33–38.
35. Borriello S.P., Wren B.W., Hyde S., Seddon S.V., Sibbons P., Krishna M.M., Tabaqchali S., Manek S. and Price A.B. (1992) Molecular, immunological, and biological characterization of a toxin A-negative, toxin B-positive strain of *Clostridium difficile. Infect Immun* **60**: 4192–4199.
36. Lyerly D.M., Barroso L.A., Wilkinis T.D., Depitre C. and Corthier G. (1992) Characterization of a toxin A-negative, toxin B-positive strain of *Clostridium difficile. Infect Immun* **60**: 4633–4639.
37. Torres J.F. (1991) Purification and characterization of toxin B from a strain of *Clostridium difficile* that does not produce toxin A. *J Med Microbiol* **35**: 40–44.
38. Banno Y., Kobayashi T. and Kano K. (1984) Biochemical characterization and biological actions of two toxins (D1 and D2) from *Clostridium difficile. Rev Infect Dis* **6**: 11–20.
39. Lyerly D.M., Roberts M.D., Phelps C.J. and Wilkins T.D. (1986) Purification and properties of toxins A and B of *Clostridium difficile. FEMS Microbiol Lett* **33**: 31–35.
40. Krivan H.C. and Wilkins T.D. (1987) Purification of *Clostridium difficile* toxin A by affinity chromatography on immobilized thyroglobulin. *Infect Immun* **55**: 1873–1877.
41. Knoop F., Martig R. and Owens M. (1990) Degradation of 2-phosphoglycerate by cytotoxin B of *Clostridium difficile. FEBS Lett* **267**: 9–12.
42. Fluit A.C., Wolfhagen M.J.H.M., Jansze M., Torensma R. and Verhoef J. Comment to Knoop et al.

(1990) FEBS Letters 267, 9–12, Toxin B of *Clostridium difficile* does not have enolase activity. *FEBS Lett* **316**: 103–104.

43. Rihn B.J., Scheftel J.M., Girardot R. and Monteil H. (1984) A new purification procedure for *Clostridium difficile* enterotoxin. *Biochem Biophys Res Commun* **124**: 690–695.
44. Pothoulakis C., Barone L.M., Ely R., Faris B., Clark M.E., Franzblau C. and LaMont J.T. (1986) Purification and properties of *Clostridium difficile* cytotoxin B. *J Biol Chem* **261**: 1316–1321.
45. Vallee B.L. and Auld D.S. (1990) Zinc coordination, function, and structure of zinc enzymes and other proteins. *Biochemistry* **29**: 5647–5659.
46. Lyerly D.M., Lockwood D.E., Richardson S.H. and Wilkins T.D. (1982) Biological activities of toxins A and B of *Clostridium difficile. Infect Immun* **35**: 1147–1150.
47. Lima A.A.M., Lyerly D.M., Wilkins T.D., Innes D.J. and Geurrant R.L. (1988) Effects of *Clostridium difficile* toxins A and B in rabbit small and large intestine *in vivo* and on cultured cells *in vitro. Infect Immun* **56**: 582–588.
48. Hecht G., Pothoulakis C., LaMont J.T. and Madara J.L. (1988) *Clostridium difficile* toxin A perturbs cytoskeletal structure and junction permeability in cultured human epithelial cells. *J Clin Invest* **82**: 1516–1524.
49. Fasano A., Bernadette B., Pumplin D.W., Wasaserman S.S., Tall B.D., Ketley J.M. and Kaper J.B. (1991) *Vibrio cholerae* produces a second enterotoxin, which affects intestinal tight junctions. *Proc Natl Acad Sci USA* **88**: 5242–5246.
50. Lyerly D.M., Saum K.E., MacDonald D.K. and Wilkins T.D. (1985) Effects of *Clostridium difficile* toxins given intragastrically to animals. *Infect Immun* **47**: 349–352.
51. Riegler M., Sedivy R., Pothoulakis C., Hamilton G., Johannes Z., Bischof G., Costentini E., Wolfgang F., Schiessel ., LaMount J.T. and Wenzl E. (1985) *Clostridium difficile* toxin B is more potent than toxin A in damaging human colonic epithelium *in vitro. J Clin Invest* **95**: 2004–2011.
52. Tucker K.D., Carrig P.E. and Wilkins T.D. (1990) Toxin A of *Clostridium difficile* is a potent cytotoxin. *J Clin Microbiol* **28**: 869–871.
53. Fiorentini C., Malorni W., Paradisi S., Guiliano M., Mastrantonio P. and Donelli G. (1990) Interaction of *Clostridium difficile* toxin A with cultured cells: cytoskeletal changes and nuclear polarization. *Infect Immun* **58**: 2329–2336.
54. Mitchell J., Laughon B.E. and Lin S. (1987) Biochemical studies on the effect of *Clostridium difficile* toxin B on actin *in vivo* and *in vitro. Infect Immun* **55**: 1610–1615.
55. Ottlinger M.E. and Lin S. (1988) *Clostridium difficile* toxin B induces reorganization of actin, vinculin and talin in cultured cells. *Exp Cell Res* **174**: 215–229.
56. Ehrich M. (1982) Biochemical and pathological effects of *Clostridium difficile* toxins in mice. *Toxicon* **20**: 983–989.
57. Arnon S.S., Mills D.C., Day P.A., Henrickson R.V., Sullivan N.M. and Wilkins T.D. (1984) Rapid death of infant rhesus monkeys injected with *Clostridium difficile* toxins A and B: physiologic and pathologic basis. *J Pediatr* **104**: 34–40.
58. Florin I. and Thelstam M. (1983) Internalization of *Clostridium difficile* cytotoxin into cultured human lung fibroblasts. *Biochim Biophys Acta* **763**: 383–392.
59. Henriques B., Florin I. and Thelestam M. (1987) Cellular internalization of *Clostridium difficile* toxin A. *Microb Pathog* **2**: 455–463.
60. Krivan H.C., Clark G.F., Smith D.F. and Wilkins T.D. (1986) Cell surface binding site for *Clostridium difficile* enterotoxin: evidence for a glycoconjugate containing the sequence Galα1-3Galβ-4GlcNAc. *Infect Immun* **53**: 573–581.
61. Tucker K. and Wilkins T.D. (1991) Toxin A of *Clostridium difficile* binds to the human carbohydrate antigens I, X, and Y. *Infect Immun* **59**: 73–78.
62. Rolfe R.D. and Song W. (1993) Purification of a functional receptor for *Clostridium difficile* toxin A from intestinal brush border membranes of infant hamsters. *Clin Infect Dis* **16** (Suppl. 4) S219–227.

63. Rolfe R.D. and Song W. (1995) Immunoglobulin and non-immunoglobulin components of human milk inhibit *Clostridium difficile* toxin A receptor binding. *J Med Microbiol* **42**: 10–19.
64. Thelestam M. and Bronnegard M. (1980) Interaction of cytopathogenic toxin from *Clostridium difficile* with cells in tissue culture. *Scand J Infect Dis* **22**: 16–29.
65. Muldrow L.L., Ibeanu G.C., Lee N.I., Bose N.K. and Johnson J. (1987) Molecular cloning of *Clostridium difficile* toxin A gene fragment in lambda/gt11. *FEBS Lett* **213**: 249–253.
66. Wren B.W., Clayton C.L., Mullany P.P. and Tabaqchali S. (1989) Molecular cloning and expression of *Clostridium difficile* toxin A in *Escherichia coli* K 12. *FEBS Lett* **225**: 82–86.
67. Eichel-Streiber C.v., Suckau D., Wachter M. and Hadding U. (1988) Cloning and characterization of overlapping DNA fragments of the toxin A gene of *Clostridium difficile. J Gen Microbiol* **135**: 55–64.
68. Frey S. and Wilkins T.D. (1992) Localization of two epitopes recognized by monoclonal antibody PCG-4 on *Clostridium difficile* toxin A. *Infect Immun* **60**: 2488–2492.
69. Koronakis V., Koronakis E. and Hughes C. (1989) Isolation and analysis of the *C*-terminal signal directing export of the *Escherichia coli* hemolysin protein across both bacterial membranes. *EMBO J* **8**: 595–605.
70. Phelps C.J., Lyerly D.M., Johnson J.L. and Wilkins T.D. (1991) Construction and expression of the complete *Clostridium difficile* toxin A gene in *Escherichia coli. Infect Immun* **59**: 150–153.
71. Eichel-Streiber C.v., Laufenberg-Feldmann R., Sartingen S., Schilze J. and Sauerborn M. (1992) Comparative sequence analysis of the *Clostridium difficile* toxins A and B. *Mol Gen Genet* **233**: 260–268.
72. Bairoch A. and Claverie J.-M. (1988) Sequence patterns in protein kinases. *Nature* **331**: 22.
73. Eichel-Streiber C.v. and Sauerborn M. (1990) *Clostridium difficile* toxin A carries a *C*-terminal repetitive structure homologous to the carbohydrate binding region of streptococcal glycosyltransferases. *Gene* **96**: 107–113.
74. Eichel-Streiber C.v., Sauerborn M. and Kuramitsu H.K. (1992) Evidence for a modular structure of the homologous repetitive *C*-terminal carbohydrate-binding sites of *Clostridium difficile* toxins and *Streptococcus mutans* glycosyltransferases. *J Bacteriol* **174**: 6707–6710.
75. Hall A. (1992) Ras-related GTP-ases and the cytoskeleton. *Mol Biol Cell* **3**: 475–479.
76. Wiegers W.I., Just I., Muller H., Hellwig A., Traub P. and Aktories K. (1991) Alteration of the cytoskeleton of mammalian cells cultured in vitro by *Clostridium botulinum* C2 toxin and C3 ADP-ribosyltransferase. *Eur J Cell Biol* **54**: 237–245.
77. Just I., Mohr C., Schallehn G., Menard L., Didsbury J.R., Vandekerckhove J., van Damme J. and Aktories K. (1992) Purification and characterization of an ADP-ribosyltransferase produced by *Clostridium limosum. J Biol Chem* **67**: 10274–10280.
78. Just I., Fritz G., Aktories K., Giry M., Popoff M.R., Boquet P., Hegenbarth S. and Eichel-Streiber C.v. (1994) *Clostridium difficile* toxin B acts on the GTP-binding protein Rho. *J Biol Chem* **269**: 10706–10712.
79. Just I., Seizer J., Eichel-Streiber C.v. and Aktories K. (1995) The low molecular mass GTP-binding protein Rho is affected by toxin A from *Clostridium difficile. J Clin Invest* **95**: 1026–1031.
80. Dillon S.T., Rubin E.J., Yakubovich M., Potholakis C., LaMont J.T., Feig L.A. and Gilbert R.J. (1995) Involvement of Ras-related Rho proteins in the mechanism of action of *Clostridium difficile* toxin A and toxin B. *Infect Immun* **63**: 1421–1426.
81. Just I., Schallehn G. and Aktories K. (1992) ADP-ribosylation of small GTP-binding proteins by *Bacillus cereus. Biochem Biophys Res Commun* **183**: 931–936.
82. Sugai M., Hashimoto K., Kikuchi A., Inoue S., Okumura H., Matsumota K., Gato Y., Ohgai H., Morishi K. and Syuto B. (1992) Epidermal cell differentiation inhibitor ADP-ribosylates small GTP-binding proteins and induces hyperplasia of epidermis. *J Biol Chem* **267**: 2600–2604.
83. Just I., Selzer J., Wilm M., Eichel-Streiber C.v., Mann M. and Aktories K. (1995) Glucosylation of Rho proteins by *Clostridium difficile* toxin B. *Nature* **375**: 500–503.

84. Just I., Wilm M., Selzer J., Rex G., Eichel Streiber C.v., Mann M. and Aktories K. (1995) The enterotoxin from *Clostridium difficile* (ToxA) monoglucosylates the Rho proteins. *J Biol Chem* **270**: 13932–13939.
85. Oswald E., Sugai M., Labigne A., Wu H.C., Fiorentini C., Boquet P. and O'Brien A.D. (1994) Cytotoxic nectrotizing factor type 2 produced by virulent *Escherichia coli* modifies the small GTP-binding proteins Rho involved in assembly of actin stress fibers. *Proc Natl Acad Sci USA* **91**: 3814–3818.
86. Barrosso L.A., Moncrief J.S., Lyerly D.M. and Wilkins T.D. (1994) Mutagenesis of the *Clostridium difficile* toxin B gene and effect on cytotoxic activity. *Microbiol Pathog* **16**: 297–303.
87. Mullany P., Wilks M., Puckey L. and Tabaqchali S. (1994) Gene cloning in *Clostridium difficile* using Tn*916* as a shuttle conjugative transposon. *Plasmid* **31**: 320–323.
88. Mullany P., Wilks M., Lamb I., Clayton C., Wren B. and Tobaqchali S. (1990) Genetic analysis of a tetracycline resistance element from *Clostridium difficile* and its conjugal transfer to and from *Bacillus subtilis*. *J Gen Microbiol* **136**: 1343–1349.
89. Mullany P., Wilks M. and Tobaqchali S. (1991) Transfer of Tn916 and Tn916ΔE into *Clostridium difficile*: Demonstration of a hot spot for these elements in the *C. difficile* genome. *FEMS Microbiol Lett* **79**: 191–194.
90. Arseculerante S.M., Panabokke R.G. and Wijesundera S. (1969) The toxins responsible for the lesions of gas gangrene. *J Med Microbiol* **2**: 37–53.
91. Yamakawa K., Nakamura S. and Nishida S. (1985) Separation of two cytotoxins of *Clostridium sordellii* strains. *Microbiol Immunol* **29**: 553–557.
92. Martinez R.D. and Wilkins T.D. (1988) Purification and characterization of *Clostridium sordellii* hemorrhagic toxin and cross-reactivity with *Clostridium difficile* toxin A (enterotoxin). *Infect Immun* **56**: 1215–1221.
93. Popoff M.R. (1986) Purification and characterization of *Clostridium sordellii* lethal toxin and cross-reactivity with *Clostridium difficile* cytotoxin. 1987. *Infect Immun* **55**: 35–43.
94. Martinez R.D. and Wilkins T.D. (1988) Purification and characterization of *Clostridium sordellii* lethal toxin and cross-reactivity with *Clostridium difficile* cytotoxicity. *Infect Immun* **56**: 1215–1221.
95. Martinez R.D. and Wilkins T.D. (1992) Comparison of *Clostridium sordellii* toxins HT and LT with toxins A and B of *C. difficile*. *J Med Microbiol* **36**: 30–36.
96. Green G.A., Schue V. and Monteil, H. (1995) Cloning and characterization of the cytotoxin L-encoding gene of *Clostridium sordellii*: homology with *Clostridium difficile* toxin B. *Gene* **161**: 57–61.
97. Giry M., Popoff M.R., Eichel-Streiber C.v. and Boquet P. (1995) Transient expression of RhoA, -B, and C GTPases in HeLa cells potentiates resistance to *Clostridium difficile* toxins A and B but not to *Clostridium sordellii* lethal toxin. *Infect Immun* **63**: 4063–4071.
98. Bette P., Frevert J., Mauler F., Suttorp N. and Habermann E. (1989) Pharmacological and biochemical studies of cytotoxicity of *Clostridium novyi* type A alpha-toxin. *Infect Immun* **57**: 2507–2513.
99. Bette P., Oksche A., Mauler F., Eichel-Streiber C.v., Popoff M.R. and Habermann E. (1991) A comparative biochemical, pharmacological, and immunological study of *Clostridium novyi* α-toxin, *C. difficile* toxin B and *C. sordellii* lethal toxin. 1991. *Toxicon* **29**: 877–887.
100. Ball D.W., Van Tassel R.L., Roberts M.D., Hahn P.E., Lyerly D.M. and Wilkins T.D. (1993) Purification and characterization of α-toxin produced by *Clostridium novyi* type A. *Infect Immun* **61**: 2912–2918.
101. Hofmann F., Herrmann E. and Eichel-Streiber C.v. (1995) Sequencing and analysis of the gene encoding the alpha-toxin of *Clostridium novyi* proves its homology to toxins A and B of *Clostridium difficile*. *Mol Gen Genet* **247**: 670–679.

Chapter 22

Molecular Mechanism of Action of *Clostridium difficile* Toxins A and B

KLAUS AKTORIES, JÖRG SELZER, FRED HOFMANN AND INGO JUST

Institut für Pharmakologie und Toxikologie, Albert-Ludwigs-Universität Freiburg, Hermann-Herder-Strasse 5, D-79104 Freiburg, Germany

I. INTRODUCTION

Various toxins and exoenzymes produced by clostridia induce dramatic alterations of the actin cytoskeleton of host cells. These clostridial cytotoxins are classified into at least three groups. One group comprises the binary actin–ADP-ribosylating toxins,[1–7] members of which are *Clostridium botulinum* C2 toxin,[8,9] *Clostridium perfringens* ι-toxin,[10,11] *Clostridium spiroforme* toxin[12,13] and an ADP-ribosyltransferase produced by *Clostridium difficile*.[14] These toxins ADP-ribosylate monomeric actin at Arg-177,[15,16] thereby inhibiting actin polymerization which eventually results in depolymerization of the microfilament network.[17,18] The second family of clostridial cytotoxins/exoenzymes are the C3-like toxins.[4,19] Among these toxins are *C. botulinum* C3 ADP-ribosyltransferase,[20–25] *Clostridium limosum* ADP-ribosyltransferase[26] and ADP-ribosyltransferases that are produced by *Bacillus cereus*[27,28] and *Staphylococcus aureus*,[29–31] respectively. All C3-like transferases have molecular masses of about 25 kDa, are basic proteins and appear to lack any specific cell-binding or transport

The Clostridia: Molecular Biology and Pathogenesis
ISBN 0-12-595020-9

subunits, a fact that brings into question the designation of these exoenzymes as toxins. However, when these bacterial exoenzymes reach the eukaryotic cytosol (possibly by non-specific up-take mechanisms *in vivo* or by microinjection *in vitro*) they are potent cytotoxins.[18,32,33] They act on the eukaryotic cytoskeleton by ADP-ribosylation of Rho proteins, which are involved in the regulation of the actin cytoskeleton.[32–36]

A third family of clostridial cytotoxins are the so-called large clostridial cytotoxins. Members of this family are *C. difficile* toxins A and B, the hemorrhagic and the lethal toxins from *Clostridium sordellii* and *Clostridium novyi* α-toxin. These toxins are characterized by molecular masses of about 250–300 kDa and are the largest bacterial toxins known. The recent delineation of the encoding genes and studies on their enzymatic activities have provided exciting insights into the molecular mechanisms of their cytotoxic effects.

II. LARGE CLOSTRIDIAL CYTOTOXINS

The best studied examples of the family of large clostridial toxins are *C. difficile* toxins A and B (see Chapter 21).[37–41] *Clostridium difficile* is suggested to be the major pathogen responsible for induction of about 20% of all cases of antibiotic-associated diarrhea and for most cases of pseudomembranous colitis (see Chapter 8).[38,40,42–44] Toxin A is designated as an enterotoxin and toxin B as a cytotoxin.[37,42] Whereas the enterotoxin induces diarrhea, inflammation and necrosis, toxin B by itself exhibits no overt enterotoxicity in animal intestine models.[45,46] It is likely that different cell surface receptors are responsible for these different effects since, in cell culture, both toxins are cytotoxic. However, toxin B is, in general, 1000-fold more potent than toxin A.[47,48] Both toxins act intracellularly and exert their cytotoxicity after receptor-mediated endocytosis.[49,50] The genes encoding these toxins have recently been cloned and sequenced.[39,51–55] The structural genes for *C. difficile* toxin A and B consist of 8130 and 7098 bp encoding proteins of 2710 and 2366 amino acids with M_r of about 308 000 and 270 000, respectively. The deduced amino acid sequences show 49% identity and 63% similarity.

Other members of the family of large clostridial cytotoxins are the lethal toxin (LT) and the hemorrhagic toxin (HT) produced by *C. sordellii*[41,56–58] and the *C. novyi* α-toxin.[41,59,60] All three toxins are important virulence factors for the induction of a gas-gangrene syndrome. The nucleotide sequence of *C. sordellii* LT was reported by Green and coworkers,[61] revealing about 76% identity and 90% homology with *C. difficile* toxin B and encoding a 270 kDa protein. Recently, the *C. novyi* α-toxin gene was cloned and sequenced and shown to encode a 250 kDa protein. At the nucleotide sequence level, α-toxin shares about 40% and 35%, respectively, homology with *C. difficile* toxins A and B.[62]

Although is has been recognized for some time that *C. difficile* toxins A and B selectively affect the actin cytoskeleton, and to a much lesser extent the microtubule system,[63] the precise molecular mode of action responsible for the effects of the large clostridial cytotoxins was unkown until quite recently.[63–66] It is now clear that *C. difficile* toxins and other large clostridial cytotoxins act on the actin cytoskeleton by modifying regulators of the microfilament system, namely the low molecular mass GTP-binding proteins of the Rho subfamily.

III. LOW MOLECULAR MASS GTP-BINDING PROTEINS AS REGULATORS OF THE CYTOSKELETON

Rho proteins belong to the superfamily or Ras-like GTPases that function as molecular switches.[67–75] In mammalian cell lines, the Rho-subfamily comprises RhoA, B, C, Racl, 2, Cdc42 (G25K), RhoE, RhoD, RhoG and TC10 proteins.[34,76,77] These proteins are at least 55% identical at the amino acid level. Four regions in the small GTPases show especially high homologies. These regions appear to be involved in GTP-binding and in coupling with effector proteins.[68]

Like other GTP-binding proteins, Rho proteins are active in the GTP bound form and inactive when GDP is bound (Figure 22.1). The active state is terminated by hydrolysis of bound GTP caused by stimulation of the intrinsic GTPase activity. The basal GTPase activity of Rho is low but several GTPase-activating proteins (GAP proteins)[78] have been shown to be involved in turning off Rho by causing several-fold stimulation of GTP hydrolysis. Guanine nucleotide exchange factors (GEFs) (also termed guanine nucleotide dissociation stimulators, GDS) are involved in the activation of Rho subtype proteins.[79,80] Finally, guanine nucleotide dissociation inhibitors (GDIs) also participate in regulation of Rho function.[81,82] GDI proteins block not only the nucleotide exchange but also appear to have important transport functions, i.e. they release Rho-subfamily proteins from the membrane location where Rho proteins are proposed to act. Furthermore, GDI stabilizes and stores the inactive GDP-bound form in the cytosol.[83]

Several lines of evidence indicate that Rho subtype proteins (RhoA, B and C) are involved in the regulation of the actin cytoskeleton.[34,77] Most important for recent advances in this field was the finding that C3-toxins, which selectively ADP-ribosylate RhoA, RhoB and RhoC, induce redistribution and depolymerization of the actin cytoskeleton.[18,32–34] C3 ADP-ribosylates Rho at Asp-41,[84] which is located in the effector region (amino acids 32–42) of the GTP-binding proteins. Modification by ADP-ribosylation is suggested to inhibit the interaction

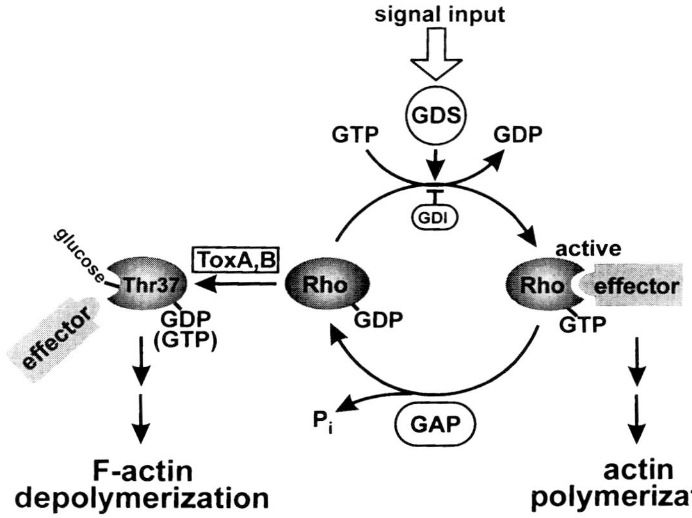

Figure 22.1 The low molecular mass GTP-binding proteins of the Rho family, which are involved in the regulation of the actin cytoskeleton, are controlled by a GTPase cycle. They are inactive in the GDP-bound form and active with GTP bound. Guanine nucleotide exchange factors (also called guanine nucleotide dissociation stimulators, GDS) promote nucleotide exchange and activate Rho proteins. Active Rho proteins interact with effectors which appear to be protein kinases that affect the polymerization of actin by an unknown mechanism. Hydrolysis of bound GTP due to intrinsic GTPase activity of Rho terminates activation of the effector. This inactivation process is largely accelerated by GTPase-activating proteins (GAPs). Guanine nucleotide dissociation inhibitor (GDI) proteins also participate in regulation of Rho proteins. GDI prevents activation and keeps the inactive form of the Rho proteins in the cytosol. It is suggested that glucosylation of Rho proteins by *C. difficile* toxins inhibits the interaction of the GTP-binding protein with their effectors, thereby inducing depolymerization of the actin microfilament system.

of Rho with its putative effector or, alternatively, it may prevent activation of Rho by guanine-nucleotide exchange factors. Further evidence for the essential role of Rho in the regulation of the actin cytoskeleton came from experiments where microinjection of the constitutively active Val14–Rho into quiescent serum-starved monolayer cells resulted in formation of stress fibers and focal adhesions.[33,85]

In addition to their well-characterized role in the regulation of the actin cytoskeleton, Rho proteins are also implicated in various signal transduction processes; e.g. in signaling via phosphoinositide-3-kinase,[86] phosphatidylinositol-4-phosphate-5-kinase,[87] phospholipase D,[88] smooth muscle contraction,[89] cell-cell contact[90] and endocytosis.[91] Moreover, the Rho subtype proteins are also involved in transcriptional activation[92,93] and may have an essential function in ras-induced transformation of cells.[94,95]

The Rac proteins (Rac1 and Rac2) are molecular switches involved in membrane ruffling and formation of lamellipodia induced by growth factors, such as platelet-derived growth factor (PDGF) and insulin.[34,96] Rac proteins also play a role in transcriptional activation[93,97] and are suggested to be important for Ras-induced transformation.[98] Moreover, Rac proteins regulate NADPH-oxidase in some cell types[99,100] and may be involved in control of phospholipase A_2.[101] Although Rac proteins are modified to some extent by C3-like transferases (maximally by about 5–10%) *in vitro*,[26] they are very poor *in vivo* substrates for C3-transferases.

A further member of the Rho family, Cdc42 (at least two mammalian isoforms exist) participates in receptor-induced formation of microspikes and filopodia.[102,103] In addition to its participation in the cortical actin filament system, Cdc42 is also involved in the activation of transcription factors.[93,97]

Based on earlier findings that the Rac and Rho signal pathways may converge, it has been suggested that the low molecular mass GTP-binding proteins of the Rho subfamily regulate cytoskeletal organization in a cascade of steps. Receptor-mediated activation of Cdc42 is followed by activation of Rac, which subsequently leads to activation of Rho.[102,103]

IV. MOLECULAR MECHANISM OF ACTION OF *CLOSTRIDIUM DIFFICILE* TOXINS A AND B

A. Rho proteins as targets of *Clostridium difficile* toxins

Clostridium difficile toxins A and B induce morphological changes which are characterized by rounding up of cells, destruction of the microfilament cytoskeleton, and dramatic retraction of the cell body.[63,65] Often, fine network-like processes between cell bodies persist. Because similar morphological changes and comparable redistribution of the actin cytoskeleton is observed with C3-like transferases,[18] studies were undertaken to determine whether *C. difficile* toxins also affect Rho proteins. These studies showed that treatment of Chinese hamster ovary cells with *C. difficile* toxins inhibits subsequent ADP-ribosylation of Rho proteins by C3 in the cell lysate.[104–107] Inhibition of ADP-ribosylation induced by toxin A or B occurs in a concentration and time-dependent manner. Moreover, the decrease in the ability of Rho to serve as a substrate for ADP-ribosylaton was shown to precede *C. difficile* toxin-induced morphological changes, suggesting a cause and effect relationship between these events. Inhibition of the C3-catalyzed ADP-ribosylation of Rho proteins by toxins A and B was also observed in a cell-free system, but this effect was strictly dependent on the presence of a cytosolic factor (see below). Gel electrophoresis studies demonstrated that Rho proteins treated with *C. difficile* toxin migrated more slowly on non-denaturing gels compared with control Rho. However, no changes were noted in the migration behaviour of *C. difficile* toxin-treated Rho using sodium dodecyl sulfate–polyacrylamide gel electrophoresis. Immunostaining of Rho from toxin-treated cells indicated that Rho proteins were not cleaved by *C. difficile* toxins. This information suggests that *C. difficile* toxins induced a covalent modification to Rho that involved a low M_r molecule.[108,109]

B. Monoglucosyltransferase activity of *Clostridium difficile* toxins

Using electrospray mass spectrometry it was demonstrated that recombinant Rho protein modified by toxin B[108] or by toxin A[109] showed an increase in molecular mass of 162 Da; this is consistent with the incorporation of hexose into this small GTP-binding protein. Further studies identified UDP-glucose as the essential cosubstrate involved in the modification of Rho proteins by toxin A and B. In contrast to the C3-like transferases that selectively ADP-ribosylate only RhoA, B, C, the *C. difficile* toxins glucosylate all members of the Rho subfamily, i.e. Rho, Rac and Cdc42. However, *C. difficile* toxins show some substrate specificity since other low molecular mass GTP-binding proteins, including Ras, Rab, Arf or Ran subfamilies or heterotrimeric G proteins, are not modified by these toxins. It was shown that a maximum of 1 mol of glucose per mol of GTPase is incorporated by the *C. difficile* toxins. In addition to glucosyltransferase activity, the *C. difficile* toxins also possess UDP-glucose hydrolase activity, i.e. in the absence of a protein substrate, they cleave UDP-glucose into UDP and glucose. However, this hydrolase activity is 10- to 100-fold lower than the glucosyltransferase activity, which is similar to the ratio of NAD glycohydrolase activity to ADP-riboslytransferase activity that has been reported for ADP-ribosylating toxins.[27]

C. Modification of threonine-37 of Rho and functional consequences

The Thr-37 residue of Rho was identified as the amino acid acceptor for glucosylation by toxin A and B. Changing this acceptor threonine residue to alanine (Thr-37AlaRho) results in a complete blockage of glucosylation indicating that Thr-37, but no other threonine residues, is modified by *C. difficile* toxins.[108,109] This finding is intriguing because a similar threonine residue is highly conserved in the effector domain of all low molecular mass GTP-binding proteins.[67,68] As can be deduced from the crystal structure of Ras protein,[110] Thr-37 is located in the effector region of Rho where coupling with the effector protein takes place (Figure 22.2). Threonine-37 in Rho (equivalent to Thr-35 in

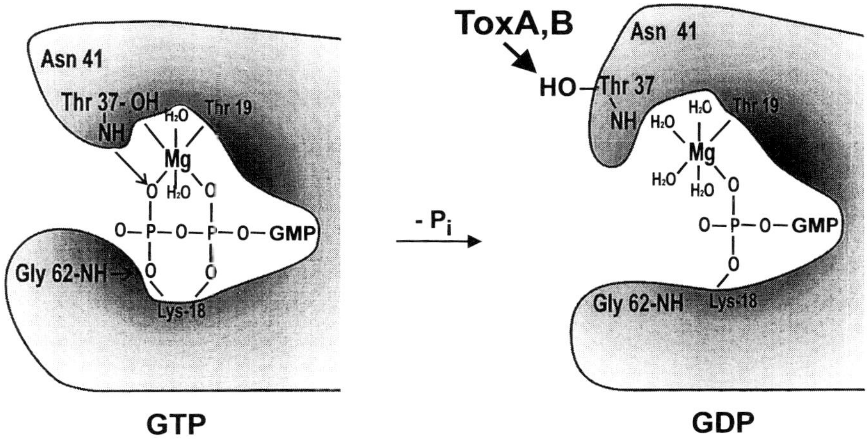

Figure 22.2 Glucosylation of Rho in the effector region. The effector region of Rho is deduced from the atomic model of Ras.[108] *Clostridium difficile* toxins A and B glucosylate Rho proteins at Thr-37. In the GTP-bound form, the hydroxyl group of Thr-37 (equivalent to Thr-35 of Rac and Cdc42, respectively), is directed to the interior of the molecule thereby participating in nucleotide and divalent cation binding. In the GDP-bound form of Rho, major conformational changes occur and the hydroxyl group of Thr-37 is oriented towards the solvent and is accessible to glucosylation by *C. difficile* toxins.

Ras, Rac and Cdc42) is involved in the binding of the nucleotide (GDP/GTP) through coordination of the magnesium cation. The region adjacent to this pivotal threonine residue undergoes major conformational changes upon activation and is, therefore, designated switch-I region.[111,112] In the active GTP-bound form, the threonine residue coordinates a Mg^{2+} ion and might participate in the binding of the γ-phosphate of the nucleotide. In the inactive GDP-bound form, the side chain of the threonine residue is directed to the solvent and is no longer available for interaction with the cation (Figure 22.2). Thus, in the case of the Rho proteins, the GDP-bound form might be a superior substrate for glucosylation to the GTP-bound form. This is supported by experimental data showing that the GTP/GTPγS-bound form of Rho protein is a poor substrate for glucosylation[108] and that in the GDP-bound form about 1 mol of glucose is incorporated per mol of GTPase.[109]

To study whether glucosylation of Rho proteins is the basis of *C. difficile* toxin-induced cytotoxic effects, recombinant toxin-glucosylated RhoA protein was microinjected into cells after toxin removal by membrane filtration. Microinjection of modified RhoA caused morphological alterations, accompanied by depolymerization of actin filaments, which resemble *C. difficile* toxin-induced effects.[108] Because glucosylated RhoA protein induced destruction of the cytoskeleton in the presence of unmodified cellular Rho, it appears that the modified protein acts in a dominant-negative manner. ADP-ribosylated RhoA protein is also dominant-negative. In contrast, microinjection of the mutant $^{\text{Thr-37Ala}}$Rho that is resistant to glucosylation did not affect the actin cytoskeleton,[33,113] indicating that the glucose moiety in the effector region is responsible for the depolymerization of the actin filaments.

V. MOLECULAR MECHANISMS OF *CLOSTRIDIUM NOVYI* α-TOXIN AND *CLOSTRIDIUM SORDELLII* LT TOXIN

Other members of the family of large clostridial cytotoxins also possess glycosyltransferase activity. Interestingly, differences in the substrate and cosubstrate specificity of these toxins have been observed. *Clostridium novyi* α-toxin modifies the Rho-subfamily proteins Rho, Rac and Cdc42 (Table 22.1). Surprisingly, the cosubstrate used by *C. novyi* α-toxin is not UDP-glucose but UDP-*N*-acetylglucosamine. Because treatment of Rho proteins with the α-toxin inhibited subsequent ADP-ribosylation by *C. difficile* toxins, it is suggested that the GTP-binding protein is modified by the toxin at the same acceptor amino acid as found for *C. difficile* toxins. In line with this notion is the finding that the Thr-37Ala mutation of Rho is not a substrate for α-toxin.[112]

Of particular interest was the study of *C. sordellii* lethal toxin. As mentioned above, this toxin exhibits about 90% amino acid sequence homology with *C. difficile* toxin B. However, it was known for some time that the morphological features induced by *C. sordellii* toxin differ from those caused by *C. difficile* toxins.[60,115,116] Studies with cell lysates performed in the presence of various activated hexoses showed that LT possesses glucosyltransferase activity and uses UDP-glucose as cosubstrate.[117] However, LT from *C. sordellii* exhibited a different protein substrate specificity (Table 22.1). The toxin glucosylates Rac protein but, in contrast to *C. difficile* toxins and *C. novyi* α-toxin, neither Rho nor Cdc42. Most interestingly, *C. sordellii* toxin was able to glucosylate recombinant

Table 22.1. Glycosylation of Rho and Ras proteins by large clostridial toxins. *Clostridium difficile* toxins A and B and *C. sordellii* LT are glucosyltransferases using UDP-glucose as a cosubstrate. The cosubstrate for *Clostridium novyi* α-toxin is UDP-*N*-acetylglucosamine. Whereas *C. difficile* toxins A and B and *C. novyi* α-toxin modify Rho, Rac and Cdc42 proteins, *Clostridium sordellii* LT glucosylates Rac and Ras proteins

	Cosubstrate	Protein substrates
Toxin A	UDP-Glc	Rho, Rac, Cdc42
Toxin B	UDP-Glc	Rho, Rac, Cdc42
α–Toxin	UDP-GlcNAc	Rho, Rac, Cdc42
LT	UDP-Glc	Rac, Ras

Ras and apparently other subtype GTP-binding proteins of the Ras family.[117] Therefore, *C. sordellii* LT was tested to see if it interfered with the Ras signal pathway in intact cells. Ras, which is activated via growth factor receptors (receptor tyrosine kinases) interacts and activates Raf kinase which subsequently activates the MAP kinase cascade by phosphorylation.[69,70] When serum-starved cells were treated with *C. sordellii* LT, addition of epidermal growth factor did not increase MAP kinase activity, indicating inhibition of Ras and blockage of the signaling pathway downstream of Ras.[117] Thus, it is suggested that the different protein substrate specificity of *C. sordellii* LT is responsible for the different morphological changes induced by this toxin compared with *C. difficile* toxins.

VI. CONCLUSION

Clostridium difficile toxins A and B have been recognized to be monoglucosyltransferases, which modify low molecular mass GTP-binding proteins of the Rho subfamily by using UDP-glucose as a cosubstrate. Modification of RhoA appears to be the underlying mechanism of the cytotoxic effects of the toxins. It is important to note that inactivation of Rho by C3-induced ADP-ribosylation was recently shown to induce alterations of the intestinal tight junctions and to increase the permeability of enterocyte monolayers.[118] Similar alterations of the functions of enterocytes were observed after treatment with *C. difficile* toxins,[119] suggesting that the pathogenic effects induced by these toxins, e.g. diarrhea and intestinal epithelial injury, are based on their ability to inactivate Rho subfamily proteins by glucosylation.

Other members of the family of large clostridial toxins also glycosylate low molecular mass GTP-binding proteins. However, it appears that the large clostridial cytotoxins, although highly similar in amino acid sequence, differ in their substrate (LT; Rac and Ras subfamily) or cosubstrate specificity (*C. novyi* toxin; UDP-*N*-acetylglucosamine).

Monoglycosylation is a novel mechanism by which bacterial toxins affect the signal cascades of eukaryotic target cells. Knowledge of the molecular mechanisms of these toxins is certainly an important prerequisite for the development of new therapeutic strategies against diseases caused by these potent virulence factors. Moreover, Rho-ADP-ribosylating C3-like toxins have proven to be useful tools to study the cellular functions of Rho proteins. Usage of C3 toxins is hampered by the fact that the toxins are unable to enter cells readily. Because the large clostridial cytotoxins are taken up by most cell lines and act at very low concentrations they will be extremely useful as cell biological tools to elucidate further the role and functions of their target proteins.

VII. REFERENCES

1. Aktories K., Wille M. and Just I. (1992) Clostridial actin-ADP-ribosylating toxins. *Curr Top Microbiol Immunol* **175**: 97–113.
2. Considine R.V. and Simpson L.L. (1991) Cellular and molecular actions of binary toxins possessing ADP-ribosyltransferase activity. *Toxicon* **29**: 913–936.
3. Aktories K. and Just I. (1990) Botulinum C2 toxin. In Moss J. and Vaughan M. (eds), *ADP-Ribosylating Toxins and G-Proteins*, pp. 79–95. American Society for Microbiology, Washington, D.C.
4. Aktories K. and Just I. (1993) GTPases and actin as targets for bacterial toxins. In Dickey B.F. and Birnbaumer L. (eds), *GTPases in Biology I*, pp. 87–112. Springer-Verlag, Berlin-Heidelberg.
5. Aktories K. and Wegner A. (1992) Mechanisms of the cytopathic action of actin-ADP-ribosylating toxins. *Mol. Microbiol* **6**: 2905–2908.
6. Aktories K. and Wegner A. (1989) ADP-ribosylation of actin by clostridial toxins. *J. Cell Biol* **109**: 1385–1387.
7. Aktories K. and Koch G. (1995) Modification of actin and of Rho proteins by clostridial ADP-ribosylating toxins. In Iglewski B., Moss J., Tu A.T. and Vaughan M. (eds) *Microbial Toxins and Virulence Factors in Disease*, pp. 491–520. Marcel Dekker, New York/Basel.
8. Aktories K., Bärmann M., Ohishi I., Tsuyama S., Jakobs K.H. and Habermann E. (1986) Botulinum C2 toxin ADP-ribosylates actin. *Nature* **322**: 390–392.
9. Ohishi I., Iwasaki M. and Sakaguchi G. (1980) Purification and characterization of two components of botulinum C2 toxin. *Infect. Immun.* **30**: 668–673.
10. Schering B., Bärmann M., Chhatwal G.S., Geipel U. and Aktories K. (1988) ADP-ribosylation of skeletal muscle and non-muscle actin by *Clostridium perfringens* iota toxin. *Eur. J. Biochem.* **171**: 225–229.
11. Simpson L.L., Stiles B.G., Zapeda H.H. and Wilkins T.D. (1987) Molecular basis for the pathological actions of *Clostridium perfringens* iota toxin. *Infect. Immun.* **55**: 118–122.
12. Popoff M.R., Milward F.W., Bancillon B. and Boquet P. (1989) Purification of the *Clostridium spiroforme* binary toxin and activity of the toxin on HEp2 cells. *Infect. Immun.* **57**: 2462–2469.
13. Simpson L.L., Stiles B.G., Zepeda H. and Wilkins T.D. (1989) Production by *Clostridium spiroforme* of an iota-like toxin that possesses mono(ADP-ribosyl)transferase activity: Identification of a novel class of ADP-ribosyltransferases. *Infect. Immun.* **57**: 255–261.
14. Popoff M.R., Rubin E.J., Gill D.M. and Boquet P. (1988) Actin-specific ADP-ribosyltransferase produced by a *Clostridium difficile* strain. *Infect. Immun.* **56**: 2299–2306.
15. Vandekerckhove J., Schering B., Bärmann M. and Aktories K. (1987) *Clostridium perfringens* iota toxin ADP-ribosylates skeletal muscle actin in Arg-177. *FEBS Lett.* **225**: 48–52.
16. Vandekerckhove J., Schering B., Bärmann M. and Aktories K. (1988) Botulinum C2 toxin ADP-ribosylates cytoplasmic β/γ-actin in arginine 177. *J Biol Chem* **263**: 696–700.
17. Reuner K.H., Presek P., Boschek C.B. and Aktories K. (1987) Botulinum C2 toxin ADP-ribosylates actin and disorganizes the microfilament network in intact cells. *Eur J. Cell Biol* **43**: 134–140.
18. Wiegers W., Just I., Müller H., Hellwig A., Traub P. and Aktories K. (1991) Alteration of the cytoskeleton of mammalian cells cultured *in vitro* by *Clostridium botulinum* C2 toxin and C3 ADP-ribosyltransferase, *Eur J Cell Biol* **54**: 237–245.
19. Aktories K. and Just I. (1995) *In vitro* ADP-ribosylation of Rho by bacterial ADP-ribosyltransferases. *Methods Enzymol.* **256**: 184–195.
20. Aktories K., Weller U. and Chhatwal G.S. (1987) *Clostridium botulinum* type C produces a novel ADP-ribosyltransferase distinct from botulinum C2 toxin. *FEBS Lett.* **212**: 109–113.
21. Aktories K., Rösener S., Blaschke U. and Chhatwal G.S. (1988) Botulinum ADP-ribosyltransferase

C3. Purification of the enzyme and characterization of the ADP-ribosylation reaction in platelet membranes. *Eur J Biochem* **172**: 445–450.

22. Rubin, E.J., Gill D.M., Boquet P. and Popoff M.R. (1988) Functional modification of a 21-Kilodalton G protein when ADP-ribosylated by exoenzyme C3 of *Clostridium botulinum*. *Mol Cell Biol* **8**: 418–426.
23. Aktories K., Mohr C. and Koch G. (1992) *Clostridium botulinum* C3 ADP-ribosyltransferase. *Curr Top Microbiol Immunol* **175**: 115–131.
24. Aktories K. (1990) Clostridial ADP-ribosyltransferases – modification of low molecular weight GTP-binding proteins and of actin by clostridial toxins. *Med Microbiol Immunol* **179**: 123–136.
25. Aktories K., Braun U., Habermann B. and Rösener S. (1990) Botulinum ADP-ribosyltransferase C3. In Moss J. and Vaughan M. (eds), *ADP-Ribosylating Toxins and G Proteins*, pp. 97–115. American Society for Microbiology, Washington, D.C.
26. Just I., Mohr C., Schallehn G., Menard L., Didsbury J.R., Vandekerckhove J., van Damme J. and Aktories K. (1992) Purification and characterization of an ADP-ribosyltransferase produced by *Clostridium limosum* *J Biol Chem* **267**: 10274–10280.
27. Just I., Selzer J., Jung M., van Damme J., Vandekerckhove J. and Aktories K. (1995) Rho-ADP-ribosylating exoenzyme from *Bacillus cereus* – purification, characterization and identification of the NAD-binding site. *Biochemistry*. **34**: 334–340.
28. Just I., Schallehn G. and Aktories K. (1992) ADP-ribosylation of small GTP-binding proteins by *Bacillus cereus*. *Biochem Biophys Res Commun* **183**: 931–936.
29. Sugai M., Hashimoto K., Kikuchi A., Inoue S., Okumura H., Matsumota K., Goto Y., Ohgai H., Moriishi K., Syuto B., Yoshikawa K., Suginaka H. and Takai Y. (1992) Epidermal cell differentiation inhibitor ADP-ribosylates small GTP-binding proteins and induces hyperplasia of epidermis. *J Biol Chem* **267**: 2600–2604.
30. Sugai M., Enomoto T., Hashimoto K., Matsumoto K., Matsuo Y., Ohgai H., Hong Y.-M., Inoue S., Yoshikawa K. and Suginaka H. (1990) A novel epidermal cell differentiation inhibitor (EDIN): Purification and characterization from *Staphylococcus aureus*. *Biochem Biophys Res Commun* **173**: 92–98.
31. Inoue S., Sugai M., Murooka Y., Paik S.-Y., Hong Y.-M., Ohgai H. and Suginaka H. (1991) Molecular cloning and sequencing of the epidermal cell differentiation inhibitor gene from *Staphylococcus aureus*. *Biochem Biophys Res Commun* **174**: 459–464.
32. Chardin P., Boquet P., Madaule P., Popoff M.R., Rubin E.J. and Gill D.M. (1989) The mammalian G protein rho C is ADP-ribosylated by *Clostridium botulinum* exoenzyme C3 and affects actin microfilament in Vero cells. *EMBO J* **8**: 1087–1092.
33. Paterson H.F., Self A.J., Garrett M.D., Just I., Aktories K. and Hall A. (1990) Microinjection of recombinant $p21^{rho}$ induces rapid changes in cell morphology. *J Cell Biol* **111**: 1001–1007.
34. Hall A. (1994) Small GTP-binding proteins and the regulation of the actin cytoskeleton. *Annu Rev Cell Biol* **10**: 31–54.
35. Aktories K., Braun U., Rösener S., Just I. and Hall A. (1989) The rho gene product expressed in *E coli* is a substrate of botulinum ADP-ribosyltransferase C3. *Biochem Biophys Res Commun* **158**: 209–213.
36. Braun U., Habermann B., Just I., Aktories K. and Vandekerckhove J. (1989) Purification of the 22 kDa protein substrate of botulinum ADP-ribosyltransferase C3 from porcine brain cytosol and its characterization as a GTP-binding protein highly homologous to the rho gene product. *FEBS Lett* **243**: 70–76.
37. Bongaerts G.P.A. and Lyerly D.M. (1994) Role of toxins A and B in the pathogenesis of *Clostridium difficile* disease. *Microb Pathog* **17**: 1–12.
38. Thielmann N.M. and Guerrant R.L. (1995) *Clostridium difficile* and its toxins. In Moss J., Iglewski B., Vaughan M. and Tu A.T. (eds), *Bacterial Toxins and Virulence Factors in Disease*, pp. 327–366. Marcel Dekker, New York/Basel.

39. Eichel-Streiber C. (1993) Molecular biology of the *Clostridium difficile* toxins. In Sebald M. (ed.), *Genetics and Molecular Biology of Anaerobic Bacteria*, pp. 264–289. Springer-Verlag, New York.
40. Kelly C.P., Pothoulakis C. and LaMont J.T. (1994) Current concepts: *Clostridium difficile* colitis. *N Engl J Med* **330**: 257–262.
41. Hatheway C.L. (1990) Toxigenic clostridia. *Clin Microbiol Rev* **3**: 66–98.
42. Borriello S.P., Davies H.A., Kamiya S. and Reed P.J. (1990) Virulence factors of *Clostridium difficile*. *Rev Infect Dis* **12**(Suppl. 2): 185–191.
43. Reinke C.M. and Messick C.R. (1994) Update on *Clostridium difficile*-induced colitis, part 2. *Am J Hosp Pharm* **51**: 1892–1901.
44. Reinke C.M. and Messick C.R. (1994) Update on *Clostridium difficile*-induced colitis, part 1. *Am J Hosp Pharm* **51**: 1771–1781.
45. Lima A.A.M., Lyerly D.M., Wilkins T.D., Innes D.J. and Guerrant R.L. (1988) Effects of *Clostridium difficile* toxins A and B in rabbit small and large intestine *in vivo* and on cultured cells *in vitro*. *Infect Immun* **56** (3): 582–588.
46. Lyerly D.M., Saum K.E., MacDonald D.K. and Wilkins T.D. (1985) Effects of *Clostridium difficile* toxins given intragastrically to animals. *Infect Immun* **47**: 349–352.
47. Rothman S.W., Brown J.E., Diecidue A. and Foret D.A. (1984) Differential cytotoxic effects of toxin A and B isolated from *Clostridium difficile*. *Infect Immun* **46**: 324–331.
48. Eichel-Streiber C., Warfolomeow I., Knautz D., Sauerborn M. and Hadding U. (1991) Morphological changes in adherent cells induced by *Clostridium difficile* toxins. *Biochem Soc Trans* **19**: 1154–1160.
49. Florin I. and Thelestam M. (1986) Lysosomal involvement in cellular intoxication with *Clostridium difficile* toxin B. *Microb Pathogen* **1**: 373–385.
50. Henriques B., Florin I. and Thelestam M. (1987) Cellular internalisation of *Clostridium difficile* toxin A. *Microb Pathogen* **2**: 455–463.
51. Eichel-Streiber C. and Sauerborn M. (1990) *Clostridium difficile* toxin A carries a *C*-terminal structure homologous to the carbohydrate binding region of streptococcal glycosyltransferase. *Gene* **96**: 107–113.
52. Eichel-Streiber C., Laufenberg-Feldmann R., Sartingen S., Schulze J. and Sauerborn M. (1992) Comparative sequence analysis of the *Clostridium difficile* toxins A and B. *Mol Gen Genet* **233**: 260–268.
53. Citi S. (1993) The molecular organization of tight junctions. *J Cell Biol* **121**: 485–489.
54. Dove C.H., Wang S.Z., Price S.B., Phelps C.J., Lyerly D.M., Wilkins T.D. and Johnson J.L. (1990) Molecular characterization of the *Clostridium difficile* toxin A gene. *Infect Immun* **58**: 480–488.
55. Sauerborn M. and Eichel-Streiber C. (1990) Nucleotide sequence of *Clostridium difficile* toxin A. *Nucl Acids Res* **18**: 1629–1630.
56. Arseculeratne S.N., Panabokke R.G. and Wijesundera S. (1969) The toxins responsible for the lesions of *Clostridium sordellii* gas gangrene. *J Med Microbiol* **2**: 37–53.
57. Martinez R.D. and Wilkins T.D. (1988) Purification and characterization of *Clostridium sordellii* hemorrhagic toxin and cross-reactivity with *Clostridium difficile* toxin A (enterotoxin). *Infect Immun* **56**: 1215–1221.
58. Nakamura S., Tanabe N., Yamakawa K. and Nishida S. (1983) Cytotoxin production by *Clostridium sordellii* strains. *Microbiol Immunol* **27**: 495–502.
59. Izumi N., Kondo H., Ohishi I. and Sakaguchi G. (1983) Purification and characterization of α-toxin of *Clostridium oedematiens* type A. *Jpn J Med Sci Biol* **36**: 135–146.
60. Bette P., Oksche A., Mauler F., Eichel-Streiber C., Popoff M.R. and Habermann E. (1991) A comparative biochemical, pharmacological and immunological study of *Clostridium novyi* α-toxin, *C. difficile* toxin B and *C. sordellii* lethal toxin. *Toxicon* **29**: 877–887.
61. Green G.A., Schué V. and Monteil H. (1995) Cloning and characterization of the cytotoxin

LT-encoding gene of *Clostridium sordellii*: Homology with *Clostridium difficile* cytotoxin B. *Gene* **161**: 57–61.

62. Hofmann F., Herrmann A., Habermann E. and von Eichel-Streiber C. (1995) Sequencing and analysis of the gene encoding the α-toxin of *Clostridium novyi* proves its homology to toxins A and B of *Clostridium difficile*. *Mol Gen Genet* **247**: 670–679.
63. Ottlinger M.E. and Lin S. (1988) Clostridium *difficile* toxin B induces reorganization of actin, vinculin, and talin in cultures cells. *Exp Cell Res* **174**: 215–229.
64. Fiorentini C., Malorni W., Paradisi S., Giuliano M., Mastrantonio P. and Donelli G. (1990) Interaction of *Clostridium difficile* toxin A with cultured cells: cytoskeletal changes and nuclear polarization. *Infect Immun* **58**: 2329–2336.
65. Fiorentini C. and Thelestam M. (1991) *Clostridium difficile* toxin A and its effects on cells. *Toxicon* **29**: 543–567.
66. Fiorentini C., Arancia G., Paradisi S., Donelli G., Giuliano M., Piemonto F. and Mastrantonio P. (1989) Effects of *Clostridium difficile* toxins A and B on cytoskeleton organization in HEp-2 cells: a comparative morphological study. *Toxicon* **27**: 1209–1218.
67. Bourne H.R., Sanders D.A. and McCormick F. (1990) The GTPase superfamily: a conserved switch for diverse cell functions. *Nature* **348**: 125–132.
68 Bourne H.R., Sanders D.A. and McCormick F. (1991) The GTPase superfamily: conserved structure and molecular mechanism. *Nature* **349**: 117–127.
69 Satoh T., Nakafuku M. and Kaziro Y. (1992) Function of Ras as a molecular switch in signal transduction. *J Biol Chem* **267**: 24149–24152.
70. McCormick F. (1994) Activators and effectors of *ras* p21 proteins. *Curr Opin Genet Dev* **4**: 71–76.
71. Wittinghofer A. and Pai E.F. (1991) The structure of *ras* protein: a model for a universal molecular switch. *Trends Biochem Sci* **16**: 382–387.
72. Balch W.E., Der C.J. and Hall A. (eds) (1995) Small GTPases and their regulators: Part A, Ras family. *Meth Enzym* **255**: 1–509.
73. Balch W.E., Der C.J. and Hall A. (eds) (1995) Small GTPases and their regulators: Part B, Rho family. *Meth Enzym* **256**: 1–366.
74. Balch W.E., Der C.J. and Hall A. (eds) (1995) Small GTPases and their regulators: Part C, proteins involved in transport. *Meth Enzym* **257**: 1–332.
75. Dickey B.F. and Birnbaumer L. (eds) (1993) *GTPases in Biology I*, pp. 1–688. Springer-Verlag, Berlin, Heidelberg, New York, Handbook of Experimental Pharmacology 108/1.
76. Nobes C. and Hall A. (1994) Regulation and function of the Rho subfamily of small GTPases. *Curr Opin Genet Dev* **4**: 77–81.
77. Takai Y., Sasaki T., Tanaka K. and Nakanishi H. (1995) Rho as a regulator of the cytoskeleton. *Trends Biochem Sci* **20**: 227–231.
78. Lamarche N. and Hall A. (1994) GAPs for rho-related GTPases. *Trends Genet* **10**: 436–440.
79. Feig L.A. (1994) Guanine-nucleotide exchange factors: A family of positive regulators of Ras and related GTPases. *Curr Opin Cell Biol* **6**: 204–211.
80. Quilliam L.A., Khosravi-Far R., Huff S.Y. and Der C.J. (1995) Guanine nucleotide exchange factors: Activators of the Ras superfamily of proteins. *Bioessays* **17**: 395–404.
81. Ueda T., Kikuchi A., Ohga N., Yamamoto J. and Takai Y. (1990) Purification and characterization from bovine brain cytosol of a novel regulatory protein inhibiting the dissociation of GDP from and the subsequent binding of GTP to *rho*B p20, a *ras* p21-like GTP-binding protein. *J Biol Chem* **265**: 9373–9380.
82. Lelias J.-M., Adra C.N., Wulf G.M., Guillemot J.-C., Khagad M., Caput D. and Lim B. (1993) cDNA cloning of a human mRNA preferentially expressed in hematopoietic cells and with homology to a GDP-dissociation inhibitor for the rho GTP-binding proteins. *Proc Natl Acad Sci USA* **90**: 1479–1483.

83. Isomura M., Kikuchi A., Ohga N. and Takai Y. (1991) Regulation of binding of rhoB p20 to membranes by its specific regulatory protein GDP dissociation inhibitor. *Oncogene* **6**: 119–124.
84. Sekine A., Fujiwara M. and Narumiya S. (1989) Asparagine residue in the rho gene product is the modification site for botulinum ADP-ribosyltransferase. *J Biol Chem* **264**: 8602–8605.
85. Ridley A.J. and Hall A. (1992) The small GTP-binding protein rho regulates the assembly of focal adhesions and actin stress fibers in response to growth factors, *Cell* **70**: 389–399.
86. Zhang J., King W.G., Dillon S., Hall A., Feig L. and Rittenhouse S.E. (1993) Activation of platelet phosphatidylinositide 3-kinase requires the small GTP-binding protein Rho. *J Biol Chem* **268**: 22251–22254.
87. Chong L.D., Traynor-Kaplan A., Bokoch G.M. and Schwartz M.A. (1994) The small GTP-binding protein Rho regulates a phosphatidylinositol 4-phosphate 5-kinase in mammalian cells. *Cell* **79**: 507–513.
88. Malcolm K.C., Ross A.H., Qiu R.-G., Symons M. and Exton J.H. (1994) Activation of rat liver phospholipase D by the small GTP-binding protein RhoA. *J Biol Chem* **269**: 25951–25954.
89. Hirata K.-I., Kikuchi A., Sasaki T., Kuroda S., Kaibuchi K., Matsuura Y., Seki H., Saida K. and Takai Y. (1992) Involvement of *rho* p21 in the GTP-enhanced calcium ion sensitivity of smooth muscle contraction. *J Biol Chem* **267**: 8719–8722.
90. Tominaga T., Sugie K., Hirata M., Morii N., Fukata J., Uchida A., Imura H. and Narumiya S. (1993) Inhibition of PMA-induced, LFA-1-dependent lymphocyte aggregation by ADP-ribosylation of the small molecular weight GTP binding protein, *rho. J Cell Biol* **120**: 1529–1537.
91. Schmalzing G., Richter H.P., Hansen A., Schwarz W., Just I. and Aktories K. (1995) Involvement of the GTP binding protein Rho in constitutive endocytosis in *Xenopus laevis* oocytes. *J Cell Biol* **130**: 1319–1332.
92. Olson M.F., Ashworth A. and Hall A. (1995) An essential role for Rho, Rac, and Cdc42 GTPases in cell cycle progression through G_1. *Science.* **269**: 1270–1272.
93. Hill C.S., Wynne J. and Treisman R. (1995) The Rho family GTPases RhoA, Rac1, and CDC42Hs regulate transcriptional activation by SRF. *Cell* **81**: 1159–1170.
94. Khosravi-Far R., Solski P.A., Clark G.J., Kinch M.S. and Der C.J. (1995) Activation of Rac1, RhoA, and mitogen-activated protein kinases is required for Ras transformation. *Mol Cell Biol* **15**: 6443–6453.
95. Qiu R.G., Chen J., McCormick F. and Symons M. (1995) A role for Rho in Ras transformation. *Proc Natl Acad Sci USA* **92**: 11781–11785.
96. Ridley A.J., Paterson H.F., Johnston C.L., Diekmann D. and Hall A. (1992) The small GTP-binding protein rac regulates growth factor-induced membrane ruffling. *Cell* **70**: 401–410.
97. Minden A., Lin A., Claret F.-X., Abo A. and Karin M. (1995) Selective activation of the JNK signaling cascade and c-Jun transcriptional activity by the small GTPases Rac and Cdc42Hs. *Cell* **81**: 1147–1157.
98. Qiu R.-G., Chen J., Kirn D., McCormick F. and Symons M. (1995) An essential role for Rac in Ras transformation. *Nature* **374**: 457–459.
99. Abo A., Pick E., Hall A., Totty N., Teahan C.G. and Segal A.W. (1991) Activation of the NADPH oxidase involves the small GTP-binding protein p21rac. *Nature* **353**: 668–670.
100. Bokoch G.M. (1994) Regulation of the human neutrophil NADPH oxidase by the Rac GTP-binding proteins. *Curr Opin Cell Biol* **6**: 212–218.
101. Peppelenbosch M.P., Qiu R.-G., De Vries-Smits A.M.M., Tertoolen L.G.J., de Laat S.W., McCormick F., Hall A., Symons M.H. and Bos J.L. (1995) Rac mediates growth factor-induced arachidonic acid release. *Cell* **81**: 849–856.
102 Nobes C.D. and Hall A. (1995) Rho, Rac, and Cdc42 GTPases regulate the assembly of multimolecular focal complexes associated with actin stress fibers, lamellipodia, and filopodia. *Cell* **81**: 53–62.
103. Kozma R., Ahmed S., Best A. and Lim L. (1995) The Ras-related protein Cdc42Hs and bradykinin

promote formation of peripheral actin microspikes and filopodia in Swiss 3T3 fibroblasts. *Mol Cell Biol* **15**: 1942–1952.

104. Just I., Fritz G., Aktories K., Giry M., Popoff M.R., Boquet P., Hegenbarth S. and Von Eichel-Streiber C. (1994) *Clostridium difficile* toxin B acts on the GTP-binding protein Rho. *J Biol Chem* **269**: 10706–10712.

105. Just I., Richter H.-P., Prepens U., Von Eichel-Streiber C. and Aktories K. (1994) Probing the action of *Clostridium difficile* toxin B in *Xenopus laevis* oocytes. *J Cell Science* **107**: 1653–1659.

106. Just I., Selzer J., Von Eichel-Streiber C. and Aktories K. (1995) The low molecular mass GTP-binding protein Rho is affected by toxin A from *Clostridium difficile*. *J Clin Invest* **95**: 1026–1031.

107. Dillon S.T., Rubin E.J., Yakubovich M., Pothoulakis C., LaMont J.T., Feig L.A. and Gilbert R.J. (1995) Involvement of Ras-related Rho proteins in the mechanisms of action of *Clostridium difficile* toxin A and toxin B. *Infect Immun* **63**: 1421–1426.

108. Just I., Selzer J., Wilm M., Von Eichel-Streiber C., Mann M. and Aktories K. (1995) Glucosylation of Rho proteins by *Clostridium difficile* toxin B. *Nature.* **375**: 500–503.

109. Just I., Wilm M., Selzer J., Rex G., Von Eichel-Streiber C., Mann M. and Aktories K. (1995) The enterotoxin from *Clostridium difficile* (ToxA) monoglucosylates the Rho proteins. *J Biol Chem* **270**: 13932–13936.

110. Pai E.F., Kabsch W., Krengel U., Holmes K.C., John J. and Wittinghofer A. (1989) Structure of the guanine-nucleotide-binding domain of the Ha-ras oncogene product p21 in the triphosphate conformation. *Nature* **341**: 209–214.

111. Kim S.-H., Privé G.G and Milburn M.V. (1993) Conformational switch and structural basis for oncogenic mutations of Ras proteins. In Dickey B.F. and Birnbaumer L. (eds), *GTPases in Biology I*, pp. 177–194. Springer-Verlag, Berlin, Heidelberg.

112. Wittinghofer A., Pai E.F. and Goody R.S. (1993) Structural and mechanistic aspects of the GTPase reaction of H-ras p21. In Dickey F. and Birnbaumer L. (eds), *GTPases in Biology I*, pp. 195–211. Springer-Verlag, Berlin, Heidelberg.

113. Nemoto Y., Namba T., Kozaki S. and Narumiya S. (1991) *Clostridium botulinum* C3 ADP-ribosyltransferase gene. *J Biol Chem* **266**: 19312–19319.

114. Selzer J., Hofman F., Rex G., Wilm M., Mann M., Just I. and Aktories K. (1996) *Clostridium novyi* α-toxin-catalyzed incorporation of GLcNAc into Rho subfamily proteins. *J Biol Chem* **271**: 25173–25177.

115. Baldacini O., Girardot R., Green G.A., Rihn B. and Monteil H. (1992) Comparative study of immunological properties and cytotoxic effects of *Clostridium difficile* toxin B and *Clostridium sordellii* toxin L. *Toxicon.* **30**: 129–140.

116. Popoff M.R. (1987) Purification and characterization of *Clostridium sordellii* lethal toxin and cross-reactivity with *Clostridium difficile* cytotoxin. *Infect Immun* **55**: 35–43.

117. Just I., Selzer J., Hofmann F., Green G.A. and Aktories K. (1996) Inactivation of Ras by *Clostridium sordellii* lethal toxin-catalyzed glucosylation. *J Biol Chem* **271**: 10149–10153.

118. Nusrat A., Giry M., Turner J.R., Colgan S.P., Parkos C.A., Carnes D., Lemichez E., Boquet P. and Madara J.L. (1995) Rho protein regulates tight junctions and perijunctional actin organization in polarized epithelia. *Proc Natl Acad Sci USA* **92**: 10629–10633.

119. Moore R., Pothoulakis C., LaMont J.T., Carlson S. and Madara J.L. (1990) *C. difficile* toxin A increases intestinal permeability and induces Cl^- secretion. *Am J Physiol* **259**: G165–G172.

H: Other Toxins and Extracellular Enzymes

Chapter 23

Extracellular Enzymes from *Clostridium perfringens* and *Clostridium histolyticum* that Damage Connective Tissue

AKINOBU OKABE AND STEWART T. COLE*

Department of Microbiology, Faculty of Medicine, Kagawa Medical University, 1750–1, Ikenobe, Miki-cho, Kita-gun, Kagawa 761–07, Japan, and

**Unité de Génétique Moléculaire Bactérienne, Institut Pasteur, 28 Rue du Docteur Roux, 75724 Paris Cedex 15, France*

I. INTRODUCTION

Infection of deep wounds with the histolytic clostridia, *Clostridium perfringens* and *Clostridium histolyticum*, often results in gas gangrene as the powerful cytotoxins that they secrete provoke tissue necrosis and may also allow the bacteria to escape from phagolysosomes by destroying membrane structure.[1] Another characteristic feature of such infections is destruction of the connective tissue, an extensive network consisting of collagen, proteoglycans and various complex polysaccharides such as hyaluronic acid, repeated units of the disaccharide *N*-acetyl-D-glucosamine-β-(1–4)-D-glucuronic acid linked by a β-(1–3) bond. Connective tissue damage is believed to result from the activity of a battery of hydrolytic enzymes, such as collagenase or κ-toxin, sialidase or neuraminidase, hyaluronidase or μ-toxin, and also proteases and endoglycosidases.[2–4]

The Clostridia: Molecular Biology and Pathogenesis
ISBN 0-12-595020-9

Destruction of the connective tissue is likely to be of major importance in gas gangrene as it allows both the spread of organisms and toxins from the initial site of infection thus leading to colonization and destruction of surrounding tissues.

Recent advances in the molecular genetics of *C. perfringens* have elucidated some of the molecular mechanisms underlying the pathogenesis of clostridial infections, particularly the roles of the α- and θ-toxins in gas gangrene.[5–7] Progress has also been made in understanding the genetics and biochemical properties of the toxins and enzymes that damage the connective tissues. This is illustrated by studies with λ-toxin, a protease produced by some *C. perfringens* types, with κ-toxin, a collagenase produced by all types of *C. perfringens* and some other clostridial species, and with μ-toxin, or hyaluronidase, an endo-β-*N*-acetylglucosaminidase. In this chapter we describe some of the characteristic properties of these toxins and also a possible role of λ-toxin in pathogenesis.

II. RELATIONSHIP BETWEEN COLLAGENASE AND GELATINASE

Collagenase of *C. perfringens* (κ-toxin) degrades collagen fiber, the most abundant matrix protein and is considered to be responsible for the pathogenesis of myonecrosis caused by this organism.[1] A single chromosomal gene, *colA*, encodes the *C. perfringens* enzyme and this has been cloned, sequenced and characterized.[8] Because large amounts of collagenase are produced by *C. histolyticum*, the biochemical properties of the enzyme from this bacterium have been most extensively studied and it has been widely used as a tissue-dispersing enzyme.[9] The multiplicity of active forms of *C. histolyticum* collagenase reported by many workers[10] was postulated to be due to gene duplication.[11] However, on Southern hybridization using the collagenase gene cloned from *C. histolyticum* JCM1403 as a probe only one band, corresponding to the cloned gene, was detected.[12] Furthermore, two polypeptides that copurified from the strain, 116 kDa collagenase and 98 kDa gelatinase showed almost identical peptide maps and possessed the same *N*-terminal amino acid sequences.[12] Therefore, the latter is considered to be produced by removal of a *C*-terminal peptide from the 116 kDa collagenase. It is possible that some, if not all, of collagenase subfractions reported for the other strains are proteolytic products of a progenitor collagenase.

A. Structure of collagenases

The primary structures of collagenases of *C. perfringens* and *Vibrio alginolyticus* are homologous to that of *C. histolyticum*. On the basis of alignment of their deduced amino acid sequences, it is postulated that these enzymes are composed of three segments.[12] Homology of each segment is statistically significant except for segment 3 of *V. alginolyticus* collagenase (data not shown). Segment 1 contains the HEXXH motif, a signature for zinc-metalloproteases. However, the sequence around the motif does not agree with an extended consensus sequence shared by all zinc-metalloproteases.[13] This raises the question whether or not HEXXH is the catalytic site of the collagenase. Substitution of both the first histidine residue with phenylalanine and the second one with arginine, by site-directed mutagenesis, resulted in complete loss of hydrolyzing activity against Pz-peptide, a synthetic substrate for collagenase.[14] This indicates that the two histidine residues are essential for the enzyme activity, and suggests that the motif is a catalytic site of the enzyme.

The 116 kDa collagenase is processed into the 98 kDa gelatinase by removal of a *C*-terminal peptide, as described above. Furthermore, *Escherichia coli* cells expressing the collagenase gene without segments 2b and 3 produced gelatinase but not collagenase activity.[15] These results suggest that segments 2 and/or 3 are involved in the interaction of collagenase with native collagen fiber. This possibility is supported by an experiment using glutathione S-transferase fused to the *C*-terminal portion of *C. histolyticum* collagenase. While the glutathione S-transferase *per se* did not bind to the collagen fiber, the hybrid protein in which the enzyme was fused via its C terminus to the peptides corresponding to segments 2b and 3, did bind thus indicating that these segments should be a collagen-binding domain.[15] Segments 2 and 3 are characteristically repeated in the collagenases of *C. perfringens* and *C. histolyticum*, respectively, while segment 3 of *V. alginolyticus* collagenase differs from those of the clostridial ones (Figure 23.1 and also data not shown). Such differences may cause difference in their substrate preference or specificity. Surprisingly, the PKD domain occurring repetitively in the amino acid sequence deduced from the *PKD1* gene, which is involved in polycystic kidney disease, shows similarity to segment 2 of *C. perfringens* collagenase and also to certain regions of some other eukaryotic and prokaryotic proteins (Figure 23.1). This has been postulated to be due to horizontal transfer of the eukaryotic gene into bacteria.[16]

B. Production of collagenase by recombinant *Bacillus subtilis* cells

Purification of the *C. histolyticum* collagenase is complicated by the presence of multiple forms and other proteases with physical and chemical characteristics similar to those of collagenase.[10] Thus, many preparations of *C. histolyticum* collagenase are often contaminated not only with various forms of collagenase and gelatinase but also with non-specific proteases. Although the use of

A

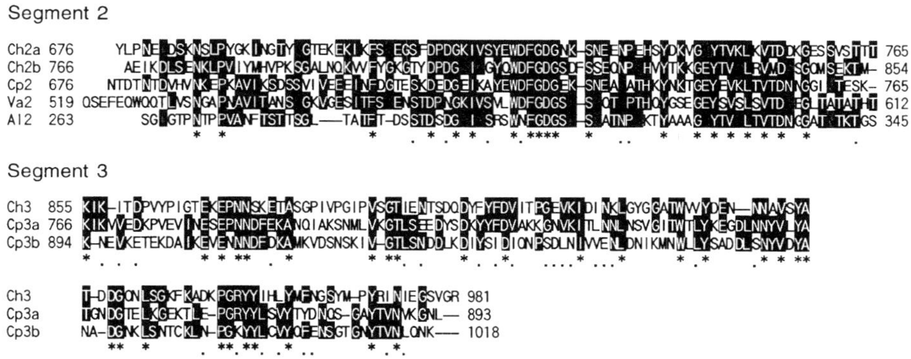

B

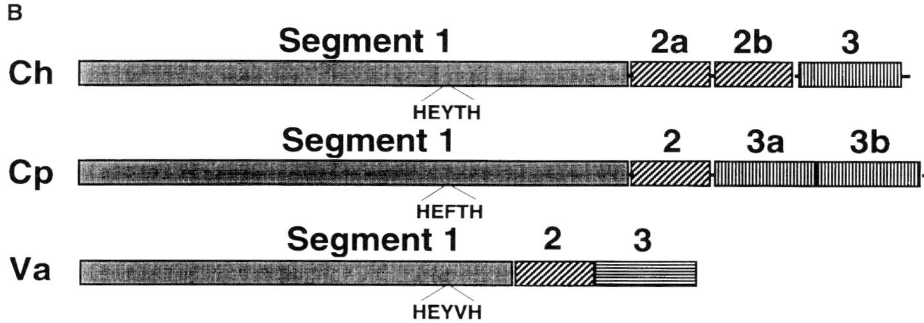

C

```
PKD1 exon15 R14   FTNRSAQFEAATS-PSPRRVAYHWDFGDGSPGQDTDEPRAEHSYLRPGDYRVQVNASNLVS-FFVAQATVTVQ
Pmel/Human        PLTFALQLHDPSGYLAEADLSYTWDFGDSSGTLISRAPVVTHTYLEPGPVTAQVVLQAAIP-LTSCGSSPVPG
Pmel/Bovine       PLTFALQLHDPSGYLAGADLSYTWDFGDSTGTLISRALTVTHTYLESGPVTAQVVLQAAIP-LTSCGSSPVPG
Nmb/Human         PIMFDVLIHDPSHFLNYSTINYKWSFGDNTGLFVSTNHTVNHTYVLNGTFSLNLTVKAAAP--GPCPPPPPPP
Coll/Vibal        VGESITFSSENSTDPNGKIVSVLWDFGDGST-STQTKPT--HQYGSEGEYSVSLSVTDSEG-LTATATHTVVI
Slpb/Meth1        TSGTAPLNVLFTDTSTGSPTTWKWNFGDGTS-STQKSPT--HAYSTAGTYTVTLTVTNSAGSNTATKTNYVTV
Slpb/Meth2        RSGIAPLTVTFKDNSSGSPTAWNWSFGDGAY-SNEKYPK--HTYTAPGSYTISLTASNAAGSNTLTKSNYIVV
Slpb/Meth4        RSGTAPLTVTFKDNSSGSPTAWNWSFGDGAY-SNEKYPK--HTYMAPGSYTISLTASNAAGSNTLIKNNYIVV
Coll/Clope        VEEEINFDGTESKDEDGEIKAYEWDFGDGEK-SNEAKAT--HKYNKTGEYEVKLTVTDNNG-GINTESKKIKV
Api/Achly         SGLTATFT-DSSTDSDGSIASRSWNFGDGST-STATNPS--KTYAAAGTYTVTLTVTDNGG-ATNTKTGSVTV
                  hh  t  h h     tttth a WDhGDGt  tt    thtH aht G h h h htN ht      h h h
```

Figure 23.1 A. Amino acid sequence alignment of the segments 2 and 3 of three homologous collagenases. Ch, Cp and Va represent collagenases of *Clostridium histolyticum*, *Clostridium perfringens* and *Vibrio alginolyticus*, respectively. Residues conserved in at least two sequences are indicated by a black background. Additional residues conserved between two sequences in segment 2 are lightly shaded. Identical and similar residues are indicated by asterisks and dots, respectively. B. Location of a zinc-metalloprotease consensus motif, HEXXH, and three-segment structure. C. PKD domains. Of 14 PKD domains (R1–R14) in PKD1 exons only R14 in the exon 15 is presented at the top of the list. Other proteins containing PKD domains are listed: human and bovine melanoma antigen Pmel17, Pmel/Human and Pmel/Bovine: human Nmb protein, Nmb/Human; *V. alginolyticus* collagenase, Coll/Vibal; three domains of the *Methanothermus fervidus* Slpb protein, Slpb/Meth1–4; *C. perfringens* collagenase, Coll/Clope; and *Achromobacter lyticus* protease, Api/Achly. The hydrophobic and turn-like or polar amino acids are denoted by h and t, respectively. Conserved amino acids are represented in the bottom line of the figure. Panels A and B were based on Yoshihara *et al.*[12], and panel C on Glücksmann-Kuis *et al*[16].

recombinant DNA technology was expected to solve this problem, expression of the collagenase gene in other hosts appears to be problematic. In general, large genes from *C. perfringens* are not well expressed in *E. coli* probably owing to the translational barrier arising from the difference in codon usage between the two organisms.[3] This translational barrier should be reduced in *Bacillus subtilis*, a Gram-positive bacterium with codon usage closer to that of *C. perfringens* than to *E. coli*.[3]

Unfortunately, problems of plasmid instability and proteolytic degradation of the gene product were encountered on transformation of *B. subtilis* with a plasmid containing the collagenase gene.[17] These were solved by selecting the host strain, culture medium and a growth condition. When *B. subtilis* DB104, a strain deficient in two major proteases, was used as the host and grown in a medium containing high concentrations of sodium succinate[18] at 35°C with gentle shaking, not only was intact plasmid maintained but sufficient quantities of collagenase were also produced, as the culture supernatant contained 116 kDa collagenase as a major product. The 116 kDa collagenase was purified to homogeneity by ammonium sulfate precipitation, gel filtration and ion exchange chromatography, as shown on sodium dodecyl sulfate (SDS)-polyacrylamide gel analysis.[17] Another clostridial exotoxin, phospholipase C, has also been purified from recombinant *B. subtilis*.[19] The plasmid containing the phospholipase C gene can be stably maintained in *B. subtilis* cells grown under standard conditions. The difference in stability of the two plasmids suggests that the longer the mRNA, the greater will be the probability that a ribosome will stall at a rare codon still present in *B. subtilis*.[3] Ribosome stalling during translation of a secreted protein would be detrimental to the membrane or secretory mechanism of the host bacterium. Measures such as those described above, might enable the difficulty associated with expressing clostridial genes in recombinant bacteria to be overcome.

III. CHARACTERIZATION OF ENZYMATIC PROPERTIES OF λ-TOXIN

It is four decades since Bidwel first identified λ-toxin in the ammonium sulfate fraction as a protease which degraded azocoll, gelatin, silk, and casein.[20] However, further characterization of the toxin has not been performed until recently. Jin *et al.*[21] purified λ-toxin from the culture supernatant of *C. perfringens* type B NCIB10691. The purified enzyme was inhibited specifically by metal chelators, and its caseinolytic activity is optimal at pH 7.5.

A. Analysis of the *lam* gene

Jin *et al.*[21] also cloned and sequenced the λ-toxin-encoding gene (*lam*) from a plasmid in the NCIB strain. The *lam* gene encodes a 553 amino acid protein with the zinc-metalloprotease consensus motif, HEXXH. The amino acid sequence at the *N*-terminus of the purified enzyme coincided with that starting at position 236 of the deduced amino acid sequence, suggesting that the *N*-terminal 235 amino acid residues are cleaved during secretion and maturation of the enzyme. The amino acid sequence contains not only an extended consensus sequence shared by all zinc-metalloproteases[13] but also the consensus sequence around the three zinc-binding sites characteristic of the thermolysin family.[22]

The *lam* gene is located on 70, 120 and 140 kb plasmids in strains NCIB10691,[21] NCTC2062 and 945P,[23] respectively, while λ-toxin was reported to be encoded by a 3.2 kb plasmid in the strain ATTC 3626B.[24] Along with the observation that the nucleotide sequence determined for the upstream region of the *lam* gene showed significant similarity to those present in IS*1151* and IS*231*,[25] this may imply that the *lam* gene is transposable from one plasmid to another.

B. Phylogenetic relationships between λ-toxin and other metalloproteases

The plasmid-borne *lam* gene appears to have evolved similarly to the chromosomal genes of *C. perfringens*, since the GC content of the *lam* gene at 30%, is close to that of the chromosomal DNA. A phylogenetic tree constructed by the neighbor-joining method (Figure 23.2) revealed that λ-toxin is closely related to the thermolysin family. The overall phylogenetic distribution of the metalloproteases is similar to relationships within bacteria, as inferred from comparative 16S rRNA sequence analysis,[27] although all enzymes listed are affiliated with only limited groups of bacteria, that is, the β- and γ-subdivisions of the proteobacteria, and high and low GC Gram-positive bacteria. It is evident that λ-toxin diverged a long time ago from the metalloproteases of Gram-positive bacteria.

The propeptide sequences of proteases are unusually long, and have been implicated in secretion, membrane anchoring, folding into the active enzyme conformation,[28,29] or in prevention of deleterious proteolytic activity in the wrong compartment.[30] Statistical analysis of prepropeptides and mature enzymes shows that the prepropeptides are less similar than the mature enzymes. The prepropeptide of λ-toxin is significantly similar to those of metalloproteases from Gram-positive bacteria but not to those from Gram-negative ones. It is conceivable that propeptides of metalloproteases have evolved, reflecting

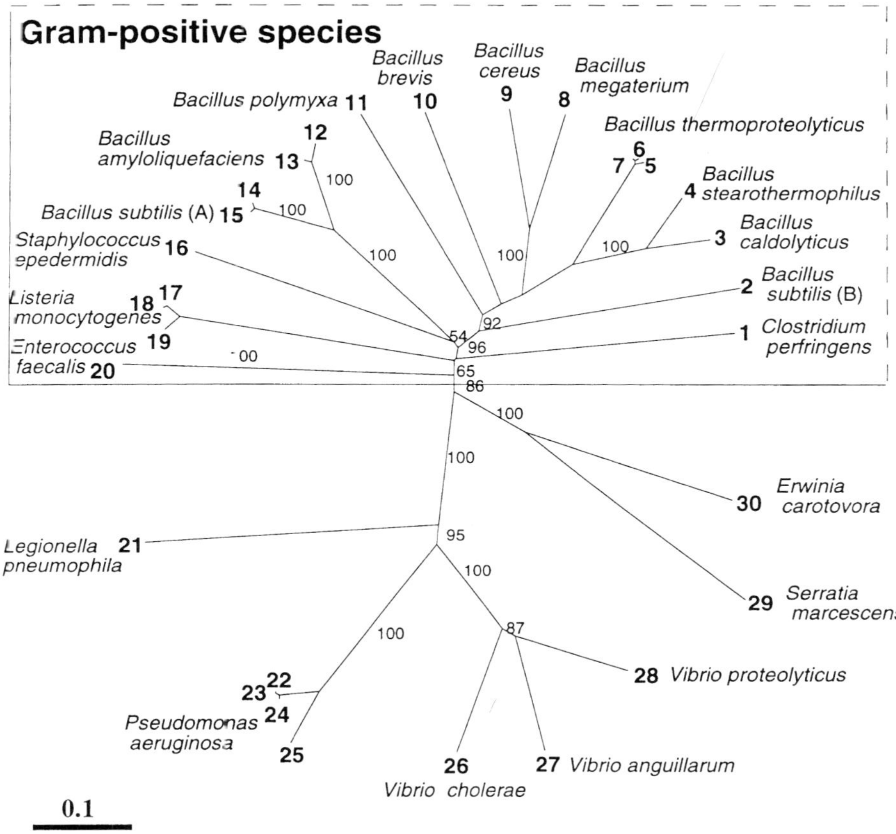

Figure 23.2 Evolutionary tree for zinc-metalloprotease amino acid sequences, constructed by the neighbor-joining method.[26] The scale bar corresponds to 0.1 estimated fixed mutation per amino acid position. Bootstrap (confidence) values shown adjacent to a node were calculated from 1000 trees. The marker bar represent 0.1 expected displacement per amino acid position. The GenBank nucleotide accession numbers for the sequences of the metalloproteases from the organisms named are as follows: 1, D45904; 2, A41042; 3, P23384; 4, P06874; 5, M21663; 6, B36706; 7, 441267; 8, S21934; 9, S22690; 10, PN0114; 11, P29148; 12, M36723; 13, P06832; 14, D10773; 15, P06142; 16, 396259; 17, C60280; 18, P23224; 19, B60280; 20, A43580; 21, P21347; 22, C38166; 23, B38166; 24, P14756; 25, A38166; 26, P24153; 27, L02528; 28, Q00971; 29, M59854; and 30, A41048. *Bacillus subtilis* (A) and *B. subtilis* (B) represent netural proteases A and B, respectively.

difference in the process of secretion or maturation of the enzymes between Gram positive and negative bacteria.

C. Pathogenic roles of λ-toxin

Many bacterial metalloproteases contribute to pathogenicity through either degrading various biologically important host components or activating inactive precursor forms of virulence factors.[31] Purified λ-toxin is capable of degrading immunoglobulin A, complement C3 component, fibrinogen, fibronectin and α_2-macroglobulin.[21] Injection of the toxin into dorsal skin of mice caused hemorrhagic edema. The toxin also rendered the injection site permeable to circulating Evans blue in a dose-dependent manner, indicating that it causes an increase in vascular permeability.[21] The dose-dependency and time course of the permeability-enhancing reaction of the λ-toxin are similar to those of *Vibrio vulnificus* metalloprotease,[32] which activates Hageman factor and prekallikrein.[33] Thus, the toxin appears to play a pathogenic role through evoking inflammatory responses when the organism invades interstitial spaces. It may also activate the other virulence factors, such as plasmid-encoded ε-toxins[34,35] and ι-toxins,[36] which are produced in precursor forms and are activated by proteolysis.[3]

IV. THE μ-TOXIN

Some, but not all, strains of *C. perfringens*[4] and other pathogenic clostridial species,[1] such as *C. difficile*, elicit a hyaluronidase also known as μ-toxin. A gene, *nagH*, encoding hyaluronidase was cloned and sequenced from *C. perfringens* and shown to secrete a 114 kDa protein endowed with *N*-acetyl glucosaminidase activity. As described above for collagenase, production of recombinant hyaluronidase was found to be highly inefficient in *E. coli*, although *B. subtilis* has not yet been tried as a host. Futher biochemical characterization performed on material produced by *C. perfringens* CPN50 convincingly demonstrated that a protein of the correct size acted as hyaluronidase, since reducing ends were released from hyaluronic acid.[37] It thus seems likely that the *nagH* gene product corresponds to the μ-toxin which, when injected intradermally, potentiates the cytolytic effect of α-toxin by facilitating its diffusion.[4]

Although the μ-toxin has *N*-acetyl glucosaminidase activity, it is clearly quite distinct, in size and sequence, from members of a family of well-characterized bacterial endo-*N*-acetyl-β-D-glucosaminidases that act on the *N-N′*-diacetylchitobiosyl core of *N*-glycosylproteins. These enzymes have molecular masses of around 30 kDa and share limited similarity in the active site region with chitinases.[38] Several other glucosaminidase activities have been detected in *C. perfringens*, that may belong to this family because they act on glycoproteins.[37]

Although full-length hyaluronidase could not be produced in *E. coli*, the expression of truncated forms of *nagH* in this host allowed some interesting conclusions to be drawn about the organization of the protein into domains.[37] The precursor form of the NagH protein is predicted to consist of 1042 amino acid residues with a typical cleavage site for leader peptidase between positions 30 and 31. The primary structure is consistent with the protein being exported, as there are no extended hydrophobic stretches, although hyaluronidase may be associated with the outer surface of the bacterial envelope, as it is only released following autolysis.[37] The processed form of the protein has an M_r of 114 392 and seems to be organized into at least two domains, an acidic *N*-terminal segment consisting of residues 1–700, and a *C*-terminal region of 300 residues, containing three copies of a 45 amino acid motif. The active site appears to be located at the *N*-terminal end because a truncated protein comprising the first *c.* 370 residues, retained enzymatic activity. There is weak but significant similarity between residues 158 and 700 of hyaluronidase and several segments of a putative 855-residue protein (T20B5.3) from *Caenorhabditis elegans.*

V. CHROMOSOMAL LOCATION OF THE GENES FOR λ- AND μ-TOXINS

In contrast to the λ-toxin gene which is located on plasmids in a limited number of strains,[23,39] both the *colA* and *nagH* genes are situated on the chromosome on the same 30 kb *Apa*I–*Sma*I fragment near the *rrnE* operon of most isolates of *C. perfringens*. The general organization of this region is depicted in Figure 23.3, although the exact order and orientation of the genes is still uncertain. It is clear, however, that a gene homologous to the *mscL* gene of *E coli*,[40] which encodes a

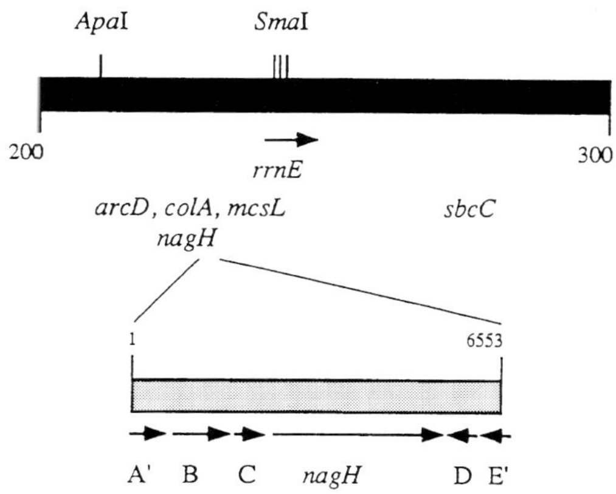

Figure 23.3: Organization of the *Clostridium perfringens* chromosome between coordinates 200 and 300 kb showing the postitions of various genes and rare restriction sites.[38] Neither the exact positions nor the order of the genes are known. The insert shows the organization of a 6.55 kb fragment from the *nagH* region (accession number M81878) and the positions of the various open reading frames (ORFs). Similarity was seen between the products of *nagH*, ORFA, ORFB, and ORFD with proteins described in the following sequences U28742 (T20B5.3), M27248 (XylR *B. subtilis*), M59232 (*B. subtilis* YCW2) and U32700 (*Haemophilus influenzae* hypothetical protein), respectively.

large-conductance mechano-sensitive protein channel (MscL), is situated downstream of *colA*.[41] Although the gene product has not been identified in the membrane of *C. perfringens*, by analogy with *E. coli* it is expected to sense environmental signals, such as oxidoreduction potential or osmotic pressure, that may also influence expression of virulence genes.

While surveying a collection of *C. perfringens* strains for the presence of virulence genes Katayama *et al.*[23] established that while the *colA* gene was always present, *nagH* was missing from four CPE-producing strains which have chromosomal *cpe* genes,[42] although the sequences flanking *nagH* were still present (see Figure 23.3). In contrast, *nagH* was always present in strains harboring *cpe* on a plasmid.

The exact distance between *nagH* and *colA* is not known but both genes appear to be organized as monocistronic operons. Attempts to detect *nagH* transcripts have been unsuccessful, probably because expression of the gene is inducible possibly by hyaluronic acid. The *nagH* gene is flanked on one side by ORFs A-C (Figure 23.3) and on the other by ORFs D and E. The proteins encoded by ORFs A and D are similar to proteins involved in the regulation of sucrose metabolism and to the YCW2 protein of *B. subtilis*, respectively.[37] The product of ORF B is strikingly similar throughout its length (279 residues) to a hypothetical protein of 299 amino acids from *Haemophilus influenzae*, and to two distinct hypothetical proteins from *E. coli* of 306 and 307 residues, although there is presently no clue to its function.

VI. REFERENCES

1. Hatheway C.L. (1990) Toxigenic clostridia. *Clin Microbiol Rev* **3**: 66–98.
2. McDonel J.L. (1986) Toxins of *Clostridium perfringens* types A, B, C, D and E, pp. 477–517. In Dorner F. and Drews J. (eds), *Pharmacology of Bacterial Toxins*. Pergamon Press, Oxford.
3. Rood J.I. and Cole S.T. (1991) Molecular genetics and pathogenesis of *Clostridium perfringens*. *Microbiol Rev* **55**: 621–648.
4. Smith L.D.S. (1979) Virulence factors of *Clostridum perfringens*.. *Rev Ind Dis* **1**: 254–260.
5. Awad M.M., Bryant A.E., Stevens D.L. and Rood J.I. (1995) Virulence studies on chromosomal α-toxin and θ-toxin mutants constructed by allelic exchange provide genetic evidence for the essential role of α-toxin in *Clostridium perfringens*-mediated gas gangrene. *Mol Microbiol* **15**: 191–202.
6. Ninomiya M., Matsushita O., Minami J., Sakamoto H., Nakano M. and Okabe A. (1994) Role of alpha-toxin in *Clostridium perfringens* infection determined by using recombinants of *C. perfringens* and *Bacillus subtilis*. *Infect Immun* **62**: 5032–5039.
7. Rood J.I. and Lyristis M. (1995) Regulation of extracellular toxin production in *Clostridium perfringens*.. *Trends Microbiol* **3**: 192–196.
8. Matsushita O., Yoshihara K., Katayama S.-I., Minami J. and Okabe A. (1994) Purification and characterization of a *Clostridium perfringens* 120–kilodalton collagenase and nucleotide sequence of the corresponding gene. *J Bacteriol* **176**: 149–156.

9. Seglen P.O. (1976) Preparation of isolated rat liver cells. *Methods Cell Biol* **13**: 29–83.
10. Peterkofsky B. (1982) Bacterial collagenase. *Methods Enzymol* **82**: 453–471.
11. Bond M.D. and van Wart H.E. (1984) Relationship between the individual collagenases of *Clostridium histolyticum*: evidence for evolution by gene duplication. *Biochemistry* **23**: 3092–3099.
12. Yoshihara K., Matsushita, O., Minami J. and Okabe A. (1994) Cloning and nucleotide sequence analysis of the *colH* gene from *Clostridium histolyticum* encoding a collagenase and a gelatinase. *J Bacteriol* **176**: 6489–6496.
13. Jongeneel C.V., Bouvier J. and Bairoch A. (1989) A unique signature identifies a family of zinc-dependent metallopeptidases. *FEBS Lett* **242**: 211–214.
14. Matsushita O., Jung C.-M., Minami J. and Okabe A. (1997) Analysis of the catalytic domain of 116-kDa *Clostridium histolyticum* collagenase. (In preparation).
15. Matsushita O., Jung C.-M., Minami J., Katayama S.-I., Nishi N. and Okabe A. (1997) A study on the collagen binding domain of 116 kDa *Clostridium histolyticum* collagenase. (In preparation).
16. Glücksmann-Kuis M.A., Tayber O., Woolf E.A. *et al.*, (1995) Polycystic kidney disease: the complete structure of the *PKD1* gene and its protein. *Cell* **81**: 289–298.
17. Jung C.-M., Matsushita O., Katayama S.-I., Minami J., Ohhira I. and Okabe A. (1996) Expression of the *colH* gene encoding *Clostridium histolyticum* collagenase in *Bacillus subtilis* and its application to enzyme purification. *Microbiol Immun* **40**: 923–929.
18. Tsutsui K., Minami J., Matsushita O., Katayama S.-I., Taniguchi Y., Nakamura S., Nishioka M. and Okabe A. (1995) Phylogenetic analysis of phospholipase C genes from *Clostridium perfringens* types A to E and *Clostridium novyi*. *J Bacteriol* **177**: 7164–7170.
19. Hirata Y., Minami J., Koyama M., Matsushita O., Katayama S.-I., Jin F., Maeta H. and Okabe A. (1995) A method for purification of *Clostridium perfringens* phospholipase C from recombinant *Bacillus subtilis* cells. *Appl Environ Microbiol* **61**: 4114–4115.
20. Bidwell E. (1950) Proteolytic enzymes of *Clostridium welchii*. *Biochem J* **46**: 589–598.
21. Jin F., Matsushita O., Katayama S.-I., Jin S., Matsushita C., Minami J. and Okabe A. (1996) Purification, characterization, and primary structure of *Clostridium perfringens* lambda-toxin, a thermolysin-like metalloprotease. *Infect Immun* **64**: 230–237.
22. Jiang W. and Bond J.S. (1992) Families of metalloendopeptidases and their relationships. *FEBS Lett* **312**: 110–114.
23. Katayama S., Dupuy B., Daube G., China B. and Cole S.T. (1996) Genome mapping of *Clostridium perfringens* strains with I-*Ceu*I shows many virulence genes to be plasmid-borne. *Mol Gen Genet* **251**: 720–726.
24. Blaschek H.P. and Solberg M. (1981) Isolation of a plasmid responsible for caseinase activity in *Clostridium perfringens* ATTC 3626B. *J Bacteriol* **147**: 262–266.
25. Mahillon J., Seurinck J., Delcour J. and Zabeau M. (1987) Cloning and nucleotide sequence of different iso-IS*231* elements and their structural association with the Tn*4430* transposon in *Bacillus thuringiensis*. *Gene* **51**: 187–196.
26. Saitou N. and Nei M. (1987) The neighbor-joining method: a new method for reconstructing phylogenetic trees. *Mol Biol Evol* **4**: 406–425.
27. Olsen G.J., Woese C.R. and Overbeek R. (1994) The winds of (evolutionary) change: breathing new life into microbiology. *J Bacteriol* **176**: 1–6.
28. Li Y. and Inouye M. (1994) Autoprocessing of prothiolsubtilisin E in which active-site serine 221 is altered to cysteine. *J Biol Chem* **269**: 4169–4174.
29. Shinde U., Li Y., Chatterjee S. and Inouye M. (1993) Folding pathway mediated by an intramolecular chaperone. *Proc Natl Acad Sci USA* **90**: 6924–6928.
30. Simonen M. and Palva I. (1993) Protein secretion in *Bacillus* species. *Microbiol Rev* **57**: 109–137.
31. Häse C.C. and Finkelstein R.A. (1993) Bacterial extracellular zinc-containing metalloproteases. *Microbiol Rev* **57**: 823–837.

32. Miyoshi N., Miyoshi S.-I., Sugiyama K., Suzuki Y., Furuta H. and Shinoda S. (1987) Activation of the plasma kallikrein–kinin system by *Vibrio vulnificus* protease. *Infect Immun* **55**: 1936–1939.
33. Miyoshi S.-I. and Shinoda S. (1992) Activation mechanism of human Hageman factor-plasma kallikrein-kinin system by *Vibrio vulnificus* metalloprotease. *FEBS Lett* **308**: 315–319.
34. Canard B., Saint-Joanis B. and Cole S.T. (1992) Genomic diversity and organization of virulence genes in the pathogenic anaerobe *Clostridium perfringens*. *Mol Microbiol* **6**: 1421–1429.
35. Hunter S.E.C., Clarke I.N., Kelly D.C. and Titball R.W. (1992) Cloning and nucleotide sequencing of the *Clostridium perfringens* epsilon-toxin gene and its expression in *Escherichia coli*. *Infect Immun* **60**: 102–110.
36. Perelle S., Gibert M., Boquet P. and Popoff M.R. (1993) Characterization of *Clostridium perfringens* iota-toxin genes and expression in *Escherichia coli*. *Infect Immun* **61**: 5147–5156.
37. Canard B., Garnier T., Saint-Joanis B. and Cole S.T. (1994) Molecular genetic analysis of the *nagH* gene encoding a hyaluronidase of *Clostridium perfringens*. *Mol Gen Genet* **243**: 215–224.
38. Karamanos Y., Bourgerie S., Barreaud J.-P., and Julien R. (1995) Are there biological functions for bacterial endo-*N*-acetyl-β-D-glucosaminidases. *Res Microbiol* **146**: 437–443.
39. Sukharev S.I., Blount P., Martinac B., Blattner F.R. and Kung C. (1994) A large-conductance mechanosensitive channel in *E. coli* encoded by *mscL* alone. *Nature* **368**: 265–268.
40. Katayama S., Dupuy B., Garnier T. and Cole S.T. (1995) Rapid expansion of the physical and genetic map of the chromosome of *Clostridium perfringens* CPN50. *J Bacteriol* **177**: 5680–5685.
41. Matsushita O., Jung C.-M. and Okabe A. (1995) Identification of the gene encoding an MscL homologue in *Clostridium perfringens*. *Gene* **165**: 147–148.
42. Cornillot E., Saint-Joanis B., Canard B., Granum P.-E., Daube G. and Cole S.T. (1995) The enterotoxin gene (*cpe*) of *Clostridium perfringens* can be chromosomal or plasmid-borne. *Mol Microbiol* **15**: 639–647.

Chapter 24

Clostridial Sialidases

PETER ROGGENTIN AND ROLAND SCHAUER
Biochemisches Institut, Christian-Albrechts-Universität, Olshausenstrasse 40, D-24098 Kiel, Germany

I. INTRODUCTION

Many clostridial species are known to secrete a variety of proteins, which were formerly named toxins and hemagglutinins because of their harmful effects on humans. Many of these proteins have been identified as enzymes, e.g. the α-toxin of *Clostridium perfringens* is a phospholipase.[1] These enzymes allow the bacteria to hydrolyze complex organic materials in their environment, which originated from higher organisms such as animals. They use the liberated fatty, amino and nucleic acids, and also carbohydrates for nutrition.

Among the proteins secreted, an enzyme named neuraminidase[2] or alternatively sialidase[3] has been found. Sialidases (EC 3.2.1.18) are glycohydrolases and cleave terminal sialic acids from sialoglycoconjugates. Sialic acids comprise a family of 36 naturally occurring derivatives of neuraminic acid, which are all *N*-acylated to form, for example, *N*-acetylneuraminic acid (Neu5Ac) which is the most common derivative of this group of monosaccharides.[4] This sugar is acidic due to a carboxyl group at C1, and may be linked via the hydroxyl group of C2 to a subterminal sugar of an oligosaccharide, sialic acid polymer, glycoprotein, glycolipid or lipopolysaccharide.

The Clostridia: Molecular Biology and Pathogenesis
ISBN 0-12-595020-9

Among the clostridia, only a few species are known to produce sialidase: *Clostridium perfringens*, *Clostridium sordellii*, *Clostridium septicum*, *Clostridium chauvoei*, and *Clostridium tertium*,[5] some strains of *Clostridium butyricum*,[6] and also strain DSM 1787 of *Clostridium nexile* (P. Roggentin, unpublished observation). The enzyme is thought to be of advantage to the bacteria, as it supplies them with a further source of energy which can be exploited if they produce *N*-acetylneuraminate lyase (EC 4.1.3.3.), an enzyme that cleaves sialic acid into pyruvic acid and *N*-acetylmannosamine. Furthermore, the release of sialic acids makes subterminal structures accessible to digestion by other hydrolytic enzymes, whose products are made available to the bacteria. Sialidases are, therefore, believed to have a mainly nutritional function for these bacteria.[7,8]

Little has been published about the role of bacterial sialidase in infections, which are frequently caused by *C. perfringens* and, to a lesser extent, by *C. septicum* and *C. sordellii*.[9] The enzyme may facilitate the degradation of tissues, and so enable a more rapid spreading of the bacteria.[10,11] On reaching the bloodstream, sialidase may suppress the biological processes mediated by sialic acids in the organism,[4,12] e.g. the masking effect on cell surfaces, or as a binding determinant in cell–cell interactions.

The following aspects made sialidase of interest for our investigations. (1) It is produced in large amounts by three clostridial species that cause severe, life-threatening infections in man and animals (clostridial myonecrosis or gas gangrene) and its presence may be used as an indicator for the growth of these bacteria in wounds at an early stage of infection. (2) Since it has become apparent that sialic acids play a central role in many regulatory and cellular processes, sialidases have become essential tools in glycobiology for the selective removal of sialic acids. For this purpose, pure, well-characterized enzymes are required. One of the aims of this work was therefore to isolate six sialidases from five clostridial species and determine their ability to hydrolyse sialic acids from typical natural substrates under optimal *in vitro* conditions. The results also allowed comparison of their enzymic properties, and the availability of pure proteins permitted genetic investigations on sialidases. (3) Sialidases are irregularly distributed in the bacterial kingdom, and within the Genus *Clostridium*. Highly related species such as *C. sordellii* and *Clostridium bifermentans* are differentiated by their ability to produce sialidase,[13] a property that can even distinguish certain strains of *C. butyricum*.[6] These observations indicate an unusual evolutionary pathway and uncommon route of genetic distribution of sialidase (an enzyme used by some microorganisms to attack a sugar) which may have first appeared in deuterostomate animals some 500 million years ago,[14] possibly as a consequence of microbial aggression.

II. SIALIDASES AS INDICATORS OF INFECTIONS CAUSED BY CLOSTRIDIA

Clostridial spores resting in dust, soil or feces may enter wounds of man and other animals. The resulting infection is called clostridial myonecrosis, as the bacteria preferentially grow in muscle tissue, or gas gangrene, because some species produce large amounts of gas in the tissue (see Chapter 9). It is often encountered after extensive injury to skeletal muscle complicated by compound fractures or foreign bodies, as a complication of surgery, or after intramuscular injections particularly into patients with diabetes or arteriosclerosis.[15] The onset of infection is facilitated by a locally reduced blood supply, leading to anaerobic conditions and induction of spore germination. The vegetative clostridial cells spread through tissue with a speed of up to 10 cm/h,[16] leading to marked systemic symptoms, which may result in circulatory collapse within 24–48 h. The rapid progression of the disease makes a rapid and sensitive diagnostic test desirable. A high degree of specificity is important to distinguish the pathogenic clostridia from those bacteria simulating gas gangrene.[17,18] Conventional tests based on the isolation of the bacteria and investigation of their biochemical properties[19, 20] are too time-consuming for early and life-saving therapy; hyperbaric oxygen treatment of the patient or resection of infected tissue are successful procedures only at an early stage of infection.

The high invasive potency of the myonecrotizing clostridia *C. perfringens* (most common in gas gangrene), *C. sordellii*, *C. septicum*, *Clostridium novyi*, *Clostridium sporogenes* and *Clostridium histolyticum* is mainly due to secreted lytic enzymes. Growing cells of the first three species listed secrete high levels of sialidase.[21] In animal experiments, sialidases have been found in resected tissues or wound fluids 8–12 h after infection with clostridial cells. It was logical, therefore, to investigate sialidase as an indicator for the presence of growing cells of one of the three clostridial species.

Antibodies were raised against the *C. perfringens*, *C. sordellii* or *C. septicum* sialidases enriched from the culture medium and were evaluated in three tests.[22,23] (1) The sialidase inhibition test measured the sialidase activity in a liquid sample in the presence or absence of one of the three polyclonal antibodies to sialidase. Inhibition of enzyme activity by one of the antisera indicated the presence of the respective clostridial species. The test can be completed within 50 min, but crossreactions with heterologous sialidases reduced the value of the test as a diagnostic tool. (2) A sandwich enzyme-linked immunosorbent assay (ELISA) was developed for each of the three sialidases. The development of a multi-step immunotest was necessary to achieve a high sensitivity; however, the test could not be completed in less than 6 h. (3) A less time-consuming, species-specific test was developed from the observation that in mutagenesis experiments of a sialidase only one antigenic site was responsible for enzyme inhibition (see

Section III C). This result suggested that other antibody-binding sites, which did not influence enzyme activity, could be used to capture sialidase in an active form. A second prerequisite for a sensitive sialidase test was the availability of the fluorogenic substrate 4-methylumbelliferyl-α-D-*N*-acetylneuraminic acid (MU-Neu5Ac).[24] The sialidase of the three clostridial species was detectable *in vitro* and in wound exudates of patients at concentrations of 0.1–1 mU/ml within 2 h. Some low-level crossreactivities, which never exceeded 10% of the homologous reaction, were observed with twelve heterologous bacterial sialidases. Clinical samples showed a high correlation between the results of the immunoassay and the bacteriological analysis of infection.

Despite the promising results obtained with the test (3), it is not yet commercially available. The reason for this low interest in producing a suitable test-kit may be due to the low number of clinical cases per year in Germany (D. Tirpitz, personal communication). This situation, however, may be different in other countries or perhaps as a consequence of catastrophes.

III. PROPERTIES OF CLOSTRIDIAL SIALIDASES

A. Properties of the enzyme proteins

The sialidases of *C. perfringens* (two isoenzymes),[25] *C. septicum*,[26] *C. sordellii*,[27] *C. chauvoei*[28] and *C. tertium*,[29] were isolated or enriched using a combination of ion-exchange chromatography, gel filtration, hydrophobic interaction chromatography and isoelectric focusing, with final purification being achieved using high-resolution chromatography. The main properties of the six sialidases are summarized in Table 24.1. The enzymes are quite different in their molecular masses which range from 43 to 150 kDa. They are monomers, except for two, which are homo-dimers (*C. chauvoei*) or homo-trimers (*C. septicum*) under native conditions. Four sialidases exhibit maximum activity at 36–37°C, while two (*C. perfringens* large isoenzyme and *C. tertium*) have optimum activity at 50–55°C. The differences in pI indicate the diversity in amino acid composition among the proteins. Only the sialidase of *C. chauvoei* required calcium ions for maximum activity. Further differences were observed with respect to stability towards freezing, and sensitivity to mercury ions. From the alignment of the protein sequences we postulate that the high resistance to Hg^{2+} of the *C. tertium* sialidase is the result of an exchange of a cysteine residue (Cys→Thr) conserved in the other enzymes (unpublished data).

To measure the substrate specificity of the sialidases, a number of natural sialylated substances were tested. They differ in three ways. (1) There are substitutions in the Neu5Ac molecule, e.g. a *N*-glycolyl-group instead of the

Table 24.1. Molecular properties of six sialidases isolated from five clostridial species

	Clostridium Species	Molecular mass (kDa)	Subunits	Location	pI [pH]	Optimum temperature [°C]	Optimum pH	Substrate Range [c]	Ca^{2+} Requirement	Stable to Freezing	50% Inhibition by Hg^{2+} (μM)	Specific Activity [U/mg]	Reference
I	*C. chauvoei*	150[a]	2	Extra Cellular	< 3.4	37	5.5	Broad	+	+	13	25	28
	C. perfringens (large isoenzyme)	73[b]	1	Extra Cellular	5.2	55	5.0	Broad	–	–	5	570	25
	C. septicum	111[b]	3	Extra Cellular	< 3.5	37	5.3	Broad	–	+	50	16	26
	C. tertium	52[a]	1	Extra Cellular	n.d.	50	5.5	Broad	–	+	5200	1	29
II	*C. perfringens* (small isoenzyme)	43[b]	1	intra Cellular	5.15	37	6.1	Narrow	–	–	0.07	470	25
	C. sordellii	45[b]	1	Extra Cellular	4.4	36	6.0	Moderate	–	+	< 10	478	27

[a] SDS-PAGE
[b] deduced from the gene
[c] See Figure 24.1
n.d., not determined

N-acetyl-group at C5, or *O*-acetyl-groups at C7, C8, C9 or C4 (hydroxylation or *O*-acetylation usually reduces or abolishes the rate of hydrolysis by bacterial sialidases).[4] (2) The nature of the sialic acid–sugar linkage: sialic acids are α2,3- or α2,6-bound to a subterminal hexose such as galactose or *N*-acetyl-galactosamine, and α2,8- or α2,9-linked to another sialic acid residue. The α2,3-linkages are usually hydrolyzed at the greatest rate. (3) The remainder of the molecule, which may be lactose, a sialic acid polymer, a glycoprotein or a glycolipid, has a considerable influence on the rate of hydrolysis, possibly by promoting or hindering secondary interactions between the enzyme and the substrate.

As shown in Figure 24.1, four of the sialidases (*C. perfringens* large isoenzyme, *C. septicum*, *C. chauvoei* and *C. tertium*) hydrolyze most of the natural substrates at high rates. Three of them (*C. septicum*, *C. chauvoei* and *C. tertium*) show a marked preference for mucins, and the fourth enzyme (*C. perfringens* large isoenzyme) for the ganglioside GD1a in the presence of a detergent. Both are substrates of high molecular mass and complex three-dimensional structure. Two sialidases (*C. perfringens* small isoenzyme and *C. sordellii*), however, are limited mostly to sialyllactoses with a preference for the α2,3-bound sialic acid. The larger substrates are hydrolyzed only poorly.

Combining the data obtained for the molecular weights, the pH-optima and the substrate specificity, two groups can be recognized (see Table 24.1): (1) relatively large enzymes hydrolyzing a broad substrate range at an optimal pH of 5.0–5.5; (2) relatively small sialidases which are highly specific for α2,3-linked sialic acids of oligosaccharides at an optimum pH of about 6.

B. Primary structures of clostridial sialidases

During the last decade we have succeeded in cloning and sequencing the genes encoding five sialidases from four clostridial species: the small[30] and large[31] isoenzyme of *C. perfringens*, named *nanH* and *nanI*, respectively, by other groups,[32] and the sialidases of *C. sordellii*,[33] *C. septicum*[34] and *C. tertium* (K. Grobe, B. Sartori, R. Schauer, P. Roggentin, unpublished data).

At a very early stage in these investigations, it became obvious that significant sequence similarities are present among these phenotypically diverse sialidases. The most important sequence motif in sialidase proteins is the 'Asp-box', which contains aspartate at a central position as the most conserved residue of this stretch of amino acids: Ser-X-Asp-X-Gly-X-Thr-Trp.[35] This motif, which is repeated four or five times in each bacterial sialidase sequence, has also been found in a viral sialidase and in all sialidases of animal origin sequenced to date.[36,37] This motif, and the occurrence of further conserved amino acids in all sialidase sequences, suggests that sialidases originated from one source.

An alignment of the deduced amino acid sequences of the five clostridial sialidases is shown schematically in Figure 24.2. From this alignment it is apparent that the larger sialidases contain *N*-terminal and/or *C*-terminal extensions and insertions that are absent in the smaller enzymes, as indicated by gaps, while other regions are present in each sialidase. The additional regions present in the larger forms explain the differences in mass. The regions conserved in all sialidases must be responsible for catalysis.

C. Site-specific mutagenesis of a clostridial sialidase

In order to gain insight into the catalytic process occurring during the hydrolysis of sialic acid linkages, the small isoenzyme of *C. perfringens* was selected for mutagenesis. The sialidase could be expressed in large amounts as a recombinant enzyme by *E. coli*, and exhibits a high specific activity and a low molecular mass (Table 24.1). Mutational studies were focused on conserved amino acids, which may be involved in substrate binding or catalysis.[38] The greatest effect was observed by the conservative substitution of Arg-37 by Lys, located in a short conserved region (Phe-Arg-Ile-Pro, named FRIP-region) upstream from the first Asp-box. The K_m, $V_{\max}$ and K_i values for the mutant enzyme, measured with MU-Neu5Ac and 2-deoxy-2,3-didehydro-Neu5Ac (Neu2en5Ac), respectively, were dramatically changed, indicating the involvement of Arg-37 in substrate binding. In addition to reduced activity, the mutant enzyme had a lower temperature optimum and was active over a more limited pH-range. The activity of this mutant enzyme was not inhibited by an antiserum raised against the wild-type enzyme.

From the three-dimensional structure of a sialidase from *Salmonella typhimurium*,[39] which has a similar size to the small *C. perfringens* enzyme, and after alignment of the two primary structures, we can conclude that Arg-37 is involved in substrate binding via the carboxyl group of sialic acids.

Two further alterations made in the second Asp-box (Asp-143→Lys; Trp-149→Ala) had no marked influence on substrate binding.

The data indicate the existence of two groups of clostridial sialidases that differ in size, optimum pH- and substrate specificity. A comparison of these sialidases at a molecular level should reveal the taxonomic relationship of these enzymes.

C. perfringens A99 (large isoenzyme)

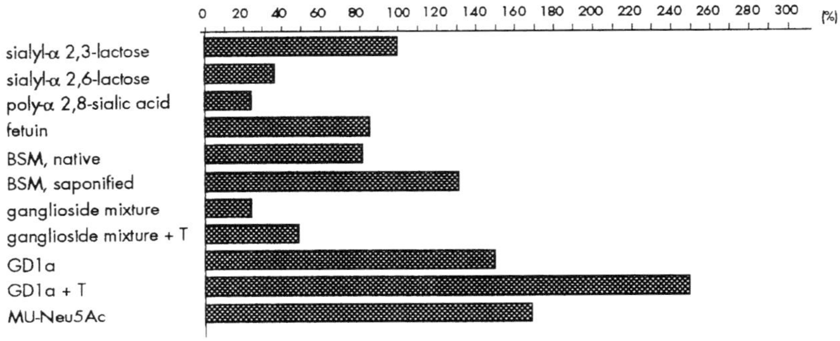

C. tertium DSM 2485

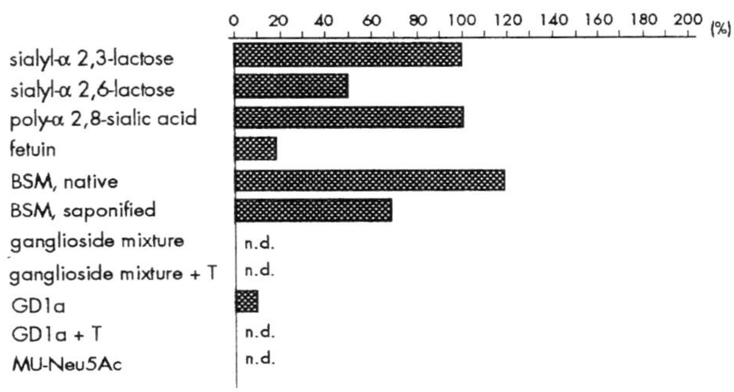

C. perfringens A99 (small isoenzyme)

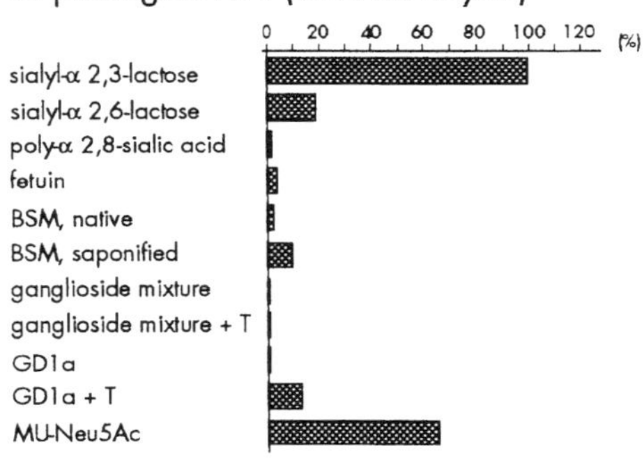

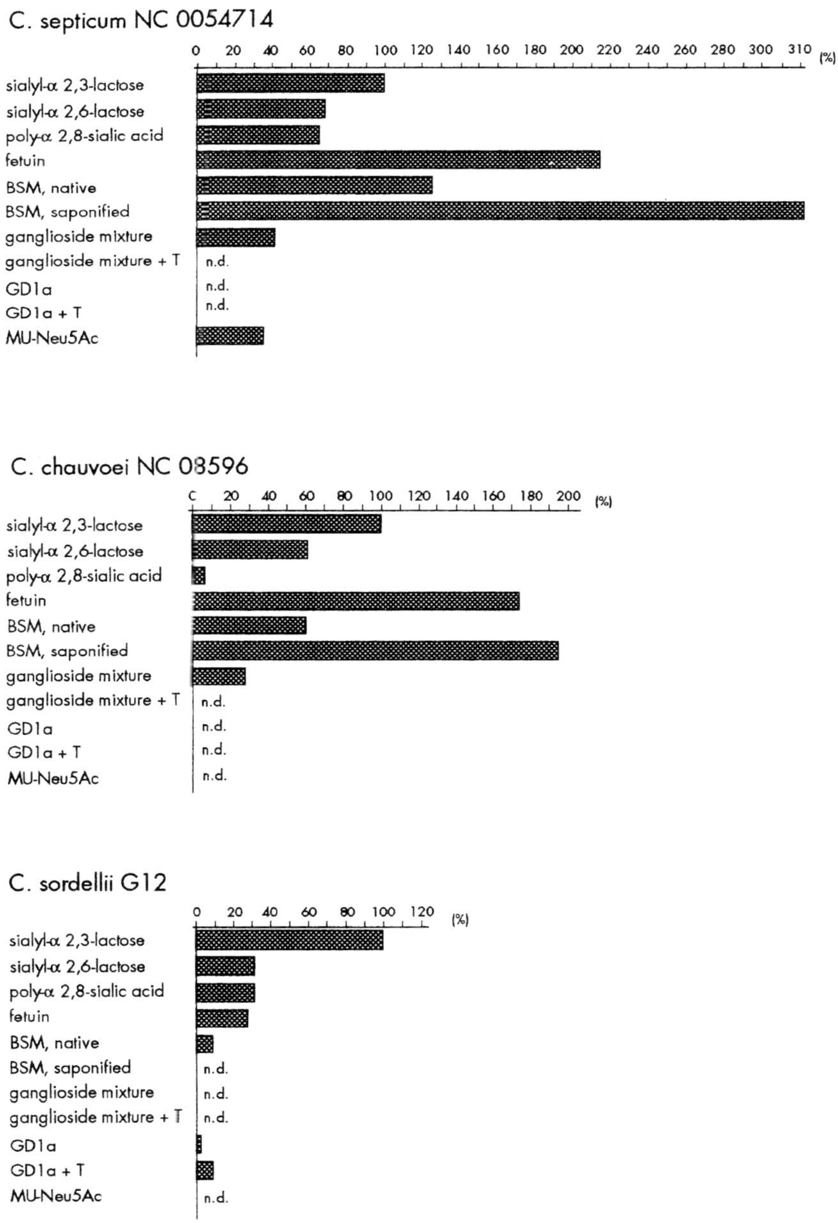

Figure 24.1 Comparison of the substrate specificity of six clostridial sialidases. The hydrolysis rates for representative sialidase substrates are shown relative to sialyl-α2,3-lactose (set at 100%). BSM, bovine submandibular gland mucin; GD1a, disialoganglioside from bovine brain; T, Triton X-100 or CF-54; n.d., not determined.

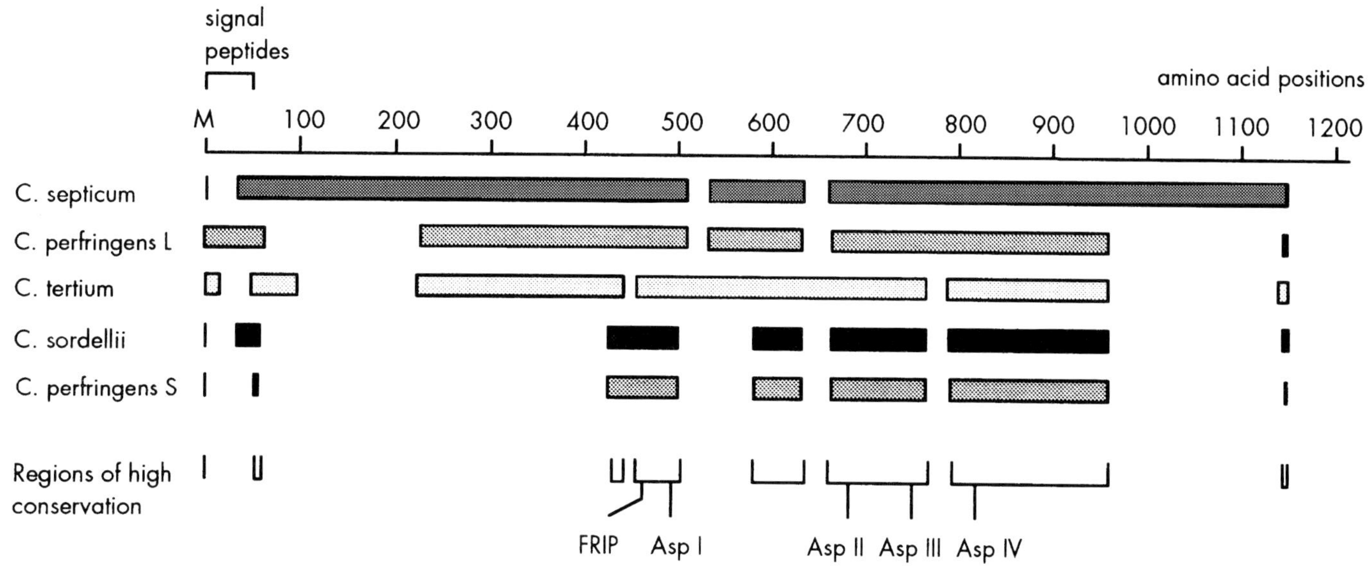

Figure 24.2 Alignment of the amino acid sequences of five clostridial sialidases. Gaps had to be introduced because of extensions and insertions in the larger proteins. The positions of conserved regions and motifs (FRIP-region, Asp-boxes) are indicated. L, large isoenzyme; S, small isoenzyme.

IV. RELATIONSHIP BETWEEN CLOSTRIDIAL SIALIDASES

The alignment of the amino acid sequences of five clostridial sialidases (Figure 24.2) allows the calculation of similarity values among each pair of sialidases compared. Only regions present in both proteins were included in the respective calculation. The number of comparable amino acids was set at 100%, and the percentage of identical amino acids was calculated for each pairwise comparison. These similarity values were combined by the average linkage method[40] visualized as a dendrogram in Figure 24.3. The five sialidases are divided into two groups at 31% similarity: (1) the sialidases of *C. perfringens* (large isoenzyme), *C. septicum* and *C. tertium*; and (2) those of *C. perfringens* (small isoenzyme) and *C. sordellii*.

From phylogenetic data derived from a comparison of 16S rRNA-sequences[41] we know that *C. perfringens* and *C. septicum* are more closely related to each other than to *C. sordellii*. This relationship was also noted in the similarity values obtained between the large sialidases, but not with the similarity values found between the small enzymes.

A comparison[36] of the G+C contents of sialidase structural genes and the respective chromosomes revealed, for example, an unexpectedly low value (40.9 mol% G+C) for the sialidase gene from *S. typhimurium* (the value for the chromosome is 50–53 mol%). The G+C content of this gene is closer to that of the small *C. perfringens* isoenzyme (31.9 mol% G+C).

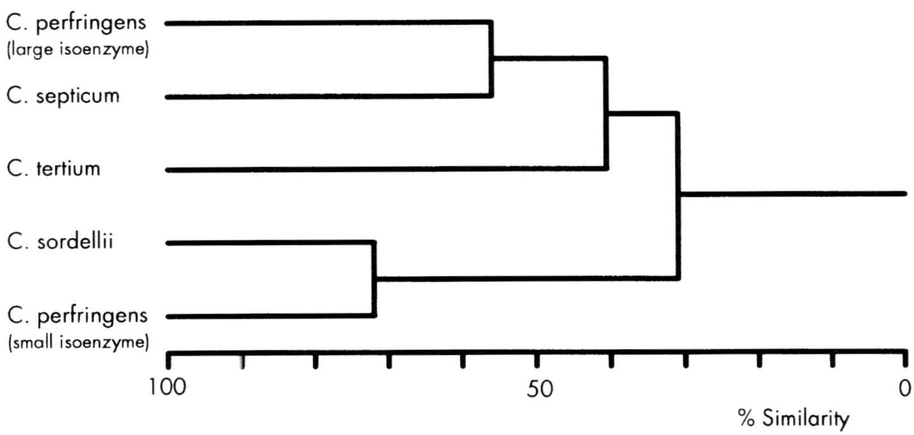

Figure 24.3 Relationship of five clostridial sialidases. The calculation is based on similarity values obtained from pairwise comparison of the amino acid sequences. The percentage content of identical amino acids was determined. The singular values were combined by the average linkage method to generate the dendrogram.

These observations indicate that some sialidases may not have evolved in parallel with their bacterial hosts.

V. CONCLUSIONS

Several arguments based on a temperature optimum of 37°C, the specificity of many bacterial sialidases for substrates common only in animals, and significant amino acid sequence homologies between bacterial and mammalian sialidases[4] have supported the assumption that sialidases, and also their substrates,[14] first evolved in animals of the deuterostomate lineage. Bacteria may have acquired the gene from this origin, during lytic processes in infections, or by fermentation of dead organisms.

Additional data supports the idea of a horizontal sialidase gene transfer within the bacterial kingdom:

(1) Sialidases are irregulary distributed among the bacteria; closely related species differ in the possession of this enzyme,[13] while similar sialidases are found in unrelated species such as *C. perfringens* and *S. typhimurium*.[36]
(2) Two bacterial sialidases (*C. perfringens* small isoenzyme, *S. typhimurium*) are located in the cells, which renders the enzymes inaccessible to their glycoconjugate substrates in the environment. This behaviour indicates a recent transfer event and is supported by the observation that only strain LT-2 among 200 other strains of *S. typhimurium* is sialidase-positive; no cryptic gene was found in the other strains of this species.[42]
(3) The possible mechanisms of gene transfer are indicated by the finding that a flanking region of the *Micromonospora viridifaciens* sialidase gene exhibits homology with a transposon sequence.[43] The involvement of phages in gene transfer has been suggested in the case of the gene encoding the small sialidase isoenzyme of *C. perfringens*, which is located close to a phage attachment site on the chromosome.[44] Evidence for an association of prophage DNA and the sialidase gene of *S. typhimurium* was derived from the reverse-strand sequence of the 3′-flanking region of the gene.[42]

Did this gene transfer event occur only once or more frequently? The following summary of physical, physiological, molecular and similarity data may provide an answer.

The characterization of several clostridial sialidases demonstrated a high diversity of properties which could be used to segregate the enzymes into two groups.

Group I enzymes were large sialidases containing additional regions in their proteins, which may not be necessary for catalysis, but may enhance the binding of large and complex substrates. Such lectin-like domains are described for the *Vibrio cholerae* sialidase.[39] The broad substrate specificity, extracellular location, and the optimum temperatures of these sialidases suggest that these enzymes are highly adapted to the requirements of the host bacterial species.

Group II enzymes include the small (43 kDa) sialidases of *C. sordellii* and *C. perfringens*, and the enzyme of *S. typhimurium*. Their substrate specificity is predominantly limited to oligosaccharides, with a significant preference for α2,3-linked sialic acids. These properties may be of little value to the bacteria, and only the *C. sordellii* enzyme appears to have undergone alterations (including a signal peptide for secretion of the enzyme and a broader substrate specificity) as an adaptation to the needs of the bacteria.

If sialidases were acquired by bacteria (e.g. from animals of the deuterostomate lineage,[14,36] then group I enzymes may have arisen from an early evolutionary event. The group II sialidases may have arisen from a later evolutionary event, in which a phenotypically limited type of sialidase was transferred, possibly being of little value to the bacterium, until adapted in the same way as the larger forms.

VI. REFERENCES

1. Cato E.P., George W.L. and Finegold, S.M. (1986) Genus *Clostridium*. In Sneath P.H.M., Mair N.S., Sharpe M.E. and Holt J.G. (eds), *Bergey's Manual of Systematic Bacteriology*, Vol 2, pp. 1141–1200. Williams & Wilkins, Baltimore, MD.
2. Gottschalk A. (1957) Neuraminidase: the specific enzyme of influenza virus and *Vibrio cholerae*. *Biochim Biophys Acta* **23**: 645–646.
3. Heimer R. and Meyer K. (1956) Studies on sialic acid of submaxillary mucoid. *Proc Natl Acad Sci USA* **42**: 728–734.
4. Schauer R., Kelm S., Reuter G., Roggentin P. and Shaw L. (1995) Biochemistry and role of sialic acids. In Rosenberg A. (ed.), *Biology of the Sialic Acids*, pp. 7–67. Plenum Press, New York.
5. Fraser A.G. (1978) Neuraminidase production by clostridia. *J Med Microbiol* **11**: 269–280.
6. Popoff M.R. and Dodin A. (1985) Survey of neuraminidase production by *Clostridium butyricum*, *Clostridium beijerinckii*, and *Clostridium difficile* strains from clinical and nonclinical sources. *J Clin Microbiol* **22**: 873–876.
7. Corfield T. (1992) Bacterial sialidases – Roles in pathogenicity and nutrition. *Glycobiology* **2**: 509–521.
8. Müller H.E. (1992) Relationships between ecology and pathogenicity of microorganisms. *BIOforum* **15**: 16–22.
9. McLennan J.D. (1962) The histotoxic clostridial infections of man. *Bacteriol Rev* **26**: 177–276.
10. Ezepchuk Y.V., Vertier Y.V. and Kostyukova N.N. (1973) Neuraminidase of *Corynebacterium diphteriae* as a factor of pathogenicity with a spreading function. *Biull Eksperiment Biol Med* **75**: 63–65.
11. Schauer R. (1982) Chemistry, metabolism, and biological functions of sialic acids. In Tipson R.S. and Horton D. (eds), *Advances in Carbohydrate Chemistry and Biochemistry*, Vol. 40, pp. 132–234. Academic Press, New York.
12. Schauer R. (1985) Sialic acids and their role as biological masks. *Trends Biochem Sci* **10**: 357–360.

13. Roggentin P., Gutschker-Gdaniec G., Schauer R. and Hobrecht R. (1985) Correlative properties for a differentiation of two *Clostridium sordellii* phenotypes and their distinction from *Clostridium bifermentans*. *Zbl Bakt Hyg A* **260**: 319–328.
14. Schauer R. and Vliegenthart J.F.G. (1982) Introduction. In Schauer R. (ed.), *Sialic Acids – Chemistry, Metabolism and Function*, Cell Biology Monographs, Vol 10, pp. 1–3. Springer, Wien.
15. Cohen R.F. and Yourofsky L.A. (1980) Gas gangrene: a postoperative complication. *J Foot Surg* **19**: 202–206.
16. Kozianka J., Kozuschek W. and Edelmann M. (1987) Der Gasbrand – eine 'wenig bekannte' Infektion mit hoher Mortalität. *Therapiewoche* **37**: 4271–4292.
17. Bessman A.N. and Wagner W. (1975) Nonclostridial gas gangrene. *JAMA* **233**: 958–963.
18. Darke S.G., King A.M. and Slack W.K. (1977) Gas gangrene and related infection: Classification, clinical features and aetiology, management and mortality. A report of 88 cases. *Br J Surg* **64**: 104–112.
19. Hobrecht R. (1975) Isolierung und Identifizierung von Gasoedemerregern aus Gewebsproben. *Wehrmed Mschr* **9**: 267–270.
20. Jancik J.M. (1980) Praktische Hinweise zur Differentialdiagnose clostridialer, histotoxischer Infektionen. *Wehrmed Mschr* **24**: 79–84.
21. Roggentin P., Gutschker-Gdaniec G.H.M., Hobrecht R. and Schauer R. (1988) Early diagnosis of clostridial gas gangrene using sialidase antibodies. *Clin Chim Acta* **173**: 251–262.
22. Roggentin P., Hobrecht R., Tirpitz D., Rothe B. and Schauer R. (1991) Application of sialidase antibodies for the diagnosis of clostridial infections. *Clin Chim Acta* **196**: 97–106.
23. Roggentin T., Kleineidam R.G., Majewski D.M., Tirpitz D., Roggentin P. and Schauer R. (1993) An immunoassay for the rapid and specific detection of three sialidase-producing clostridia causing gas gangrene. *J Immunol Meth* **157**: 125–133.
24. Potier M., Mameli L., Belislé M., Dallaire L. and Melançon S.B. (1979) Fluorometric assay of neuraminidase with a sodium (4-methylumbelliferyl-α-D-*N*-acetylneuraminate) substrate. *Anal Biochem* **94**: 287–296.
25. Roggentin P., Kleineidam R.G. and Schauer R. (1995) Diversity in the properties of two sialidase isoenzymes produced by *Clostridium perfringens*. *Biol Chem Hoppe-Seyler* **376**: 569–575.
26. Zenz K.I., Roggentin P. and Schauer R. (1993) Isolation and properties of the natural and the recombinant sialidase from *Clostridium septicum* NC 0054714. *Glycoconj J* **10**: 50–56.
27. Roggentin P., Berg W. and Schauer R. (1987) Purification and characterization of sialidase from *Clostridium sordellii* G12. *Glycoconj J* **4**: 349–359.
28. Heuermann D., Roggentin P., Kleineidam R.G. and Schauer R. (1991) Purification and characterization of a sialidase from *Clostridium chauvoei* NC08596. *Glycoconj J* **8**: 95–101.
29. Sartori B. (1993) Charakterisierung der Sialidase aus *Clostridium tertium*. Diploma Thesis, Kiel.
30. Roggentin P., Rothe B., Lottspeich F. and Schauer R. (1988) Cloning and sequencing of a *Clostridium perfringens* sialidase gene. *FEBS Lett* **238**: 31–34.
31. Traving C., Schauer R. and Roggentin P. (1994) Gene structure of the 'large' sialidase isoenzyme from *Clostridium perfringens* A99 and its relationship with other clostridial *nanH* proteins. *Glycoconj J* **11**: 141–151.
32. Katayama S-I., Dupuy B., Garnier T. and Cole S.T. (1995) Rapid expansion of the physical and genetic map of the chromosome of *Clostridium perfringens* CPN50. *J Bacteriol* **177**: 5680–5685.
33. Rothe B., Roggentin P., Frank R., Blöcker H. and Schauer R. (1989) Cloning, sequencing and expression of a sialidase gene from *Clostridium sordellii* G12. *J Gen Microbiol* **135**: 3087–3096.
34. Rothe B., Rothe B., Roggentin P. and Schauer R. (1991) The sialidase gene from *Clostridium septicum*: cloning, sequencing, expression in *Escherichia coli* and identification of conserved sequences in sialidases and other proteins. *Mol Gen Genet* **226**: 190–197.
35. Roggentin P., Rothe B., Kaper J.B., Galen J., Lawrisuk L., Vimr E.R. and Schauer R. (1989) Conserved sequences in bacterial and viral sialidases. *Glycoconj J* **6**: 349–353.

36. Roggentin P., Schauer R., Hoyer L.L. and Vimr E.R. (1993) The sialidase superfamily and its spread by horizontal gene transfer. *Mol Microbiol* **9**: 915–921.
37. Saito M. and Yu R.K. (1995) Biochemistry and function of sialidases. In Rosenberg A. (ed.), *Biology of the Sialic Acids*, pp. 261–313. Plenum Press, New York.
38. Roggentin T., Kleineidam R.G., Schauer R. and Roggentin P. (1992) Effects of site-specific mutations on the enzymatic properties of a sialidase from *Clostridium perfringens. Glycoconj J* **9**: 235–240.
39. Taylor G., Vimr E.R., Garman E. and Laver G. (1992) Purification, crystallization and preliminary crystallographic study of neuraminidase from *Vibrio cholerae* and *Salmonella typhimurium* LT2. *J Mol Biol* **226**: 1287–1290.
40. Anderberg M.R. (ed.) (1973) *Cluster Analysis for Application.* Academic Press, New York.
41. Dorsch M. (1990) Ein Beitrag zur Phylogenie der Gram-positiven Eubakterien. Doctoral Thesis, Kiel.
42. Hoyer L.L., Hamilton A.C., Steenbergen S.M. and Vimr E.R. (1992) Cloning, sequencing and distribution of the *Salmonella typhimurium* LT-2 sialidase gene, *nanH*, provides evidence for interspecies gene transfer. *Mol Microbiol* **6**: 873–884.
43. Sakurada K., Ohta T. and Hasegawa M. (1992) Cloning, expression, and characterization of the *Micromonospora viridifaciens* neuraminidase gene in *Streptomyces lividans. J Bacteriol* **174**: 6896–6903.
44. Canard B. and Cole S.T. (1990) Lysogenic phages of *Clostridium perfringens*: Mapping of the chromosomal attachment sites. *FEMS Microbiol Lett* **66**: 323–326.

Chapter 25

The *Clostridium perfringens* ε-Toxin

DEAN PAYNE AND PETRA OYSTON
Chemical and Biological Defence Establishment, Porton Down, Salisbury, Wiltshire, SP4 0JQ, UK

I. INTRODUCTION

ε-Toxin is the most potent clostridial toxin after botulinum and tetanus neurotoxins. The toxin is produced by *Clostridium perfringens* types B and D but not by types A, C or E.[1] *Clostridium perfringens* types B and D have a limited host range, being mainly isolated from sheep and lambs, only occasionally from goats and cattle, and rarely from man.[2] They are responsible for producing severe and rapidly fatal enterotoxaemia: *C. perfringens* type B enterotoxaemia infection of lambs causes lamb dysentery while type D enterotoxaemia causes pulpy kidney disease in sheep and lambs (see Chapter 10). Mortality rates in both diseases can be as high as 100% and their outbreak is of great economic importance wherever animals are raised intensively.[3] Neither diesease is transmissible, but sporadic outbreaks occur when the microbial balance of the gut is disrupted, for example after antibiotic treatment or by changes in diet. Pulpy kidney disease is often associated with a change from a poor to a rich diet, accompanied by excessive eating.[4] Such overeating causes considerable quantities of undigested, starch-rich food to pass from the rumen into the small intestine. The nutritious, anaerobic environment generated allows the multiplication of *C. perfringens*,

The Clostridia: Molecular Biology and Pathogenesis
ISBN 0-12-595020-9

resulting in up to 10^9 viable clostridia per gram of ileal contents and the production of high levels of ε-toxin.[5] Several vaccines exist for the prevention of *C. perfringens* enterotoxaemia. The vaccines are based on formaldehyde-treated cell filtrates or whole-cell cultures. The vaccines confer a high degree of protection in animals.[6] However, the immunogenicity of the ε-toxin in the preparations has been reported to be variable and a more defined and consistent vaccine may be preferable. Immunity to a single epitope on the toxin has been shown to be sufficient to protect against purified ε-toxin and *C. perfringens* type D infection.[7]

II. PHYSICAL AND CHEMICAL PROPERTIES

ε-Toxin is produced by *C. perfringens* types B and D as a relatively inactive prototoxin of 311 amino acids with a molecular weight of 32 700.[8] Proteolytic cleavage of 13 or 14 basic amino acids from the amino-terminal end of the prototoxin results in the production of the mature toxin with a molecular weight of 31 200.[8,9] This also causes a marked shift in pI from 8.02 (prototoxin) to either 5.36 (fully active toxin) or 5.74 (partially active toxin) and a significant change in conformation.[8,10] Activation of the prototoxin appears to produce several isoforms with a range of specific activities between that of the prototoxin and the mature toxin[8,11] and it has been found that the toxin itself also has several isoforms.[9] Thus, activation of ε-prototoxin may be a multistep process, possibly with multiple proteolytic cleavages and post-translation modifications such as deamination and phosphorylation resulting in the production of the heterogeneous mature toxin.[9]

ε-Toxin for use in experimental studies has usually been isolated from cultures of *C. perfringens* type D. Several methods, including methanol precipitation, ammonium sulphate precipitation, column chromatography, size exclusion and various forms of ion exchange chromatogrphy are used to purify the ε-toxin.[12–15]

III. DETECTION AND MEASUREMENT OF ACTIVITY

Several methods have been employed for the detection of ε-toxin and the identification of the disease in animals. Immunological assays to test for the presence of ε-toxin have been based on single radial diffusion,[16] micro-complement fixation,[17] reverse-phase passive haemagglutination,[18] enzyme-linked immunosorbent assay (ELISA)[19–21] or latex agglutination (LAT).[22] The ELISA and LAT assays were able to detect ε-toxin in buffers, culture supernatants and also in intestinal contents. A polymerase chain reaction (PCR) test has been used to demonstrate the presence of genes encoding ε-toxin production in laboratory cultures of *C. perfringens* type B and D strains.[23]

Traditionally, the activity of purified ε-toxin was determined using mouse lethality tests;[13,24] the toxin has an LD_{50} in mice of 78 ng when administered intravenously.[15] However, it was found that the Madin Darby canine kidney (MDCK) cell line was sensitive to *C. perfringens* type D culture filtrates and this could be used as the basis of an alternative assay for ε-toxin activity (Figure 25.1).[25] Subsequently, a new cytotoxicity assay was developed for the determination of the activity of *C. perfringens* type D ε-toxin *in vivo*, based on the sensitivity of this cell line.[15] As little as 15 ng/ml of purified ε-toxin was sufficient to reduce the viability of MDCK cells by 50% and as little as 8 ng/ml sufficient to cause a significant reduction in the viability of the MDCK cells.[15]

IV. GENETICS

The ε-toxin *etx* gene is carried on an episome distinct from the 3.6 Mb chromosome.[26] The genes from *C. perfringens* types B and D have been cloned and nucleotide sequences determined (Figure 25.2). The *etxB* gene encoded a protein of M_r *c.* 32 981.[9] Neither *etxB* nor the derived protein showed homology with other proteins in the DDBJ, EMBL or GenBank databases. Comparison of the *etxB* and *etxD* genes revealed two nucleotide differences within the open reading frame resulting in a single amino acid substitution.[23] The promoters for the genes were not homologous, with different putative −10 and −35 sequences (Figure 25.2). This allowed the development of ε-toxin-specific PCR primers to produce a system for typing B and D strains of *C. perfringens*. ε-Toxin was expressed from the *etxB* promoter in *Escherichia coli*, although only at low levels.[9] ε-Prototoxin is preceded by a signal peptide resulting in the export of the prototoxin from *C. perfringens*. The recombinant protein accumulated in the periplasmic space of *E. coli*, expressing *etxB*.[9,27] The recombinant prototoxin expressed in *E. coli* and activated with trypsin was shown to have identical biochemical and biological properties to those of the native toxin activated in a similar manner and should provide an easily purified source of ε-toxin.[9]

Chemical modification experiments have demonstrated the importance of certain amino acid residues for the lethality of ε-toxin. One tryptophan,[28] one histidine,[29] one tyrosine[30] and three or four aspartic or glutamic acid[31] residues were shown to be essential for the lethal activity of ε-toxin (Figure 25.2). The histidine residue appeared to be required for the active site, while the single tryptophan and tyrosine residues appeared to be necessary for the binding site at least. Eight lysine residues have also been shown to be important in activity, but

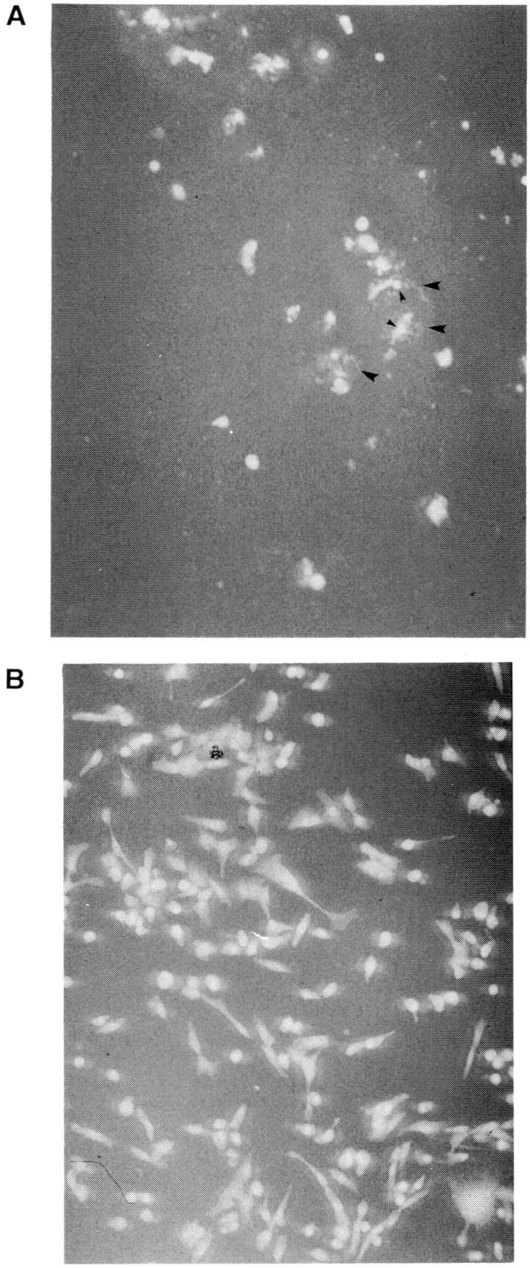

Figure 25.1 MDCK cells (A) were exposed to 4 μg/ml of ε-toxin for 4 h at 37°C. Cells were stained with a mixture of ethidium bromide and acridine orange and examined by fluorescence microscopy (× 400 magnification). On exposure to the toxin MDCK cells rounded up and the cytoplasmic membrane ruptured, scattering the cytoplasmic contents (large arrowheads). Despite swelling, nuclei remain intact (small arrowheads). Control cells are shown in (B). (See colour plate.)

```
A  ATGAAAAAAAATCTTGTAAAAAGTTTAGCAATCGCATCAGCGGTGATATCCATCTATTCAAT
    M  K  K  N  L  V  K  S  L  A  I  A  S  A  V  I  S  I  Y  S  I

   AGTTAATATTGTTTCACCAACTAATGTAATAGCTAAGGAAATATCTAATACAGTATCTAATG
    V  N  I  V  S  P  T  N  V  I  A  K  E  I  S  N  T  V  S  N

   AAATGTCCAAAAAAGCTTCTTATGATAATGTAGATACATTAATTGAGAAAGGAAGATATAAT
   E  M  S  K  K  A  S  Y* D  N  V  D  T  L  I  E  K  G  R  Y* N

   ACAAAATATAATTACTTAAAGAGAATGGAAAAATATTATCCTAATGCTATGGCATATTTTGA
    T  K  Y* N  Y* L  K  R  M  E  K  Y* Y* P  N  A  M  A  Y* F  D

   TAAGGTTACTATAAATCCACAAGGAAATGATTTTTATATTAATAATCCTAAAGTTGAATTAG
    K  V  T  I  N  P  Q  G  N  D  F  Y* I  N  N  P  K  V  E  L

   ATGGAGAACCATCAATGAATTATCTTGAAGATGTTTATGTTGGAAAAGCTCTCTTAACTAAT
   D  G  E  P  S  M  N  Y* L  E  D  V  Y* V  G  K  A  L  L  T  N

   GATACTCAACAAGAACAAAAATTAAAATCACAATCATTCACTTGTAAAAATACTGATACAGT
    D  T  Q  Q  E  Q  K  L  K  S  Q  S  F  T  C  K  N  T  D  T  V

   AACTGCAACTACTACTCATACTGTGGGAACTTCGATACAAGCAACTGCTAAGTTTACTGTTC
    T  A  T  T  T  H* T  V  G  T  S  I  Q  A  T  A  K  F  T  V

   CTTTTAATGAAACAGGAGTATCATTAACTACTAGTTATAGTTTTGCAAATACAAATACAAAT
   P  F  N  E  T  G  V  S  L  T  T  S  Y* S  F  A  N  T  N  T  N

   ACTAATTCAAAAGAAATTACTCATAATGTCCCTTCACAAGATATACTAGTACCAGCTAATAC
    T  N  S  K  E  I  T  H* N  V  P  S  Q  D  I  L  V  P  A  N  T

   TACTGTAGAAGTAATAGCATATTTAAAAAAAGTTAATGTTAAAGGAAATGTAAAGTTAGTAG
    T  V  E  V  I  A  Y* L  K  K  V  N  V  K  G  N  V  K  L  V

   GACAAGTAAGTGGAAGTGAATGGGGAGAGATACCTAGTTATTTAGCTTTTCCTAGGGATGGT
   G  Q  V  S  G  S  E  W* G  E  I  P  S  Y* L  A  F  P  R  D  G

   TATAAATTTAGTTTATCAGATACAGTAAATAAGAGTGATTTAAATGAAGATGGTACTATTAA
    Y* K  F  S  L  S  D  T  V  N  K  S  D  L  N  E  D  G  T  I  N

   TATTAATGGAAAAGGAAATTATAGTGCAGTTATGGGAGATGAGTTAATAGTTAAGGTTAGAA
    I  N  G  K  G  N  Y* S  A  V  M  G  D  E  L  I  V  K  V  R

   ATTTAAATACAAATAATGTACAAGAATATGTAATACCTGTAGATAAAAAAGAAAAAAGTAAT
   N  L  N  T  N  N  V  Q  E  Y* V  I  P  V  D  K  K  E  K  S  N

   GATTCAAATATAGTAAAATATAGGAGTCTTTCTATTAAGGCACCAGGAATAAAATAA
    D  S  N  I  V  K  Y* R  S  L  S  I  K  A  P  G  I  K  .

     -35                    -10
B  TTTGTATATTAAATAATTTTATTTATATTATTTACTTTTTTTAAAAAATATAGAAAAATATA
   GGCCTATATTAAATAATTTTATTTTTATTATTTACTTTTTTAAAAAAATATAGTAAAATATA

                        SD        start
   GAAAAATATATTAATGAAAGGGTGGTTTTATG
   GAAAATTATGTTAATGAAAGGGTGGTTTTATG
```

Figure 25.2 (A) Nucleotide and derived amino acid sequence of the *etxB* gene. The signal sequence is shown in italic type and the prototoxin that is cleaved by trypsin is shown underlined. Tyrosine (Y), tryptophan (W) and histidine (H) residues have been identified as essential for the lethal activity of ε-toxin, and these are indicated by an asterisk. (B) The putative Shine–Dalgarno (SD), −10 and −35 sequences of *etxB* (upper sequence) are shown, and the sequence from *etxD* (lower sequence) is aligned for comparison.

are probably involved in maintaining conformation rather than integral to an active site.[32]

V. BIOLOGICAL ACTIVITY

A. *In vivo*

ε-Prototoxin produced in the gut of animals is activated by proteolytic enzymes present in intestinal fluid.[33] The toxin increases intestinal permeability and enters the blood supply.[4,27,34] In sheep, the toxin becomes concentrated in the brain and, to a lesser extent, in the kidneys.[14] Within 2–3 min of a high toxin dose, the toxin causes a widespread disturbance in the permeability balance of the brain by disrupting vascular endothelia.[35,36] As the degenerative changes progress, serum proteins and eventually red cells leak from the vasculature, astrocyte end feet rupture and edema ensues.[37] In acute cases of enterotoxemia characteristic lesions occur at specific sites in the brain.[36,38,39]

B. *In vitro*

Buxton,[40] investigated the effect of ε-toxin on a wide range of isolated cells from a number of species. Nonsensitive cells included pulmonary alveolar macrophages from guinea pigs and rabbits, lymphocytes from guinea pigs and sheep and peritoneal macrophages and Lan Schultz ascites-tumour cells from mice. However, guinea pig peritoneal macrophages were susceptible and showed gross morphological changes such as swelling, blistering, noncontinuous nuclear and cytoplasmic membranes and featureless cytoplasms. Of those cell lines examined by Knight *et al.*[41] and Payne *et al.*[15] only the MDCK cell line was susceptible to ε-toxin. Unlike guinea pig peritoneal macrophages, only the plasma membranes of MDCK cells are disrupted (Figure 25.1).[15]

VI. MODE OF ACTION

The mode of action of ε-toxin is not known, but several observations have suggested that it acts upon the central nervous system. ε-Intoxication is characterized by elevated levels of both cAMP and glucose in serum[38] which is thought to be caused by toxin-induced brain edema.[42] This edema is also thought to be responsible for the increase in levels of the catecholamines adrenaline and noradrenaline observed in the lamb model system.[42] Increased catecholamine production was thought to activate adenyl cyclase, which in turn raised the cAMP level.[42] In the mouse model system[43] no significant changes in the levels of either adrenaline or noradrenaline in the brain were observed 30 min after intravenous injection of ε-toxin. However, the level of dopamine was found to drop significantly. It was found that dopamine antagonists or any drugs which directly or indirectly inhibit dopamine release or activity protected mice from the lethal effects of ε-toxin.

No drug has yet been found which directly inhibits the binding of ε-toxin. However, prior injection of prototoxin or formalinized ε-toxin can alleviate the effects of ε-intoxication *in vivo*,[14,36] indicating that a specific receptor is required for toxin binding before the toxin can act on its target. Considering the known effect of the toxin on the vascular endothelium, the evidence would suggest that the ε-toxin receptor is probably located on this endothelial surface. Nagahama and Sakurai[44] prepared rat brain homogenates and synaptosomal membrane fractions and found that radiolabeled ε-toxin bound with single binding phases and dissociation constants of 2.5 nM and 3.3. nM, respectively. Treatment of synaptosomal membrane fractions with pronase and neuraminidase, but not with trypsin and phospholipase C, lowered the binding of the labelled toxin. Heating of the synaptosomal fractions also reduced the binding of the toxin while treatment with lipase completely abolished binding. The authors suggested that ε-toxin binding was facilitated by a sialoglycoprotein and that a lipid component or environment was also important; it has been found that the major component that ε-toxin binds to in crude rat brain homogenates is a 26 kDa sialoglycoprotein (unpublished data). This data supports the proposal that ε-toxin binds to specific sites in brain tissue.[44]

Very little data is available on the mechanism by which ε-toxin alters vascular permeability. It is possible that the binding of ε-toxin results in the opening of tight junctions which alters fluid dynamics and vascular permeability.[37] It is not known whether ε-toxin binds to a receptor and a signal is then transduced across the membrane or whether the toxin is internalized before interacting directly with an intracellular component.

VII. REFERENCES

1. Brooks M.E., Sterne M. and Warrack G.H. (1957) A reassessment of the criteria used for type differentiation of *Clostridia perfringens*. *J Pathol Bacteriol* **74**: 185–195.
2. Smith L.D. and Williams B.L. (1984) *The Pathogenic Anaerobic Bacteria*, 3rd edn. C. Thomas, Springfield, IL.
3. Buxton A. and Fraser G. (1977) *Animal Microbiology*, Vol. 1. Blackwell Scientific Publications Ltd., Oxford.

4. Bullen J.J. (1970) Role of toxins in host-parasite relationships. In Ajl, S., Kadis, S. and Montie, T.C. (eds) *Microbial Toxins*, Vol. 1, pp. 233–276. Academic Press, New York.
5. Bullen J.J. and Scarisbrick R. (1957) Enterotoxaemia of sheep: experimental reproduction of the disease. *J Pathol Bacteriol* **73**: 494–509.
6. Stephen J. and Pietrowski R.A. (1986) *Bacterial Toxins*, 2nd edn. van Nostrand Reinhold (UK) Co. Ltd., Wokingham, England.
7. Percival D.A., Shuttleworth A.D., Williamson E.D. and Kelly D.C. (1990) Anti-idiotypic antibody-induced protection against *Clostridium perfringens* type D. *Infect Immun* **58**: 2487–2492.
8. Worthington R.W. and Mulders M.S. (1977) Physical changes in the epsilon prototoxin molecule of *Clostridium perfringens* during enzymatic activation. *Infect Immun* **18**: 549–551.
9. Hunter S.E., Clarke I.N., Kelly D.C. and Titball R.W. (1992) Cloning and nucleotide sequencing of the *Clostridium perfringens* epsilon toxin gene and its expression in *Escherichia coli*. *Infect Immun* **60**: 102–110.
10. Habeeb A.F., Lee C.L. and Atassi M.Z. (1973) Conformational studies on modified proteins and peptides. VII. Conformation of epsilon-prototoxin and epsilon-toxin from *Clostridium perfringens*. Conformational changes associated with toxicity. *Biochem Biophys Acta* **322**: 245–250.
11. Habeeb A.F. (1975) Studies on epsilon-prototoxin of *Clostridium perfringens* type D. Physiochemical and chemical properties of epsilon-prototoxin. *Biochem Biophys Acta* **412**: 62–69.
12. Verwoerd D.W. (1960) Isolation van die protoksien van *Clostridium welchii* type D. *J S Afr Vet Med Assoc* **31**: 195–203.
13. Habeeb A.F. (1969) Studies on epsilon prototoxin of *Clostridium perfringens* type D. I. Purification methods: evidence for multiple forms of epsilon-prototoxin. *Arch Biochem Biophys* **130**: 430–440.
14. Worthington R.W., Mulders M.S. and van Rensburg J.J. (1973) *Clostridium perfringens* type D epsilon prototoxin. Some chemical, immunological and biological properties of a highly purified prototoxin. *Onderstepoort J Vet Res* **40**: 143–149.
15. Payne D.W., Williamson E.D., Havard H., Modi N. and Brown J. (1994) Evaluation of a new cytotoxicity assay for *Clostridium perfringens* type D epsilon toxin. *FEMS Microbiol Lett* **116**: 161–167.
16. Mancini G., Carbonara A.O. and Heremans J.F. (1965) Immunochemical quantitation of antigens by single radial immunodiffusion. *Immunochemistry* **2**: 235–254.
17. Marucci A.A. and Fuller T.C. (1971) Quantitative microcomplement fixation test. *Appl Microbiol* **21**: 260–264.
18. Beh K.J. and Buttery S.H. (1978) Reverse phase passive haemagglutination and single radial immunodiffusion to detect epsilon antigen of *Clostridium perfringens* type D. *Aust Vet J* **54**: 541–544.
19. Naylor R.D., Martin P.K. and Sharpe R.T. (1987) Detection of *Clostridium perfringens* epsilon toxin by ELISA. *Res Vet Sci* **42**: 255-256.
20. Nagahama M., Kobayashi K., Ochi S. and Sakurai J. (1991) Enzyme-linked immunosorbent assay for rapid detection of toxins from *Clostridium perfringens*. *FEMS Microbiol Lett* **68**: 41–44.
21 el Idrissi A.H. and Ward G.E. (1992) Evaluation of enzyme-linked immunosorbent assay for diagnosis of *Clostridium perfringens* enterotoxaemias. *Vet Microbiol* **31**: 389–396.
22 Martin P.K. and Naylor R.D. (1994) A latex agglutination test for the qualitative detection of *Clostridium perfringens* epsilon toxin. *Res Vet Sci* **56**: 259–261.
23 Havard H.L., Hunter S.E. and Titball R.W. (1992) Comparison of the nucleotide sequence and development of a PCR test for the epsilon toxin gene of *Clostridium perfringens* type B and type D. *FEMS Microbiol Lett* **76**: 77–81.
24 Worthington R.W., Mulders M.S. and van Rensburg J.J. (1973) Enzymatic activation of *Clostridium perfringens* epsilon prototoxin and some biological properties of activated toxin. *Onderstepoort J Vet Res* **40**: 151–154.

25 Knight P.A., Burnett C., Whitaker A.M. and Queminet J. (1986) The titration of clostridial toxoids and antisera in cell culture. *Develop Biol Standard* **64**: 129–136.

26 Canard B., Saint Joanis B. and Cole S.T. (1992) Genomic diversity and organization of virulence genes in the pathogenic anaerobe *Clostridium perfringens. Mol Microbiol* **6**: 1421–1429.

27 Bullen J.J. and Batty I. (1956) The effect of *Clostridium welchii* type D culture filtrates on the permeability of the mouse intestine. *J Pathol Bacteriol* **71**: 311–323.

28 Sakurai J. and Nagahama M. (1985) Role of one tryptophan residue in the lethal activity of *Clostridium perfringens* epsilon toxin. *Biochem Biophys Res Commun* **128**: 760–766.

29 Sakurai J. and Nagahama M. (1987) Histidine residues in *Clostridium perfringens* epsilon toxin. *FEMS Microbiology Letters* **41**: 317–319.

30 Sakurai J. and Nagahama M. (1987) The inactivation of *Clostridium perfringens* epsilon toxin by treatment with tetranitromethane and *N*-acetylimidazole. *Toxicon* **25**: 279–284.

31 Sakurai J. and Nagahama M. (1987) Carboxyl groups in *Clostridium perfringens* epsilon toxin. *Microb Pathog* **3**: 469–474.

32 Sakurai J. and Nagahama M. (1986) Amino groups in *Clostridium perfringens* epsilon prototoxin and epsilon toxin. *Microb Pathog* **1**: 417–423.

33 Niilo L. (1965) Bovine enterotoxaemia. III. Factors affecting the stability of the toxins of *Clostridium perfringens* types A, C and D. *Can Vet J* **6**: 38–42.

34 Jansen B.C. (1967) The production of a basic immunity against pulpy kidney disease. *Onderstepoort J Vet Res* **34**: 65–80.

35 Finnie J.W. (1984) Ultrastructural changes in the brain of mice given *Clostridium perfringens* type D epsilon toxin. *J Comp Pathol* **94**: 445–452.

36 Buxton D. (1976) Use of horseradish peroxidase to study the antagonism of *Clostridium welchii* (*Cl. perfringens*) type D epsilon toxin in mice by the formalinized epsilon prototoxin. *J Comp Pathol* **86**: 67–72.

37 Buxton D. and Morgan K.T. (1967) Studies of the lesions produced in the brains of colostrum-deprived lambs by *Clostridium welchii* (*Clostridium perfringens*) type D toxin. *J Comp Path* **86**: 435–447.

38 McDonel J.L. (1986) Toxins of *Clostridium perfringens* types A, B, C, D and E. In Dorner F. and Drews J. (eds), *Pharmacology of Bacterial Toxins*, pp. 477–517. Pergamon Press, Oxford.

39 Hartley W.J. (1956) A focal symmetrical encephalomalacia of lambs. *NZ Vet J* **4**: 129–135.

40 Buxton D. (1978) *In vitro* effects of *Clostridium welchii* type-D epsilon toxin on guinea-pig, mouse, rabbit and sheep cells. *J Med Microbiol* **11**: 299–302.

41 Knight P.A., Queminet J., Blanchard J.H. and Tilleray J.H. (1990) *In vitro* tests for the measurement of clostridial toxins, toxoids and antisera. II. Titration of *Clostridium perfringens* toxins and antitoxins in cell culture. *Biologicals* **18**: 263–270.

42 Worthington R.W., Bertschinger H.J. and Mulders M.S. (1979) Catecholamine and cyclic nucleotide response of sheep to the injection of *Clostridium welchii* type D epsilon toxin. *J Med Microbiol* **12**: 497–501.

43 Nagahama M. and Sakurai J. (1993) Effect of drugs acting on the central nervous system on the lethality in mice of *Clostridium perfringens* epsilon toxin. *Toxicon* **31**: 427–435.

44 Nagahama M. and Sakurai J. (1992) High-affinity binding of *Clostridium perfringens* epsilon-toxin to rat brain. *Infect Immun* **60**: 1237–1240.

I: Regulation of Virulence

Chapter 26

Regulation of Toxin Production in *Clostridium Perfringens*

TOHRU SHIMIZU, AKINOBU OKABE* AND JULIAN I. ROOD†

*Department of Microbiology, Institute of Basic Medical Sciences, University of Tsukuba, Tsukuba, Ibaraki 305, Japan; *Department of Microbiology, Kagawa Medical School, Kita-gun, Kagawa 761–07, Japan, and †Department of Microbiology, Monash University, Clayton, VIC 3168, Australia.*

I. INTRODUCTION

Clostridium perfringens produces many extracellular toxins, several of which have been postulated to be associated with its ability to cause gas gangrene or clostridial myonecrosis.[1,2] The most significant of these toxins are the α-toxin (phospholipase C), θ-toxin (perfringolysin O) and κ-toxin (collagenase). The genes encoding these toxins and other extracellular enzymes from *C. perfringens* have been cloned and sequenced, which has greatly facilitated their molecular analysis.[3–5]

Since genetic methods for the use of recombinant DNA technology in *C. perfringens* have become available,[5] it has become possible to analyze genes directly in this organism. In particular, the role of α-toxin and θ-toxin in the pathogenesis of *C. perfringens* infections has been examined using alellic exchange mutagenesis of the chromosomal structural genes.[6] Examination of these isogenic mutants in a mouse myonecrosis model has shown that α-toxin is

The Clostridia: Molecular Biology and Pathogenesis
ISBN 0–12–595020–9

an essential *C. perfringens* virulence factor and suggested that α-toxin and θ-toxin may act synergistically *in vivo*.

To aid in the prevention of the severe wound infections caused by *C. perfringens*, it is important to understand how this organism, which is a member of the normal intestinal flora of both animals and humans, controls and regulates the production of its toxins. Because the expression of many of these toxins appears to increase in the exponential growth phase and to decrease in stationary phase,[7] it is reasonable to assume that *C. perfringens* has mechanisms by which it can regulate toxin production in response to environmental or growth phase changes. Indeed, recent studies have led to the discovery of a two-component regulatory system, consisting of the *virR* and *virS* genes, which regulate the production of at least five extracellular toxins or enzymes (α-toxin, θ-toxin, κ-toxin, protease, and sialidase).[8,9]

In many bacteria, it has become clear that cell-to-cell communication plays an important role in the regulation of physiological processes or in the expression of virulence factors.[10,11] The production of *N*-acyl homoserine lactones is widespread in Gram-negative bacteria and these molecules act as diffusible chemical communication signals which may regulate bioluminescence, antibiotic production, plasmid conjugal transfer and the synthesis of extracellular enzymes and virulence factors.[10] In Gram-positive bacteria, oligopeptides play an important role in the mating processes of *Enterococcus faecalis*[12] and in transformation competence and the differentiation of *Bacillus* sp.[11] In *C. perfringens* there have been reports of a diffusible signalling substance, substance A, which appears to be involved in the regulation of toxin production.[13–15] It is possible that the production of substance A may reflect changes in cell density and may act to signal the *virR–virS* two-component regulatory system, finally resulting in activation of toxin production.

II. EVIDENCE FOR AN INTERCELLULAR SIGNALLING SUBSTANCE

Higashi and co-workers used treatment with *N*-methyl-*N′*-nitro-*N*-nitrosoguanidine (NTG) to isolate pleiotropic mutants of *C. perfringens* strain PB6K which were unable to produce several extracellular toxins, especially θ-toxin.[13] These θ-toxin mutants were classified into two groups (*a* and *b*) based upon complementation studies. Complementation of θ-toxin production was detected by cross-streaking combinations of θ-toxin-negative mutants on sheep blood agar. Production of θ-toxin was only observed from a mutant in group *b* when a mutant in group *a* was cross-streaked. No complementation was detected when combinations of two mutants from the same group were tested. The same complementation phenomenon was detected using mixed broth cultures containing one mutant strain from each group. Using antibiotic resistance as a marker, it was shown that no genetic transfer events were involved in this phenomenon. Because complementation did not require direct contact between the cells from the two groups, it was concluded that cells of group *b* produce θ-toxin after being activated by a product diffusing from group *a* cells. The diffusible signalling product could activate θ-toxin production in group *b* mutants through a cellophane membrane, indicating that it was a relatively small compound.

In further studies of this complementation phenomenon separate mutants unable to produce λ-toxin (protease), κ-toxin, or hemaggulutinin (HA), were isolated from strain PB6K.[15] As for the θ-toxin mutants, these mutants could be classified into groups *a* and *b* by complementation analysis. All of the group *b* mutants were pleiotropically negative or leaky with respect to θ-toxin, λ-toxin, κ-toxin, and HA production, suggesting that this group of mutants had a mutation in a gene or genes which regulate the production of these four extracellular proteins. It was proposed that these gene(s) enable the production of a diffusible factor, designated as substance A. It was also suggested that the group *a* mutants could produce the diffusible stimulating factor but had separate mutations in the structural genes encoding the extracellular activities. The ability of the group *b* mutants and the PB6K parent strain to produce other extracellular enzymes, including sialidase, DNase, RNase, hyaluronidase, and β-galactosidase, was compared, but no significant differences were observed. These results indicate that the production of these enzymes might not be regulated by the same mechanism.

Further mutational analysis was carried out to characterize the genetic background of the group *b* mutants.[14] Group *a* mutants could be converted to the group *b* phenotype by a second NTG treatment, indicating that the group *a* strains now had mutations in a gene or genes encoding a substance that operates as an activator. It was postulated that the group *b* strains still have wild-type structural genes for the four extracellular toxins but have lost the capacity to produce substance A, which means that they are in an inactivated state with respect to their ability to produce these extracellular proteins. Group *b* strains produce these proteins only when they obtain substance A from a group *a* strain (Figure 26.1). Because complementation occurred through a membrane whose molecular size cut-off was 1000–2000, substance A is a very small molecule. Further studies showed that substance A activity was unlikely to involve cyclic AMP biosynthesis.[14]

More recently, Imagawa and Higashi partially characterized substance A.[16] Substance A activity appeared in exponentially growing cultures of group *a* mutants and disappeared in stationary phase. It stimulated the production of θ-toxin from group *b* mutants in a dose-dependent manner. Its activity was most

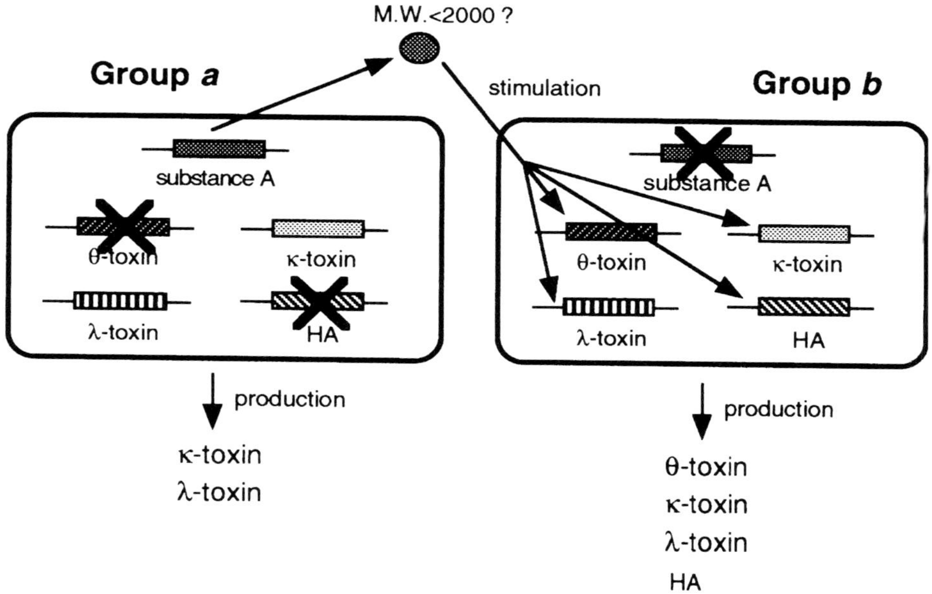

Figure 26.1 Schematic representation of the *Clostridium perfringens* complementation phenomenon. Examples of group *a* and group *b* cells are shown, together with selected structural genes. Group *a* mutants have mutations in the structural genes for one or more toxins but have an intact substance A structural gene. Group *b* mutants have a mutation in the substance A structural gene but have wild-type toxin structural genes. Group *b* mutants only produce toxins when they obtain substance A which has diffused from a group *a* mutant. HA, hemagglutinin. Based on data described previously.[13–15]

stable at pH 5.0 and at 0°C, and even increased for 5 h upon storage, gradually decreasing during the following 15 h. The reason why activity increased at 0°C is unknown. It also was confirmed that substance A somehow functions in the *de novo* synthesis of θ-toxin but not in the secretion of the mature toxin.

Substance A was the first global regulatory factor involved in extracellular toxin production to be reported from *C. perfringens*. The evidence indicates that it is part of a complex regulatory system which controls toxin production in this organism. To date, none of the genes or factors involved in the substance A system have been identified. The precise mechanism by which this regulatory system operates therefore remains to be clarified.

Other studies carried out before modern methods of genetic analysis were available also reported the isolation of pleiotropic mutants of *C. perfringens*.[17] Although some of these mutants may have originated from multiple NTG-derived mutations, the mutagenesis of global regulatory genes now appears to be the more likely hypothesis.

III. EVIDENCE FOR THE REGULATION OF SPECIFIC TOXIN GENES

A. Regulation of the α-toxin gene (*plc*)

Although the cloning and sequence analysis of the α-toxin (phospholipase C) structural gene (*plc* or *cpa*) was reported from several laboratories,[18–22] little information was obtained regarding the genetic regulation of the *plc* gene in *C. perfringens. Clostridium perfringens* produces α-toxin under all of the conditions examined to date, leading to general belief that the *plc* gene is expressed constitutively. However, specific regulatory mechanisms may be responsible for observed differences in α-toxin production between type A strains and other toxin types, and between different type A strains. Several *plc* genes have been cloned from strains that produce different amounts of α-toxin. Although these α-toxins have various amino acid substitutions, their specific activities are very similar.[23] This finding suggests that the variable α-toxin levels observed in different *C. perfringens* isolates results from differences in toxin production. The level of *plc* mRNA in these *C. perfringens* strains varies up to 40-fold,[23] despite the fact that the promoter sequences are almost identical. Therefore, it seems possible that the high and low α-toxin-producing strains differ in an as yet unidentified gene region which affects *plc* gene expression.[23,24] This conclusion is further strengthened by evidence that a natural *C. perfringens* variant which was deficient in α-toxin and θ-toxin production, did not produce α-toxin after transformation with a recombinant plasmid carrying a cloned *plc* gene.[25]

At least two proteins have been shown to bind to the *plc* gene. One protein appears to be specific for a DNA fragment containing the *plc* promoter, the other binds to a region within the *plc* gene. Both proteins have been shown by gel retardation assays to be present in cell extracts from *C. perfringens* type A NCTC8237, a high α-toxin-producing strain.[24] The former protein, but not the latter, is present in extracts from *C. perfringens* type C NCIB10662, a low α-toxin-producing strain.[24] This result suggests that the protein which binds to the *plc* coding region and was only seen in NCTC8237, may play an important role in the positive regulation of *plc* gene expression (Figure 26.2). Since this putative regulatory protein has not yet been purified or characterized, its precise role in *plc* gene expression remains unknown.

Several attempts have been made to identify a gene, or nucleotide sequence in the vicinity of the *plc* gene, which affects *plc* expression. There is an open reading frame, designated ORF2, located upstream of the *plc* gene. Two putative ORF2 promoters, and the *plc* promoter, are oriented head to head and are separated by a 90-bp intervening sequence which includes a direct repeat. This divergent promoter was suspected to be involved in the regulation of the *plc* gene. However, deletion of ORF2 and its promoters, did not affect the

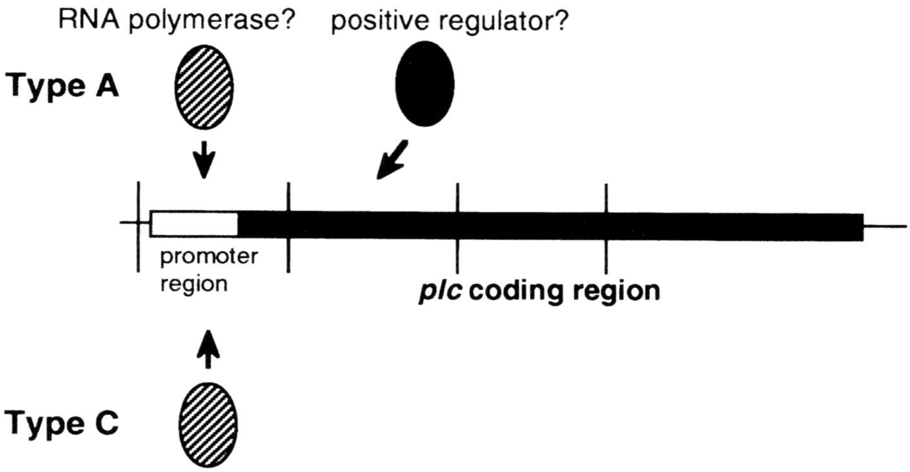

Figure 26.2 Diagrammatic representation of *plc* gene regulation. The *plc* structural gene is shown together with the putative regulatory proteins which are produced by type A and type C strains of *Clostridium perfringens* and which bind to the *plc* gene region. Based on data of Katayama *et al.*[24]

transcriptional activity of the *plc* promoter in *Escherichia coli*[24] or in *C. perfringens* (S.-I. Katayama, unpublished results).

Located upstream of the *plc* gene, between positions 66 and 40, are three poly(A) tracts with a 10–11 bp periodicity (Figure 26.3). Fragments containing these tracts showed anomalously slow mobility in an electric field at low temperature, indicating that they form a region of sequence-dependent DNA curvature.[26] Deletion of the poly(A) tracts from a plasmid containing *plc* significantly increased *plc* gene expression in *E. coli*,[26] but decreased *plc* expression in *C. perfringens*.[27] However, since a stimulatory effect of the curved DNA on *plc* promoter activity was shown in an *in vitro* transcription experiment using purified *C. perfringens* RNA polymerase,[27] the curved DNA *per se* seems to be an element which stimulates the *plc* promoter like many curved DNA sequences located upstream of promoters.[28] The discrepancy between the two hosts seems to be due to the difference in the host–vector system. One possible explanation is that *C. perfringens* RNA polymerase fits a DNA structure formed by the promoter and its upstream curved DNA region but *E. coli* RNA polymerase does not. Alternatively, the curved DNA present in the *bla* gene of the *E. coli* vector might indirectly affect the expression of the cloned *plc* gene, as is the case for the curved DNA present in the *luxAB* reporter system.[29] The complex formed by the curved DNA upstream of the −35 region in *Bacillus subtilis* bacteriophage promoters and RNA polymerase is stable enough to give a positive gel retardation assay.[30] Thus, the protein previously observed by gel retardation to bind specifically to the promoter region may be the *C. perfringens* RNA polymerase

```
                                                  -35
GAATTGTATTCAAAAATATTTTAAAAAATATTCAAAAATTTAGTGAGCTTAT
             -10         +1
GGTAATTATATGGTATAATTTCAGTGCAAGTGTTAATCGTTATCAAAAAAGG

GGAGATTAATACTTGAAAAAATTAACGGGGGATATAAAAAATGAAAAGAAAG
                                          M  K  R  K
```

Figure 26.3 Nucleotide sequence of the upstream regulatory region of the *plc* gene. The poly(dA-dT) tracts[26] are boxed and the consensus − 35 and −10 promoter sequences and the mRNA start site (+1) are indicated. The putative ribosome-binding site is shown by the dotted line and the α-toxin coding region is in bold-type.

(Figure 26.3). *Clostridium perfringens* can express the *plc* gene efficiently even at low temperature, which may be related to the function of the curved DNA since the stimulatory effect of the curved DNA on *plc* gene expression is more prominent at low temperature than at high temperature.[27]

B. Regulation of the θ-toxin gene (*pfoA*)

The structural gene encoding the θ-toxin or perfringolysin O (*pfoA*) has been cloned and sequenced, and the similarity of the toxin to other thiol-activated cytolysins determined[31,32] (see Chapter 13). In subsequent studies, other workers found that the region upstream of the *pfoA* gene is involved in regulating *pfoA* expression in *E. coli*.[33] When the upstream region is deleted, θ-toxin production in *E. coli* decreases 20-fold. Sequence analysis revealed that 591 bp upstream of the *pfoA* gene there is an ORF, designated *pfoR*, which potentially encodes a 343 amino acid protein (Figure 26.4). When the *pfoR* and *pfoA* genes are located on different replicons, no activation of θ-toxin production is observed. In contrast, limited activation occurs when both genes are located on the same plasmid, especially when these genes are located in the reverse orientation (Figure 26.4). Inactivation of *pfoR* by deletion of a 87-bp internal fragment results in the loss of *pfoA* activation (Figure 26.4). Based on these observations, it is concluded that the *pfoR* gene encodes a *cis*-dominant activator of the *pfoA* gene (Figure 26.4).

The deduced amino acid sequence of the putative PfoR protein contains motifs which are characteristic of DNA-binding regulatory proteins. In addition, there are many palindromic structures and directly repeated sequences located in the intergenic region between the *pfoR* and *pfoA* genes, some of which may represent the PfoR-binding sites. The putative PfoR protein is highly hydrophobic and

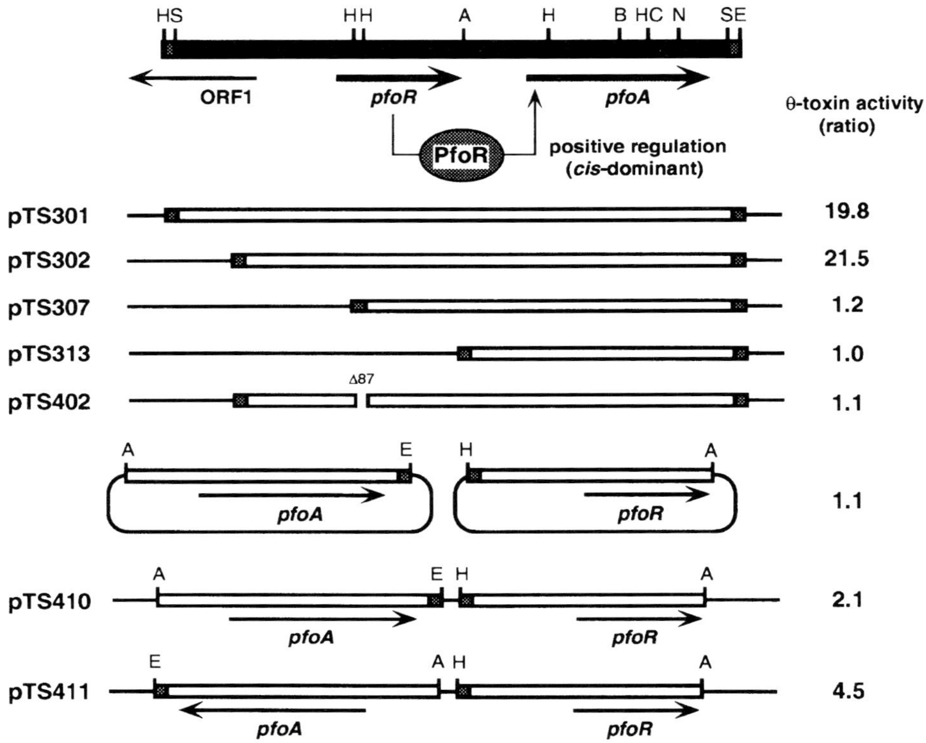

Figure 26.4 Deletion analysis of the *pfoR–pfoA* genes in *Escherichia coli.*[33] The various plasmid constructs are shown together with the ratio of the resultant θ-toxin activities in *E. coli*, relative to that of pTS313. A model for the proposed mode of action of the *pfoR* product is shown at the top of the diagram. The shaded boxes represent multiple cloning sites of pUC19 vector. Restriction endonuclease sites are shown as *Hin*dIII (H), *Sma*I (S), *Apa*LI (A), *Bgl*II (B), *Hin*cII (HC), *Nde*I (N), *Eco*RI(E). Based on a diagram of Shimizu *et al.*[33]

appears to be fairly toxic for *E. coli* cells since *pfoR* deletions frequently occur when multicopy plasmids carrying *pfoR* are maintained in *E. coli* (M. Tamaki and T. Shimizu, unpublished results).

All of the work concerning the *pfoR* gene has been performed in *E. coli*, and it is possible that the expression of the *pfoA* gene is regulated in a different manner in *C. perfringens*. Genetic analysis of the *pfoR* gene in *C. perfringens* is therefore required to determine the precise function of the PfoR protein. Specifically, it is necessary to generate a chromosomal *pfoR* mutation by allelic exchange or homologous recombination and to examine the effect of the resultant mutation on θ-toxin production.

IV. IDENTIFICATION OF A TWO-COMPONENT SYSTEM FOR THE GLOBAL REGULATION OF TOXIN PRODUCTION

Bacterial two-component regulatory systems generally comprise a response regulator protein and a sensor histidine kinase.[34] The genes encoding these proteins are often linked in an operon. In many such systems either an environmental or growth-phase change induces an alteration in the conformation of a transmembrane membrane component of the sensor kinase. The resultant structural modification induces the autophosphorylation of a histidine residue in a cytoplasmic domain of the protein. The activated histidine kinase then acts as a phosphodonor to activate the response regulator protein. The modified response regulator generally then either activates or represses the transcription of one or more target genes. In this manner, an environmental change specifically induces the production of proteins which enable the bacterium to respond to that change or represses the production of proteins that no longer need to be synthesized.[34]

Although early mutational analysis suggested that a global system for regulating toxin production existed in *C. perfringens*,[13–15,17] little was known about the precise mechanism. Recent studies in two of our laboratories have led independently to the discovery that the expression of many of the important extracellular *C. perfringens* toxin genes is coordinately regulated by the products of a two-component signal transduction system.[8,9]

NTG-mutagenesis of the transformable *C. perfringens* strain 13 led to the isolation of mutants altered in their ability to produce θ-toxin.[9] Transformation with a recombinant plasmid carrying the *pfoA* gene was used to divide these mutants into two groups. In the first group of mutants, typified by SI107, θ-toxin production was restored, indicating that the mutation was in the *pfoA* gene. The other group of mutants, typified by SI112, could not be complemented and remained unable to produce θ-toxin. These results suggested that SI112, which also produced reduced levels of κ-toxin and HA, had a mutation in a global regulatory gene.

A chromosomal DNA library was constructed in an *E. coli–C. perfringens* shuttle vector and used to transform SI112. A recombinant plasmid, pBT404, which restored the ability to produce θ-toxin in SI112 was obtained (Table 26.1). The 2.5 kb *Hin*dIII fragment carried by this plasmid was shown to be responsible for the activation of *pfoA* expression at the transcriptional level (Figure 26.5). The same plasmid also resulted in the activation of κ-toxin and HA production (Table 26.1), supporting the hypothesis that a global regulatory gene had been identified. Sequence analysis revealed that the recombinant plasmid carried an ORF, designated *virR*.[9] The deduced amino acid sequence of the putative VirR protein had similarity to response regulators of bacterial two-component regulatory systems. Complementation analysis of a chromosomal *virR* mutant

Table 26.1. Activities of toxins in *Clostridium perfringens* strains

Strain	Perfringolysin O (U/μg of protein)	Collagenase (OD_{320} units/mg of protein)	Hemagglutinin
13(pJIR418)	4.1±0.2	15.2±0.3	+
SI112(pJIR418)	<0.3±0.1	5.1±0.2	−
SI112(pBT404)	2.3±0.1	18.6±0.5	+
TS133(pJIR418)	<0.3±0.1	5.4±0.3	−
TS133(pBT404)	2.6±0.2	19.5±0.5	+

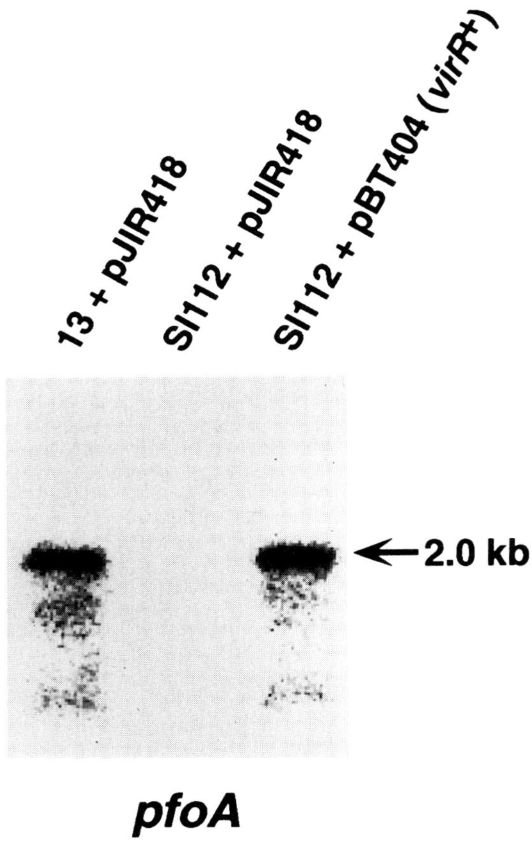

Figure 26.5 Effect of the *virR* mutation on the expression of *pfoA* mRNA in *Clostridium perfringens*. Total RNA of *C. perfringens* was prepared from logarithmic-phase culture and 3 μg of each RNA preparation was denatured, separated by electrophoresis on a formaldehyde denaturing gel, and blotted onto a nylon filter membrane. A ^{32}P-labeled *pfoA* fragment was used for a probe. pJIR418 is a *Escherichia coli–C. perfringens* shuttle vector and pBT404 is pJIR418 carrying the *virR* gene. The expected size of the *pfoA* transcript (2.0 kb) is indicated by an arrow. Modified from Shimizu *et al.*[9]

constructed by homologous recombination confirmed that *virR* encoded a positive activator of θ-toxin, κ-toxin and HA production.[9]

In an independent laboratory a strain 13 mutant which was unable to produce θ-toxin and produced reduced levels of α-toxin, sialidase, and protease, was isolated by Tn *916* mutagenesis.[8] Inverse PCR was used to isolate the region into which Tn*916* had inserted, this region was subsequently cloned. The end result of these experiments was the isolation of a recombinant plasmid, pJIR872, which could complement the Tn*916*-derived mutant JIR4000 and restore its ability to produce all of the affected extracellular toxins (Table 26.2). Sequence analysis revealed that the 4.3 kb *Pst*I fragment carried by pJIR872 contained the *virR* gene and a downstream overlapping gene, designated *virS* (Figure 26.6). The Tn*916* chromosomal insertion was located within *virS*. The *virS* mutant showed significantly decreased virulence in a mouse myonecrosis model, indicating that this gene plays an important role in the pathogenesis of histotoxic *C. perfringens* infections.[8]

The deduced amino acid sequences of the putative VirR and VirS proteins contained motifs commonly found in response regulator proteins and sensor histidine kinases, respectively, of bacterial two-component regulatory systems (Figure 26.7). Response regulators generally have two conserved Asp residues and a conserved Lys residue which form an acidic pocket that is the acceptor site which receives the phosphate residue from the sensor kinase. The amino-terminal region of VirR has these conserved amino acids and it is postulated that Asp-73 represents the site of phosphorylation.[8]

Sequence similarity between sensor histidine kinases is usually located in the carboxy-terminal domains, with the amino-terminal regions generally containing transmembrane domains that act to sense an environmental stimulus. The amino-terminal region of the VirS protein has six or seven transmembrane domains and therefore appears to function in a similar manner. The carboxy-terminal,

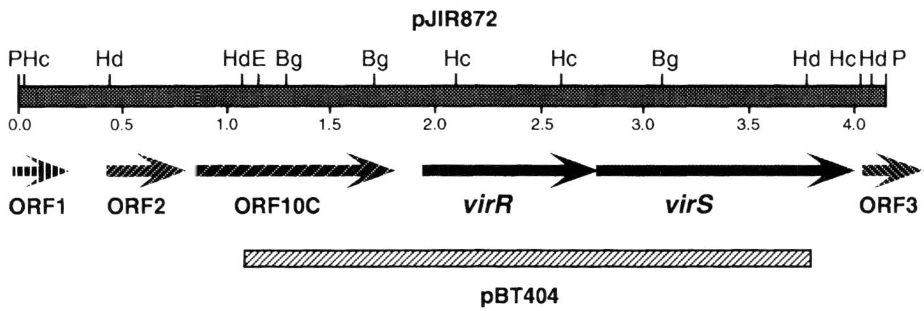

Figure 26.6 Genetic and physical map of the *virR–virS* gene region. Open reading frames (ORFs) are shown by the arrows. Restriction sites are indicated as follows, *Pst*I (P), *Hinc*II (Hc), *Hind*III (Hd), *Eco*RI(E), *Bgl*II (Bg) Based on data obtained by sequence analysis of pJIR872[8] and pBT404.[9]

Table 26.2. Extracellular enzyme levels of JIR4025 and its derivatives

Strain	Description	Phospholipase C (units/mg protein)	Perfringolysin O ($\log_2$(titer))	Protease (units/mg protein)	Sialidase (units/mg protein)
JIR4025	Parent	$(11.9 \pm 1.4) \times 10^{-3}$	9.0 ± 0.0	0.089 ± 0.005	0.31 ± 0.09
JIR4000	JIR4125::Tn*916*	$(2.1 \pm 0.4) \times 10^{-3}$	$<1.0 \pm 0.0$	0.060 ± 0.007	0.06 ± 0.03
JIR4086	JIR4000(pJIR751)	$(2.2 \pm 0.5) \times 10^{-3}$	$<1.0 \pm 0.0$	0.065 ± 0.005	0.09 ± 0.04
JIR4087	JIR4000(pJIR870)	$(6.0 \pm 1.3) \times 10^{-3}$	7.7 ± 0.5	0.094 ± 0.009	0.35 ± 0.12
JIR4089	JIR4000(pJIR872)	$(11.4 \pm 2.4) \times 10^{-3}$	10.3 ± 0.5	0.190 ± 0.060	0.85 ± 0.09

A

VirR	-19-	IALCEDN	-42-	MLFLDIQMGE-LTG	-15-	IIFITA	-10-	VRAFRYLIKP-
AgrA	-2-	IFICEDD	-45-	CYFLDIQLSTDING	-15-	IIFVTS	-12-	VAAMDFIFKD-
RprY	-9-	ILLCEDD	-37-	LCVFDVMMPK-KDG	-15-	IIFLTA	-12-	IGADDYITKP-
MrkE	-58-	VIIVEDE	-39-	AIFLDINIOS-LDG	-15-	IVFITA	-10-	LEAFDYILKP-
PhoB	-4-	ILVVEDE	-38-	LILLDWMLPG-GSG	-17-	VVMLTA	-12-	TGADDYITKP-
AlgR	-2-	VLIVDDE	-40-	IVLLDIRMPG-LDG	-15-	VIFCTA	-10-	VSAVGYLVKP-
CheY	-7-	FLVVDDF	-38-	FIISDWNMPN-MDG	-17-	VLMLTA	-12-	AGASGYVVKP-
OmpR	-6-	NLVVDDD	-37-	LMVLDLMLPG-EDG	-15-	IIMVTA	-12-	IGADDYIPKP-
TctD	-2-	LLLAEDN	-37-	LAVLDINMPG-MDG	-15-	VLLLTA	-12-	VGADDYLPKP-
CopR	-2-	LLVAEDA	-37-	LLILDVMMPG-LDG	-15-	VLFLTA	-12-	LGADDYLVKP-

B

VirS	-251-	RLYHDIKNHIA	-69-	YFDICTIFSNCIDNAIEA	-50-	FLHGIGLKNIRLALEKYNG-
SakB	-236-	KFFHDYKNLLL	-80-	IFDCIRILGILIDNAIEA	-49-	GHSGLGLNTIQEFNQKFPN-
NtrB	-146-	MLAHEIKNPLS	-89-	QDQLIQVFLNLVKNAAEA	-66-	TGSGLGLALVAKIVGDHGG-
VirA	-470-	GIAHEFNNILG	-91-	PLELQQVLINICKNASQA	-66-	GGTGLGLASVHGHISAFAG-
PhoR	-210-	NVSHELRTPLT	-93-	EDQLRSAISNLVYNAVNH	-55-	KGFGLGLAIVKHAVNHHES-
RprX	-305-	NMTHEFKTPIS	-98-	EMHITNVIFNLMDNAVKY	-33-	KGFGLGLAYVKKIIQDHKG-
EnvZ	-240-	GVSHDLRTPLT	-83-	PLSIKRAVANMVVNAARY	-51-	SGTGLGLAIVQRIVDNHNG-

Figure 26.7 Comparative analysis of the amino acid sequence of the VirR and VirS proteins. A, Comparison of the sequence of the VirR protein with the response regulators, AgrA,[35] RprY,[36] MrkE,[37] PhoB,[38] AlgR,[39] CheY,[40] OmpR,[41] TctD,[42] and CopR.[43] B, Comparison of the sequence of the VirS protein with the sensor histidine kinases, SakB,[44] NtrB,[45] VirA,[46] PhoR,[47] RprX,[36] and EnvZ.[48] Modified from Lyristis *et al.*[8]

hydrophilic, and presumably cytoplasmic, region of VirS contains domains that have been shown to be involved in nucleotide binding and kinase activity and are conserved in sensor histidine kinases (Figure 26.7). Based on sequence comparisons it is postulated that the conserved His-255 residue represents the site of autophosphorylation of VirS.[8]

The finding that the chromosomal *virS* mutant could be complemented by a recombinant *virS* gene was as expected. However, this mutation could also be partly complemented by a recombinant *virR* gene.[8] This result can be explained by postulating that in this situation some phosphorylation of the VirR protein occurs by cross-talk with other sensor kinases or by phosphorylation from a small molecular weight phosphodonor.

Transcriptional analysis of the *virR*/*virS* regulon was recently performed.[49] The results showed that the *virR* and *virS* genes were transcribed on a single mRNA molecule and were organized in an operon. Therefore, the *virR* mutant TS133, which was previously isolated by insertional inactivation,[9] was also unable to produce VirS owing to polarity effects. A multicopy *virR* gene was able to partially restore the phenotypic changes present in TS133 (Table 26.1), providing independent evidence that the VirR protein can be at least partly activated in the absence of a functional VirS protein.

Transcription of the *pfoR* gene can only be detected at a very low level but does appear to be regulated by the *virR/virS* system, suggesting that a cascade mechanism may exist for the activation of the *pfoA* gene.[49] That is, the VirR protein may activate the *pfoR* gene, the product of which may in turn activate the *pfoA* gene. However, the activated VirR protein may also activate *pfoA* directly. As stated earlier, it is necessary to construct a *pfoR* mutant in *C. perfringens* to determine the precise role of this gene in the regulation of *pfoA* expression.

The production of θ-toxin is completely dependent upon a functional *virR–virS* system, since mutation of either the *virR* or *virS* genes eliminates θ-toxin production.[8,9] As expected, no *pfoA*-specific transcripts were detected in either the TS133 or JIR4000 mutants.[49] In contrast, the production of κ-toxin and α-toxin was only partly affected by the *virR* and *virS* mutations and transcription of the respective *colA* and *plc* genes was reduced but not eliminated in these mutants.

Primer extension analysis was used to identify the *pfoA*, *plc*, and *colA* promoters.[49] The *pfoA* gene had both a major and a minor promoter, with the major promoter being *virR/virS*-dependent. The *colA* gene had two major promoters, only one of which was *virR/virS*-dependent. In contrast, the *plc* gene had only one promoter, which was shown to be partially regulated by the *virR* and *virS* genes. Comparative analysis of the *virR*-dependent transcription start sites of each of these genes did not reveal any significant common sequences that could act as VirR-binding sites. These results suggest that other unidentified regulatory factors including PlcR, the proposed regulator of the *plc* gene,[8] might mediate the signal from the *virR–virS* system to the target genes in *C. perfringens*. Alternatively, VirR may resemble the OmpR response regulator of *E. coli*, which does not have a single consensus binding sequence.[50]

V. CONCLUSIONS AND FUTURE DIRECTIONS

A model which describes the regulatory network mediated by the *virR–virS* system has been proposed[8] and subsequently modified.[51] An updated version of this model is shown in Figure 26.8. It is proposed that the *virR* and *virS* genes encode a response regulator and sensor histidine kinase, respectively. The amino-terminal domain of the VirS protein is postulated to be located in the cell membrane and to be activated by an environmental or growth-phase stimulus, which leads to the autophosphorylation of the His-255 residue, which is

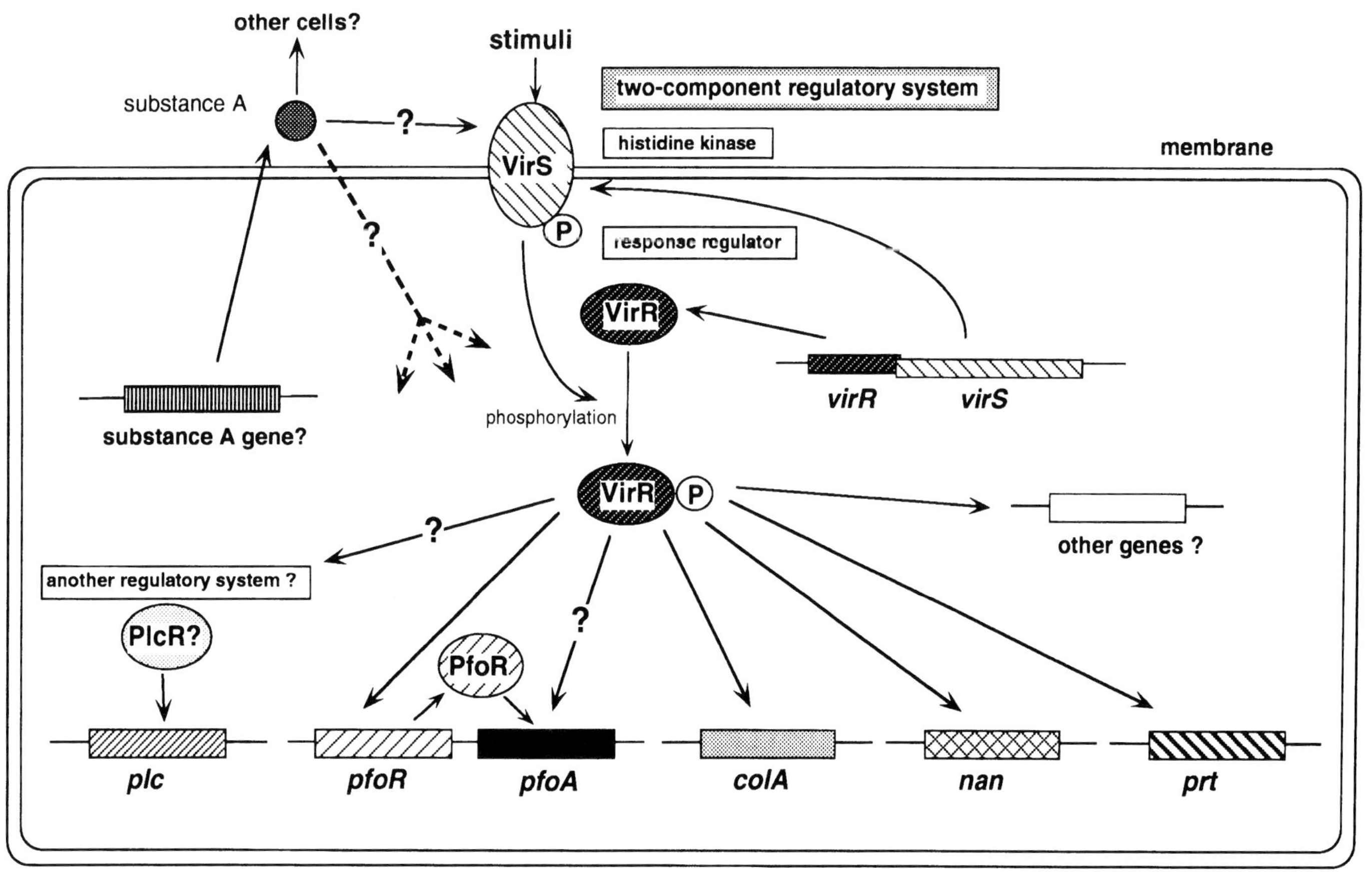

Figure 26.8 Schematic model for the regulation of extracellular toxin production in *C. perfringens.* The postulated interaction between substance A, the VirR and VirS proteins, and the various toxin structural genes is shown. *nan* and *prt* stand for the sialidase and protease genes, respectively. Based on previous diagrams.[8, 51]

located in the cytoplasmic carboxy-terminal domain. The phosphorylated VirS protein then presumably acts as a phosphodonor and leads to the phosphorylation of the Asp-73 residue of the VirR protein resulting in the activation of VirR. The activated VirR protein then either directly activates transcription of the toxin structural genes or secondary regulatory genes, such as *pfoR* or the putative *plcR* gene. For the latter, a regulatory cascade would lead to the increased level of transcription of the toxin structural genes. Therefore, the *virR–virS* system may contribute to the pathogenesis of *C. perfringens* infections by regulating the production of extracellular toxins which may then act synergistically on the infected tissues.[6]

Although from the available data we can now gain a limited but exciting picture of the regulation of the toxin production in *C. perfringens*, we still have many more questions than answers.[51] For example, what are the environmental or growth-phase stimuli which activate the VirS sensor protein? Because many changes in growth conditions or nutritional factors may affect the production of extracellular toxins, the factors which activate the VirS protein must be identified.

How is substance A involved in the regulation of toxin production? Because the toxins whose production is stimulated by substance A are almost the same as those regulated by the *virR–virS* system, except for sialidase production,[8] it is possible that the two regulatory systems are involved in the same regulatory cascade. Different results in sialidase production observed in previous studies[8,15] may have originated from the difference in *C. perfringens* strains used. Perhaps substance A is directly responsible for triggering the autophosphorylation of VirS. Moreover, increasing knowledge about the intercellular communication between bacteria suggests that such communication might play an important role in pathogenesis.[10] In *C. perfringens* infections, substance A may be responsible for signalling cells to produce toxins during the course of an infection. The genetic analysis and molecular characterization of the genes responsible for the synthesis of substance A are needed to elucidate its precise role in the regulation of toxin production.

What is the precise mechanism of action of the VirR protein? The determination of the binding sequence of the activated VirR protein and the analysis of its mode of activation are prerequisites for understanding its mechanism of action. It will be necessary to analyze the promoter region of several *virR–virS*-regulated genes to identify the putative VirR-boxes, or the regulatory sequences to which the VirR protein binds. These studies may also require the isolation and analysis of other regulatory genes that are involved in the putative VirR cascade.

What is the role of the PfoR protein in the *virR–virS* system? Because the PfoR protein appears to be a *cis*-dominant activator of the *pfoA* gene, it is important to determine its functional role in regulating θ-toxin production.

That is, to resolve whether the *virR–virS* system regulates *pfoA* gene expression directly or by regulating *pfoR*. These studies necessitate the isolation of a chromosomal *pfoR* mutant of *C. perfringens* by allelic exchange.

Are there any other genes which are regulated by the *virR–virS* system? The VirR/VirS regulon involves many extracellular toxin genes and it is possible that there are other genes that are also regulated by this system. The identification of such genes is essential if we are to have a precise picture of the functional significance of the role of the *virR–virS* system in *C. perfringens*.

Finally, are there any other genes involved in the regulation of toxin production in *C. perfringens*? It is important to determine whether other unknown regulatory genes are mediating regulation by the *virR/virS* system, such as the putative *plcR* gene.

In conclusion, with the development of methods for the genetic and molecular analysis of *C. perfringens*, considerable advances have been made in our understanding of how the production of extracellular toxins is regulated. However, much remains to be determined and we can expect significant developments in this exciting research area over the next few years. Studies will involve determining the precise molecular mechanisms by which *C. perfringens* cells control the production of the extracellular toxins that are involved in virulence, and will make a major contribution to our understanding of *C. perfringens* infections.

VI. ACKNOWLEDGMENTS

Work in the University of Tsukuba laboratory was supported by a Grant-in-Aid for scientific research from the Ministry of Education, Science and Culture of Japan and a grant from University of Tsukuba Research Projects. Work at Monash University was supported by grants from the Australian National Health and Medical Research Council and CSL Ltd. Work at Kagawa Medical School was supported by a Grant-in-Aid for scientific research from the Ministry of Education, Science and Culture of Japan.

VII. REFERENCES

1. McDonel J.L. (1980) *Clostridium perfringens* toxins (Type A, B, C, D, E). *Pharmacol Ther* **10**: 617–635.
2. Hatheway C.L. (1990) Toxigenic clostridia. *Clin Microbiol Rev* **3**: 66–98.
3. Matsushita O., Yoshihara K., Katayama S.-I., Minami J. and Okabe A. (1994) Purification and characterization of a *Clostridium perfringens* 120-kilodalton collagenase and nucleotide sequence of the corresponding gene. *J Bacteriol* **176**: 149–156.

4. Canard B., Garnier T., Saint-Joanis B. and Cole S.T. (1994) Molecular genetic analysis of the *nagH* gene encoding a hyaluronidase of *Clostridium perfringens. Mol Gen Genet* **243**: 215–224.
5. Rood J.I. and Cole S.T. (1991) Molecular genetics and pathogenesis of *Clostridium perfringens. Microbiol Rev* **55**: 621–648.
6. Awad M.M., Bryant A.E., Stevens D.L. and Rood J.I. (1995) Virulence studies on chromosomal α-toxin and θ-toxin mutants constructed by allelic exchange provide genetic evidence for the essential role of α-toxin in *Clostridium perfringens*-mediated gas gangrene. *Mol Microbiol* **15**: 191–202.
7. Smyth C.J. and Arbuthnott J.P. (1974) Properties of *Clostridium perfringens* (welchii) type-A alpha-toxin (phospholipase C) purified by electrofocusing. *J Med Microbiol* **7**: 41–67.
8. Lyristis M., Bryant A.E., Sloan J., Awad M.M., Nisbet I.T., Stevens D.L. and Rood J.I. (1994) Identification and molecular analysis of a locus that regulates extracellular toxin production in *Clostridium perfringens. Mol Microbiol* **12**: 761–777.
9. Shimizu T., Ba-Thein W., Tamaki M. and Hayashi H. (1994) The *virR* gene, a member of a class of two-component response regulators, regulates the production of the perfringolysin O, collagenase, and hemagglutinin in *Clostridium perfringens. J Bacteriol* **176**: 1616–1623.
10. Salmond G.P.C., Bycroft B.W., Stewart G.S.A.B. and Williams P. (1995) The bacterial 'enigma': cracking the code of cell–cell communication. *Mol Microbiol* **16**: 615–624.
11. Kaiser D. and Losick R. (1993) How and why bacteria talk to each other. *Cell* **73**: 873–885.
12. Clewell D.B. and Weaver K.E. (1989) Sex pheromones and plasmid transfer in *Enterococcus faecalis. Plasmid* **21**: 175–184.
13. Higashi Y., Chazono M., Inoue K., Yanagase Y., Amano T. and Shimada K. (1973) Complementation of theta toxinogenicity between mutants of two groups of *Clostridium perfringens. Biken J* **16**: 1–9.
14. Imagawa T., Tatsuki T., Higashi Y. and Amano T. (1981) Complementation characteristics of newly isolated mutants from two groups of strains of *Clostridium perfringens. Biken J* **24**: 13–21.
15. Tatsuki T., Imagawa T., Kitajima H., Higashi Y., Chazono M., Yanagase Y. and Amano T. (1981) Further studies on complementation between mutants of *Clostridium perfringens. Biken J* **24**: 1–11.
16. Imagawa T. and Higashi Y. (1992) An activity which restores theta toxin activity in some theta toxin-deficient mutants of *Clostridium perfringens. Microbiol Immunol* **36**: 523–527.
17. Rood J.I. and Wilkinson R.G. (1975) Isolation and characterization of *Clostridium perfringens* mutants altered in both hemagglutinin and sialidase production. *J Bacteriol* **123**: 419–427.
18. Leslie D., Fairweather N., Pickard D., Dougan G. and Kehoe M. (1989) Phospholipase C and haemolytic activities of *Clostridium perfringens* alpha-toxin cloned in *Escherichia coli*: sequence and homology with a *Bacillus cereus* phospholipase C. *Mol Microbiol* **3**: 383–392.
19. Okabe A., Shimizu T. and Hayashi H. (1989) Cloning and sequencing of a phospholipase C gene of *Clostridium perfringens. Biochem Biophys Res Commun* **160**: 33–39.
20. Saint-Joanis B., Garnier T. and Cole S.T. (1989) Gene cloning shows the alpha toxin of *Clostridium perfringens* to contain both sphingomyelinase and lecithinase activities. *Mol Gen Genet* **219**: 453–460.
21. Titball R.W., Hunter S.E., Martin K.L., Morris B.C., Shuttleworth A.D., Rubidge T., Anderson D.W. and Kelly D.C. (1989) Molecular cloning and nucleotide sequence of the alpha-toxin (phospholipase C) of *Clostridium perfringens. Infect Immun* **57**: 367–376.
22. Tso J.Y. and Siebel C. (1989) Cloning and expression of the phospholipase C gene from *Clostridium perfringens* and *Clostridium bifermentans. Infect Immun* **57**: 468–476.
23. Tsutsui K., Minami J., Matsushita O., Katayama S.-I., Taniguchi Y., Nakamura S., Nishioka M. and Okabe A. (1995) Phylogenetic analysis of phospholipase C genes from *Clostridium perfringens* types A to E and *Clostridium novyi. J. Bacteriol* **177**: 7164–7170.
24. Katayama S.-I., Matsushita O., Minami J., Mizobuchi S. and Okabe A. (1993) Comparison of the alpha-toxin genes of *Clostridium perfringens* type A and C strains: evidence for extragenic regulation of transcription. *Infect Immun* **61**: 457–463.

25. Kameyama K., Matusushita O., Katayama S.-I., Minami J., Maeda M., Nakamura S. and Okabe A. (1995) Analysis of the phospholipase C gene of *Clostridium perfringens* KZ1340 isolated from Antarctic soil. *Microbiol Immunol* **40**: 255–263.
26. Toyonaga T., Matsushita O., Katayama S.-I., Minami J. and Okabe A. (1992) Role of the upstream region containing an intrinsic DNA curvature in the negative regulation of the phospholipase C gene of *Clostridium perfringens. Microbiol Immunol* **36**: 603–613.
27. Matsushita C., Matsushita O., Katayama S.-I., Minami J., Takai K. and Okabe A. (1996) An upstream activating sequence containing curved DNA involved in activation of the *Clostridium perfringens plc* promoter. *Microbiol* **142**: 2561–2566.
28. Pérez-Martín J., Rojo F. and Lorenzo V. de. (1994) Promoters responsive to DNA bending: a common theme in prokaryotic gene expression. *Microbiol Rev* **58**: 268–290.
29. Forsberg A.J., Pavitt G.D. and Higgins C.F. (1994) Use of a transcriptional fusions to monitor gene expression: a cautionary tale. *J Bacteriol* **176**: 2128–2131.
30. Nickerson C.A. and Achberger E.C. (1995) Role of curved DNA in binding of *Escherichia coli* RNA polymerase to promoters. *J Bacteriol* **177**: 5756–5761.
31. Tweten R.K. (1988) Nucleotide sequence of the gene for perfringolysin O (theta-toxin) from *Clostridium perfringens*: significant homology with the genes for streptolysin O and pneumolysin. *Infect Immun* **56**: 3235–3240.
32. Tweten R.K. (1988) Cloning and expression in *Escherichia coli* of the perfringolysin O (theta-toxin) gene from *Clostridium perfringens* and characterization of the gene product. *Infect Immun* **56**: 3228–3234.
33. Shimizu T., Okabe A., Minami J. and Hayashi H. (1991) An upstream regulatory sequence stimulates expression of the perfringolysin O gene of *Clostridium perfringens. Infect Immun* **59**: 137–142.
34. Stock J.B., Ninfa A.J. and Stock A.M. (1989) Protein phosphorylation and regulation of adaptive responses in bacteria. *Microbiol Rev* **53**: 450–490.
35. Peng H.-L., Novic R.P., Kreiswirth B., Kornblum J. and Schlievert P. (1988) Cloning, characterization, and sequencing of an accessory gene regulator (*agr*) in *Staphylococcus aureus. J Bacteriol* **170**: 4365–4372.
36. Rasmussen B.A. and Kovacs E. (1993) Cloning and identification of a two-component signal-transducing system from *Bacteriodes fragilis. Mol Microbiol* **7**: 765–776.
37. Allen B.L., Gerlach G.-F. and Clegg S. (1991) Nucleotide sequence and functions of *mrk* determinants necessary for expression of type 3 fimbriae in *Klebsiella pneumoniae. J Bacteriol* **173**: 916–920.
38. Makino K., Shinagawa H., Amemura M. and Nakata A. (1986) Nucleotide sequence of the *phoB* gene, the positive regulatory gene for the phosphate regulon of *Escherichia coli* K-12. *J Mol Biol* **190**: 37–44.
39. Deretic V., Dikshit R., Konyecsni W.M., Chakrabarty A.M. and Misra T.K. (1989) The *algR* gene, which regulates mucoidity in *Pseudomonas aeruginosa*, belongs to a class of environmentally responsive genes. *J Bacteriol* **171**: 1278–1283.
40. Stock A., Chen T., Welsh D. and Stock J. (1988) CheA protein, a central regulator of bacterial chemotaxis, belongs to a family of proteins that control gene expression in response to changing environmental conditions. *Proc Natl Acad Sci USA* **85**: 1403–1407.
41. Lijestroem P., Laamanen I. and Palva E.T. (1988) Structure and expression of the *ompB* operon, the regulatory locus for the outer membrane porin regulon in *Salmonella typhimurium. J Mol Biol* **201**: 663–673.
42. Widenhom K.A., Somers J.M. and Kay W.W. (1989) Genetic regulation of the tricarboxylate transport operon (*tct*) of *Salmonella typhimurium. J. Bacteriol* **171**: 4436–4441.
43. Mellano M.A. and Cooksey D.A. (1988) Nucleotide sequence and organization of copper resistance genes from *Pseudomonas syringae* pv. *tomato. J Bacteriol* **170**: 2879–2883.
44. Axelsson L., Holck A., Birkeland S., Aukrust T. and Blom H. (1993) Cloning and nucleotide

sequence of a gene from *Lactobacillus sake* Lb706 necessary for sakacin A production and immunity. *Appl Environ Microbiol* **59**: 2868–2875.

45. Nixon T.B., Ronson C.W. and Ausubel F.M. (1986) Two-component regulatory systems responsive to environmental stimuli share strongly conserved domains with the nitrogen assimilation regulatory gene *ntrB* and *ntrC*. *Proc Natl Acad Sci USA* **83**: 7850–7854.
46. Leroux B., Yanofsky M.F., Winans S., Ward J.E., Ziegler S.F. and Nester E.W. (1987) Characterization of the *virA* of *Agrobacterium tumefaciens*: a transcriptional regulator and host range determinant. *EMBO J* **6**: 849–856.
47. Makino K., Shinagawa H., Amemura M. and Nakata A. (1986) Nucleotide sequence of the *phoR* gene, a regulatory gene for the phosphate regulon of *Escherichia coli*. *J. Mol Biol* **192**: 549–556.
48. Mizuno T., Wurtzel E.T. and Inouye M. (1982) Cloning of the regulatory genes (*ompR* and *envZ*) for the matrix proteins of the *Escherichia coli* outer membrane. *J Bacteriol* **150**: 1462–1466.
49. Ba-Thein W., Lyristis M., Ohtani K., Nisbet I.T., Hayashi H., Rood J.I. and Shimizu T. (1996) The *virR*/*virS locus* regulates the transcription of genes encoding extracellular toxin production in *Clostridium perfringens*. *J Bacteriol* **178**: 2514–2520.
50. Pratt L. and Silhavy T.J. (1995) Identification of base pairs important for OmpR-DNA interaction. *Mol Microbiol* **17**: 565–573.
51. Rood J.I. and Lyristis M. (1995) Regulation of extracellular toxin production in *Clostridium perfringens*. *Trends Microbiol* **3**: 192–196.

Chapter 27

Regulation of Enterotoxin Production in *Clostridium perfringens*

STEPHEN B. MELVILLE, RENÉE E. COLLIE*
AND BRUCE A. McCLANE[†]

*Department of Microbiology and Immunology, University of Tennessee, School of Medicine, 858 Madison Avenue, Memphis, TN 38163, USA, and *Department of Molecular Genetics and Biochemistry, University of Pittsburgh School of Medicine, E1240 Biomedical Science Tower, Pittsburgh, PA 15261, USA.*

I. INTRODUCTION

As described in greater detail in Chapters 8 and 19, the course of *Clostridium perfringens* type A food poisoning is unusual in that the cell developmental process of spore production is a prerequisite for synthesis of *C. perfringens* enterotoxin (CPE), which is believed to be the causative factor for this food poisoning. This link between sporulation and enterotoxin production raises many interesting questions as to how these two processes may be coregulated. Some aspects of enterotoxin synthesis have also been the subject of much debate in the literature, primarily owing to conflicting reports. The application of recently developed molecular biological techniques for *C. perfringens* is now providing some answers concerning the regulation of enterotoxin synthesis; these recent results are the primary focus of this chapter.

The Clostridia: Molecular Biology and Pathogenesis
ISBN 0-12-595020-9

II. PHYSIOLOGICAL AND GENETIC CORRELATION BETWEEN SPORULATION AND ENTEROTOXIN FORMATION

Soon after Duncan and Strong[1] developed a medium (DS medium) that gave consistent sporulation for some enterotoxigenic *C. perfringens* strains, evidence began to accumulate that sporulating cells were the primary producers of enterotoxin. Specifically, Duncan and Strong[2] and Hauschild *et al.*[3] independently demonstrated that sporulating, but not vegetative, cultures of *C. perfringens* produce a factor (later identified by both groups[4,5] as CPE, a single, 35 kDa polypeptide) that causes diarrhea in human volunteers and fluid accumulation in ileal loops of many mammalian species.

These early studies led to investigations into the physiological and genetic basis for enterotoxin production by sporulating *C. perfringens* cells. Using DS medium, Duncan[6] constructed a fairly precise kinetic analysis of sporulation, enterotoxin synthesis, and release of the mature spore by the mother cell during the growth and sporulation cycle of *C. perfringens* type A strain NCTC 8798 (Figure 27.1). This study demonstrated that no heat-resistant spores or enterotoxin could be detected during the exponential growth phase. However, after the onset of stationary phase (i.e., after 2–3 h of growth in DS), Duncan noted the

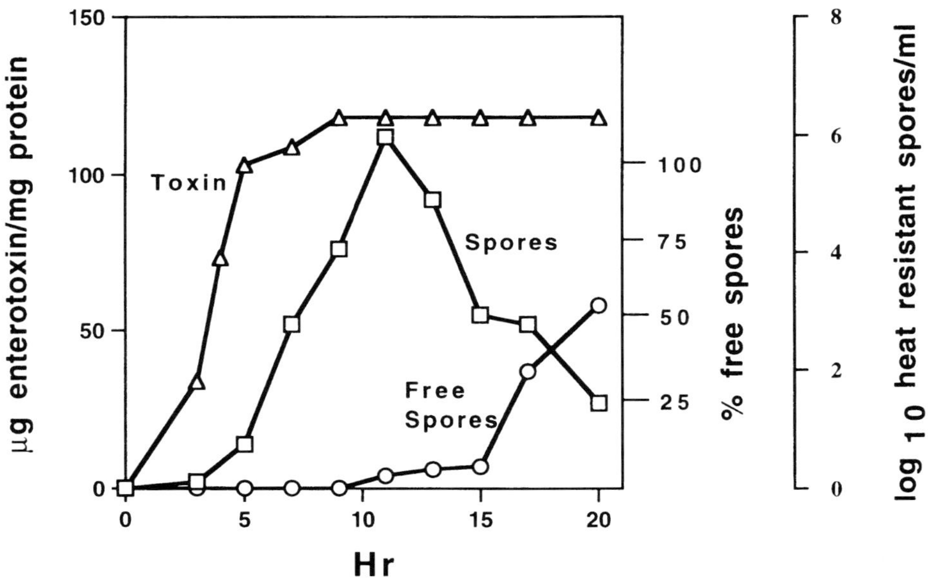

Figure 27.1 Kinetics of enterotoxin formation, endospore formation and the release of spores from the sporangium by *Clostridium perfringens* type A strain NCTC 8798 grown in DS medium. Derived from Duncan (1973).[6]

following order of events: (1) heat-resistant spore formation; (2) intracellular synthesis of CPE protein; (3) release of mature spores; and, (4) the appearance of extracellular CPE protein in the culture medium. These, and other studies indicate that since (1) maximal expression of CPE occurs after the appearance of heat-resistant spores, and, (2) many *C. perfringens* strains unable to express CPE are capable of sporulating, CPE synthesis is not required to produce an apparently normal spore. The absence of CPE protein in the culture medium prior to lysis of the mother cells also indicates that CPE is not actively secreted from the cell (later cloning/sequencing studies have shown that *cpe* does not encode an *N*-terminal signal peptide).

Genetic evidence in support of the coregulation of sporulation and CPE synthesis was provided by the analysis of spontaneous and chemical sporulation mutants of CPE+ *C. perfringens* type A strain NCTC 8798 (Table 27.1).[7] (For the sake of clarity, we refer to strains as CPE+ if the enterotoxin protein has been detected and CPE− if it has not, and to *cpe*+ if the enterotoxin gene has been detected and *cpe*− if it has not.) The stage at which sporulation was blocked for each of these mutants was determined by electron microscopic analysis of ultrastructural changes, and the ability of each mutant to express CPE was determined following its growth in DS. As shown in Table 27.1, mutants blocked at stage 0 were unable to express CPE, while mutants blocked at stages III, IV, and V continued to do so, although in reduced amounts. This pattern of CPE expression indicates that CPE synthesis is induced during stage I or II of the sporulation cycle, a result which is temporally consistent with the kinetics of CPE synthesis shown in Figure 27.1. Duncan *et al.*[7] also isolated sporulation revertants for one stage 0 mutant and showed that these sporulation revertants regained the ability to synthesize CPE, indicating that a single mutation was probably responsible for the loss of both sporulation and CPE synthesis by this mutant.

Table 27.1. Synthesis of enterotoxin and formation of paracrystalline inclusions by sporulation mutants of *Clostridium perfringens*

Strain	Sporulation stage blocked	Enterotoxin made	Paracrystalline inclusion formed
NCTC 8798	Spo+ (wild type)	+	+
8–7	Stage III	+	?
8–20	Stage IV	+	+
8–17	Stage V	+	+
8–1	Stage 0	−	−
R–8, R–42, R–43	Spo+, revertants of mutant 8–1	+	+

Data derived from Duncan *et al.*[8]

Despite this early evidence closely linking sporulation and CPE expression, there have been occasional reports of CPE expression by 'vegetative' (nonsporulating) *C. perfringens* cultures. In at least some cases, these reports are very likely artifactual – one study[9] reported vegetative expression of CPE by *C. perfringens* strain ATCC 3624, while later studies from several groups have clearly demonstrated that this strain does not carry the *cpe* gene.[10,11] In the absence of the *cpe* gene, it is difficult to imagine how the CPE protein could be detected; this may reflect crossreactivity of the antibodies used in this study with a 48 kDa nonenterotoxigenic protein (see Section VIII). The relationship between CPE expression and sporulation has recently been probed using highly sensitive and specific Western immunoblots that have conclusively demonstrated that vegetative cultures of *C. perfringens* food poisoning isolate NCTC 8239 are able to produce very small amounts of CPE in the apparent absence of sporulation.[12] However, this trace-level vegetative CPE expression is probably a result of 'leaky' regulation since sporulating cultures of this same isolate routinely express greater than 1500-fold more CPE.

III. CLONING AND SEQUENCING OF THE *cpe* GENE

In 1989, three separate groups reported the cloning and sequencing of different *cpe* gene fragments from strain NCTC 8239.[13–15] Notably, van Damme-Jongsten *et al.*[13] obtained overlapping partial *cpe* clones that generated the sequence of the entire *cpe* structural gene.[13] Because none of the three groups initially isolated the intact *cpe* coding region, questions were raised as to whether the *cpe* gene product may be toxic for *Escherichia coli*.[16] This concern was resolved by a publication describing the cloning of the intact *cpe* gene in *E. coli*.[12]

The DNA sequence derived from the complete *cpe* gene predicts that CPE is a polypeptide of 319 amino acids and 35 317 Da;[12] based upon the few *cpe* genes sequenced to date, the CPE sequence appears to be conserved among different enterotoxigenic isolates, although there may be a few microsequence variations. Initial *cpe* sequence comparisons revealed no significant levels of homology to other proteins. Recently, however, an open reading frame with significant homology to *cpe*, designated *Antp70/C*1, was identified in the *bont/C*1 locus of *C. botulinum*.[17] The *Antp70/C*1 gene shows 27.6% identity and 47.4% similarity to the 259 *N*-terminal amino acids of CPE.[17] The functional role of the Antp70/C1 protein and the significance of its homology with CPE have not been determined. Based upon its coding sequence, CPE (and perhaps Antp70/C1) can be considered a distinct member of the bacterial toxin family (see Chapter 17).

If the three published sequences for the *cpe* gene from strain NCTC 8239 are compared, several minor differences are notable in the region immediately upstream of the *cpe* structural gene, usually involving an insertion or substitution of a single base pair.[11] These differences suggest heterogeneity in the *cpe* flanking sequence purportedly derived from a single strain, perhaps owing to continued subculturing in individual laboratories.[12] Moreover, when the sequences directly upstream of the *cpe* gene were compared for six *cpe*+ *C. perfringens* strains,[11] a 45 bp insertion about 265 nucleotides upstream of the start of the *cpe* gene was detected in half of these strains, indicating that there is a mixed population of upstream regions in naturally enterotoxigenic isolates.[11]

Immediately downstream from the coding sequence is a region that can potentially form a hairpin structure, followed by a string of thymidine residues. This structure is characteristic of rho-independent transcriptional terminators, and may also play a role in message stability.[12]

Recent studies suggest that the *cpe* gene may lie within or adjacent to a mobile genetic element.[13–20] Mapping studies of the *C. perfringens* genome of strain 8–6 (a derivative of strain NCTC 8798) located its *cpe* gene on a 5 kb *Nru*I fragment within a variable region of the chromosome.[18] In addition, sequencing of regions upstream of the *cpe* gene in strain NCTC 8239 has identified three open reading frames (ORFs), one of which, ORF3, shows strong homology to an ORF from the *Salmonella typhimurium* insertion sequence IS*200* (Figure 27.2).[19] Interestingly, ORF3 has been found to precede the *cpe* ORF in all but one of the enterotoxigenic *C. perfringens* strains examined to date[19,20] but the significance of this is not yet understood.

Very recently, Cornillot *et al.*[20] studied 29 *cpe*+ *C. perfringens* isolates and demonstrated that the *cpe* gene can reside either on the chromosome or on a large plasmid ($\geqslant$ 100 kb). All of the episomally located *cpe* genes lacked a 5 kb *Nru*I fragment found at the chromosomal *cpe* location, suggesting significant rearrangements had occurred. Further, in seven of the 25 strains shown to contain a plasmid-borne *cpe* gene, there was an insertion sequence element, IS*1151* (Figure 27.2),[21] associated with the toxin gene, but this element was never associated with the chromosomally located genes (see Chapter 4).[20]

IV. TRANSCRIPTIONAL CONTROL OF *cpe* EXPRESSION DURING THE GROWTH CYCLE

During the sporulation cycle CPE synthesis appears to be regulated at the transcriptional level, as shown by several lines of evidence. First, when *C. perfringens* strain NCTC 10240 was inoculated into DS sporulation medium (Figure 27.3A),[11] only very low levels of *cpe* mRNA and no CPE protein could be detected in the exponential growth phase (0–4 h of growth in DS). However, the levels of both *cpe* mRNA and protein rose rapidly, and in synchrony, about

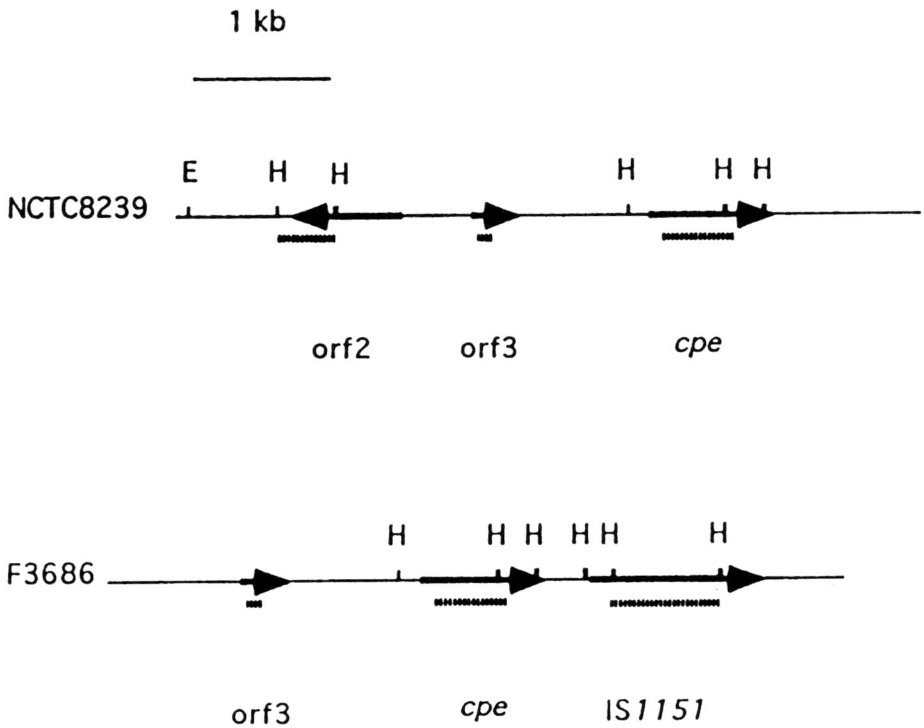

Figure 27.2 Organization of the respective *cpe* loci discussed in the text. The upper part shows the organization of the *cpe* region of the chromosome of strain NCTC 8239 while the lower part shows the *cpe* locus present on a plasmid in strain F3686. Sites for restriction enzymes *Eco*RI (E) and *Hin*dIII (H) are indicated, as are the positions of orf2, orf3, *cpe* and IS*1151*. From Brynestad *et al.*[19] and Cornillot *et al.*[20]

1 h after the cells entered stationary phase (about 5 h after inoculation into DS), indicating that the increase in CPE expression as cells enter sporulation results from increased transcription of the *cpe* gene.

Further evidence of transcriptional control was seen in a translational fusion made between DNA containing the putative *cpe* promoter region and 5′ end of the *cpe* ORF from strain NCTC 10240 and the coding region of the *gusA* gene from *E. coli*.[22] This fusion was cloned into the *C. perfringens–E. coli* shuttle

---

Figure 27.3 (Opposite) A, Synthesis of *cpe* mRNA (squares), and CPE protein (solid circles) during the growth cycle (OD_{600}, open circles) of *Clostridium perfringens* NCTC 10240 grown in DS medium. From reference by Melville *et al.*[11] B, Expression of NCTC 10240 *cpe–gusA* (solid circles) and the vector lacking the *cpe* promoter region (solid squares) in *C. perfringens* NCTC 8798 grown in DS medium. OD_{600} of NCTC 10240 *cpe–gusA* (open circles) and the vector alone (open squares). From Melville *et al.*[11]

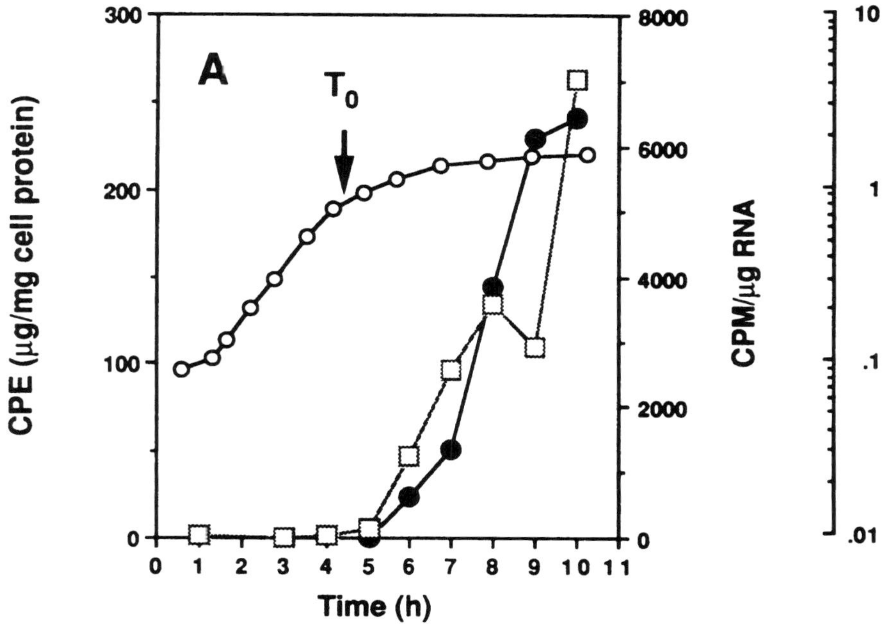
A
T0
CPE (μg/mg cell protein)
CPM/μg RNA
OD600
Time (h)

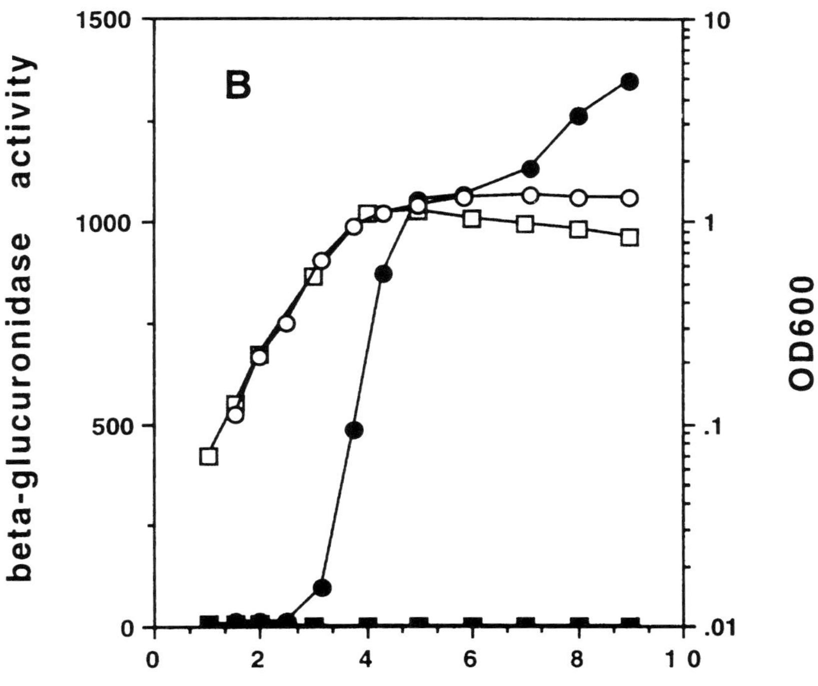
B
beta-glucuronidase activity
OD600

vector pJIR750 and introduced into strain NCTC 8798 by electroporation. Consistent with the direct measurements of *cpe* mRNA and protein, only very low levels of β-glucuronidase were detected during the exponential phase however, expression of this enzyme from the *cpe* promoter rose dramatically about 1 h after the cells entered stationary phase (Figure 27.3B).

Northern blot analysis of *cpe* mRNA synthesis also supports the conclusion that *cpe* is regulated at the level of transcription. Czeczulin *et al.*[23] have recently demonstrated that a *cpe*-specific probe hybridizes to a single, *c.* 1.2 kb species present in RNA samples extracted from sporulating, but not vegetative cultures, of strain NCTC 8239. These results also suggest that *cpe* is transcribed as a monocistronic message.

The molecular basis for transcriptional regulation of CPE expression remains unknown. However, during their analysis of *cpe* flanking regions, Brynestad *et al*,[19] also identified three pairs of perfect Hpr protein (a transition-state regulator in *Bacillus subtilis*) consensus-binding sites in the region between ORF3 and the *cpe* gene and another perfect Hpr consensus binding site 3′ to *cpe*. These findings led the authors to suggest that an Hpr-like homolog in *C. perfringens* may be involved in the transcriptional regulation of CPE expression.

V. *cpe* PROMOTER(S)?

Because of the abundance of A/T base pairs in intergenic regions of *C. perfringens* DNA, it is futile to attempt to identify promoters based on sequence analysis alone.[24] However, comparing sequences upstream of transcriptional start sites that have been identified by experimental means (such as primer extension and RNase protection assays) should help identify the *cpe* promoter.

Primer extension experiments with *cpe* mRNA isolated from sporulating cells of strains NCTC 8798 and NCTC 10240 indicated the presence of several 5′ ends *in vivo*, with the longest extension product mapping to a position about −200, relative to the start of the *cpe* gene coding sequence, and two clusters of 5′ ends located at positions −69 and −55.[11] Only the presence of 5′ ends in the clusters at positions −69 and −55 could be confirmed by RNase T2-protection assay (S.B. Melville and A.L. Sonnenshein, unpublished results); in addition, the DNA sequence just upstream of the 5′ end of the longest extension product does not show any striking homology to previously identified promoter-recognition sequences assigned to σ-factors from Gram-positive organisms.[11,25] Therefore, it remains uncertain whether the longest extension product represents the major promoter driving *cpe* expression. Mutational analysis of the regions upstream of the *cpe* gene will be required to confirm the precise location of the *cpe* promoter.

As described in Section III, DNA sequences upstream of the *cpe* structural gene contain more strain-specific differences than have been seen within the coding region of the *cpe* gene itself.[11] Different CPE+ strains, sporulating to the same degree, appear to produce widely different levels of CPE protein, ranging from barely detectable amounts to up to 30–40% of the total protein in cell-free extracts (R. Collie and B.A. McClane, unpublished results).[26] It is noteworthy that in several CPE+ strains, CPE accumulates to high enough concentrations in the mother cell to form well-defined inclusion bodies that are readily visible by electron microscopy (Table 27.1).[8,27] Whether these differences in CPE expression levels result from differences in promoter strength or the presence of additional regulatory factors in some strains has not been determined.

Some minor differences in the timing of CPE expression between two strains of *C. perfringens* have been observed; strain NCTC 10240 shows induction of CPE synthesis 1 h after entering stationary phase (Figure 27.3), while strain NCTC 8798 exhibits a sharp increase in CPE expression 1 h before entering into stationary phase.[11] This becomes especially interesting when one considers that a translational fusion containing the putative *cpe* promoter DNA derived from strain NCTC 10240 is expressed at the same relative time (i.e. after 1 h in stationary phase) in both strain NCTC 10240 and the heterologous strain NCTC 8798 (Figure 27.3),[11] suggesting that the timing of the initiation of *cpe* expression may be more dependent on promoter DNA sequences than the presence of strain-specific regulatory factors (see next section).

VI. CPE EXPRESSION IN *cpe*− STRAINS

To assess the distribution of *cpe* regulatory factors among *C. perfringens* (most of which are known to lack the *cpe* gene; see Section IX), Czeczulin *et al.*[23] recently introduced the *cpe* ORF into three naturally *cpe*− *C. perfringens* isolates (ATCC 3624, a type A strain; PS49, a type B strain; and CN 5383, a type C strain) and then checked whether these isolates could produce CPE and, if so, whether this expression occurred during vegetative growth or sporulation. Results from these studies indicated that all three naturally *cpe*− isolates expressed CPE in a sporulation-dependent manner similar to the naturally enterotoxigenic *C. perfringens* strains. This suggests that some of the regulatory factor(s) involved in CPE expression are widely distributed among *C. perfringens*, including type B isolates, which are not known to naturally carry the *cpe* gene. This distribution appears consistent with these regulatory factor(s) involved in CPE expression also being involved in the regulation of other *C. perfringens* genes.

During these studies, the level of CPE expression by each naturally *cpe*− transformant was observed to correlate directly with its sporulation levels, a result that further supported the close relationship between sporulation and CPE expression discussed in Section II. This conclusion has been independently confirmed by Melville and Sonenshein (unpublished data), who could not detect sporulation-dependent *cpe* expression when a *cpe–lacZ* translational fusion (using the *lacZ* gene from *E. coli*) was introduced into the naturally *cpe*− *C. perfringens* strain 13, a strain which produces only a very low number of spores in sporulation media such as DS. The limited spore-forming ability of strain 13 and its easy transformability may make this strain valuable as a transformation host for experiments designed to identify *cpe* regulatory factors.

VII. CPE EXPRESSION IN *ESCHERICHIA COLI* AND *BACILLUS SUBTILIS*

Despite recent advances in the understanding of the molecular genetics of *C. perfringens* (see Chapters 4 and 5), it is still not nearly as well understood as the genetics of *E. coli* and *B. subtilis*. Therefore, expression of CPE in *E. coli* could be useful as a heterologous host system for identifying gene products that regulate *cpe* expression. Czeczulin *et al.*[12] observed 'moderate' levels of CPE expression from a clostridial promoter when a 5.7 kb *Xba*I *C. perfringens* DNA fragment containing the *cpe* gene from strain NCTC 8239 was cloned into *E. coli* on the multicopy plasmid pUC19. When the same *cpe*-containing *Xba*I fragment was introduced into *E. coli* on a lower copy-number shuttle plasmid, however, no CPE expression was detected by these *E. coli* transformants, even though naturally nonenterotoxigenic *C. perfringens* transformants carrying the same shuttle plasmid were able to produce significant levels of CPE; this suggests that expression of the *cpe* gene in *E. coli* involves a strong copy number effect.[23] When synthesized in *E. coli*, the CPE protein has identical immunological and biological properties to that made by *C. perfringens*,[12] indicating that there is no proteolytic processing of the CPE polypeptide in *E. coli*. CPE accumulates intracellularly in *C. perfringens*, and so it is interesting that most of the toxin produced in *E. coli* localizes to the periplasm, suggesting that CPE is 'transported' across the cytoplasmic membrane, but not the outer membrane, of this Gram-negative bacterium. The mechanism responsible for the export of CPE to the periplasm of *E. coli* remains unknown, but may be related to the membrane-active biochemical properties of CPE (see Chapter 23).

Because CPE expression is strongly sporulation-associated in *C. perfringens*, it was also of interest to determine whether transformants of *B. subtilis*, another Gram-positive, spore-forming bacterium, would regulate *cpe* expression in the same manner as *C. perfringens*. If so, the powerful tools of *B. subtilis* genetics (including many well-defined sporulation and stationary phase regulatory protein mutants) could then be applied to determine which gene products are responsible for *cpe* regulation. To this end, transcriptional[11] and translational (S.B. Melville and A.L. Sonnenshein, unpublished data) fusions between the putative *cpe* promoter DNA from strains NCTC 8798 and NCTC 10240 and the *E. coli gusA* gene were placed on the *B. subtilis* chromosome in single copy by homologous recombination at the *amyE* locus. Unfortunately, the pattern of *cpe* expression seen in *B. subtilis* was quite different from that in *C. perfringens* (Figure 27.3B): there was low-level expression during exponential growth, followed by only a threefold induction as the cells entered stationary phase (this contrasts with the >100-fold induction of expression of this reporter construct observed for *C. perfringens* as the cells entered stationary phase). In addition, mutations in *B. subtilis* genes that encoded well-studied sporulation and stationary phase regulatory proteins such as Spo0A, AbrB, SigB, SigE, and SigH did not significantly affect expression of these fusions during the *B. subtilis* growth cycle (S.B. Melville and A.L. Sonneshein, unpublished data).[11] A likely explanation for why *B. subtilis* fails to regulate expression of these fusions in the same fashion as *C. perfringens* is that *B. subtilis* appears to utilize a different *cpe* promoter from *C. perfringens*.[11]

VIII. THE EXPRESSION OF CPE BY ENTEROTOXIGENIC *CLOSTRIDIUM PERFRINGENS*

As mentioned in Section I, there are several controversial aspects of CPE regulation and biological function. In addition to the issue of vegetative CPE expression addressed in Section II, these subjects include the localization of CPE,[28,29] reports of CPE synthesis in presumptive CPE− strains,[9,30] and the possible synthesis of a 52 kDa precursor of the mature 34 kDa enterotoxin protein.[31]

In the early 1970s, Frieben and Duncan[28,29] showed that some extracted coat proteins from mature *C. perfringens* spores reacted with 'CPE-specific' antibodies, and suggested that CPE may be an integral spore coat protein. In two later papers, Ryu and Labbe[32,33] used more specific exraction methods to isolate spore coat proteins from mature spores of CPE+ and CPE− *C. perfringens* strains, and clearly demonstrated by Western blotting that the spore coat of CPE+ strains contains two proteins that react with 'CPE-specific' antibodies. One of these proteins is 34 kDa, and is presumably the CPE protein, while the other protein migrates as a 48 kDa species. Interestingly, spores from four out of seven CPE− strains also contained an immunoreactive 48 kDa species. Since the cloned *cpe* gene is not large enough to encode a 48 kDa protein,[12,19] this species probably represents a spore-associated protein that shares an epitope(s) with CPE and thus exhibits crossreactivity in immunological assays. Ryu and Labbe[33] also concluded that the amount of 34 kDa CPE associated with spore coats is small, suggesting that this CPE may be simply trapped during spore assembly.[32] Consistent with this hypothesis, immunostaining of thin sections in electron microscopy studies has indicated that the vast majority of CPE synthesis occurs in the mother cell.[27] Therefore, contrary to Frieben and Duncan's early speculation, CPE does not appear to be a major structural component of spore coats. Further, since many *cpe* strains form normal spores, it is obvious that CPE is not an essential structural component of the spore coat.

This 48 kDa crossreactive protein may account for the occasional reports (see Section II) describing CPE synthesis in strains later shown to be *cpe*−, especially since, in these reports, CPE expression was assayed by serological techniques such as immunoelectrophoresis or gel diffusion assays that would not specifically demonstrate the presence of a 34 kDa protein. The use of highly specific CPE Western blotting techniques,[10,33] which separate proteins according to size should, in future, eliminate such confusion.

Using an *in vitro* protein synthesis system containing polysomes isolated from sporulating *C. perfringens* cultures, Smith and McDonel[31] identified a 52 kDa protein that was crossreactive with CPE antibodies. A 34 kDa protein that reacted with CPE antibodies (presumably CPE), was also detected in this system, leading these investigators to hypothesize that the 52 kDa species may be a precursor of CPE. It is now apparent that this hypothesis cannot be correct since, as mentioned above, the cloned *cpe* gene is too small to encode a protein of 52 kDa. Perhaps this species is the same 48 kDa spore-associated protein discussed above or, alternatively, it could represent aggregated enterotoxin, since one of the aggregates sometimes formed during sodium dodecyl sulfate–polyacrylamide gel electrophoresis (SDS–PAGE) of purified CPE is a 52 kDa species.[34]

IX. MOLECULAR EPIDEMIOLOGY

van Damme-Jongsten *et al.*[13] used *cpe*-specific probes to test for the presence of the *cpe* gene in 98 strains of *C. perfringens* isolated from various animal sources and found that DNA from only 6% of these isolates hybridized with the probes, suggesting that most strains, at least from animal sources, lack the *cpe* gene. Hybridization of these *cpe* probes with DNA from food-poisoning strains gave a 60% positive response;[35] this 60% figure seems somewhat low considering that CPE is believed to be the causative factor of *C. perfringens* food poisoning. However, some of the isolates tested in the van Damme-Jongsten *et al.* study may have been *cpe*−, normal flora *C. perfringens* isolates (only a single isolate was tested from each outbreak). Other studies[12] have found sequence differences in *cpe* regions used to design some of the probes used by van Damme-Jongsten *et al.*; these differences could potentially interfere with probe hybridization and produce an underestimation of the number of *cpe*+ isolates for the food poisoning and animal isolates. In response to this concern, polymerase chain reaction (PCR) and gene probe assays based on the *cpe* cloning/sequencing work of Czeczulin *et al.* have also been developed; studies using these assays[10] have confirmed that *cpe*+ isolates appear to represent only a small minority of the *C. perfringens* population since only 16/454 (*c.* 4%) of veterinary (and some environmental) isolates tested in this study were found to carry the *cpe* gene.

Gene detection methods may improve the diagnosis of *C. perfringens* type A food poisoning outbreaks. Because the traditional bacteriological methods for identifying this food poisoning are nonspecific and suffer limitations,[36] there is interest in updating the diagnostic criteria for *C. perfringens* type A food poisoning to include demonstrating whether food or fecal *C. perfringens* isolates are enterotoxigenic. Given that CPE expression is sporulation-associated, demonstration of the enterotoxigenicity of *C. perfringens* isolates by serologic techniques requires that an isolate sporulate *in vitro*; because many *C. perfringens* strains sporulate poorly *in vitro*, a significant percentage of enterotoxigenic isolates will be missed by serologic assays. However, since *in vitro* sporulation is not required to detect the *cpe* gene, gene detection assays give a significantly improved rate of detection of potentially enterotoxigenic *C. perfringens* isolates compared with these serologic assays (Table 27.2).

It is becoming increasingly apparent that merely detecting the presence of *cpe*+ *C. perfringens* isolates in food or feces will not suffice to establish a *C. perfringens* type A food poisoning outbreak. As mentioned in Section III, Cornillot *et al.*[20] recently found that some *C. perfringens* isolates contain a plasmid-borne copy of *cpe*, while others contain a chromosomal copy. Interestingly, all strains with a chromosomal copy of the *cpe* gene in this study were known to be the cause of human enteritis outbreaks, while all the strains with plasmid-borne *cpe* genes were isolated from livestock. If confirmed, these results would suggest that there are differences in pathogenicity between *C. perfringens* isolates carrying chromosomal versus plasmid copies of *cpe*, and that it may be important to demonstrate that food or fecal isolates carry a chromosomal *cpe* gene in order to establish an outbreak of *C. perfringens* type A food poisoning.

The presence of the *cpe* gene on an episome (i.e., a plasmid or episomally replicating phage) begs the question of whether this episome is transferable between strains of *C. perfringens* or even across species lines. There is some evidence that *C. perfringens* has transferred genetic material with other Gram-positive organisms, such as *Enterococcus faecalis* and *Clostridium difficile*,[37] although there is currently no evidence that any isolate of these two organisms is able to express CPE. If the *cpe* gene is readily transferred, even within the *C.*

Table 27.2. Comparison of polymerase chain reaction (PCR) and *TaqI–StyI* gene probe results with Western immunoblot results for previously uncharacterized *C. perfringens* isolates

Strain	Sporulation[a]		Western blot[b]		*TaqI–StyI* gene probe[b]	PCR[b]	Isolate origin
	DS[c]	RC[d]	DS[c]	RC[d]			
5	Low	Low	+	+	+	+	Human
153	High	Low	+	+	+	+	Porcine
155	Med	Med	+	+	+	+	Porcine
222	High	Med	+	+	+	+	Canine
382	Low	Med	+	+	+	+	Canine
452	Low	Low	+	+	+	+	Canine
458	Med	Low	+	+	+	+	Canine
CP-1	sp−	sp−	−	−	+	+	Porcine
157	sp−	sp−	−	−	+	+	Canine
455	sp−	sp−	−	−	+	+	Canine
456	sp−	sp−	−	−	+	+	Canine
457	sp−	sp−	−	−	+	+	Canine
75	Med	Med	−	−	−	−	Avian
406	sp−	Low	−	−	−	−	Porcine
424	High	Low	−	−	−	−	Canine
425	Low	sp−	−	−	−	−	Bovine
427	High	Low	−	−	−	−	Bovine
428	Low	Med	−	−	−	−	Feline
125	sp−	sp−	−	−	−	−	Human
126	sp−	sp−	−	−	−	−	Human
127	sp−	sp−	−	−	−	−	Human
150	sp−	sp−	−	−	−	−	Avian
414	sp−	sp−	−	−	−	−	Ovine
416	sp−	sp−	−	−	−	−	Equine

[a] Degree of culture sporulation (as assessed by phase-contrast microscopy): sp−, no sporulation; low, <25% of cells in culture; Med, 25 to 75% of cells in culture; High, >75% of cells in culture (average of four experiments for each strain).
[b] Results shown are based upon at least three independent determinations. Consistent results were obtained for all samples tested by each assay.
[c] Cells grown in Duncan–Strong sporulation medium.
[d] Cells grown in raffinose–caffeine-modified Duncan–Strong sporulation medium.
Data from Kokai-Kun *et al.*[10]

perfringens population, this would have important implications for understanding the epidemiology of the disease and indicate an increased role in the future for molecular approaches to *cpe* detection.

X. THE FUTURE

This chapter has described several features of the regulation of enterotoxin synthesis during sporulation in *C. perfringens*. While some aspects of this regulation have become clearer in recent years, many fascinating questions remain to be answered. For example, what are the transcriptional regulatory factors that control *cpe* expression, and are they also directly involved in the sporulation process? One possibility is that sporulation-dependent σ-factors are involved in the control of *cpe* expression such that CPE is expressed only at the proper stage in the sporulation cycle. A recent report[38] describing the cloning and sequencing of the genes for the sporulation sigma factors SigE, SigG and SigK from *Clostridium acetobutylicum* (a close relative of *C. perfringens*) should make it relatively straightforward to clone these same genes from *C. perfringens*. Making mutations in the genes that code for these σ-factors will help define the stage at which these sigma factors are active during sporulation, and these mutants can then be tested for altered *cpe* expression.

Another area which remains to be explored is the effect enterotoxin synthesis has on the growth and survival of *C. perfringens* in the animal gastrointestinal tract. The development of isogenic mutants will allow clearer determination of the importance of CPE in enteric illnesses associated with *C. perfringens* type A, and may address questions such as whether CPE production somehow enhances the survival of the bacterium in the gastro-intestinal tract. These questions are currently being studied in several laboratories.

XI. REFERENCES

1. Duncan C.L. and Strong D.H. (1968) Improved medium for sporulation of *Clostridium perfringens*. *Appl Microbiol* **16**: 82–89.
2. Duncan C.L. and Strong D.H. (1969) Ileal loop fluid accumulation and production of diarrhea in rabbits by cell-free products of *Clostridium perfringens*. *J Bacteriol* **100**: 86–94.
3. Hauschild A.H.W., Niilo L. and Dorward W.J. (1970) Enteropathogenic factors of food-poisoning *Clostridium perfringens* type A. *Can J Microbiol* **16**: 331–338.
4. Hauschild A.H.W. and Hilsheimer R. (1971) Purification and characteristics of the enterotoxin of *Clostridium perfringens* type A. *Can J Microbiol* **17**: 1425–1433.
5. Stark R.L. and Duncan C.L. (1972) Purification and biochemical properties of *Clostridium perfringens* type A enterotoxin. *Infect Immun* **6**: 662–673.
6. Duncan C.L. (1973) Time of enterotoxin formation and release during sporulation of *Clostridium perfringens* type A. *J Bacteriol* **113**: 932–936.
7. Duncan C.L., Strong D.H. and Sebald M. (1972) Sporulation and enterotoxin production by mutants of *Clostridium perfringens*. *J Bacteriol* **110**: 378–391.
8. Duncan C.L., King G.J. and Frieben W.R. (1973) A paracrystalline inclusion formed during sporulation of enterotoxin-producing strains of *Clostridium perfringens* type A. *J Bacteriol* **114**: 845–859.
9. Goldner, S.B., Solberg M., Jones S. and Post L.S. (1986) Enterotoxin synthesis by nonsporulating cultures of *Clostridium perfringens*. *Appl Environ Microbiol* **52**: 407–412.
10. Kokai-Kun J.F., Songer J.G., Czeczulin J.R., Chen F. and McClane B.A. (1994) Comparison of Western immunoblots and gene detection assays for identification of potentially enterotoxigenic isolates of *Clostridium perfringens*. *J Clin Microbiol* **32**: 2533–2539.
11. Melville S.B., Labbe R. and Sonenshein A.L. (1994) Expression from the *Clostridium perfringens* *cpe* promoter in *C. perfringens* and *Bacillus subtilis*. *J Bacteriol* **62**: 5550–5558.

12. Czeczulin J.R., Hanna P.C. and McClane B.A. (1993) Cloning, nucleotide sequencing, and expression of the *Clostridium perfringens* enterotoxin gene in *Escherichia coli. Infect Immun* **61**: 3429–3439.
13. van Damme-Jongsten M., Werners K. and Notermans S. (1989) Cloning and sequencing of the *Clostridium perfringens* enterotoxin gene. *Antonie Leeuwenhoek* **56**: 181–190.
14. Hanna P.C., Wnek A.P. and McClane B.A. (1989) Molecular cloning of the 3′ half of the *Clostridium perfringens* enterotoxin gene and demonstration that this region encodes receptor-binding activity. *J Bacteriol* **171**: 6815–6820.
15. Iwanejko L.A., Routledge M.N. and Stewart G.S.A.B. (1989) Cloning in *Escherichia coli* of the enterotoxin gene from *Clostridium perfringens* type A. *J Gen Microbiol* **135**: 903–909.
16. Granum P.E. and Stewart G.S.A.B. (1993) Molecular biology of *Clostridium perfringens* enterotoxin. In Sebald M. (ed.), Molecular Biology of Anaerobic Bacteria, pp. 235–247. Springer-Verlag, New York.
17. Hauser D., Eklund M.W., Boquet P. and Popoff M.R. (1994) Organization of the botulinum neurotoxin C1 gene and its associated non-toxic protein genes in *Clostridium botulinum* C 468. *Mol Gen Genet* **243**: 631–640.
18. Canard B., Saint-Joanis B. and Cole S.T. (1992) Genomic diversity and organization of virulence genes in the pathogenic anaerobe *Clostridium perfringens. Mol Microbiol* **6**: 1421–1429.
19. Brynestad S., Iwanejko L.A., Stewart G.S.A.B. and Granum P.E. (1994) A complex array of Hpr consensus DNA recognition sequences proximal to the enterotoxin gene in *Clostridium perfringens* type A. *Microbiol* **140**: 97–104.
20. Cornillot E., Saint-Joanis B., Daube G., Katayama S., Granum P.E., Canard B. and Cole S.T. (1995) The enterotoxin gene (*cpe*) of *Clostridium perfringens* can be chromosomal or plasmid-borne. *Mol Microbiol* **15**: 639–647.
21. Daube G., Simon P. and Kaeckenbeeck A. (1993) IS*1151*, an IS-like element of *Clostridium perfringens. Nucl Acids Res* **21**: 352.
22. Jefferson R.A., Burgess S.M. and Hirsh D. (1986) β-Glucuronidase from *Escherichia coli* as a gene-fusion marker. *Proc Natl Acad Sci USA* **83**: 8447–8451.
23. Czeczulin J.R., Collie R.E. and McClane B.A. (1996) Regulated expression of the *Clostridium perfringens* enterotoxin gene (*cpe*) after its electroporation into naturally *cpe*-negative type A, B, and C isolates of *C. perfringens. Infect Immun* **64**: 3301–3309.
24. Rood J.L. and Cole S.T. (1991) Molecular genetics and pathogenesis of *Clostridium perfringens. Microbiol Rev* **55**: 621–648.
25. Haldenwang W.G. (1995) The sigma factors of *Bacillus subtilis. Microbiol Rev* **59**: 1–30.
26. Labbe R.G. and Rey D.K. (1979) Raffinose increases sporulation and enterotoxin production by *Clostridium perfringens* type A. *Appl Env Microbiol* **37**: 1196–1200.
27. Roper G., Short J.A. and Walker D.A. (1976) The ultrastructure of *Clostridium perfringens* spores. In Baker A.M., Wold J., Ellar D.J., Dring G.H. and Gould G.W. (eds) *Spore Research*, pp. 279–296. Academic Press, London.
28. Frieben W.R. and Duncan C.L. (1973) Homology between enterotoxin protein and spore structural protein in *Clostridium perfringens* type A. *Eur J Biochem* **39**: 393–401.
29. Frieben W.R. and Duncan C.L. (1973) Heterogeneity of enterotoxin-like protein extracted from spores of *Clostridium perfringens* type A. *Eur J Biochem* **55**: 455–463.
30. Granum P.E., Telle W. and Stavn O. (1984) Enterotoxin formation by *Clostridium perfringens* during sporulation and vegetative growth. *Int J Food Microbiol* **1**: 43–49.
31. Smith W.P. and McDonel J.L. (1980) *Clostridium perfringens* type A: *in vitro* system for sporulation and enterotoxin synthesis. *J Bacteriol* **144**: 306–311.
32. Ryu S. and Labbe R.L. (1989) Coat and enterotoxin-related proteins in *Clostridium perfringens* spores. *J Gen Microbiol* **135**: 3109–3118.

33. Ryu S. and Labbe R.L. (1993) A 48 kilodalton enterotoxin-related protein from *Clostridium perfringens* vegetative and sporulating cells. *Int J Food Microbiol* **17**: 343–348.
34. McClane B.A., Wnek A.P., Hulkower K.I. and Hanna P.C. (1988) Divalent cation involvement in the action of *Clostridium perfringens* enterotoxin: early events in enterotoxin action are divalent cation independent. *J Biol Chem* **263**: 2433–2435.
35. van Damme-Jongsten M., Rodhouse J., Gilbert R.J. and Notermans S. (1990) Synthetic DNA probes for detection of enterotoxigenic *Clostridium perfringens* strains isolated from outbreaks of food poisoning. *J Clin Microbiol* **28**: 131–133.
36. McClane B.A. (1992) *Clostridium perfringens* enterotoxin: structure, action and detection. *J Food Safety* **12**: 237–252.
37. Young M. and Cole S.T. (1993) *Clostridium.* In Sonenshein A.L., Hoch J.A. and Losick R. (eds), Bacillus subtilis *and Other Gram-positive Bacteria: Biochemistry, Physiology and Molecular Genetics*, pp. 35–52. American Society for Microbiology, Washington, D.C.
38. Sauer U., Treuner A., Buchholz M., Santangelo J.D. and Durre P. (1994) Sporulation and primary sigma factor genes in *Clostridium acetobutylicum. J Bacteriol* **176**: 6572–6582.

J: Control and Diagnosis

Chapter 28

Molecular and Immunological Methods for the Diagnosis of Clostridial Diseases

J. GLENN SONGER
Department of Veterinary Science, The University of Arizona, Tucson, AZ 85721, USA

I. INTRODUCTION

The application of molecular methods is making a significant impact on the accuracy and rapidity of diagnosis of bacterial diseases of humans and domestic animals, including those diseases caused by clostridia. The majority of work has concentrated on three species, *Clostridium difficile*, *Clostridium botulinum*, and *Clostridium perfringens*, and these will be the emphasis of this chapter. However, it should be noted that clostridial diseases remain a predominant diagnostic challenge, particularly in domestic animals.[1] Application of molecular methods which are both more sensitive and more specific than those currently used will improve the rapidity and accuracy of diagnosis, and results may challenge dogma regarding the etiology, natural history, epidemiology, and pathogenesis of clostridial diseases.

II. *CLOSTRIDIUM BOTULINUM*

Polymerase chain reaction (PCR), nucleic acid probe-based methods, and immunological assays have been applied to detection of *C. botulinum* toxin genes and

The Clostridia: Molecular Biology and Pathogenesis
ISBN 0-12-595020-9

Table 28.1. Summary of PCR assays for detection of *Clostridium botulinum, Clostridium difficile* and *Clostridium perfringens*

Species	Target	Comments	Reference(s)
C. botulinum	Neurotoxin genes	Sensitivity to 3–45 fg purified DNA, 10–1000 cfu in food or soil; unexpected PCR positives from tissues	2,3,5,6,8–11
C. difficile	Toxin A/B genes, 16S rRNA gene	≈ 100 × more sensitive than culture	44,45
C. perfringens	*cpe*	Useful for isolates not readily sporulating *in vitro*	90,91
	cpa (*plc*)	Sensitivity ≈ 5 $\times 10^2$ cfu g feces	85
	Major toxin genes	Single reactions or multiplex; ≈ 100% correlation with *in vivo* assay	79,80,82,83,86,87, 89,98–101

toxins. PCR is rapid,[2] sensitive, and specific,[3–8] and should be very convenient for direct testing of individual isolates for the presence of toxin genes (Table 28.1).[2] Degenerate primers derived from the nucleotide sequence of the portion of the neurotoxin gene that encodes the heavy chain amplified a fragment from neurotoxin-producing strains of *C. botulinum* toxin types, and from *C. barati* and *Clostridium butyricum*, but not from nontoxigenic strains of *C. botulinum*.[11] The detection limit for strains of various toxin types is in the range 3–45 fg of purified DNA (≈ 1–15 bacteria),[2,3,9,10] but from 10 to 1000 bacteria in artificially contaminated foods.[3,11] Specificity can also be excellent because primers can be designed for a given toxin type; furthermore, products are not detected after amplification of template from other clostridia[2] or from other bacteria that may be commonly encountered.

Samples of milk, infant formula, infant feces, various foods and soil were inoculated with 10, 100, or 1000 spores of *C. botulinum* types A, B, or E, and examined by PCR and mouse bioassay. The correlation of results obtained by PCR and the mouse bioassay was more than 95%; a few samples of soil and feces were PCR positive but bioassay negative,[6] suggesting that template was present but that toxin, if produced, was not present in detectable amounts. PCR was comparable in sensitivity to the mouse bioassay when used to examine environmental samples and clinical material from a foal. Botulinum toxin type B was detected by mouse bioassay in enrichments of serum, spleen, feces, and intestinal contents, while examination by PCR revealed type B neurotoxin genes in these samples and in liver enrichments.[7]

Investigation of an outbreak of avian botulism revealed a 100% correlation between results of the mouse bioassay and PCR (with enrichment of soil or tissue samples by bacteriological culture). Type C toxin was found by bioassay in only one of six samples of avian tissue examined without enrichment, but the gene was found in all six by PCR.[2]

Despite the usefulness of PCR, the importance of toxin detection by some *in vivo* or *in vitro* method cannot be safely overlooked. The appropriate genes were detected by PCR in a group of strains producing neurotoxins A, B, or E (n = 207).[5] but apparently silent type B toxin genes (based upon failure to detect biologically active type B toxin) were found in more than half of the toxin type A cultures. Thus, specificity problems may occur if strains which apparently do not produce toxin *in vitro* also fail to do so *in vivo*. In addition, sequence variation, at least among type F strains, was the source of false negative results.[8]

Immunoassays have also been applied to detection of botulinum toxin.[12–16] Many of these assays are in a standard enzyme-linked immunoassay (ELISA) format, although some have adopted unique detection systems. A chemiluminescence immunosorbent assay was about tenfold more sensitive than standard ELISA for detection of *C. botulinum* antigens.[17] Another ELISA detected *C. botulinum* toxins A, B, and E at mouse bioassay levels (< 10 pg/ml, or < 0.07 pM).[19,20] In this procedure, toxin is captured by immobilized equine antibodies and detection is by an antibody conjugated to a factor X activator from the venom of Russell's viper. The toxin–conjugate complex is detected by the action of the activator on a substrate containing coagulation factors plus alkaline phosphatase-labeled fibrinogen and solid-phase fibrinogen, with deposition of solid-phase, enzyme-labeled fibrin indicating a positive. A modification of this assay has been used to titrate antitoxic antibodies.[21] The sensitivity of a sandwich ELISA procedure was 1–2 mouse bioassay units, although the correlation between these two assays varied among toxin preparations from different manufacturers.[22] Use of a detection system employing a fiber optic-based biosensor allowed rapid detection of toxin captured by affinity-purified horse polyclonal antitoxin A antibodies (as little as 5 ng toxin/ml detected in less than 1 min).[18]

Results of these and other development efforts suggest that replacement of the mouse bioassay method by immunoassay methods and/or PCR is justified in many instances. However, results must be interpreted with caution in the light of unexpected findings (e.g. apparently 'silent' toxin genes and PCR positives from tissues not normally colonized by *C. botulinum*).

III. *CLOSTRIDIUM DIFFICILE*

Noncommercialized,[23,24] and commercial enzyme immunoassays for diagnosis of *C. difficile* infections by detection of toxin A, toxin B, or both have been described. Numerous clinical tests of these enzyme immunoassays have been performed and, while results are not identical from test-to-test, several trends are worth noting. First, bacteriological culture (accompanied by toxin testing of isolates) and tissue culture cytotoxicity assays, used alone or together, remain more sensitive than enzyme immunoassays developed to date.[25–38] Specificity, on the other hand, is comparable, between enzyme immunoassays and culture/tissue culture cytotoxicity and among immunoassays.[25–33] Thus, the negative predictive values for immunoassays tend to be high (approaching or at 100%), while the positive predictive values are somewhat lower (65–95%).[25,31,32,37] The major advantages of the enzyme immunoassays are rapidity (2–4 h, versus 1–2 days for cytotoxicity assays) and simplicity (in not requiring maintenance of cell culture facilities). In general, latex agglutination tests are favored for their ease of use and low hands-on time in the clinical laboratory,[39] but are of low sensitivity and specificity.[36,38,40]

A variety of PCR assays have been developed for diagnosis of *C. difficile* infections, some based upon amplification of the toxin A gene, some on the toxin B gene, and some on both (Table 28.1). Template preparation for detection of *C. difficile* genes has been accomplished by ion exchange after phenol chloroform extraction of stools,[41] by heating enrichment cultures in a microwave oven,[42] and by immunomagnetic separation of the organism from stools prior to PCR.[43] Using the tissue culture cytotoxicity assay as the gold standard, the sensitivity and specificity of PCR assays has been very close to 100% in most cases, whether the assay is applied to stools or cultures.[41,43–48] Five to forty per cent of cytotoxin-negative stool samples have been found PCR positive (for the toxin A gene, toxin B gene, or both).[44,48] In four of seven recovered colitis patients who eventually experienced a relapse of symptoms, fecal toxigenic *C. difficile* were detected by PCR before the relapse, even though *C. difficile* was not cultured from, and toxin B was not detected in, fecal samples collected at the same time. Therefore, PCR appears to be much more sensitive than standard bacteriologic culture (perhaps 100-fold more sensitive)[45] and is apparently of similar or greater sensitivity as cytotoxin testing.[44] PCR has also been used for ribotyping of *C. difficile*,[49,50] and although randomly primed PCR has been used, it was reported to be inferior to ribotyping as a method for classification of isolates.[51]

Nucleic acid probes for detection of *C. difficile* have also been reported. A probe based upon the toxin B gene was moderately sensitive (83%) and highly specific (100%) when compared with the tissue culture cytotoxicity assay. When used for direct examination of nearly 200 fecal samples, the correlation of results of the two assays was 96%.[52] The sensitivity of another hybridization assay evaluated on fecal samples after cultural enrichment was 100%, and the specificity 96.7%. The positive and negative predictive values were 86.7% and 100%, respectively, compared with the tissue culture cytotoxicity assay.[53]

In summary, although PCR and nucleic acid probes hold promise, at this time culture and/or toxin detection remain the tests of choice for diagnosis of *C. difficile* infections.

IV. *CLOSTRIDIUM PERFRINGENS*

The key components in diagnostic systems for diseases caused by *C. perfringens* are evaluation of clinical signs and gross and microscopic lesions, bacteriologic culture of appropriate specimens, and detection of toxins in pathologic specimens and in supernatant fluids of pure cultures (Table 28.2).[1,54,55] Molecular methods for detection of *C. perfringens* are mainly enzyme immunoassays for detection of toxins or gene probes or PCR for detection of toxin genes. *Clostridium perfringens* enterotoxin (CPE) detection methods include enzyme-linked immunosorbent assay (ELISA) and reverse passive latex agglutination (RPLA).[56–62] The standard assay currently used for CPE detection is the commercial RPLA test, although ELISA may be more reliable.[63,64] Indeed, results of clinical studies comparing RPLA with ELISA (for which commercialization is being sought) may make the latter the preferred method. An ELISA for sialidase was useful in examining specimens from patients with clostridial myonecrosis.[65]

Immunoassays for enterotoxemia-associated *C. perfringens* toxins have also been reported, including ε-toxin detection by immunoelectrophoresis,[66–68] latex agglutination,[69] immunodiffusion,[70] and ELISA.[71–78] ELISA detected $\geqslant 0.005$ U of α-toxin[74] and 1–8 ng of β-toxin/ml from cultures,[71,76] and 30 ng β-toxin/g of gut contents.[72] Results with ELISA for detection of ε- and ι-toxin have been much the same.[71,72,76] Sensitivity and specificity are 90–100% compared with the mouse protection test.[72,75,77,79]

Use of gene probes or PCR assays for detection of toxigenic *C. perfringens* is becoming more common.[80–92] Aimed at CPE detection, such methods limit or eliminate the need to maintain cell cultures for cytotoxicity testing. CPE expression is strongly sporulation-associated and many clinical isolates of *C.*

Table 28.2. Toxin production by types of *Clostridium perfringens*

	Major toxins				
Type	α	β	ε	ι	Enterotoxin
A	+				±[a]
B	+	+	+		±
C	+	+			±
D	+		+		±
E	+			+	±

[a] Enterotoxin may be produced by any of the toxinogenic phenotypes.

perfringens sporulate poorly in laboratory sporulation media. False positives (isolates with *cpe* but unable to produce CPE) may occur, but probe or PCR assays avoid false negatives from isolates that sporulate and produce CPE *in vivo* but cannot be induced to sporulate *in vitro*.

The *cpe* gene can be detected with oligonucleotide probes[83,84,92–94] or PCR assays.[86,87,89–91] A *cpe* probe detected enterotoxigenic *C. perfringens* in artificially contaminated raw beef, with a sensitivity of ≤ 10 colony forming units (cfu)/g.[95] Extensive use has been made of oligonucleotide probes for detection of *cpe*-containing *C. perfringens* in domestic animals,[92,94,96] and to detect genes for α-, β-, ε-, ι-, θ-, and μ-toxins, as well as CPE and sialidase.[84] All of a group of 750 bovine enterotoxemia strains from Belgium hybridized with probes for α-toxin and sialidase genes, most with a probe for the μ-toxin gene, and few with the *cpe* probe. A more recent study by the same group identified 11 pathologic types, based upon hybridization with probes for the same genes.[97] There were no strains of types B or E, but five strains of enterotoxigenic type D were found in sheep and deer. Such extended toxin genotyping may have value for studies of pathogenesis, or in diagnosis and prophylaxis, but this has not been confirmed.

PCR assays are sensitive and specific predictors of CPE production by individual isolates (Table 28.2).[90,91] The *cpe* gene was detected in feces after heat-shocking the specimen, culturing in sporulation medium, and examining culture supernatant fluids by PCR, and there was good correlation among bacteriological culture and assays for *cpe* and CPE.[91] The *cpe* gene was also reliably detected by both PCR and hybridization with a probe, and most of these strains produced CPE. Strains which were *cpe*+ but Cpe$^-$ did not sporulate *in vitro*.[90]

Primers derived from the sequences of the *cpa*, *cpb*, *etx*, *iA*, and *iB* genes were used in PCR assays to detect these toxin genes (Tables 28.1 and 28.2).[79,80,82,86,87,89,98] Amplification of template from nearly 500 isolates allowed them to be readily placed into genotype A (86.7%), genotype C (4%), and genotype D (9.3%).[80–82] Strains of types B and D were readily differentiated by PCR from strains of types A, C, and E.[88] The α- and β-toxin genes were detected in cultures of *C. perfringens* from 20 piglets with necrotizing enteritis, while 16 piglets with lesions other than necrotizing enteritis yielded isolates positive for the α-toxin gene only.[98]

Detection of *C. perfringens* by *cpa* amplification has been reported,[85] with a detection limit of *c.* 5×10^2 bacteria/g of feces. PCR genotyping methods have also been adapted to detection of the organism in direct enrichments of intestinal contents of affected animals.[79,83] A multiplex PCR system produced similar results,[83] although these investigators were unable to amplify all five major toxin genes in a single reaction. Meer and Songer[99–101] also developed a multiplex PCR method and amplified genes for α-, β-, ε-, and ι-toxins and CPE in a single reaction mixture. Concurrence between genotype and phenotype was

100% on a group of 86 strains (phenotypes A–E, and also enterotoxigenic type A). Eighty-three per cent of 361 field isolates were nonenterotoxigenic genotype A, 12% were enterotoxigenic genotype A, 3% were genotype C, and the remainder included strains of genotypes B, enterotoxigenic C, D, enterotoxigenic D (n = 1 each), and enterotoxigenic E (n = 3).

The 16S rDNA sequences have also been the target of PCR-based methods for detection of *C. perfringens* in food or soft drinks.[102,103] A potential drawback to these methods is lack of specificity owing to the ubiquity of *cpe* negative strains which would be 16S rDNA probe or PCR positive.

V. FUTURE DIRECTIONS

The opportunities to apply molecular methods to the diagnosis of clostridial disease are nearly limitless. Successes to date are important, and continued experience in the clinical setting will better establish the most useful procedures and techniques. For many clostridial infections, however, little, if any, effort has yet been invested in development of rapid and simple *in vitro* diagnostic methods. Examples include detection of *C. septicum* and *C. chauvoei* in myonecrosis in domestic ruminants, which have traditionally been diagnosed by use of a fluorescent antibody test. While valuable in the past, an unreliable supply of conjugate has compromised the continuing usefulness of the test. Other organisms, such as *Clostridium spiroforme*, *Clostridium colinum*, and *Clostridium piliforme*, have received little meaningful attention. Even with significant progress on diagnosis of *C. perfringens* enteric infections, few diagnostic laboratories employ these molecular methods on a routine basis. Investments in these areas are likely to provide benefits in improved health care for both humans and domestic animals.

VI. REFERENCES

1. Songer J.G. (1996) Clostridial enteric diseases of domestic animals. *Clin Microbiol Rev* **9**: 216–234.
2. Franciosa G., Fenicia L., Caldiani C. and Aureli P. (1996) PCR for detection of *Clostridium botulinum* type C in avian and environmental samples. *J Clin Microbiol* **34**: 882–885.
3. Fach P., Hauser D., Guillou J.P. and Popoff M.R. (1993) Polymerase chain reaction for the rapid identification of *Clostridium botulinum* type A strains and detection in food samples. *J Appl Bacteriol* **75**: 234–239
4. Ferreira J.L., Baumstark B.R., Hamdy M.K. and McCay. (1993) Polymerase chain reaction for detection of type A *Clostridium botulinum* in foods. *J Food Protect* **56**: 18–20.
5. Franciosa G., Ferreira J.L. and Hatheway C.L. (1994) Detection of type A, B, and E botulism

neurotoxin genes in *Clostridium botulinum* and other *Clostridium* species by PCR: evidence of unexpressed type B toxin genes in type A toxigenic organisms. *J Clin Microbiol* **32**: 1911–1917.

6. Szabo E.A., Pemberton J.M., Gibson A.M., Eyles M.J. and Desmarchelier P.M. (1994) Polymerase chain reaction for detection of *Clostridium botulinum* types A, B and E in food, soil and infant faeces. *J Appl Bacteriol* **76**: 539–545.
7. Szabo E.A., Pemberton J.M., Gibson A.M., Thomas R.J., Pascoe R.R. and Desmarchelier P.M. (1994) Application of PCR to a clinical and environmental investigation of a case of equine botulism. *J Clin Microbiol* **32**: 1986–1991.
8. Ferreira J.L., Hamdy M.K., McCay S.G., Hemphill M., Kirma N. and Baumstark B.R. (1994) Detection of *Clostridium botulinum* type F using the polymerase chain reaction. *Mol Cell Probes* **8**: 365–373.
9. Szabo E.A., Pemberton J.M. and Desmarchelier P.M. (1993) Detection of the genes encoding botulinum neurotoxin types A to E by the polymerase chain reaction. *Appl Environ Microbiol* **59**: 3011–3020.
10. Fach P., Gibert N., Griffais R., Guillou J.P. and Popoff M.R. (1995) PCR and gene probe identification of botulinum neurotoxin A-, B-, E-, F-, and G-producing *Clostridium* spp. and evaluation in food samples. *Appl Environ Microbiol* **61**: 389–392.
11. Campbell K.D., Collins M.D. and East A.K. (1993) Gene probes for identification of gene and specific identification of the botulinal types B, E, and F. *J Clin Microbiol* **31**: 2255–2262.
12. Dezfulian M. and Bartlett J.G. (1985) Detection of *Clostridium botulinum* type B toxin in the presence of a lethal substance interfering with toxin neutralization. *Diagn Microbiol Infect Dis* **3**: 105–112.
13. Dezfulian M., Hatheway C.L., Yolken R.H. and Bartlett J.G. (1984) Enzyme-linked immunosorbent assay for detection of *Clostridium botulinum* type A and type B toxins in stool samples of infants with botulism. *J Clin Microbiol* **20**: 379–383.
14. Dezfulian M., Yolken R.H. and Bartlett J.G. (1985) Rapid diagnosis of a case of infant botulism by enzyme immunoassay. *Pediatr Infect Dis* **4**: 399–401.
15. Jubb T.F., Ellis T.M. and Gregory A.R. (1993) Diagnosis of botulism in cattle using ELISA to detect antibody to botulinum toxins. *Aust Vet J* **70**: 226–227.
16. Thomas R.J. (1991) Detection of *Clostridium botulinum* types C and D toxin by ELISA. *Aust Vet* **68**: 111–113.
17. Ligieza J., Reiss J. and Michalik M. (1994) Chemiluminescence immunosorbent assay (CLISA) and a possibility of the specific detection of soluble antigens of *Clostridium botulinum* type A. *Arch Immunol Ther Ep (Warsz)* **42**: 129–133.
18. Ogert R.A., Brown J.E., Singh B.R., Shriver-Lake L.C. and Ligler F.S. (1992) Detection of *Clostridium botulinum* toxin A using a fiber optic-based biosensor. *Anal Biochem* **205**: 306–312.
19. Doellgast G.J., Triscott M., Beard G.A., Bottoms J.D., Cheng T., Roh B.H., Roman M.G., Hall P.A. and Brown J.E. (1993) Sensitive enzyme-linked immunosorbent assay for detection of *Clostridium botulinum* neurotoxins A, B, and E using signal amplification via enzyme-linked coagulation assay. *J Clin Microbiol* **31**: 2402–2409.
20. Doellgast G.J., Beard G.A., Bottoms J.D., Cheng T., Roh B.H., Roman M.G., Hall P.A. and Triscott M. (1994) Enzyme-linked immunosorbent assay and enzyme-linked coagulation assay for detection of *Clostridium botulinum* neurotoxins A, B, and E and solution phase complexes with dual-label antibodies. *J Clin Microbiol* **32**: 105–111.
21. Doellgast G.J., Triscott M., Beard G.A. and Bottoms J.D. (1994) Enzyme-linked immunosorbent assay-enzyme linked coagulation assay for detection of antibodies to *Clostridium botulinum* neurotoxins A, B, and E and solution-phase complexes. *J Clin Microbiol* **32**: 851–853.
22. Ekong T.A., McLellan K. and Sesardic D. (1995) Immunological detection of *Clostridium botulinum* toxin type A in therapeutic preparations. *J Immunol Methods* **180**: 181–191.
23. Muller F., Stiegler C. and Hadding U. (1992) Monoclonal antibodies specific for *Clostridium difficile* toxin B and their use in immunoassays. *J Clin Microbiol* **30**: 1544–1550.

24. Depitre C., Avesani V., Delmee M. and Corthier G. (1993) Detection of *Clostridium difficile* toxins in stools. Comparison between a new enzyme immunoassay for toxin A and other routine tests. *Gastroenterol Clin Biol* **17**: 283–286.

25. Merz C.S., Kramer C., Forman M., Gluck L., Mills K., Senft K., Steiman I., Wallace N. and Charache P. (1994) Comparison of four commercially available rapid enzyme immunoassays with cytotoxin assay for detection of *Clostridium difficile* toxin(s) from stool specimens. *J Clin Microbiol* **32**: 1142–1147.

26. Shanholtzer C.J., Willard K.E., Holter J.J., Olson M.M., Gerding D.N. and Peterson L.R. (1992) Comparison of the VIDAS *Clostridium difficile* toxin A immunoassay with *C. difficile* culture and cytotoxin and latex tests. *J Clin Microbiol* **30**: 1837–1840.

27. Barbut F., Kajzer C., Planas N. and Petit J.C. (1993) Comparison of three enzyme immunoassays, a cytotoxicity assay, and toxigenic culture for diagnosis of *Clostridium difficile*-associated diarrhea. *J Clin Microbiol* **31**: 963–967.

28. Knapp C.C., Sandin R.L., Hall G.S., Ludwig M.D., Rutherford I. and Washington J.A. (1993) Comparison of vidas *Clostridium difficile* toxin-A assay and premier *C. difficile* toxin-A assay to cytotoxin-B tissue culture assay for the detection of toxins of *C. difficile*. *Diagn Microbiol Infect Dis* **17**: 7–12.

29. Altaie S.S., Meyer P. and Dryja D. (1994) Comparison of two commercially available enzyme immunoassays for detection of *Clostridium difficile* in stool specimens. *J Clin Microbiol* **32**: 51–53.

30. Schue V., Green G.A. and Monteil H. (1994) Comparison of the ToxA test with cytotoxicity assay and culture for the detection of *Clostridium difficile*-associated diarrhoea disease. *J Med Microbiol* **41**: 316–318.

31. Arrow S.A., Croese L., Bowman R.A. and Riley T.V. (1994) Evaluation of three commercial enzyme immunoassay kits for detecting faecal *Clostridium difficile* toxins. *J Clin Pathol* **47**: 954–956.

32. DeGirolami P.C., Hanff P.A., Eichelberger K., Longhi L., Teresa H., Pratt J., Cheng A., Letourneau J.M. and Thorne G.M. (1992) Multicenter evaluation of a new enzyme immunoassay for detection of *Clostridium difficile* enterotoxin A. *J Clin Microbiol* **30**: 1085–1088.

33. Delmee M., Mackey T. and Hamitou A. (1992) Evaluation of a new commercial *Clostridium difficile* toxin A enzyme immunoassay using diarrhoeal stools. *Eur J Clin Microbiol Infect Dis* **11**: 246–249.

34. Doern G.V., Coughun R.T. and Wu L. (1992) Laboratory diagnosis of *Clostridium difficile*-associated gastrointestinal disease: comparison of a monoclonal antibody enzyme immunoassay for toxins A and B with a monoclonal antibody enzyme immunoassay for toxin A only and two cytotoxicity assays. *J Clin Microbiol* **30**: 2042–2046.

35. Barrett U., Silva E., Hirsch D. and Guerrant R.L. (1995) Concurrence of *Clostridium difficile* toxin A enzyme-linked immunosorbent assay, fecal lactoferrin assay, and clinical criteria with *C. difficile* cytotoxin titer in two patient cohorts. *J Clin Microbiol* **33**: 1755–1759.

36. Gilligan P.H., Walden T.P., Kelly W.F., Wait K.J., Kraft J.A. and Willis D.H. (1993) The use of a commercially available enzyme immunoassay for the detection of *Clostridium difficile* toxin A. *Arch Pathol Lab Med* **117**: 507–510.

37. Bowman R.A., Arrow S., Croese L. and Riley T.V. (1994) Evaluation of an enzyme immunoassay kit for the detection enterotoxin. *Pathology* **26**: 480–481.

38. Mattia A.R., Doern G.V., Clark J., Holden J., Wu L. and Ferraro M.J. (1993) Comparison of four methods in the diagnosis of *Clostridium difficile* disease. *Eur J Clin Microbiol Infect Dis* **12**: 882–886.

39. Riederer K.M., Lawson P., Held M.S., Petrylka K., Briski L.E. and Khatib R. (1995) Diagnosis of *Clostridium difficile*-associated diarrhea. *Can J Microbiol* **41**: 88–91.

40. Kelly W.F., Wait K.J. and Gilligan P.H. (1992) Evaluation of the latex agglutination test for detection of *Clostridium difficile*. *Arch Pathol Lab Med* **116**: 517–520.

41. Kato N., Ou C.U., Kato H., Bartley S.L., Luo C.C., Killgore G.E. and Ueno K. (1993) Detection of

toxigenic *Clostridium difficile* in stool specimens by the polymerase chain reaction. *J Inf Dis* **167**: 455–458.
42. Ratcliff R.M., Goodwin A.A. and Lanser J.A. (1994) Use of gene amplification to detect *Clostridium difficile* in clinical specimens. *Pathology* **26**: 477–479.
43. Wolfhagen M.J., Fluit A.C., Torensma R., Poppelier N.J. and Verhoef J. (1994) Rapid detection of toxigenic *Clostridium difficile* in fecal samples by magnetic immuno PCR assay. *J Clin Microbiol* **32**: 1629–1633.
44. Boondeekhun H.S., Gurtler V., Odd M.L., Wilson V.A. and Mayall B.C. (1993) Detection of *Clostridium difficile* enterotoxin gene in clinical specimens by the polymerase chain reaction. *J Med Microbiol* **38**: 384–387.
45. Kuhl S.J., Tang Y.J., Navarro L., Gumerlock P.H. and Silva J. Jr. (1993) Diagnosis and monitoring of *Clostridium difficile* infections with the polymerase chain reaction. *Clin Infect Dis* **16**: (Suppl. 4): S234–S238.
46. Gumerlock P.H., Tang Y.J., Weiss J.B. and Silva J. Jr. (1993) Specific detection of toxigenic strains of *Clostridium difficile* in stool specimens. *J Clin Microbiol* **31**: 507–511.
47. McMillin D.E., Muldrow L.L. and Laggette S.J. (1992) Simultaneous detection of toxin A and toxin B genetic determinants of *Clostridium difficile* using the multiplex polymerase chain reaction. *Can J Microbiol* **38**: 81–83.
48. Tang Y.J., Gumerlock P.H., Weiss J.B. and Silva J. Jr. (1994) Specific detection of *Clostridium difficile* toxin A gene sequences in clinical isolates. *Mol Cell Probes* **8**: 463–467.
49. Gurtler V. (1993) Typing of *Clostridium difficile* strains by PCR-amplification of variable length 16S–23S rDNA spacer regions. *J Gen Microbiol* **139**: 3089–3097.
50. Martirosian G., Kuipers S., Verbrugh H., van Belkum A. and Meisel-Mikolajczyk F. (1995) PCR ribotyping and arbitrarily primed PCR for typing strains of *Clostridium difficile* from a Polish maternity hospital. *J Clin Microbiol* **33**: 2016–2021.
51. Collier M.C., Stock F., De Girolami P.C., Samore M.H. and Cartwright C.P. (1996) Comparison of PCR-based approaches to molecular epidemiologic analysis of *Clostridium difficile. J Clin Microbiol* **34**: 1153–1157.
52. Green G.A., Riot B. and Monteil H. (1994) Evaluation of an oligonucleotide probe and an immunological test for direct detection of toxigenic *Clostridium difficile* in stool samples. *Eur J Clin Microbiol Infect Dis* **13**: 576–581.
53. Wolfhagen M.J., Fluit A.C., Jansze M., Rademaker C.M. and Verhoef J. (1993) Detection of toxigenic *Clostridium difficile* in fecal samples by colony blot hybridization. *Eur J Clin Microbiol Infect Dis* **12**: 463–466.
54. Hoefling D.C. (1989) Recognizing diarrhea caused by *Clostridium perfringens* type C. *Vet Med* **84**: 437–440.
55. Smith L. (1980) Clostridial infections, In Hitchner S.B., Domermuth C.H., Purchase H.G. and Williams J.E. (eds), *Isolation and Identification of Avian Pathogens*, pp. 33–35. Am Assoc Avian Pathol, Kennet Square, PA.
56. McClane B.A. and Strouse R.J. (1984) Rapid detection of *Clostridium perfringens* type A enterotoxin by enzyme-linked immunosorbent assay. *J Clin Microbiol* **19**: 112–115.
57. Berry P.R., Rodhouse J.C., Hughes S., Bartholomew B.A. and Gilbert R.J. (1988) Evaluation of ELISA, RPLA, and Vero cell assays for detecting *Clostridium perfringens* enterotoxin in faecal specimens. *J Clin Pathol* **41**: 458–461.
58. Harmon S.M. and Kautter D.A. (1986) Evaluation of a reversed passive latex agglutination test kit for *Clostridium perfringens* enterotoxin. *J Food Protect* **49**: 523–525.
59. Hunter S.E.C., Brown J.E., Oyston P.C.F., Sakurai J. and Titball R.W. (1993) Molecular genetic analysis of beta-toxin of *Clostridium perfringens* reveals sequence homology with alpha-toxin, gamma-toxin, and leukocidin of *Staphylococcus aureus. Infect Immun* **61**: 3958–3965.
60. Jackson S.G., Yip-Chuck D.A. and Brodsky M.H. (1986) Evaluation of the diagnostic application of

an enzyme immunoassay for *Clostridium perfringens* type A enterotoxin. *Appl Environ Microbiol* **52**: 969–970.

61. Notermans S., Heuvelman C., Beckers H. and Uemura T. (1984) Evaluation of the ELISA as tool in diagnosing *Clostridium perfringens* enterotoxins. *Zentralbl Bakteriol Mikrobiol Hyg B* **179**: 225–234.
62. Uemura T., Yoshitake S., Hu D and Kajikawa T. (1992) A highly sensitivie enzyme-linked immunosorbent assay for *Clostridium perfringens* enterotoxin. *Lett Appl Microbiol* **15**: 23–25.
63. Mpamugo O., Donovan T. and Brett M.M. (1995) Enterotoxigenic *Clostridium perfringens* as a cause of sporadic cases of diarrhoea. *J Med Microbiol* **43**: 442–445.
64. Brett M.M., Rodhouse J.C., Donovan T.J., Tebbutt G.M. and Hutchinson D.N. (1992) Detection of *Clostridium perfringens* and its enterotoxin in cases of sporadic diarrhoea. *J Clin Pathol* **45**: 609–611.
65. Roggentin T., Kleineidam R.G., Majewski D.M., Tirpitz D., Roggentin P. and Schauer R. (1993) An immunoassay for the rapid and specific detection of three sialidase-producing clostridia causing gas gangrene. *J Immunol Methods* **157**: 125–133.
66. Henderson T.G. (1984) The detection of *Clostridium perfringens* type D enterotoxin in the intestinal contents of animals by counterimmunoelectrophoresis. *NZ J Sci* **27**: 423–426.
67. Kumar A.A. and Uppal P.K. (1988) Quick detection and quantitation of *Clostridium perfringens* type D epsilon toxin by counterimmunoelectrophoresis. *Ind J Comp Microbiol Immunol Infect Dis* **9**: 221–222.
68. Tripathi B.N., Parihar N.S., Singh K.P. and Kumar A.A. (1992) Immunodiagnostic tests for the detection of *Clostridium perfringens* type C infection. *Ind J Animal Sci* **62**: 607–610.
69. Martin P.K. and Naylor R.D. (1994) A latex agglutination test for the qualitative detection of *Clostridium perfringens* epsilon toxin. *Res Vet Sci* **56**: 259–261.
70. Beh K.J. and Buttery S.H. (1978) Reverse-phase passive haemagglutination and single radial immunodiffusion to detect epsilon antigen of *Clostridium perfringens* type D. *Aust Vet J* **54**: 541–544.
71. El-Idrissi A.H. and Ward G.E. (1992) Development of double sandwich ELISA for *Clostridium perfringens* beta and epsilon toxins. *Vet Microbiol* **31**: 89–99.
72. El-Idrissi A.H. and Ward G.E. (1992) Evaluation of enzyme-linked immunosorbent assay for diagnosis of *Clostridium perfringens* enterotoxemia. *Vet Microbiol* **31**: 389–396.
73. El-Idrissi A.H., Ward G.E., Johnson D.W., Benkirane A. and Fassi-Fehri M.M. (1992) Bacteriological investigation of sudden sheep mortality in Morocco. *Prev Vet Med* **12**: 35–46.
74. Holdsworth R.J. and Parratt D. (1994) Development of an ELISA assay for *Clostridium perfringens* phospholipase C (alpha toxin). *J Immunosassay* **15**: 293–304.
75. Martin P.K., Naylor R.D. and Sharpe R.T. (1988) Detection of *Clostridium perfringens* beta toxin by enzyme-linked immunosorbent assay. *Res Vet Sci* **44**: 270–271.
76. Nagahama M., Kobayashi K., Ochi S. and Sakurai J. (1991) Enzyme-linked immunosorbent assay for rapid detection of toxins from *Clostridium perfringens. FEMS Microbiol Lett* **68**: 41–44.
77. Naylor R.D., Martin P.K. and Sharpe R.T. (1987) Detection of *Clostridium perfringens* epsilon toxin by ELISA. *Res Vet Sci* **42**: 255–256.
78. Sojka M.G., White V.J., Thorns C.J. and Roeder P.L. (1989) The detection of *Clostridium perfringens* epsilon antitoxin in rabbit serum by monoclonal antibody based competition ELISA. *J Biol Stand* **17**: 117–124.
79. Møller K., Ahrens P. and Meyling A. (1995) Comparison of toxicity neutralization, ELISA, and PCR tests for typing of *Clostridium perfringens* and for detection of the enterotoxin gene. *International Conference on the Molecular Genetics and Pathogenesis of the Clostridia, 1995.* (Abst. B16), p. 23. American Society for Microbiology, Washington, D.C.
80. Songer J.G. (1995) Diagnosis of clostridial infections of domestic animals. *International Conference*

on the Molecular Genetics and Pathogenesis of the Clostridia, 1995. p. 9. American Society for Microbiology, Washington, D.C.

81. Songer J.G., Hunter S.E.C. and Chen F. (1993) PCR method for detection of alpha, beta, and epsilon toxin genes of *Clostridium perfringens*, 74th Ann. Mtg. Conf. Res. Workers Animal Dis., 1993. (Abstr. 71). Conf Res Workers Anim Dis, Chicago IL.
82. Songer J.G. and Meer R.R. (1996) Genotyping of *Clostridium perfringens* by polymerase chain reaction is a useful adjunct to diagnosis of clostridial enteric disease in animals. *Anaerobe* **2** 197–203.
83. Daube G., China B., Simon P., Hvala K. and Mainil J. (1994) Typing of *Clostridium perfringens* by *in vitro* amplification of toxin genes. *J Appl Bacteriol* **77**: 650–655.
84. Daube G., Simon P., Limbourg B., Renier K. and Kaeckenbeeck A. (1994) Molecular typing of *Clostridium perfringens. Ann Med Veterinaire* **138**: 183–191.
85. Fach P. and Guillou J.P. (1993) Detection by *in vitro* amplification of the alpha-toxin (phospholipase C) gene from *Clostridium perfringens. J Appl Bacteriol* **74**: 61–66.
86. Fach P. and Popoff M.R. (1995) Development of a rapid and sensitive multiplex PCR assay for detection of enterotoxigenic *C. perfringens* in food and fecal samples. International Conference on the Molecular Genetics and Pathogenesis of the Clostridia, 1995. (Abstr. B1), p. 21. American Society for Neurobiology, Washington, DC.
87. Fach P., Delbart M.O., Schlachter A., Poumeyrol M. and Popoff M.R. (1993) Use of the polymerase chain reaction for the diagnosis of *Clostridium perfringens* foodborne diseases. *Méd Mal Infect* **23**: 70–77.
88. Havard H.L., Hunter S.E.C. and Titball R.W. (1992) Comparison of the nucleotide sequence and development of a PCR test for the epsilon toxin gene of *Clostridium perfringens* type B and type D. *FEMS Microbiol Lett* **97**: 77–82.
89. Kadra B., Fach P. and Guillou J.P. (1995) Utilisation of PCR (Polymerase chain reaction) for typing of *Clostridium perfringens* and for determination of its enterotoxin. International Conference on the Molecular Genetics and Pathogenesis of the Clostridia, 1995. (Abstr. B2), p. 19. American Society for Neurobiology, Washington, DC.
90. Kokai-Kun J.F., Songer J.G., Czeczulin J.R., Chen F. and McClane B.A. (1994) Comparison of western immunoblots and gene detection assays for identification of potentially enterotoxigenic isolates of *Clostridium perfringens. J Clin Microbiol* **32**: 2533–2539.
91. Saito M., Matsumoto M. and Funabashi M. (1992) Detection of *Clostridium perfringens* enterotoxin gene by the polymerase chain reaction amplification procedure. *Int J Food Microbiol* **17**: 47–55.
92. Tschirdewahn B., Notermans S., Wernars K. and Untermann F. (1992) The presence of enterotoxigenic *Clostridium perfringens* strains in faeces of varous animals. *Int J Food Microbiol* **14**: 175–178.
93. Schlapp T., Blaha I., Bauerfeind R., Wieler L.H., Schoepe H., Weiss R. and Baljer G. (1995) Synthesis and evaluation of a non-radioactive gene probe for the detection of *C. perfringens* alpha toxin. *Mol Cell Probes* **9**: 101–109.
94. van Damme-Jongsten M., Rodhouse J., Gilbert R. and Notermans S. (1990) Synthetic DNA probes for detection of enterotoxigenic *Clostridium perfringens* strains isolated from outbreaks of food poisoning. *J Clin Microbiol* **28**: 131–133.
95. Baez L.A. and Juneja V.K. (1995) Nonradioactive colony hybridization assay for detection and enumeration of enterotoxigenic *Clostridium perfringens* in raw beef. *Appl Environ Microbiol* **61**: 807–810.
96. van Damme-Jongsten M., Haagsma J. and Notermans S. (1990) Testing strains of *Clostridium perfringens* type A isolated from diarrhoeic piglets for the presence of the enterotoxin gene. *Vet Rec* **126**: 191–192.
97. Daube G., Simon P., Limbourg B., Manteca C., Mainil J. and Kaeckenbeeck A. (1996) Hybridization

of 2,659 *Clostridium perfringens* isolates with gene probes for seven toxins (alpha, beta, epsilon, iota, theta, mu, and enterotoxin) and for sialidase. *Am J Vet Res* **57**: 496–501.

98. Buogo C., Capaul S., Hani H., Frey J and Nicolet J. (1995) Diagnosis of *Clostridium perfringens* type C enteritis in pigs using a DNA amplification technique (PCR). *Zbl Vet Med (B)* **42**: 51–58.
99. Meer R.R. and Songer J.G. (1995) Multiplex PCR method for genotyping of *Clostridium perfringens*, 76th Ann. Mtg. Conf. Res. Workers Animal Dis., 1995. (Abstr. 28). Conf Res Workers Animal Dis, Chicago, IL.
100. Meer R.R. and Songer J.G. (1996) PCR genotyping of field isolates of *Clostridium perfringens*. Ann. Mtg. Arizona Branch Am. Soc. Microbiol., 1996. (Abstr. B1). American Society for Microbiology (Arizona branch), Tucson, Az.
101. Meer R.R. and Songer J.G. (1996) Multiplex PCR method for genotyping of *Clostridium perfringens*. *Am J Vet Res*, (in press).
102. Wang R.F., Cao W.W., Franklin W., Campbell W. and Cerniglia C.E. (1994) A 16S rDNA-based PCR method for rapid and specific detection of *Clostridium perfringens* in food. *Mol Cell Probes* **8**: 131–137.
103. Galindo I., Rangel-Aldao R. and Ramirez J.L. (1993) A combined polymerase chain reaction-colour development hybridization assay in a microtitre format for the detection of *Clostridium* spp. *Appl Microbiol Biotechnol* **39**: 553–557.

Chapter 29

Molecular Approaches to Novel Vaccines for the Control of Clostridial Toxemias and Infections

J. Edward Brown and E. Diane Williamson*

*US Army Medical Research Institute of Infectious Diseases, Fort Detrick, MD 21703, USA, and *Chemical and Biological Defence Sector, DERA, Porton Down, Salisbury, Wiltshire SP4 0JQ, UK*

I. INTRODUCTION

This chapter addresses the application of vaccines to prevent the toxaemias and associated disease induced by three major species of pathogenic clostridia: *Clostridium tetani*, *Clostridium botulinum* and *Clostridium perfringens*.

II. TETANUS VACCINE

A. Toxoid vaccine

Use of tetanus toxoid vaccine to eradicate tetanus from most of the world's population ranks as one of the outstanding achievements in infectious diseases in the twentieth century.[1] The tetanus vaccine has proved so successful in the industrialized world in eliminating tetanus that the World Health Organization (WHO) Expanded Program on Immunizations has focused on the prevention of

The Clostridia: Molecular Biology and Pathogenesis
ISBN 0-12-595020-9

neonatal tetanus as a world-wide goal. With over 500 000 newborn deaths from tetanus in 1991, the WHO/United Nations Development Program (UNDP) for Vaccine Development has established a goal of developing a tetanus vaccine requiring only one administration.[2,3] In industrialized countries, where concern with mortality from tetanus has been removed, attention is focused on the side reactions associated with vaccine boosters of the aluminum-salt adjuvantized toxoid, such as toxicity to aluminum, reactions to toxoid contaminants and hyperimmunization.

The current tetanus toxoid is a formaldehyde-inactivated preparation of partially purified toxin using aluminum hydroxide as an adjuvant. Modification of the composition, scheduling, dosage, or adjuvant of this toxoid continues in a search to improve the vaccine. Alum-adsorbed vaccine has been known for a number of years to induce increased IgE levels, usually indicative of allergy. However, a recent study comparing adsorbed to fluid phase toxoid booster immunization demonstrated that both preparations caused significant local reactions[4] and increased IgE levels.[5] In these studies, IgG responses to fluid-phase toxoid were excellent. Another approach to reducing adverse reactions might be to administer lower booster doses of toxoid. In 66 Swedish adult vaccinees, booster doses of 0.75 or 1.9 Limits of flocculation (Lf) units reduced rates of adverse reactions and provided satisfactory immunity.[6] A third modification of the toxoid preparation might be to eliminate nonessential components by toxin purification before toxoiding. In mice and guinea pigs, toxoid prepared from pure toxin had similar immunogenicity to conventional toxoid.[7] That study also showed that, after the second immunization, calcium phosphate and stearyl tyrosine adjuvants were as effective as aluminum phosphate adjuvant, without inducing higher IgE levels. In pregnant women, the antibody response after immunization with toxoid prepared from pure toxin and adsorbed to calcium phosphate was actually much higher than that after immunization with conventional toxoid.[8]

For a single dose parenteral vaccine, WHO/UNDP has focused on development of a toxoid encapsulated in polylactide-co-polyglycolide (PLA:PGA) derived polymers.[2] Microencapsulated tetanus toxoid has been compared to conventional alum-adsorbed toxoid in mice.[9] Serum antibody titers after the second injection were significantly higher in animals given microencapsulated toxoid than in animals given the conventional toxoid, demonstrating that the immunogenicity of the toxoid could be maintained during the preparation process. In studies using chromatographically purified toxoid, single doses of toxoid encapsulated in PLA:PGA (50:50) polymers produced neutralization titers significantly higher than 100% PLA-encapsulated toxoid and similar to alum-adsorbed toxoid.[10] Moreover, toxoid simply adsorbed onto PLA microspheres and administered by the nasal route produced high systemic titers after a booster, whereas free toxoid produced no significant response.[11] Finally, after a single injection, mixtures of toxoid-containing microspheres containing, respectively, 50, 75, and 100% PLA induced long-term antibody responses in mice similar to those obtained with three injections of toxoid-alum.[12] Human studies of the immune response to microencapsulated tetanus toxoid would appear warranted on the basis of these results.

B. Recombinant vaccine

Clostridial neurotoxins are synthesized as a single inactive 150 kDa peptide and are released *in situ* by bacterial lysis (see Chapters 17 and 18). Proteolytic cleavage results in the active dimer molecule, consisting of a heavy chain (100 kDa) and a light chain (50 kDa) joined by a single disulfide bond.[13]

More than two decades ago, it was shown that a fragment obtained by trypsin digestion of tetanus toxin, which was derived from the *C*-terminus of the toxin heavy chain could be used to induce protective immunity against tetanus.[14] The protective nontoxic fragment C (now termed H_C) is a 50 kDa *C*-terminal peptide of tetanus toxin (Figure 29.1). In 1986, Fairweather and colleagues[15] at Wellcome Biotechnology Ltd, who had previously sequenced the entire tetanus toxin gene, were able to identify the region coding for H_C. A recombinant plasmid was constructed which expressed a fusion of *Escherichia coli trpE* protein and tetanus H_C at around 1% of the total cell protein.[16] The purified fusion protein, when used to immunize mice, protected against 10 LD_{50} doses of toxin. The first step toward improving yield was made by construction of plasmid pTETtac2, where the initial 43 residues and terminal 11 residues of H_C were encoded from synthetic gene sequences so that the codon bias was optimized for expression in *E. coli*.[17] The yield was 3% of the total cell protein and the recombinant protein was similar in biochemical and immunogenic properties to the authentic fragment C. Expression of this hybrid synthetic-native gene sequence in a

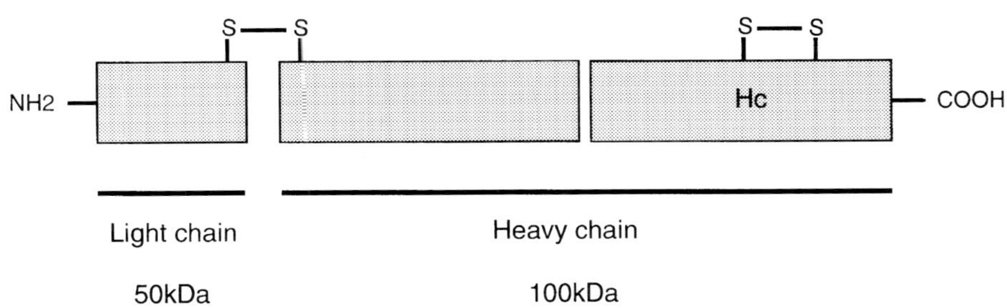

Figure 29.1 Generalized toxin structure for tetanus and botulinum toxins.

Salmonella typhimurium aroA mutant under control of the *tac* promoter resulted in a live vaccine vehicle that produced protective immunity when administered to mice either orally or intravenously.[18] When the same structural gene was incorporated into a baculovirus expression system and expressed in *Spodoptera frugiperda* cells, soluble H_C could be recovered from cell lysates as 10–12% of the total cellular protein.[19] Purified protein protected mice against at least 50 LD_{50} doses of tetanus toxin.

The immunogenic potential of tetanus H_C produced using genetic engineering has been confirmed by others. Using polymerase chain reaction (PCR), the fragment encoding H_C was amplified from *C. tetani* and cloned into plasmid pSS1261 under control of the *tac* promoter. The protein produced in *E. coli* conferred protective immunity in mice.[20] This gene fragment was cloned and expressed in BCG vaccine strain and the recombinant vaccine induced increasing antibody levels in mice for at least 12 weeks.[21] The native H_C-encoding fragment was also expressed in the Gram-positive bacterium, *Lactococcus lactis*, where up to 22% of the soluble cell protein was obtained as H_C.[22] Mice immunized with the bacterium were protected from lethal tetanus toxin challenge. Finally, the gene fragment from pSS1261 has been ligated to a mutated gene for pertussis toxin S1 subunit.[23] The 75-kDa fusion protein, expressed in *E. coli*, induced strong protective immunity to tetanus toxin and some protection against *Bordetella pertussis* infection.

Because the codon usage in *Clostridium* sp. is AT-rich, the Wellcome group chose to prepare a series of gene constructs with increasing portions of synthetic gene sequence base on an *E. coli* codon bias.[24] Complete sequence and improved promoter strength resulted in plasmid pTETtac115, where soluble expression from *E. coli* was about 11–14% cellular protein. The recombinant antigen could be incorporated into DPT vaccines.[25] This nearly synthetic gene, where rare *E. coli* codons were excluded, has been used for a number of subsequent studies in other expression systems.

The synthetic gene was used to direct both intracellular and secreted expression from *Saccharomyces cerevisiae*.[26] Fragment C that had been produced intracellularly was effective in immunizing mice against toxin challenge, but the secreted form of the protein was inactive. Deglycosylation rendered the material fully immunogenic. To achieve higher expression levels, the synthetic gene was inserted into the methylotrophic yeast *Pichia pastoris* under the control of the promoter from the methanol-induced alcohol oxidase gene, AOX1.[27] Multiple integrated copies of the H_C expression cassette led to yields of 27% of total cell protein in high biomass fermentations. One liter yielded approximately 10^5 vaccine doses and the material was as effective as native H_C at immunizing mice. More recently, the immunogenicity of smaller derivatives of fragment C, expressed as fusions with maltose-binding protein in *E. coli*, have been examined.[28] Subfragments that induce protective immunity but do not bind to neurons, such as $TetC_{1-180}$, were identified. Consideration of the use of these derivatives for vaccines was recommended because the consequences of the immunogen binding to neurons are unknown, especially if delivered by live vectors. When a series of synthetic peptides was tested for their ability to induce protective immunity, a peptide which induced complete protection was located within H_C.[29] This neutralizing epitope occurs within the same minimum protective region ($TetC_{80-180}$) deduced above.[28]

Live *Salmonella* vaccines would make excellent candidates for expression of heterologous antigens, such as tetanus fragment C. Using the synthetic fragment C gene, candidate oral tetanus vaccines have been constructed in attenuated *Salmonella typhi*[30] and *S. typhimurium*,[31] each with double mutations in *aro* genes. A single oral dose of about 5×10^9 organisms of the *S. typhimurium* construct induced protection against 50 LD_{50} doses of tetanus toxin. Further advances have produced novel vectors expressing either subunits or peptides as guest antigen fusions with tetanus H_C in *Salmonella*.[32,33] Prior immunization against tetanus toxoid appears to have no adverse effect on response to either *Salmonella* or a third antigen in these constructs.[34] Two recent reviews of these efforts are available.[35,36]

III. BOTULINUM VACCINES

A. Toxoid vaccines

The currently available toxoid vaccine for botulism induces immunity against five serotypes (types A–E) of *C. botulinum*. Several recent reviews have described the immunogenicity of this pentavalent vaccine.[37,38] Although generally accepted as a safe and effective vaccine,[39] dissatisfaction with the current formulation has increased over the years, not only because of the cost of production, but also because of the small amount, as a percentage, of individual purified toxoids incorporated and because it does not cover all seven serotypes.

The pentavalent vaccine resulted from an effort at the US Army Biological Laboratories starting during the 1940s. After a number of animal studies, initial immunization experiments in man with crude type A and type B toxoids were reported in 1947.[40] Immune responses to fluid toxoid and alum-precipitated toxoid administered by a variety of schedules were evaluated. Three injections at 3–4 week intervals were found to be the most effective. Following improvements in both culture media and toxin isolation, immunity to bivalent AB botulinum toxoid produced by Parke, Davis and Co. was reexamined in 1962.[41] An immunization schedule of 0–2–10 weeks was found to produce

100% immunity after the primary series and to result in the most favorable maintenance of long-term immunity after an annual booster. No systemic reactions were observed in approximately 800 injections, and local reactions were less frequent and milder than observed by Reames. Prior immunization with type A toxoid was noted to depress the response to the type B component of bivalent AB toxoid. In 1963, Fiock reported on the immunogenicity in humans of pentavalent ABCDE botulinum toxoid.[42] An immunization schedule of 0–2–12 weeks was adopted and immune titers were measured at 14 weeks and 52 weeks after the primary injection and 8 weeks after an annual booster. Titers were judged to be satisfactory to all five components, although the proportion of satisfactory titers varied with serotype and between vaccine batches. A few individuals had measurable titers at 52 weeks, but satisfactory titers to all five components were observed after the annual booster in at least 86% of individuals immunized. These studies of various schedules indicate that timing of injections will influence the magnitude of the positive response, so that an increased spacing reduces the need for the amount of toxoid necessary for satisfactory immunization. For example, two immunizations on either a 0–8- or 0–10-week schedule produced titers as high as four doses on a 0–2–4–6-week schedule.[41] The economic benefits of increased timings between immunizations should not be forgotten when establishing immunization schedules for expensive vaccines.

The neutralization titer of sera, quantified using a mouse lethality assay, is used as a measure of the immune response in man. The relationship between antitoxin titer and protection against botulinum intoxication has been examined in animals.[43] The immunogenicity and protective efficacy of each of the toxoids A to E, was evaluated in mice, rabbits, and guinea pigs. Animals were immunized with single doses of monovalent toxoid and challenged intraperitoneally with homologous toxin. On the day before challenge, serum was taken for the measurement of neutralization titer. As expected, the level of resistance increased with antitoxin titer. However, at approximately twice the lowest measureable neutralization titer, animals could withstand 10^4–10^5 guinea pig intraperitoneal LD_{50} doses. Other experiments with the pentavalent toxoid A–E, demonstrated that animals given single immunizations and challenged by aerosol inhalation had 80% survival to 2000 guinea pig LD_{50} doses of type D toxin and 92% survival to 30 000 guinea pig LD_{50} doses of type A toxin.[43]

Unpublished work has demonstrated the efficacy of pentavalent toxoid in nonhuman primates.[44] In the early 1960s, 16 rhesus monkeys (approximately 5 kg) were immunized with Lot No. 1 (Parke Davis & Co.) using the same 0–2–12-week schedule recommended for humans. Eight received a mean inhaled dose of 23 500 mouse (i.p.) LD_{50} ($MIPLD_{50}$) of type A toxin and eight received a dose of 250 000 $MIPLD_{50}$. All survived with no evidence of a change in their temperature, feeding habits or appearance throughout the 15-day observation period. These doses were calculated to be from 24 to 252 monkey LD_{50} doses, indicating a vigorous protective immunity to aerosol challenge with botulinum toxin. We have recently extended these studies to examine whether an abbreviated immunization schedule with pentavalent toxoid would elicit protective immunity.[45] Monkeys were immunized with a normal human dose of pentavalent toxoid at 0 and 2 weeks, then challenged at 5 weeks with 25 monkey LD_{50} aerosol doses of botulinum toxin type A. Control animals were immunized with DT vaccine. All pentavalent toxoid-immunized animals survived and none showed clinical signs or behavior distinguishable from the unexposed animals. All individuals had neutralizing serum antibody titers before exposure. These studies suggest that pentavalent toxoid can provide protective immunity when given with less than three doses.

In 1992, it became possible to evaluate the immunogenicity of pentavalent botulinum toxoid in military volunteers whose immunization schedule had been interrupted (P.R. Pittman, M.H. Sjogren, D. Hack, D. Franz, R.S. Makuch and J. Arthur, 1995, unpublished data). A group of volunteers had 18–24 months previously received either three, two or one injections before military deployment. Of those given two or three doses, 11% retained detectable titers while none given one dose retained a measurable titer. After boosting, 98% of those who had completed the primary series had measurable titers, whereas 97% of those who had previously received two primary injections and 70% of those had previously received one primary injection developed neutralization titers. More striking, however, was the titer of those receiving a delayed booster immunization. For example, those who previously received two injections had a mean titer after booster immunization above 7 IU/ml – 350 times above the minimum protective serum titer. One might speculate that long intervals between initial and second injections might routinely offer the best mechanism to develop a strong immune protection to botulism.

Insights from efforts to improve tetanus vaccine suggest other avenues for obtaining improved botulism protection with the toxoids A–E in the immediate future, such as further purification of the antigen, use of alternative adjuvants, microencapsulation, or optimization of antigen dose. Batches of toxoid are now being formulated to contain toxoid produced from the seven purified (>90%) individual neurotoxins and this is likely to create the second generation of botulinum vaccines.

A significant step forward in botulinum vaccines has been the production and testing of a type F toxoid vaccine. The toxoid was developed and produced from purified (>90%) toxin under Good Manufacturing Practice (GMP) conditions by the Centre for Applied Microbiology and Research, Porton Down, UK under US Army contract. Under Investigational New Drug (IND) status, the toxoid was tested in a phase I trial for safety and immunogenicity (P.R. Pittman *et al.*, 1994, unpublished data). Single doses of 2, 5 and 10 μg were tested for reactogenicity, and also two and three doses at 5 μg each. The toxoid was very well tolerated and, after three doses, produced satisfactory titers for a year. In a parallel study, we evaluated the usefulness of ELISA to track serum titers in these volunteers over time.[46] Individuals receiving two or three doses had serum ELISA titers to pure toxin 10- to 100-fold higher than from nonimmune controls and were easily distinguishable. After the second immunization, the seroconversion rate was 85%; after three doses, 90%. ELISA titers correlate well with neutralization data in sera taken after the second immunization ($r^2 = 0.86$). ELISA values were greater than 1.0 in 70% of volunteers for at least 180 days. The data indicate ELISA is a satisfactory screening tool for serum immune titers.

B. Recombinant vaccine

Much of the work on a recombinant botulinum vaccine has been patterned on the approach laid out for a recombinant tetanus vaccine. After cloning and sequencing of the fragments of the gene for botulinum toxin A,[47] selection of the gene sequence encoding the H_C fragment was feasible. The first approach was to use the native gene fragments to construct gene segments designed to produce a native H_C polypeptide or an H_C polypeptide fused to *E. coli* maltrose-binding protein.[48] The gene insert for the native protein appeared to be unstable, and the expressed fusion product was recognizable by ELISA but appeared to be in inclusion bodies. Immunization trials in mice with extracts containing fusion product conferred protection against challenge with native botulinum A.

Because work with the tetanus H_C suggested that native clostridial gene sequences would not be translated well in *E. coli*, it was decided to construct a synthetic gene encoding the botulinum type A H_C-containing codons utilized preferentially in *E. coli*. Oligomers of 60–70 nucleotides were ligated together to produce a synthetic gene encoding H_C, a protein fragment of approximately 50 kDa, of botulinum neurotoxin A.[49] When expressed in *E. coli*, the full-sized product could be observed by immunoblot analysis. Crude extracts were used to immunize mice, and animals given three subcutaneous doses were protected against 10^6 LD_{50} doses of toxin given intraperitoneally. When mice were immunized with purified H_C at various doses mixed with alum, the degree of protection was as efficacious as for mice immunized with pentavalent toxoid. The recombinant protein did not protect against three LD_{50} doses of botulinum toxin B or 10 LD_{50} doses of botulinum toxin E.

Because of the need to develop seven recombinant vaccines, several studies with type A H_C were performed to determine if the 50 kDa H_C polypeptide was the minimum immunogen required to elict satisfactory immunity. Using native gene fragments isolated from pCBA2 and pCBA3,[47] polypeptides of 26–34 kDa were produced which spanned the entire neurotoxin protein.[50] A fragment encoding residues 573–830 elicited weak protective immunity. Two other fragments, whose sequence is part of the H_C domain but lacking its amino-terminal end, also induced weak protective immunity. In a more detailed study, PCR was used to generate 10 overlapping native gene fragment constructs covering the entire neurotoxin sequence.[51] The polypeptides encoded by these fragments were expressed in *E. coli* and purified by sodium dodecyl sulfate– polyacrylamide gel electrophoresis. Two polypeptides conferred protection ($H_{455-661}$ and $H_{1150-1289}$). However, protection was weak compared with that resulting from immunization with the ent re H_C polypeptide. Both studies indicate the importance of amino- and carboxy-terminal sequences of the H_C domain and suggest that multiple epitopes are required for vigorous protective immunity.

These studies have provided encouragement that improved vaccines for each of the botulinum serotypes might be possible based on recombinant H_C domains of each. The synthetic H_C gene for type A neurotoxin has been expressed in baculovirus[38] and in *Pichia pastoris* (L. Smith, personal communication). Expression of the type A H_C protein in baculovirus-infected cells yielded a soluble secreted protein of the correct size that was recognized by serotype A-specific antiserum. Immunization with this material protected mice from 10^6 LD_{50} doses of the toxin. However intracellularly expressed protein failed to induce protective immunity. As seen with tetanus fragment C, botulinum type A H_C protein produced in *Pichia pastoris* yielded a product that protected mice to the same extent as the recombinant protein from *E. coli* or baculovirus as expression systems.

Synthetic genes for the H_C domain have been constructed for botulinum neurotoxins from type B (J.L. Middlebrook, personal communication), type E (L. Smith, personal communication), and type F (R. Titball, personal communication). Lysates of *Pichia pastoris* containing recombinant H_C protein encoded by synthetic genes for both type B and type E (L. Smith, personal communication) produced protective immunity when injected in mice. As seen with tetanus fragment C, secretion of type B H_C resulted in a nonprotective product, probably from glycosylation. Methods for purification to homogeneity from *E. coli* constructs have been devised for the H_C protein for type A and type B (Ophidian Pharmaceuticals, Madison, WI, USA); phase I trials of the type A H_C protein would verify its safety and immunogenicity in humans and be a prudent step toward full-scale development of a seven-component recombinant botulinum vaccine.

IV. PERFRINGENS VACCINES

A. Toxoid vaccines for α- and ε-toxins

Clostridium perfringens is an obligate anaerobe which is widespread in the environment and is found in soil, decaying organic matter and as a member of the gut flora in man and animals.[52] Different strains of *C. perfringens* can be assigned to one of five biotypes (A–E) depending on the spectrum of toxins produced.[53] *Clostridium perfringens* type A is the most important causative agent of gas gangrene in man. In total, *C. perfringens* produces at least 12 different toxins,[52,54] four of which, α, β, ι and ε, are considered major toxins. For the purposes of this chapter, only the α- and ε-toxins will be considered further because these are the only two for which there is currently a prospect of vaccination.

A crude *C. perfringens* toxoid vaccine was demonstrated to induce protection against experimentally induced gas gangrene as long ago as 1937[55] and subsequent studies suggested that the effective component of this vaccine was derived from the α-toxin.[56–58] Despite these favorable results no vaccine has been developed for use in humans and current treatment for gas gangrene usually involves the removal of the affected limb or tissues.

Vaccines containing toxoided α- and/or ε-toxins are available for veterinary use as multivalent formulations protecting against the five major clostridial diseases, namely, blackleg, tetanus, malignant oedema, pulpy kidney and black disease. However, it might be desirable to minimize the toxoids incorporated in a veterinary vaccine down to the protective subunits in order to enrich for the latter and remove unecessary (for protection) and potentially reactogenic components. In this way, a well-defined vaccine could be obtained which would not cause side-effects in economically important livestock.

B. Subunit vaccine for gas gangrene

The first bacterial toxin to be identified as an enzyme, the α-toxin, is a zinc-metallophospholipase C, catalyzing the hydrolysis of phosphatidylcholine to phosphorylcholine and 1,2-diglyceride.[59] It also has lethal, necrotizing, haemolytic and cytolytic properties (see Chapter 14).[60] The cloning and sequencing of the *C. perfringens* α-toxin as a 43 kDa protein[61] led to study of the structure–function relationships of the protein.[62,63] As part of these studies, the location of some antibody epitopes was identified[64] and, in particular, the epitope recognized by a monoclonal antibody which inhibits the phospholipase C (PLC) activity of the toxin.[64] This neutralizing epitope is located in the amino-terminal region of the molecule at position 193–198 in the primary amino acid sequence. Since the amino-terminal region of the α-toxin molecule shows some sequence homology with the amino-terminal region of *Bacillus cereus*[65] which also possesses PLC activity, while the carboxy-terminal region of the toxin shows homology with an enzyme in the human arachidonic acid pathway,[62] it seemed possible that the alpha-toxin might have two functional domains. To examine this hypothesis, genetically engineered amino- and carboxy-terminal fragments of the toxin of size 28 kDa and 14 kDa, respectively, were produced as soluble proteins in *E. coli*.[66] The amino-terminal fragment comprised the primary amino acid sequence 1–249 and was denoted Cpa_{1-249} while the carboxy-terminal fragment spanned amino acid sequence 247–370 ($Cpa_{247-370}$). Study of the biological properties of each fragment compared with the intact cloned toxin (Cpa) revealed that Cpa_{1-249} retained PLC activity, but was devoid of haemolytic and lethal activity with vastly reduced sphingomyelinase activity, while $Cpa_{247-370}$ had no intrinsic biological activity.[67] Neither fragment was toxic in an experimental animal model.

However, combination of the two fragments *in vitro* resulted in the restoration of haemolytic activity, albeit at a reduced level.[67] These results indicated that $Cpa_{247-370}$ conferred hemolytic and (possibly also sphingomyelinase) activity on the amino-terminal region of the molecule, which had intrinsic PLC activity. This conclusion was confirmed by demonstration of the potent inhibitory activity of an antiserum raised to $Cpa_{247-370}$ on the haemolytic activity of the holotoxin.[66] Surprisingly, the anti-$Cpa_{247-370}$ also inhibited the PLC activity of the toxin, suggesting that anti-$Cpa_{247-370}$ inhibits interaction of the PLC moiety of the α-toxin with the cell membrane. Thus, these studies identified the $Cpa_{247-370}$ as a candidate subunit vaccine to protect against α-toxin. To test the efficacy of the vaccine, $Cpa_{247-370}$ or Cpa_{1-249} or the formal toxoid were used in an emulsion with Freund's incomplete adjuvant to immunize Balb/c mice[66] (Table 29.1).

Mice hyperimmune to $Cpa_{247-370}$ were challenged either with 50 mouse lethal doses (MLD) of purified Cpa or with α-toxin-producing *C. perfringens*. Challenge of untreated control mice with the organism resulted in a fulminating gas gangrene characterized by edema and cyanosis of the affected limb within 10 h

Table 29.1. Antibody titer to α-toxin induced by α-toxin fragments

Candidate vaccine	Anti-toxin titer (dilution $\times 10^3$)
$Cpa_{247-370}$	12.8
Cpa_{1-249}	51.2
Formol toxoid	51.2

Table 29.2. Protection against challenge with either purified cloned toxin or live organism

Candidate vaccine	Survivors[a] after toxin challenge	Mean TTD[b] ± SEM (h)	Survivors of challenge[c] with *Clostridium perfringens*	Mean TTD ± SEM (h)
$Cpa_{247-370}$	6/6		6/6	
Cpa_{1-249}	0/6	4.8 ± 0.27	Not tested	
Formol toxoid	6/6		6/6	
Untreated control	0/6	1.83 ± 0	0/6	6.83 ± 0.44

[a] The formol toxoid and $Cpa_{247-370}$ groups were challenged intraperitoneally with 50 μg purified Cpa; while the Cpa_{1-249} and untreated control groups were challenged with 5 μg purified Cpa.
[b] Mean time to death (TTD) ± SEM in hours.
[c] All groups except Cpa_{1-249} were challenged intramuscularly with 10^9 viable *C. perfringens* organisms.

of a challenge of 10^8 or more organisms. However, $Cpa_{247-370}$ vaccinees were protected against either toxin or organism challenge, without development of symptoms.[66] In contrast, mice hyperimmune to Cpa_{1-249} were not protected against challenge either by toxin or by the organism (Table 29.2)

These results indicated that $Cpa_{247-370}$ is a protective subunit vaccine which is as effective as the formol toxoid in preventing the development of experimental gas gangrene. The subunit $Cpa_{247-370}$ was derived from the α-toxin gene from a human gas gangrene isolate of *C. perfringens* (NCTC 8237 strain). The sequence of NCTC 8237 α-toxin was subsequently found to be atypical,[68] in that α-toxins produced by most other isolates of *C. perfringens* contained point mutations variously at sequence positions 174, 177, 335 and 345. It is hypothesized that some of these mutations, particularly at positions 174, 177 and 335, adjacent to aromatic amino acids, have evolved to provide chymotrypsin resistance to α-toxins produced by enteric isolates of *C. perfringens*.[68] However, the $Cpa_{247-370}$ subunit, based on the sequence of NCTC 8237-derived α-toxin, provided complete protection not only against the homologous toxin, but also against α-toxin from an enteric isolate containing all four point mutations.[68] These results underline the common protective capacity of the $Cpa_{247-370}$ subunit vaccine against α-toxins derived from a number of sources.

The recombinant subunit vaccine described above represents a vaccine with application for use in man. The vaccine would have utility to prevent the development of gas gangrene in at-risk groups within the population. The subunit vaccine could also be substituted for the formol toxoid currently incorporated in a multivalent veterinary vaccine.

C. Vaccine strategies to prevent ε enterotoxemia

ε-Toxin is the major toxin produced by *C. perfringens* type D,[69] although other toxins are produced in smaller amounts. This toxin induces a rapidly fatal enterotoxemia in sheep, cattle and goats and has marked dermonecrotic, edematic and neurotoxic activity in a range of animal species,[70] lending weight to the suspicion that it may be toxic in man. It is produced as a relatively inactive prototoxin which is converted to the toxin after cleavage by proteolytic enzymes such as trypsin and chymotrypsin.[71]

Little is known of the mode of action of ε-toxin, following binding to cell membrane receptors in target tissues (see Chapter 25). Studies towards an antidote were facilitated by the identification of a monoclonal antibody (mAb), raised to the epsilon prototoxoid, with neutralizing activity for the mature toxin.[72] This mAb, denoted A5C12, provided the starting point for the derivation of an anti-idiotypic antibody (anti-Id) approach to protect against the toxin.[72]

In this approach to vaccination, anti-Id are derived by raising antibodies that recognize the variable regions within the antigen-binding Fab fragment of the neutralizing primary mAb.[73] For ε-toxin, since the primary mAb, A5C12, has a binding site which is a mirror image of the neutralizing epitope, then the anti-Id has an internal image of the neutralizing epitope on the toxin. The internal image anti-Id can, in turn, be used as an active immunogen to induce a polyclonal serum containing ε-toxin-neutralizing antibodies. In this way, the anti-Id is being used as a surrogate antigen for epsilon-toxin.

D. Derivation of anti-Id to ε-toxin

In order to prepare anti-Id for use as a vaccine, Balb/c mice were immunized with keyhole limpet haemocyanin (KLH)-conjugated A5C12, and the polyclonal anti-Id was purified from hyperimmune mouse serum by affinity chromatography against A5C12.[72] This anti-Id provided significant protection to susceptible guinea pig peritoneal macrophages (GPPM) from the cytotoxic effects of epsilon-toxin, when preincubated with the cells overnight prior to addition of activated toxin at a concentration of 1 μg/ml (Table 29.3).

Evidence has accumulated to suggest that ε-toxin binds to a cell surface receptor.[74] Results from this study suggest that the epitope for A5C12 is co-located with the cell-binding site of the toxin[72] and that internal image anti-Id, which probably constitutes only a fraction of the total anti-Id, is protective by competing with ε-toxin for cell surface receptor sites on GPPM. The ability of purified anti-Id, conjugated to KLH, to induce an anti-ε-toxin response in two animal species, namely mouse and rabbit, was tested.[72] Balb/c mice and a number

Table 29.3. Inhibitory effect of anti-Id on *in vitro* cytotoxicity of ε–toxin

Treatment	Percentage viability of GPPMs
None[a]	100
ε-Toxin[b]	< 1.0
ε-Toxin + irrelevant mouse IgG[c]	< 1.0
ε-Toxin + anti-Id[d]	50 ± 5

[a] GPPM (5 × 10^6/ml) were incubated in tissue culture medium.
[b] Epsilon toxin was added to GPPMs (5 × 10^6/ml) to a final concentration of 1 μg/ml
[c] GPPMs (5 × 10^6 /ml) were preincubated with 100 μg/ml of an irrelevant mouse IgG prior to the addition of epsilon toxin (1 μg/ml).
[d] GPPMs (5 × 10^6/ml) were preincubated with 100 μg/ml anti-Id prior to the addition of ε-toxin (1 μg/ml).

Table 29.4. Detection of Ab_3 specific for ε-toxin in anti-Id immunized animals

Vaccinees (haplotype)	Anti-ε titers (× 10^3)[a]
Balb/c mice (H-2^d)	17.8
DBA/2 mice (H-2^d)	0.79
CBA mice (H-2^d)	1.0
C3H mice (H-2^k)	1.78
NZB mice (H-2^k)	0.63
NZW rabbit	31.62

[a] Serum titers were determined as total mouse immunoglobulin specific for ε-prototoxoid

of homologous (for haplotype) and heterologous mouse strains and a New Zealand white rabbit were immunized with the anti-Id–KLH conjugate. In recipients of the anti-Id, antibody titers to the ε-prototoxoid were detected, indicating that an internal image component within the anti-Id is able to induce a specific immune response in syngeneic, allogeneic and xenogeneic animals, by mimicking the epitope on ε-toxin which is responsible for cell binding[72] (Table 29.4).

The protective efficacy of anti-Id vaccination was tested by injecting mouse and rabbit vaccines with 10 MLD (16 μg/kg) of activated toxin (Table 29.5). Vaccinees were afforded significant protection against the challenge. Balb/c mice and rabbits were completely protected, while the other mouse strains showed at least a three-fold increase in TTD. Protection against toxin challenge correlated with the mean serum titer of antibody specific for ε-toxin in anti-Id vaccinees.

Thus, a protective anti-ε response was established in anti-Id-vaccinated animals which had not at any time been exposed to the toxin itself.[72] Further, the basis of protection resided in the single epitope on ε-toxin recognized by

Table 29.5. Protective efficacy of anti-Id vaccination against *in vivo* challenge with 10 MLD ε-toxin

Vaccinees (haplotype)	Mean TTD ± SEM (min)
Balb/c mice (H-2^d)	Survived
DBA/2 mice (H-2^d)	300 ± 25
CBA mice (H-2^d)	320 ± 5
C3H mice (H-2^k)	380 ± 20
NZB mice (H-2^k)	350 ± 5
NZW rabbit	Survived

A5C12, and this was sufficient to protect vaccinees against acute toxic challenge. This illustrates the potential of the anti-Id approach to vaccination.

For the anti-Id approach to be a viable option, it would be necessary to derive a monoclonal anti-Id preparation to substitute for the polyclonal reagent described above. This would ensure product consistency, reproducibility and secure a continuous supply of anti-Id vaccine. Further understanding of structure–function relationships in the toxin would facilitate alternative approaches to the derivation of a vaccine. The ε-toxin encoding gene has been cloned and sequenced[75] and the primary amino acid sequence of the toxin has been epitope-mapped using a panel of mAbs including A5C12.[76] No sequential binding sites for mAbs have been identified, suggesting that the epitopes in ε-toxin are largely, if not entirely, conformational.[76] Thus, the preparation of fragments of the toxin in an attempt to identify an active site, can be carried out only on the basis of predicted antigenicity from secondary-structure modelling of the molecule. However, amino acid residues essential for toxicity have been identified within the toxin.[77–81] This information may be useful in deriving genetically engineered toxoids as vaccine candidates.

V. FUTURE PROSPECTS

It can be seen that there exist efficacious recombinant subunit vaccines protective against tetanus, botulism and gas gangrene. As more knowledge is acquired about the mode of action of ε-toxin, it may be possible to produce a genetically engineered toxoid with which to vaccinate against ε-enterotoxemia.

Subunit vaccines are amenable to formulation in a number of delivery systems. This means that injected delivery of the vaccine is not mandatory and that oral or intra-nasal delivery might be alternatives. The last two routes offer the advantage of convenience and allow self-administration; they also preferentially stimulate mucosal immunity over systemic immunity. Encapsulation of subunit vaccines in liposomes or polymeric microspheres may have several advantages.

It gives scope for targetting of the vaccine by adsorption or co-encapsulation of ligands which bind receptors at epithelial cell surfaces and it facilitates the co-delivery of immunoenhancers, for example, adjuvants and cytokines, with the vaccine. Microcapsules form a slow-release depot system for the vaccine, which not only protects the vaccine, but may also remove the need for boosting.

Currently, there is much interest in genetic vaccination, a technique in which immunization is effected with naked DNA encoding the sequence of a protective immunogen. It is conceivable that DNA encoding the sequences for these toxin subunits could be incorporated into a mammalian expression vector and used in genetic immunization against clostridial infections and toxaemias. Development of recombinant vaccines has opened a number of avenues toward development of new generation vaccines for these clostridial diseases.

VI. REFERENCES

1. Galazka A. and Gasse F. (1995) The present status of tetanus and tetanus vaccination. In Montecucco C. (ed.), *Clostridial Neurotoxins. The Molecular Pathogenesis of Tetanus and Botulism*, pp. 31–53. Springer-Verlag, Berlin.
2. Aguado M.T. and Lambert P.H. (1992) Controlled-release vaccines – biodegradable polylactide/polyglycolide (PL/PG) microspheres as antigen vehicles. *Immunobiol* **184**: 113–125.
3. Aguado M.T. (1993) Future approaches to vaccine development: single-dose vaccines using controlled-release delivery systems. *Vaccine* **11**: 596–597.
4. Mark A. and Granstrom M. (1994) The role of aluminum for adverse reactions and immunogenicity of diphtheria–tetanus booster vaccine. *Acta Paediatr* **83**: 159–163.
5. Mark A., Bjorksten B. and Granstrom M. (1995) Immunoglobulin E responses to diphtheria and tetanus toxoids after booster immunizations with alum-adsorbed and fluid DT-vaccines. *Vaccine* **13**: 669–673.
6. Bjorkholm B., Wahl M., Granstrom M. and Hagberg L. (1994) Immune status and booster effects of low doses of tetanus toxoid in Swedish medical personnel. *Scand J Infect Dis* **26**: 471–475.
7. Gupta R.K. and Siber G.R. (1994) Comparison of adjuvant activities of aluminum phosphate, calcium phosphate and stearyl tyrosine for tetanus toxoid. *Biologicals* **22**: 53–63.
8. Relyveld E., Bengounia A., Huet M. and Kreeftenberg J.G. (1991) Antibody response of pregnant women to two different adsorbed tetanus toxoids. *Vaccine* **9**: 369–372.
9. Esparza I., and Kissel T. (1992) Parameters affecting the immunogenicity of microencapsulated tetanus toxoid. *Vaccine* **10**: 714–720.
10. Alonso M.J., Gupta R.K., Min C., Siber G.R. and Langer R. (1994) Biodegradable microspheres as controlled-release tetanus toxoid delivery systems. *Vaccine* **12**: 299–306.
11. Almeida A.J., Alpar H.O. and Brown M.R. (1993) Immune response to nasal delivery of antigenically intact tetanus toxoid associated with poly (L-lactic acid) microspheres in rats, rabbits and guinea-pigs. *J Pharm Pharmacol* **45**: 198–203.
12. Men Y., Thomasin C., Merkle H.P., Gander B. and Corradine G. (1995) A single administration of tetanus toxoid in biodegradable microsphers elicits T cell and antibody responses similar or superior to those obtained with aluminum hydroxide. *Vaccine* **13**: 683–689.
13. DasGupta B.R. and Sugiyama H. (1972) A common subunit structure in *Clostridium botulinum* type A, B and E toxins. *Biochem Biophys Res Comm* **48**: 108–112.

14. Helting T.B. and Zwisler O. (1977) Structure of tetanus toxin I. Breakdown of the toxin molecule and discrimination between polypeptide fragments. *J Biol Chem* **252**: 187–193.

15. Fairweather N.F., Lyness V.A., Pickard D.J., Allen G. and Thomson R.O. (1986) Cloning, nucleotide sequencing, and expression of tetanus toxin fragment C in *Escherichia coli. J Bacteriol* **165**: 21–27.

16. Fairweather N.F., Lyness V.A. and Maskell D.J. (1987) Immunization of mice against tetanus with fragments of tetanus toxin synthesized in *Escherichia coli. Infect Immun* **55**: 2541–2545.

17. Makoff A.J., Ballantine S.P., Smallwood A.E. and Fairweather N.F. (1989) Expression of tetanus toxin fragment C in *E coli*: its purification and potential use as a vaccine. *Bio/Technology* **7**: 1043–1046.

18. Fairweather N.F., Chatfield S.N., Makof A.J., Strugnell R.A., Bester J., Maskell D.J. and Dougan G. (1990) Oral vaccination of mice against tetanus by use of a live attenuated *Salmonella* carrier. *Infect Immun* **58**: 1323–1326.

19. Charles I.G., Rodgers B.C., Makoff A.J., Chatfield S.N., Slater D.E. and Fairweather N.F. (1991) Synthesis of tetanus toxin fragment C in insect cells by use of a baculovirus expression system. *Infect Immun* **59**: 1627–1632.

20. Halpern J.L., Habig W.H., Neale E.A. and Stibitz S. (1990) Cloning and expression of functional fragment C of tetanus toxin. *Infect Immun* **58**: 1004–1009.

21. Stover C.K., de la Cruz V.F., Fuetst T.R., Burlein J.E., Benson L.A., Bennett L.T., Bansal G.P., Young J.F., Lee M.H., Hatfull G.F., Snapper S.B., Barletta R.G., Jacobs Jr. W.R. and Bloom B.R. (1991) New use of BCG for recombinant vaccines. *Nature* **351**: 456–460.

22. Wells J.M., Wilson P.W., Norton P.M., Gasson M.J. and Le Page R.W. (1993). *Lactococcus lactis*: high-level expression of tetanus toxin fragment C and protection against lethal challenge. *Mol Microbiol* **8**: 1155–1162.

23. Bouche P., Sato H., Sato Y. and Locht C. (1994) Neutralizing antibodies and immunoprotection against pertussis and tetanus obtained by use of a recombinant pertussis toxin-tetanus toxin fusion protein. *Infect Immun* **62**: 449–456.

24. Makoff A.J., Oxer M.D., Romanos M.A., Fairweather N.F. and Ballantine S. (1989) Expression of tetanus toxin fragment C in *Eschericia coli*: high level expression by removing rare codons. *Nucl Acids Res* **17**: 10191–10202.

25. Makoff A.J., Charles I.G. and Fairweather N.F. (1994) Recombinant antigens as components of a diphtheria-tetanus-pertussis vaccine. *Bioprocess Technol* **19**: 205–231.

26. Romanos M.A., Makoff A.J., Fairweather N.F., Beesley K.M., Slater D.E., Rayment F.B., Payne M.M. and Clare J.J. (1991) Expression of tetanus toxin fragment C in yeast: gene synthesis is required to eliminate fortuitous polyadenylation sites in AT-rich DNA. *Nucl Acids Res* **19**: 1461–1467.

27. Clare J.J., Rayment F.B., Ballantine S.P., Sreekrishna K. and Romanos M.A. (1991) High-level expression of tetanus toxin fragment C in *Pichia pastoris* strains containing multiple tandem integrations of the gene. *Bio/Technology* **9**: 455–460.

28. Figueiredo D., Turcotte C., Frankel G., Li Y., Dolly O., Wilkin G., Marriott D., Fairweather N. and Dougan G. (1995) Characterization of recombinant tetanus toxin derivatives suitable for vaccine development. *Infect Immun* **63**: 3218–3221.

29. Fischer P.M. and Howden M.E. (1994) Synthetic peptide antigens of tetanus toxin. *Mol Immunol* **31**: 1141–1148.

30. Chatfield S.N., Fairweather N., Charles I., Pickard D., Levine M., Hone D., Posada, M., Strugnell, R.A. and Dougan G. (1992b) Construction of a genetically defined *Salmonella typhi* Ty2 aroA, aroC mutant for the engineering of a candidate oral typhoid-tetanus vaccine. *Vaccine* **10**: 53–60.

31. Chatfield S.N., Charles I.G., Makoff A.J., Oxer M.D., Dougan G., Pickard D., Slater D. and Fairweather N.F. (1992a) Use of the *nirB* promoter to direct the stable expression of heterologous antigens in *Salmonella* oral vaccine strains: development of a single-dose tetanus vaccine. *Bio/Technology* **10**: 888–892.

32 Khan C.M.A., Villarreal-Ramos B., Pierce R.J., de Hormaeche R.D., McNeill H., Ali T., Chatfield S.,

Capron A., Dougan G. and Hormaeche C.E. (1994) Construction, expression, and immunogenicity of multiple tandem copies of the *Schistosoma mansoni* peptide 115–131 of the P28 glutathione S-transferase expressed as *C*-terminal fusions to tetanus toxin fragment C in a live *aro*-attenuated vaccine strain of *Salmonella. J Immunol* **153**: 5634–5642.

33. Khan C.M.A., Villarreal-Ramos B., Pierce R.J., Riveau G., de Hormaeche R.D., McNeill H., Ali T., Fairweather N., Chatfield S., Capron A., Dougan G. and Hormaeche C.E. (1994) Construction, expression, and immunogenicity of the *Schistosoma mansoni* P28 glutathione S-transferase as a genetic fusion to tetanus toxin fragment C in a live *aro*-attenuated vaccine strain of *Salmonella. Proc Natl Acad Sci USA* **91**: 11261–11265.
34. Chabalgoity J.A., Villareal-Ramos B., Khan C.M.A., Chatfield S.N., de Hormaeche R.D. and Hormaeche C.E. (1995) Influence of preimmunization with tetanus toxoid on immune responses to tetanus toxin fragment C-guest antigen fusions in a *Salmonella* vaccine carrier. *Infect Immun* **63**: 2564–2569.
35. Chatfield S., Roberts M., Li J., Starns A. and Dougan G. (1994) The use of live attenuated *Salmonella* for oral vaccination. *Dev Biol Stand* **82**: 35–42.
36. Jackson R.J., Staats H.F., Xu A.J., Takahashi I., Kiyono H., Hudson M.E., Gilley R.M., Chatfield S.N. and McGhee J.R. (1994) Oral vaccine models: multiple delivery systems employing tetanus toxoid. *Ann NY Acad Sci* **730**: 217–234.
37. Hatheway C.L. and Dang C. (1994) Immunogenicity of the neurotoxins of *Clostridium botulinum*. In Jankovoc J. and Hallett M. (eds), *Therapy with botulinum toxins*, pp. 93–107. Marcel Dekker, New York.
38. Middlebrook J.L. and Brown J.E. (1995) Immunodiagnosis and immunotherapy of tetanus and botulinum neurotoxins. In Montecucco C. (ed.), *Clostridial Neurotoxins. The Molecular Pathogenesis of Tetanus and Botulism*, pp. 89–122. Springer-Verlag, Berlin.
39. Anderson J.H.J. and Lewis G.E. Jr. (1981) Clinical evaluation of botulinum toxoids. In Lewis G.E. Jr. (ed.), *Biomedical Aspects of Botulism*, pp. 233–246. Academic Press, London.
40. Reames H.R., Kadull P.J., Housewright R.D. and Wilson J.B. (1947) Studies on botulinum toxoids, types A and B. III. Immunization of man. *J Immunol* **55**: 309–324.
41. Fiock M.A., Devine L.F., Gearinger N.F., Duff J.T., Wright G.G. and Kadull P.J. (1962) Studies on immunity to toxins of *Clostridium botulinum*. VIII. Immunological response of man to purified bivalent AB botulinum toxoid. *J Immunol* **88**: 277–283.
42. Fiock M.A., Cardella M.A. and Gearinger N.F. (1963) Studies on immunity to toxins of *Clostridium botulinum*. IX. Immunological response of man to purified pentavalent ABCDE botulinum toxoid. *J Immunol* **90**: 697–702.
43. Cardella M.A. (1964) Botulinum toxoid. In Lewis K.H. and Cassel K.J. (eds), *Botulism*, pp. 113–130. US Department of Health, Education and Welfare Public Health Service, Cincinnati, OH.
44. Demchak P., Miller W.S., Tonik E.J. and Jemski J.V. (1963) *Response of Various Animal Species to Botulinum Toxin Administered by Different Routes (U)*. Technical Memorandum No. 29. US Army Biological Laboratories, Fort Detrick.
45. Brown J.E., Parker G.W., Pitt L.M., Swauger J.E., Elliott J.J., Ruble D.L. and Wilhelmsen C. (1995) Protective efficacy in monkeys of pentavalent botulinum toxoid vaccine on an abbreviated immunization schedule. In Rood J., McClane B., Songer J.G. and Titball R.W. (eds), *Abstracts of the ASM First International Conference on the Molecular Genetics and Pathogenesis of the Clostridia*, p. 23. American Society for Microbiology, Washington D.C.
46. Montgomery V.A., Makuch R.S., Pittman P.R. and Brown J.E. (1995) Evaluation of botulinum type F vaccine by ELISA. In *Annual Meeting of the American Society for Microbiology*, p. 289. American Society for Microbiology, Washington, D.C.
47. Thompson D.E., Brehm J.K., Oultram J.D., Swinfield T.J., Shone C.C., Atkinson T., Melling J. and Minton N.P. (1990) The complete amino acid sequence of the *Clostridium botulinum* type A

neurotoxin, deduced by nucleotide sequence analysis of the encoding gene. *Eur J Biochem* **189**: 73–81.

48. LaPenotierre H.F., Clayton M.A., Brown J.E. and Middlebrook J.L. (1993) Development of molecular engineered vaccine for *C botulinum* neurotoxins. In DasGupta B.R. (ed.), *Botulinum and Tetanus Neurotoxins*, pp. 463–466. Plenum Press, New York.
49. Clayton M.A., Clayton J.M., Brown D.R. and Middlebrook J.L. (1995) Protective vaccination with a recombinant fragment of *Clostridium botulinum* neurotoxin serotype A expressed from a synthetic gene in *Escherichia coli. Infect Immun* **63**: 2738–2742.
50. Smith L., Zuber M. and Hoover T. (1994) Studies on the protective efficacy of recombinant truncated H_C from *Clostridium botulinum.* In *Abstracts of the 11th World Congress on Animal, Plant and Microbial Toxins*, p. 127. Tel Aviv, Israel.
51. Dertzbaugh M.T. and West M.W. (1996) Mapping of protective and cross-reactive domains of the type A neurotoxin of *Clostridium botulinum. Vaccine* **14**: 1538–1544.
52. McDonel J.L. (1986) Toxins of *Clostridium perfringens* types A, B, C, D and E. In Dorner, F. and Drews, J. (eds), *Pharmacology of Bacterial Toxins*, p. 477. Pergamon Press, Oxford.
53. Hatheway C.L. (1990) Toxigenic clostridia. *Clin Micro Rev* **3**: 66–98.
54. Worthington R.W., Mulders M.S.G. and van Rensburg J.J. (1973) Enzymatic activation of *Clostridium perfringens* epsilon prototoxin and some biological properties of activated toxin. *Onderstepoort J Vet Res* **40**: 153–156.
55. Penfold W.J. and Tolhurst J.C. (1937) Formol-toxoids in the prophylaxis of gas gangrene. *Med J Aust* **1**: 982–989.
56. Kameyama S., Sato H. and Murata R. (1975) The role of α-toxin of *Clostridium perfringens* in experimental gas gangrene in guinea pigs. *Jap J Med Sci Biol* **25**: 200.
57. Robertson M. and Keppie J. (1943) Gas gangrene. Active immunisation by means of concentrated toxoids. *Lancet* **ii**: 311–314.
58. Boyd N.A., Thomson R.O. and Walker P.D. (1972) The prevention of experimental *Clostridium novyi* and *Clostridium perfringens* gas gangrene in high-velocity missile wounds by active immunisation. *J Med Microbiol* **5**: 467–472.
59. MacFarlane M.G. and Knight B.C.J.G. (1941) The biochemistry of bacterial toxins. I. The lecithinase activity of *Cl. welchii toxins. Biochem J* **35**: 884–902.
60. Evans D.G. (1943) The protective properties of the alpha antitoxin and the theta haemolysin in *Cl. welchii* type A antiserum. *Br J Exp Pathol* **24**: 81–88.
61. Titball R.W., Hunter S.E.C., Martin K., Morris B., Shuttleworth A., Rubidge T., Anderson D. and Kelly D. (1989) Molecular cloning and nucleotide sequence of the alpha-toxin (phospholipase C) of *Clostridium perfringens. Infect Immun* **57**: 367–376.
62. Titball R.W., Leslie D.L., Harvey S. and Kelly D. (1991) Hemolytic and sphingomyelinase activities of *Clostridium perfringens* alpha-toxin are dependent on a domain homologous to that of an enzyme from the human arachidonic acid pathway. *Infect Immun* **59**: 1872–1874.
63. Titball R.W. and Rubidge T. (1990) The role of histidine residues in the alpha toxin of *Clostridium perfringens. FEMS Microbiol Lett* **68**: 261–266.
64. Logan A.J., Williamson E.D., Titball R.W., Percival D.A., Shuttleworth A.D., Conlan J.W. and Kelly D.C. (1991) Epitope mapping of the alpha-toxin of *Clostridium perfringens. Infect Immun* **59**: 4338–4342.
65. Leslie D., Fairweather N., Pickard D., Dougan G. and Kehoe M. (1989) Phospholipase C and haemolytic activities of *Clostridium perfringens* alpha-toxin cloned in *Escherichia coli*: sequence and homology with a *Bacillus cereus* phospholipase C. *Mol Microbiol* **3**: 383–392.
66. Williamson E.D. and Titball R.W. (1993) A genetically engineered vaccine against the alpha-toxin of *Clostridium perfringens* protects mice against experimental gas gangrene. *Vaccine* **11**: 1253–1258.
67. Titball R.W., Fearn A.M. and Williamson E.D. (1993) Biochemical and immunological properties of

the *C*-terminal domain of the alpha-toxin of *Clostridium perfringens. FEMS Microbiol Lett* **110**: 45–50.

68. Ginter A., Williamson E.D., Dessy F., Coppe Ph., Bullifent H., Howells A. and Titball R.W. (1996) Molecular variation between the α-toxins from the type strain (NCTC 8237) and clinical isolates of *Clostridium perfringens* associated with disease in man and animals. *Microbiology* **142**: 191–198.
69. Glenny A.T., Barr M., Llewellyn-Jones M., Dalling T. and Ross H.E. (1933). Multiple toxins produced by some organisms of the *Clostridium welchii* group. *J Pathol* **37**: 53–74.
70. Hauschild A.H.W. (1971) *Clostridium perfringens* toxin types B, C, D and E. In Kadis, S., Monti, T.C. and Ajl, S.J. (eds), *Microbial Toxins 2A: Bacterial Protein Toxins*, pp. 159–188. Academic Press Inc., New York.
71. Niko L. (1965) Bovine enterotoxaemia. III. Factors affecting the stability of the toxins of *Clostridium perfringens* types A, C and D. *Can Vet J* **6**: 38–42.
72. Percival D.A., Shuttleworth A.D., Williamson E.D. and Kelly D.C. (1990) Anti-idiotypic antibody-induced protection against *Clostridium perfringens* type D. *Infect Immun* **58**: 2487–2492.
73. Nisonoff A. and Lamoyi E. (1981) Implications of the presence of internal image of the antigen in anti-idiotypic antibodies: possible application to vaccine production. *Clin Immunol Immunopathol* **21**: 397–406.
74. Payne D.W., Williamson E.D., Havard H., Modi N.K. and Brown J. (1994) Evaluation of a new cytotoxicity assay for *Clostridium perfringens* type D epsilon-toxin. *FEMS Microbiol Lett* **116**: 161–168.
75. Hunter S.E.C., Clarke I.N., Kelly D.C. and Titball R.W. (1991) Cloning and nucleotide sequencing of the *Clostridium perfringens* epsilon-toxin gene and its expression in *Escherichia coli. Infect Immun* **60**: 102–110.
76. Hunter S.E.C. (1992) Molecular genetics of *Clostridium perfringens* epsilon-toxin. PhD Thesis. Faculty of Medicine, Southampton University, UK.
77. Sakurai J. and Nagahama M. (1985) Role of one tryptophan residue in the lethal activity of *Clostridium perfringens* epsilon-toxin. *Biochem Biophys Res Commun* **128**: 760–766.
78. Sakurai J. and Nagahama M. (1986) Amino groups in *Clostridium perfringens* prototoxin and epsilon toxin. *Microb Pathog* **1**: 417–423.
79. Sakurai J. and Nagahama M. (1987) Histidine residues in *Clostridium perfringens* epsilon toxin. *FEMS Microbiol Lett* **41**: 317–319.
80. Sakurai J. and Nagahama M. (1987) The inactivation of *Clostridium perfringens* epsilon toxin by treatment with tetranitromethane and *N*-acetylamidazole. *Toxicon* **25**: 279–284.
81. Sakurai J. and Nagahama M. (1987) Carboxyl groups in *Clostridium perfringens* epsilon toxin. *Microb Pathog* **3**: 469–474.

Index

Note: Page references in *italic* refer to Figures; those in **bold** refer to Tables